D0606087

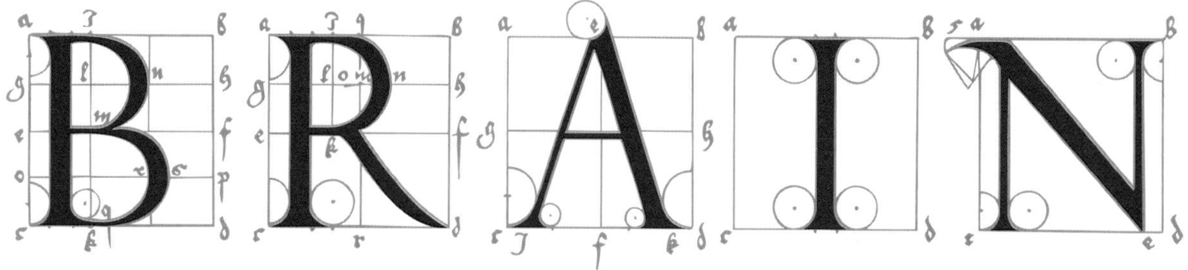
BRAIN

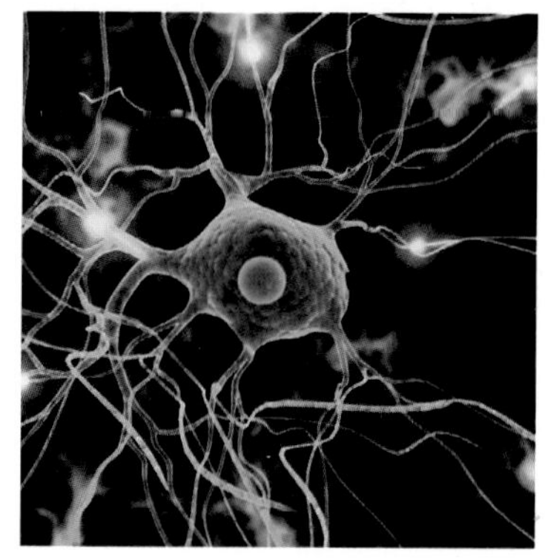

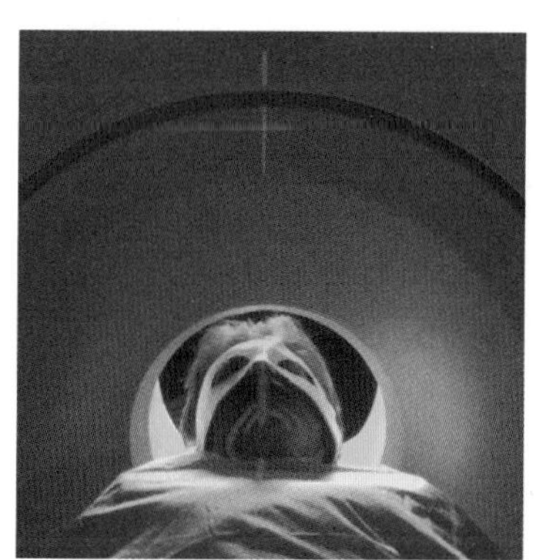

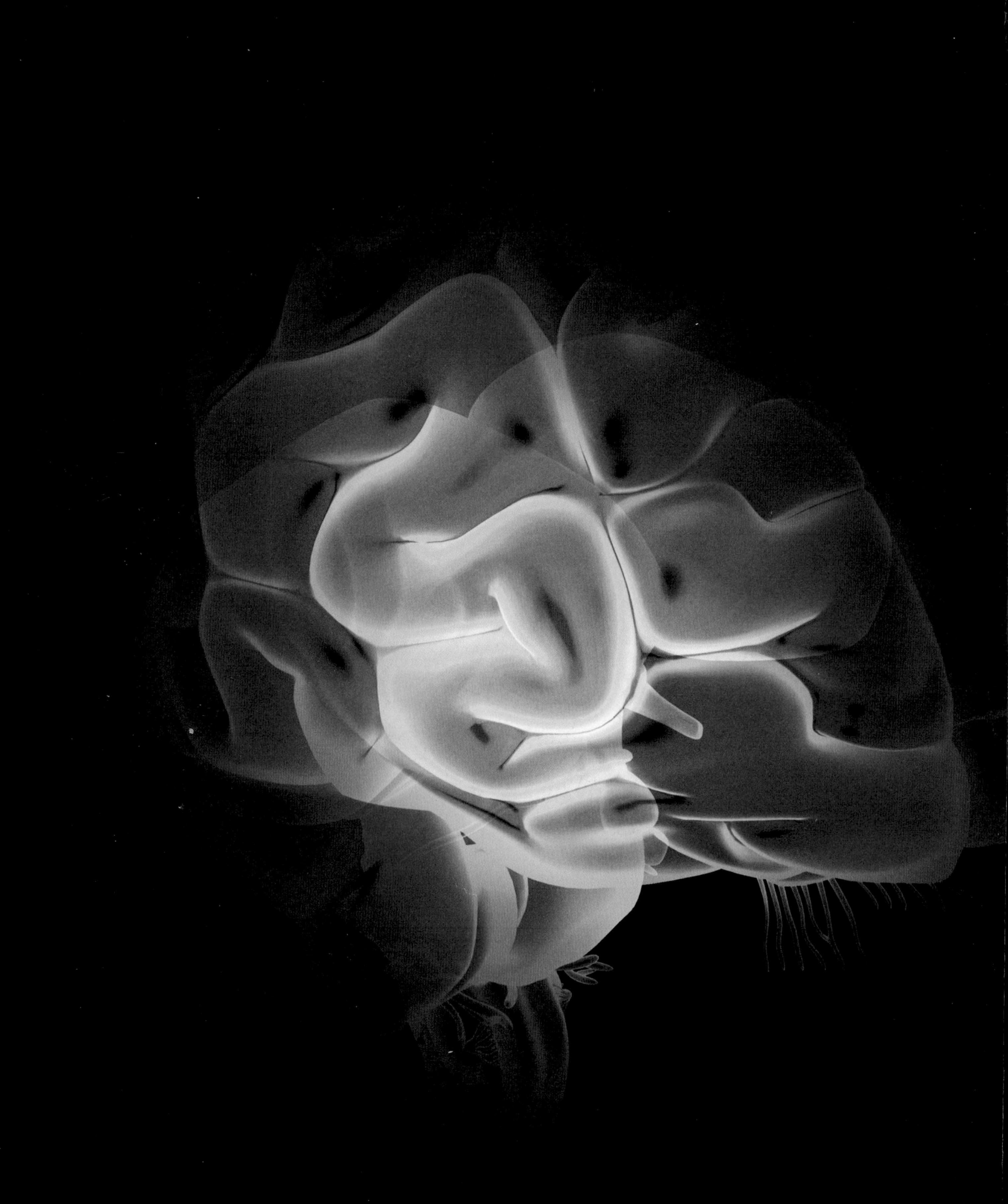

BRAIN

THE COMPLETE MIND

HOW IT DEVELOPS, HOW IT WORKS, AND HOW TO KEEP IT SHARP

MICHAEL S. SWEENEY
Foreword by Richard Restak, M.D.

WASHINGTON, D.C.

CONTENTS

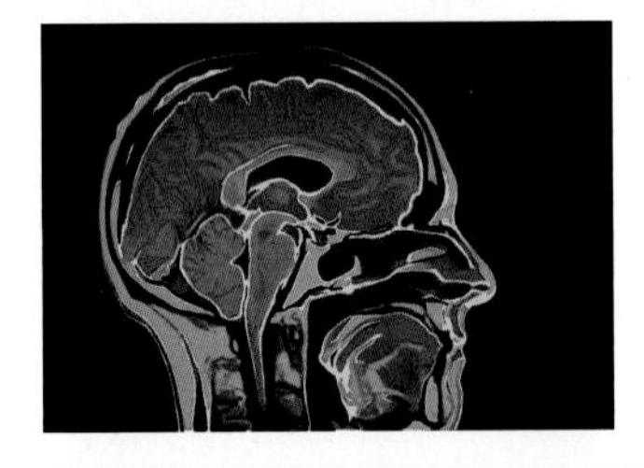

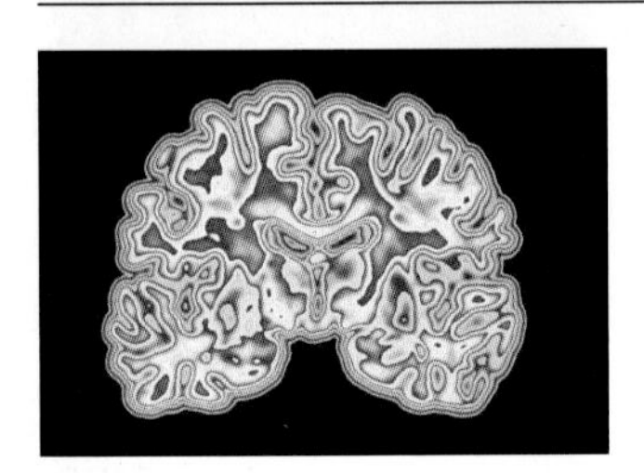

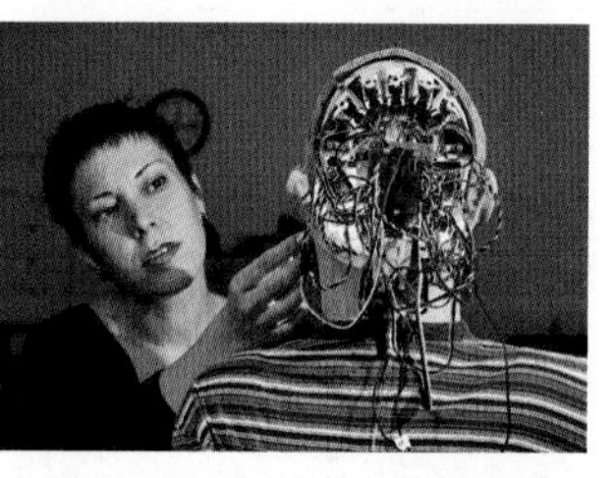

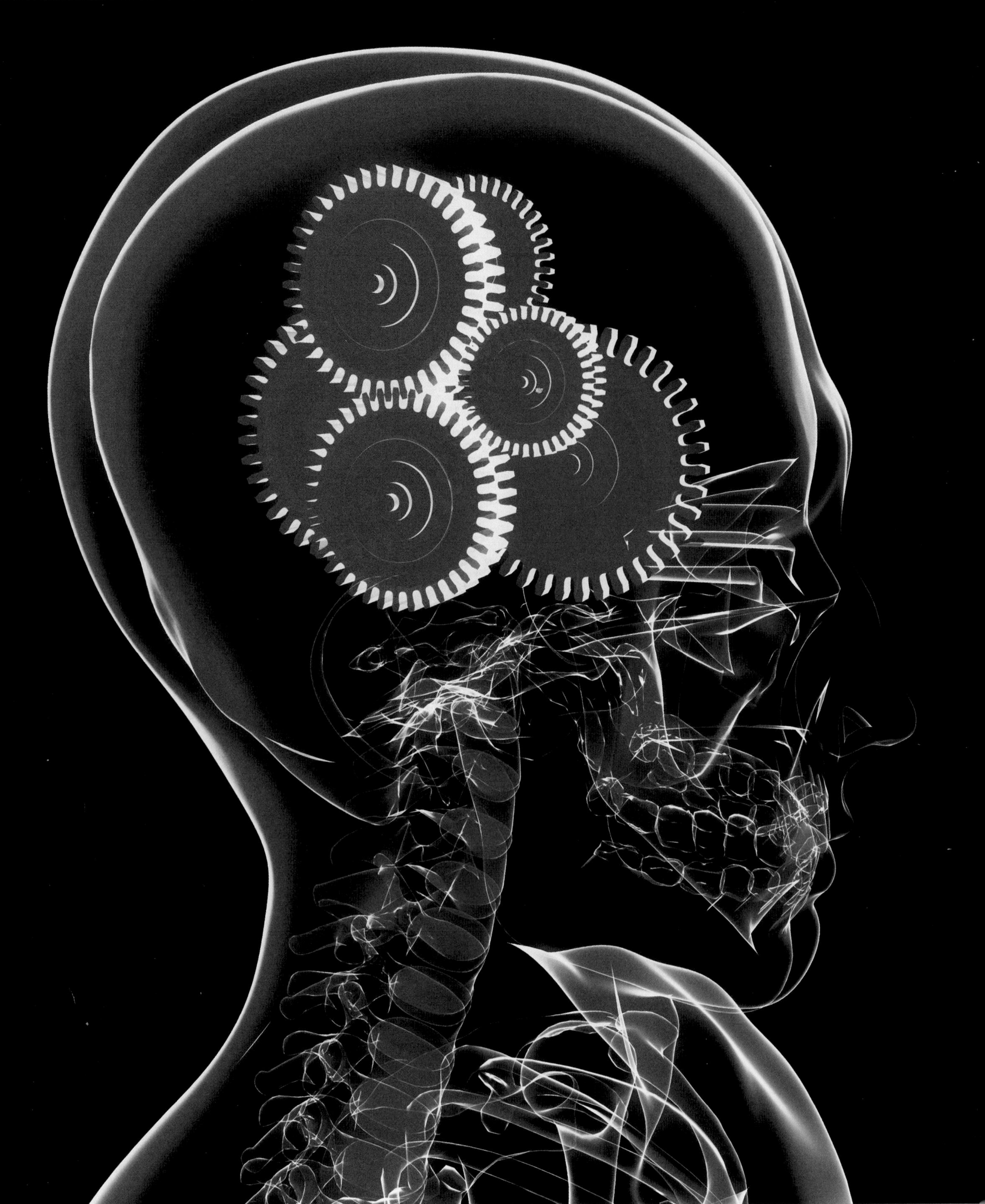

FOREWORD

BY RICHARD RESTAK, M.D.

BESTRIDING, Colossus-like, both our inner and outer worlds, the brain is the essential conduit for everything we learn. Yet despite its pivotal importance, we've only recently begun exploring it. That's because the risks and side effects of available technology formerly precluded the study of the *normal* brain. As a result, we knew more about diseased and injured brains than properly functioning ones.

Now thanks to new and safer technologies we're able to visualize the healthy brain in color-coded three dimensional images. These images have provoked widespread excitement and research. As a result, we've learned more about the brain during the last 20 years than the previous 500. Yet our knowledge is still limited.

Neuroscientists are feverishly pursuing the so far elusive goal of an overarching explanation of how the brain works. At times the search is reminiscent of the childhood game of trying to jump on one's own shadow: We're learning about the brain via the perspective provided by our brain.

While this self-referential aspect of neuroscience doesn't present an insurmountable obstacle, it does require us to employ a kind of double bookkeeping system. On one hand we must study the brain with the objectivity befitting any other area of inquiry. On the other hand we have to remain alert that our brain's ways of perceiving, thinking, and acting may lead us to incorrect conclusions.

For instance, when we sleep our conscious world temporarily ceases. Based on this nightly experience, it's tempting to assume that our brain goes into a corresponding temporary suspension of activity. But our brain doesn't shut down; it continues to consolidate and enhance the knowledge acquired during our waking hours. This insight awaited the development of the electroencephalogram (EEG) in the 1920s followed by the demonstration, 10 years later, that EEG patterns change dramatically over the course of a night's sleep.

Throughout our lives, we're establishing pathways within the brain composed of millions of nerve cells. As we mature these pathways increase in complexity—a process similar to the branching of a tree as it grows. Our cognitive abilities evolve as an accompaniment to this anatomical and functional brain complexity.

But here's the most inspiring of insights about the brain: We can enhance our brain's performance by our own efforts. Thus learning about the brain provides a wonderful mix of instruction, amazement, and self improvement. As you gain knowledge, you're in a better position to improve its functioning and thereby increase the quality of your life.

Which brings me to *Brain,* that rarest of treasures: an easily readable book on a difficult and complex subject that is understandable, encompassing, entertaining, and just plain fun to read. (And after writing 20 books on the brain I trust I can claim, without seeming immodest, that I recognize a good "brain book" when I read it.)

Michael Sweeney has done an admirable job here in fashioning a one-volume synthesis of an incredible amount of information. For the reader new to the subject, *Brain* provides a valuable and rewarding introduction. And for those readers who are already "hooked" on the brain, they will find in this volume a marvelous compendium of the current state of our knowledge, as well as provocative suggestions about the future direction of brain research.

OPPOSITE: *Whirring cogs and spinning gears represent the flurry of activity driven and controlled by the brain.*

HOW TO USE THIS BOOK

THE NINE chapters found in *Brain: The Complete Mind* are packed with riveting information. Complementing the engaging narrative—which covers the brain's anatomy, its myriad functions, and its interactions with the world—are hardworking reference elements that pepper every page with facts and figures, amazing stories of breakthrough developments and the pioneering thinkers behind them, descriptions of ailments and maladies, and helpful strategies to keeping the brain at its best. Each feature brings out the fascinating dimensions of the human brain.

❶ SUBSECTIONS: Divide a chapter into segments on the major subjects
❷ DIAGRAMS: Show the inner workings and anatomy of the brain, its processes, and its functions
❸ FAST FACTS: Present fascinating bits of information and figures
❹ TABLES: Organize key information into a quickly understood format
❺ FACT BOXES: Feature entertaining explanations and interesting anecdotes that you didn't know about the brain
❻ CROSS-REFERENCES: Make valuable connections to related information in areas throughout the book
❼ WHAT CAN GO WRONG SIDEBARS: Discuss what happens to the brain when an injury or an illness occurs, various treatments, and current research
❽ FLOW CHARTS: Illustrate processes and functions in an easy to understand format
❾ CHAPTER GLOSSARIES: Define key terms found within each chapter
❿ BREAKTHROUGH SIDEBARS: Chronicle the amazing discoveries that deepen our understanding of the brain
⓫ STAYING SHARP SIDEBARS: Document smart practices and strategic tactics for keeping the brain healthy
⓬ HISTORY SIDEBARS: Reveal the stories behind historical neuroscience beliefs and practices and the men and women who shaped them

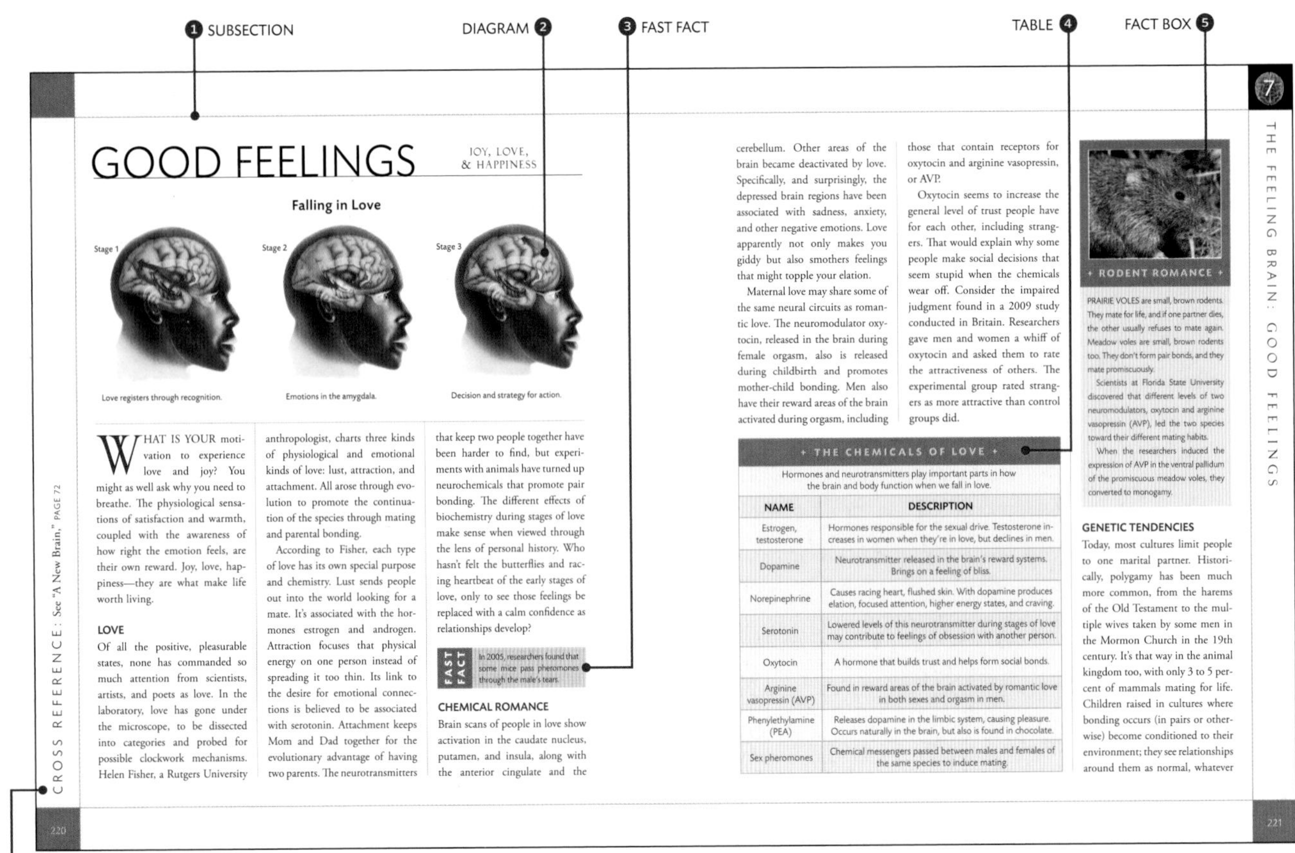

GOOD FEELINGS

JOY, LOVE, & HAPPINESS

Falling in Love

Stage 1 — Love registers through recognition.

Stage 2 — Emotions in the amygdala.

Stage 3 — Decision and strategy for action.

WHAT IS YOUR motivation to experience love and joy? You might as well ask why you need to breathe. The physiological sensations of satisfaction and warmth, coupled with the awareness of how right the emotion feels, are their own reward. Joy, love, happiness—they are what make life worth living.

LOVE

Of all the positive, pleasurable states, none has commanded so much attention from scientists, artists, and poets as love. In the laboratory, love has gone under the microscope, to be dissected into categories and probed for possible clockwork mechanisms. Helen Fisher, a Rutgers University anthropologist, charts three kinds of physiological and emotional kinds of love: lust, attraction, and attachment. All arose through evolution to promote the continuation of the species through mating and parental bonding.

According to Fisher, each type of love has its own special purpose and chemistry. Lust sends people out into the world looking for a mate. It's associated with the hormones estrogen and androgen. Attraction focuses that physical energy on one person instead of spreading it too thin. Its link to the desire for emotional connections is believed to be associated with serotonin. Attachment keeps Mom and Dad together for the evolutionary advantage of having two parents. The neurotransmitters that keep two people together have been harder to find, but experiments with animals have turned up neurochemicals that promote pair bonding. The different effects of biochemistry during stages of love make sense when viewed through the lens of personal history. Who hasn't felt the butterflies and racing heartbeat of the early stages of love, only to see those feelings be replaced with a calm confidence as relationships develop?

FAST FACT: In 2005, researchers found that some mice pass pheromones through the male's tears.

CHEMICAL ROMANCE

Brain scans of people in love show activation in the caudate nucleus, putamen, and insula, along with the anterior cingulate and the cerebellum. Other areas of the brain became deactivated by love. Specifically, and surprisingly, the depressed brain regions have been associated with sadness, anxiety, and other negative emotions. Love apparently not only makes you giddy but also smothers feelings that might topple your elation.

Maternal love may share some of the same neural circuits as romantic love. The neuromodulator oxytocin, released in the brain during female orgasm, also is released during childbirth and promotes mother-child bonding. Men also have their reward areas of the brain activated during orgasm, including those that contain receptors for oxytocin and arginine vasopressin, or AVP.

Oxytocin seems to increase the general level of trust people have for each other, including strangers. That would explain why some people make social decisions that seem stupid when the chemicals wear off. Consider the impaired judgment found in a 2009 study conducted in Britain. Researchers gave men and women a whiff of oxytocin and asked them to rate the attractiveness of others. The experimental group rated strangers as more attractive than control groups did.

+ THE CHEMICALS OF LOVE +

Hormones and neurotransmitters play important parts in how the brain and body function when we fall in love.

NAME	DESCRIPTION
Estrogen, testosterone	Hormones responsible for the sexual drive. Testosterone increases in women when they're in love, but declines in men.
Dopamine	Neurotransmitter released in the brain's reward systems. Brings on a feeling of bliss.
Norepinephrine	Causes racing heart, flushed skin. With dopamine produces elation, focused attention, higher energy states, and craving.
Serotonin	Lowered levels of this neurotransmitter during stages of love may contribute to feelings of obsession with another person.
Oxytocin	A hormone that builds trust and helps form social bonds.
Arginine vasopressin (AVP)	Found in reward areas of the brain activated by romantic love in both sexes and orgasm in men.
Phenylethylamine (PEA)	Releases dopamine in the limbic system, causing pleasure. Occurs naturally in the brain, but also is found in chocolate.
Sex pheromones	Chemical messengers passed between males and females of the same species to induce mating.

+ RODENT ROMANCE +

PRAIRIE VOLES are small, brown rodents. They mate for life, and if one partner dies, the other usually refuses to mate again. Meadow voles are small, brown rodents too. They don't form pair bonds, and they mate promiscuously.

Scientists at Florida State University discovered that different levels of two neuromodulators, oxytocin and arginine vasopressin (AVP), led the two species toward their different mating habits.

When the researchers induced the expression of AVP in the ventral pallidum of the promiscuous meadow voles, they converted to monogamy.

GENETIC TENDENCIES

Today, most cultures limit people to one marital partner. Historically, polygamy has been much more common, from the harems of the Old Testament to the multiple wives taken by some men in the Mormon Church in the 19th century. It's that way in the animal kingdom too, with only 3 to 5 percent of mammals mating for life. Children raised in cultures where bonding occurs (in pairs or otherwise) become conditioned to their environment; they see relationships around them as normal, whatever

CROSS REFERENCE: See "A New Brain," PAGE 72

7 THE FEELING BRAIN: GOOD FEELINGS

220

221

8 FLOW CHART

9 GLOSSARY

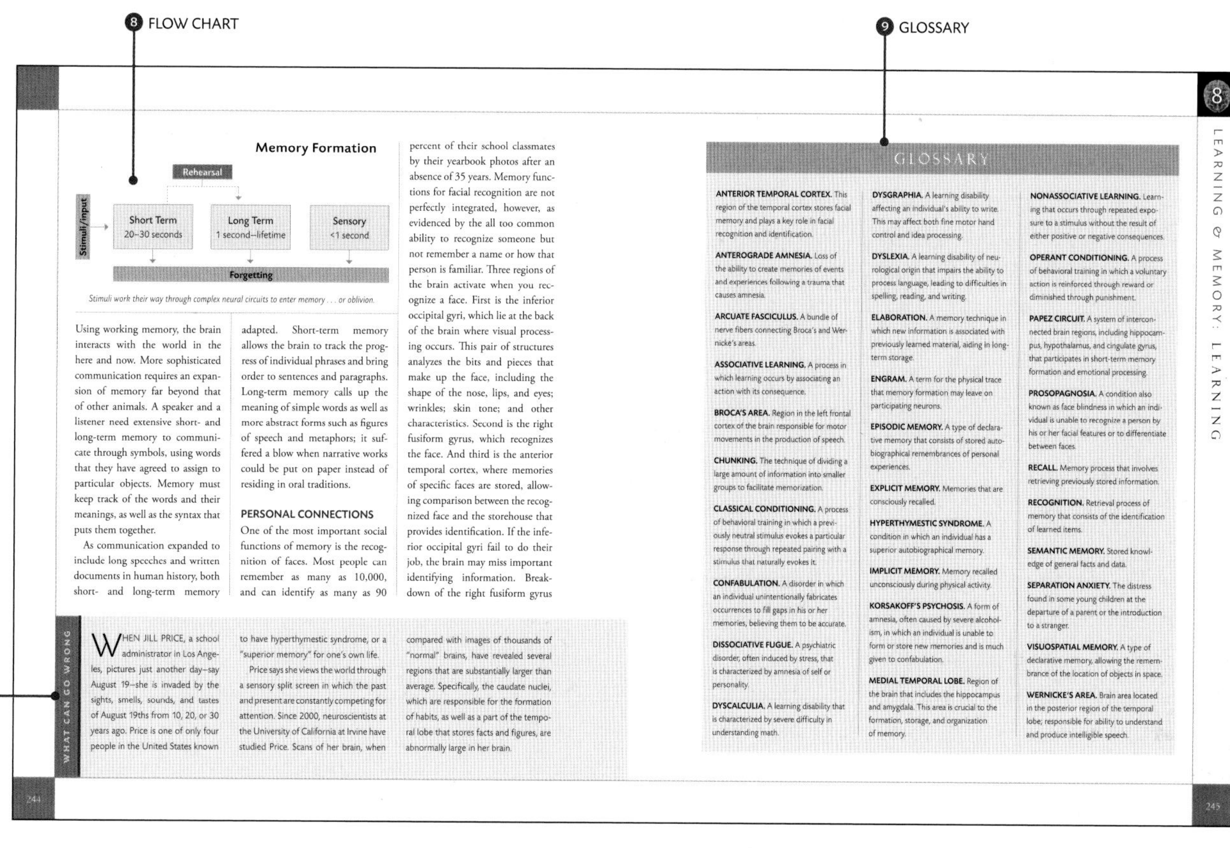

Stimuli work their way through complex neural circuits to enter memory . . . or oblivion.

Using working memory, the brain interacts with the world in the here and now. More sophisticated communication requires an expansion of memory far beyond that of other animals. A speaker and a listener need extensive short- and long-term memory to communicate through symbols, using words that they have agreed to assign to particular objects. Memory must keep track of the words and their meanings, as well as the syntax that puts them together.

As communication expanded to include long speeches and written documents in human history, both short- and long-term memory adapted. Short-term memory allows the brain to track the progress of individual phrases and bring order to sentences and paragraphs. Long-term memory calls up the meaning of simple words as well as more abstract forms such as figures of speech and metaphors; it suffered a blow when narrative works could be put on paper instead of residing in oral traditions.

PERSONAL CONNECTIONS

One of the most important social functions of memory is the recognition of faces. Most people can remember as many as 10,000, and can identify as many as 90 percent of their school classmates by their yearbook photos after an absence of 35 years. Memory functions for facial recognition are not perfectly integrated, however, as evidenced by the all too common ability to recognize someone but not remember a name or how that person is familiar. Three regions of the brain activate when you recognize a face. First is the inferior occipital gyri, which lie at the back of the brain where visual processing occurs. This pair of structures analyzes the bits and pieces that make up the face, including the shape of the nose, lips, and eyes; wrinkles; skin tone; and other characteristics. Second is the right fusiform gyrus, which recognizes the face. And third is the anterior temporal cortex, where memories of specific faces are stored, allowing comparison between the recognized face and the storehouse that provides identification. If the inferior occipital gyri fail to do their job, the brain may miss important identifying information. Breakdown of the right fusiform gyrus

WHAT CAN GO WRONG

WHEN JILL PRICE, a school administrator in Los Angeles, pictures just another day—say August 19—she is invaded by the sights, smells, sounds, and tastes of August 19ths from 10, 20, or 30 years ago. Price is one of only four people in the United States known to have hyperthymestic syndrome, or a "superior memory" for one's own life.

Price says she views the world through a sensory split screen in which the past and present are constantly competing for attention. Since 2000, neuroscientists at the University of California at Irvine have studied Price. Scans of her brain, when compared with images of thousands of "normal" brains, have revealed several regions that are substantially larger than average. Specifically, the caudate nuclei, which are responsible for the formation of habits, as well as a part of the temporal lobe that stores facts and figures, are abnormally large in her brain.

244

GLOSSARY

ANTERIOR TEMPORAL CORTEX. This region of the temporal cortex stores facial memory and plays a key role in facial recognition and identification.

ANTEROGRADE AMNESIA. Loss of the ability to create memories of events and experiences following a trauma that causes amnesia.

ARCUATE FASCICULUS. A bundle of nerve fibers connecting Broca's and Wernicke's areas.

ASSOCIATIVE LEARNING. A process in which learning occurs by associating an action with its consequence.

BROCA'S AREA. Region in the left frontal cortex of the brain responsible for motor movements in the production of speech.

CHUNKING. The technique of dividing a large amount of information into smaller groups to facilitate memorization.

CLASSICAL CONDITIONING. A process of behavioral training in which a previously neutral stimulus evokes a particular response through repeated pairing with a stimulus that naturally evokes it.

CONFABULATION. A disorder in which an individual unintentionally fabricates occurrences to fill gaps in his or her memories, believing them to be accurate.

DISSOCIATIVE FUGUE. A psychiatric disorder, often induced by stress, that is characterized by amnesia of self or personality.

DYSCALCULIA. A learning disability that is characterized by severe difficulty in understanding math.

DYSGRAPHIA. A learning disability affecting an individual's ability to write. This may affect both fine motor hand control and idea processing.

DYSLEXIA. A learning disability of neurological origin that impairs the ability to process language, leading to difficulties in spelling, reading, and writing.

ELABORATION. A memory technique in which new information is associated with previously learned material, aiding in long-term storage.

ENGRAM. A term for the physical trace that memory formation may leave on participating neurons.

EPISODIC MEMORY. A type of declarative memory that consists of stored autobiographical remembrances of personal experiences.

EXPLICIT MEMORY. Memories that are consciously recalled.

HYPERTHYMESTIC SYNDROME. A condition in which an individual has a superior autobiographical memory.

IMPLICIT MEMORY. Memory recalled unconsciously during physical activity.

KORSAKOFF'S PSYCHOSIS. A form of amnesia, often caused by severe alcoholism, in which an individual is unable to form or store new memories and is much given to confabulation.

MEDIAL TEMPORAL LOBE. Region of the brain that includes the hippocampus and amygdala. This area is crucial to the formation, storage, and organization of memory.

NONASSOCIATIVE LEARNING. Learning that occurs through repeated exposure to a stimulus without the result of either positive or negative consequences.

OPERANT CONDITIONING. A process of behavioral training in which a voluntary action is reinforced through reward or diminished through punishment.

PAPEZ CIRCUIT. A system of interconnected brain regions, including hippocampus, hypothalamus, and cingulate gyrus, that participates in short-term memory formation and emotional processing.

PROSOPAGNOSIA. A condition also known as face blindness in which an individual is unable to recognize a person by his or her facial features or to differentiate between faces.

RECALL. Memory process that involves retrieving previously stored information.

RECOGNITION. Retrieval process of memory that consists of the identification of learned items.

SEMANTIC MEMORY. Stored knowledge of general facts and data.

SEPARATION ANXIETY. The distress found in some young children at the departure of a parent or the introduction to a stranger.

VISUOSPATIAL MEMORY. A type of declarative memory, allowing the remembrance of the location of objects in space.

WERNICKE'S AREA. Brain area located in the posterior region of the temporal lobe; responsible for ability to understand and produce intelligible speech.

245

7 WHAT CAN GO WRONG SIDEBAR

11 STAYING SHARP SIDEBAR

12 HISTORY SIDEBAR

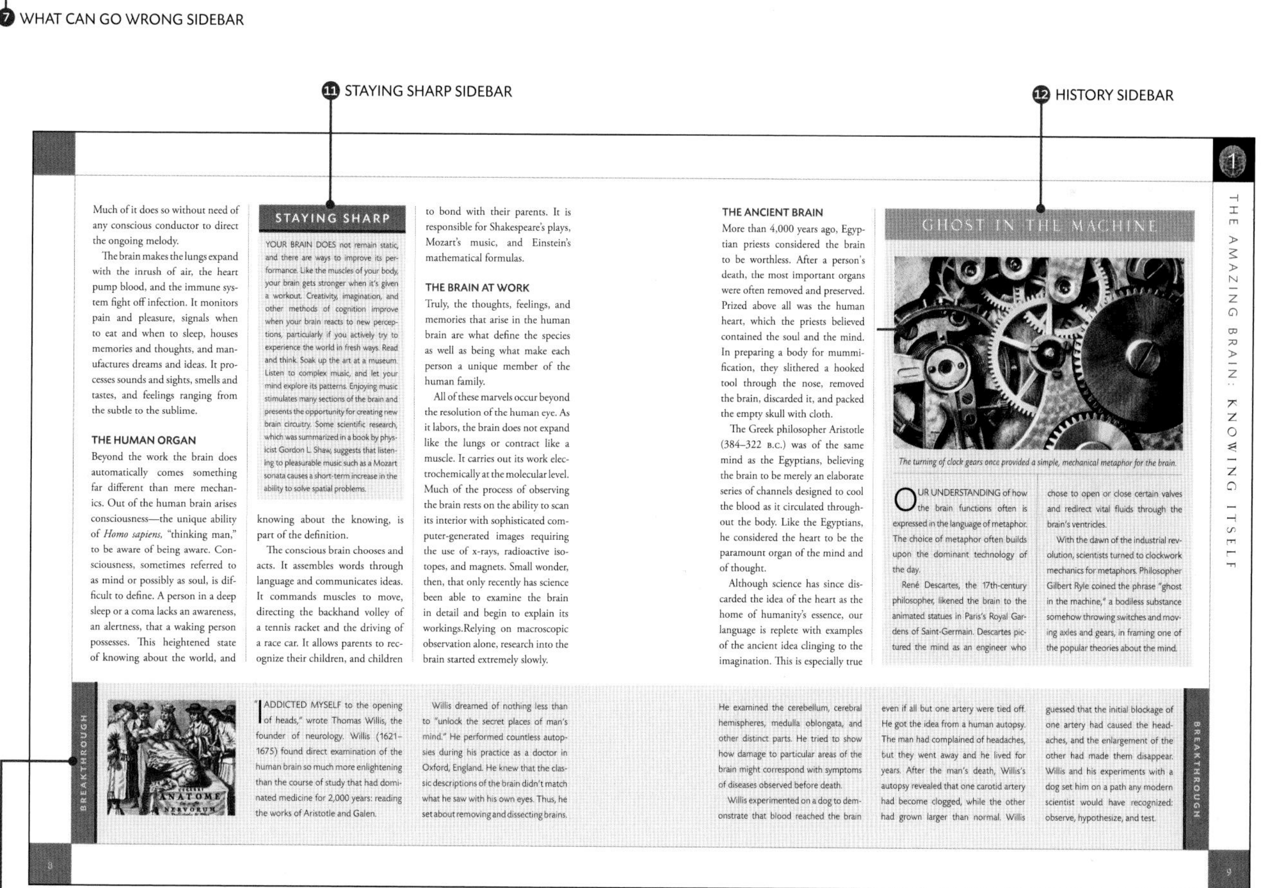

Much of it does so without need of any conscious conductor to direct the ongoing melody.

The brain makes the lungs expand with the inrush of air, the heart pump blood, and the immune system fight off infection. It monitors pain and pleasure, signals when to eat and when to sleep, houses memories and thoughts, and manufactures dreams and ideas. It processes sounds and sights, smells and tastes, and feelings ranging from the subtle to the sublime.

THE HUMAN ORGAN

Beyond the work the brain does automatically comes something far different than mere mechanics. Out of the human brain arises consciousness—the unique ability of *Homo sapiens,* "thinking man," to be aware of being aware. Consciousness, sometimes referred to as mind or possibly as soul, is difficult to define. A person in a deep sleep or a coma lacks an awareness, an alertness, that a waking person possesses. This heightened state of knowing about the world, and knowing about the knowing, is part of the definition.

The conscious brain chooses and acts. It assembles words through language and communicates ideas. It commands muscles to move, directing the backhand volley of a tennis racket and the driving of a race car. It allows parents to recognize their children, and children to bond with their parents. It is responsible for Shakespeare's plays, Mozart's music, and Einstein's mathematical formulas.

STAYING SHARP

YOUR BRAIN DOES not remain static, and there are ways to improve its performance. Like the muscles of your body, your brain gets stronger when it's given a workout. Creativity, imagination, and other methods of cognition improve when your brain reacts to new perceptions, particularly if you actively try to experience the world in fresh ways. Read and think. Soak up the art at a museum. Listen to complex music, and let your mind explore its patterns. Enjoying music stimulates many sections of the brain and presents the opportunity for creating new brain circuitry. Some scientific research, which was summarized in a book by physicist Gordon L. Shaw, suggests that listening to pleasurable music such as a Mozart sonata causes a short-term increase in the ability to solve spatial problems.

THE BRAIN AT WORK

Truly, the thoughts, feelings, and memories that arise in the human brain are what define the species as well as being what make each person a unique member of the human family.

All of these marvels occur beyond the resolution of the human eye. As it labors, the brain does not expand like the lungs or contract like a muscle. It carries out its work electrochemically at the molecular level. Much of the process of observing the brain rests on the ability to scan its interior with sophisticated computer-generated images requiring the use of x-rays, radioactive isotopes, and magnets. Small wonder, then, that only recently has science been able to examine the brain in detail and begin to explain its workings.Relying on macroscopic observation alone, research into the brain started extremely slowly.

BREAKTHROUGH

"I ADDICTED MYSELF to the opening of heads," wrote Thomas Willis, the founder of neurology. Willis (1621–1675) found direct examination of the human brain so much more enlightening than the course of study that had dominated medicine for 2,000 years: reading the works of Aristotle and Galen.

Willis dreamed of nothing less than to "unlock the secret places of man's mind." He performed countless autopsies during his practice as a doctor in Oxford, England. He knew that the classic descriptions of the brain didn't match what he saw with his own eyes. Thus, he set about removing and dissecting brains. He examined the cerebellum, cerebral hemispheres, medulla oblongata, and other distinct parts. He tried to show how damage to particular areas of the brain might correspond with symptoms of diseases observed before death.

Willis experimented on a dog to demonstrate that blood reached the brain even if all but one artery were tied off. He got the idea from a human autopsy. The man had complained of headaches, but they went away and he lived for years. After the man's death, Willis's autopsy revealed that one carotid artery had become clogged, while the other had grown larger than normal. Willis guessed that the initial blockage of one artery had caused the headaches, and the enlargement of the other had made them disappear. Willis and his experiments with a dog set him on a path any modern scientist would have recognized: observe, hypothesize, and test.

8

THE ANCIENT BRAIN

More than 4,000 years ago, Egyptian priests considered the brain to be worthless. After a person's death, the most important organs were often removed and preserved. Prized above all was the human heart, which the priests believed contained the soul and the mind. In preparing a body for mummification, they slithered a hooked tool through the nose, removed the brain, discarded it, and packed the empty skull with cloth.

The Greek philosopher Aristotle (384–322 B.C.) was of the same mind as the Egyptians, believing the brain to be merely an elaborate series of channels designed to cool the blood as it circulated throughout the body. Like the Egyptians, he considered the heart to be the paramount organ of the mind and of thought.

Although science has since discarded the idea of the heart as the home of humanity's essence, our language is replete with examples of the ancient idea clinging to the imagination. This is especially true

GHOST IN THE MACHINE

The turning of clock gears once provided a simple, mechanical metaphor for the brain.

OUR UNDERSTANDING of how the brain functions often is expressed in the language of metaphor. The choice of metaphor often builds upon the dominant technology of the day.

René Descartes, the 17th-century philosopher, likened the brain to the animated statues in Paris's Royal Gardens of Saint-Germain. Descartes pictured the mind as an engineer who chose to open or close certain valves and redirect vital fluids through the brain's ventricles.

With the dawn of the industrial revolution, scientists turned to clockwork mechanics for metaphors. Philosopher Gilbert Ryle coined the phrase "ghost in the machine," a bodiless substance somehow throwing switches and moving axles and gears, in framing one of the popular theories about the mind.

9

10 BREAKTHROUGH SIDEBAR

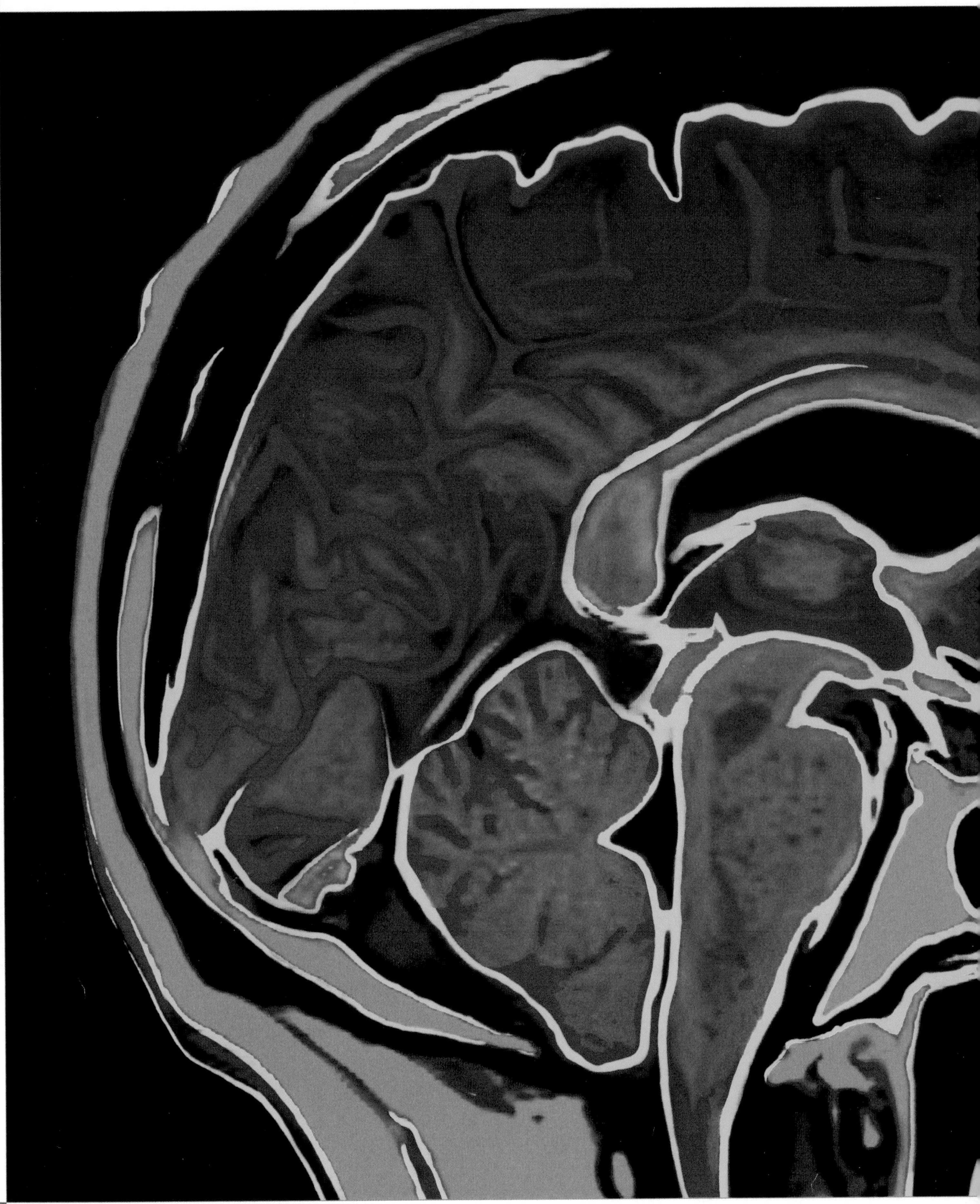

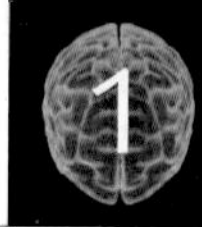

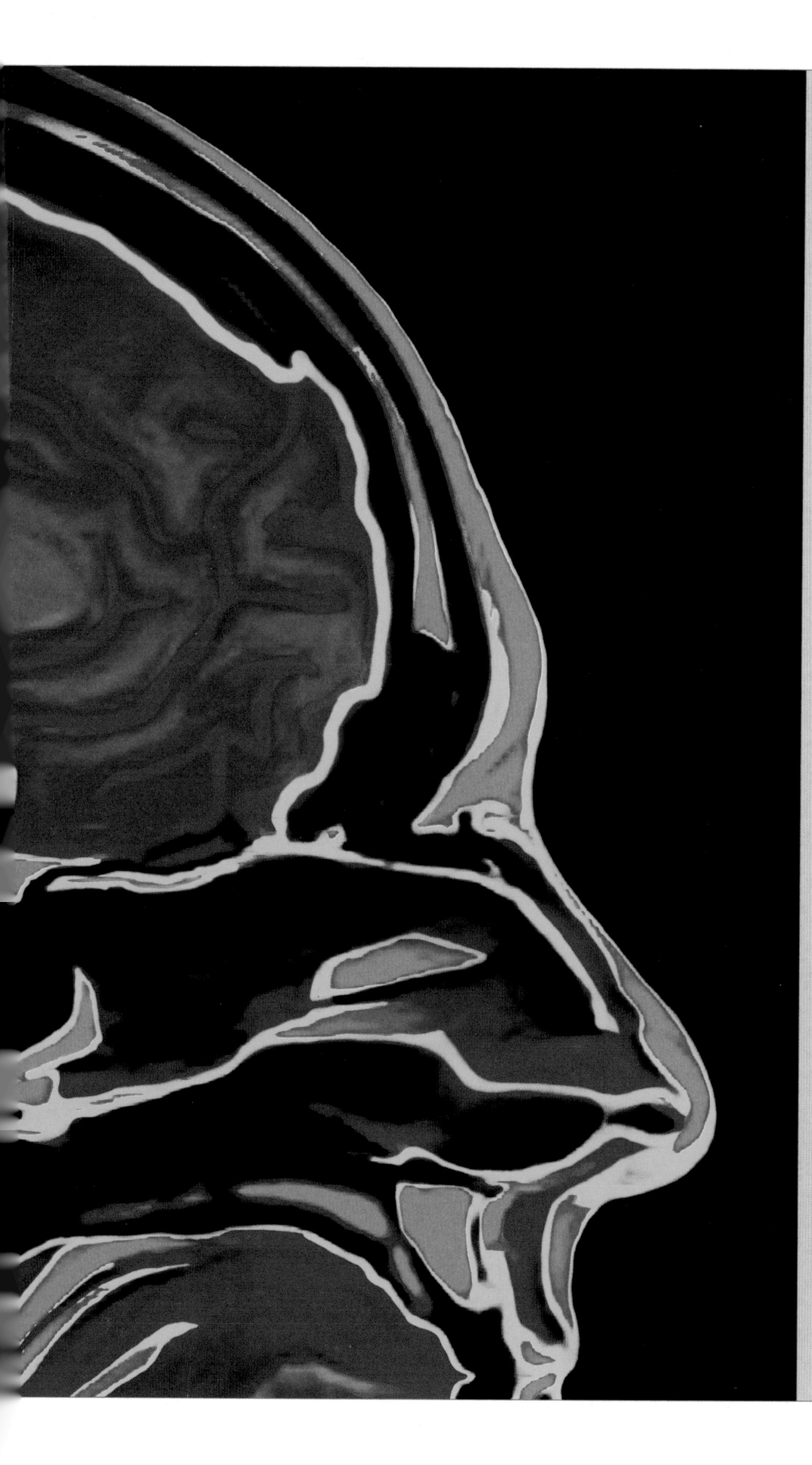

CHAPTER ONE

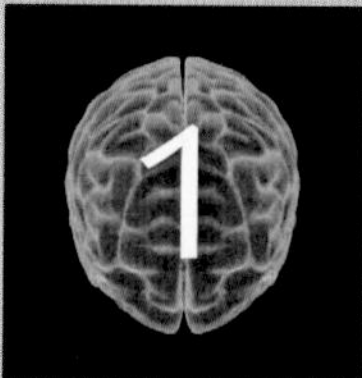

THE AMAZING BRAIN

THE BRAIN should need no introduction. You should know it intimately. After all, the brain is what makes you you. But it's a paradox that the organ that lets you know and connect with the world understands so little about itself. Now, thanks to stunning research building upon decades—no, centuries—of investigation, science is peeling away the layers of mystery to reveal how three pounds of flesh create an entire universe inside your head.

A magnetic resonance image of the human brain reveals its complex internal structure.

KNOWING ITSELF

UNDERSTANDING THE BRAIN

IT'S NOT MUCH to look at. Hippocrates, the Greek healer identified with the birth of medicine more than 2,000 years ago, thought it was made of moist phlegm. English philosopher Henry More, writing in the 1600s, compared it to bone marrow, a bowl of curds, or a cake of suet. Modern-day neurologist Richard Restak says it resembles nothing so much as a large, wrinkly, squishy walnut. Looks can be deceiving.

The brain, a three-pound chunk of organic matter, is not only the body's most marvelous organ, it is the most complicated object known. It is "wider than the sky," wrote poet Emily Dickinson. "For, put them side by side, / The one the other will contain / With ease, and you beside."

It is humbling to consider the brain and all that it does in every moment of our lives. In this corrugated mass of flesh, a staggeringly complex symphony of electrochemical reactions plays out every second of every day. Much of it does so without need of any conscious conductor to direct the ongoing melody.

The brain makes the lungs expand with the inrush of air, the heart pump blood, and the immune system fight off infection. It monitors pain and pleasure, signals when to eat and when to sleep, houses memories and thoughts, and manufactures dreams and ideas. It processes sounds and sights, smells and tastes, and feelings ranging from the subtle to the sublime.

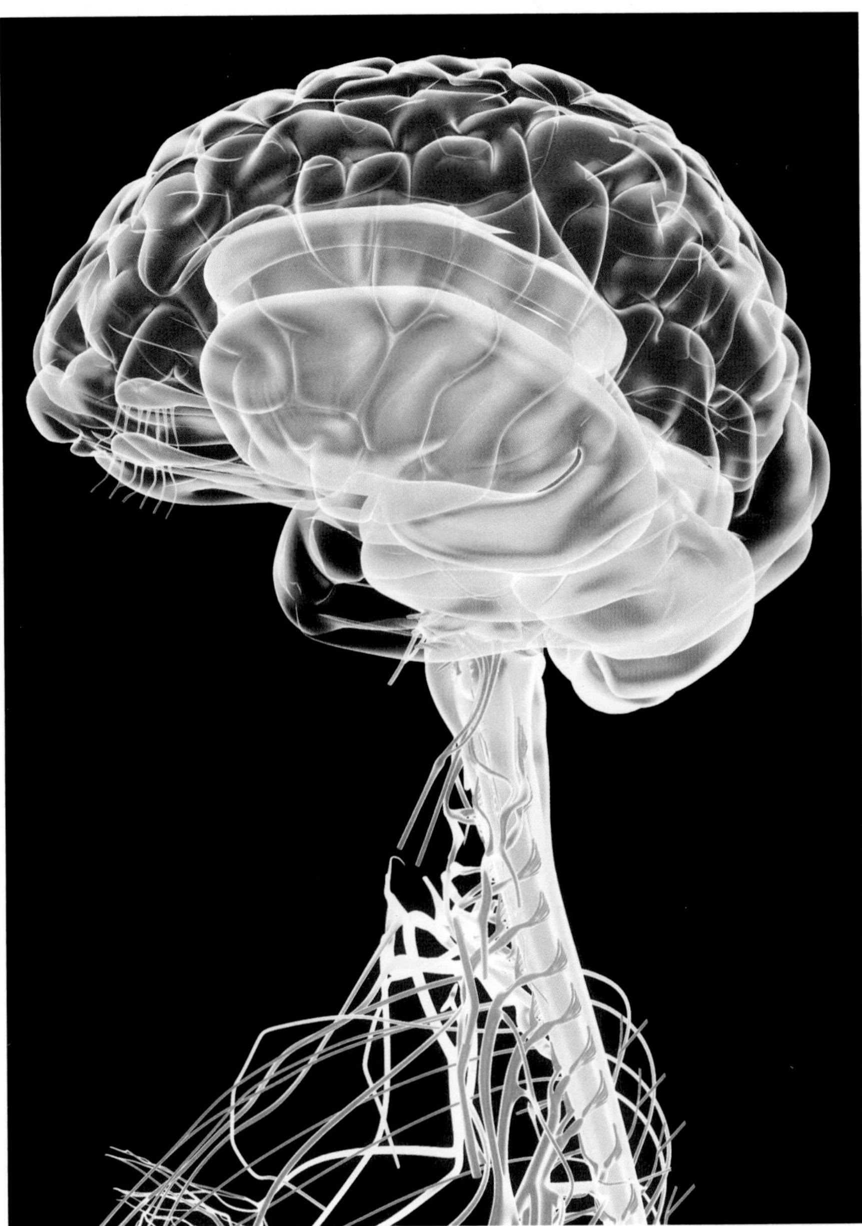

Nerves and spinal cord fan out from the brain in this stylized view of the neck and head.

THE HUMAN ORGAN

Beyond the work the brain does automatically comes something far different than mere mechanics. Out of the human brain arises consciousness—the unique ability of *Homo sapiens,* "thinking man," to be aware of being aware. Consciousness, sometimes referred to as mind or possibly as soul, is difficult to define. A person in a deep sleep or a coma lacks an awareness, an alertness, that a waking person possesses. This heightened state of knowing about the world, and knowing about the knowing, is part of the definition.

The conscious brain chooses and acts. It assembles words through language and communicates ideas. It commands muscles to move, directing the backhand volley of a tennis racket and the driving of a race car. It allows parents to recognize their children, and children to bond with their parents. It is responsible for Shakespeare's plays, Mozart's music, and Einstein's mathematical formulas.

THE BRAIN AT WORK

Truly, the thoughts, feelings, and memories that arise in the human brain are what define the species as well as being what make each person a unique member of the human family.

All of these marvels occur beyond the resolution of the human eye. As it labors, the brain does not expand like the lungs or contract like a muscle. It carries out its work electrochemically at the molecular level. Much of the process of observing the brain rests on the ability to scan its interior with sophisticated computer-generated images requiring the use of x-rays, radioactive isotopes, and magnets. Small wonder, then, that only recently has science been able to examine the brain in detail and begin to explain its workings. Relying on macroscopic observation alone, research into the brain started extremely slowly.

FAST FACT External stimuli can physically alter the brain. For example, stress weakens the encoding of memories.

THE ANCIENT BRAIN

More than 4,000 years ago, Egyptian priests considered the brain to be worthless. After a person's death, the most important organs were often removed and preserved. Prized above all was the human heart, which the priests believed contained the soul and the mind. In preparing a body for mummification, they slithered a hooked tool through the nose, removed the brain, discarded it, and packed the empty skull with cloth.

The Greek philosopher Aristotle (384–322 B.C.) was of the same mind as the Egyptians, believing the brain to be merely an elaborate series of channels designed to cool the blood as it circulated throughout the body. Like the Egyptians, he considered the heart to be the paramount organ of the mind and of thought.

Although science has since discarded the idea of the heart as the home of humanity's essence, our language is replete with examples of the ancient idea clinging to the imagination. This is especially true in love and romance. We speak of losing a heart to a loved one, suffering a broken heart, and being heartsick. In reality, falling in and out of love is a matter of losing our brain—or perhaps, as anyone insane with romance could tell you, our mind.

STAYING SHARP

YOUR BRAIN DOES not remain static, and there are ways to improve its performance. Like the muscles of your body, your brain gets stronger when it's given a workout. Creativity, imagination, and other methods of cognition improve when your brain reacts to new perceptions, particularly if you actively try to experience the world in fresh ways. Read and think. Soak up the art at a museum. Listen to complex music, and let your mind explore its patterns. Enjoying music stimulates many sections of the brain and presents the opportunity for creating new brain circuitry. Some scientific research, which was summarized in a book by physicist Gordon L. Shaw, suggests that listening to pleasurable music such as a Mozart sonata causes a short-term increase in the ability to solve spatial problems. Neurologist Richard Restak concurs with this finding. He believes that listening to Mozart for a few minutes each day may boost your cognition across many levels, from simple perceptions to deeper thoughts. *Eine kleine Nachtmusik,* anyone?

THE GREEKS & THE BRAIN

The Greeks were the first to begin to recognize the brain's paramount status. About 2,500 years ago, a Pythagorean philosopher named Alcmaeon of Croton favored the head over the heart as the home of sensory awareness. Consciousness arose in the brain, he said. Alcmaeon is reported to have peered into the skull of a dead animal after removing the eyes. He speculated on the possibility of life-giving spirits moving through open channels in the body, such as those he saw leading backward from the eye cavities. However, he probably did little to examine the human brain directly, given the Greek taboo against dissection.

Hippocrates (circa 460 B.C.–377 B.C.) took a similar view of the brain's importance a half century later. "The eyes and ears and tongue and hands and feet do whatsoever the brain determines," he wrote. "It is the brain that is the messenger to the understanding [and] the brain that interprets the understanding." Furthermore, the brain gives rise to joys, sorrows, griefs, and all other emotions, he said.

Hippocrates saw the brain as the potential generator of madness, depression, and other illnesses. He believed that four "humors"—black bile, yellow bile, blood, and phlegm—governed the body's health, as well as imbalances that led to illness. The brain was phlegm, he said, and if it became too wet its condition might lead to disorders such as epilepsy.

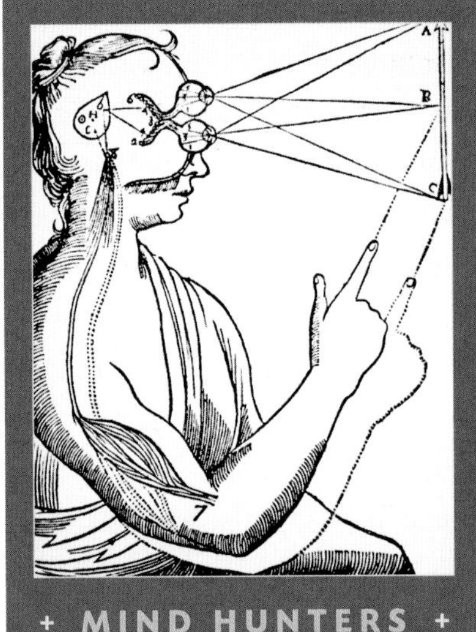

+ MIND HUNTERS +

HISTORICALLY, GREAT thinkers have placed the mind—sometimes referred to as the soul or the psyche—at various places in the human body. Some candidates across time:

+ Aristotle: The heart. It's in the center of the body and is the first organ to be discerned in an embryo.
+ Thomas Aquinas: The ventricles, or empty spaces, of the brain. Being pure spirit, it survives beyond death.
+ René Descartes: The pineal gland.

A PLATONIC VIEW

Aristotle's teacher, Plato, reasoned that the mind had to exist inside the brain because of geometry and pure logic. The brain was round, he said, and close to the perfect roundness of the sphere. It also inhabited the part of the human body closest to heaven.

Plato and other Greek philosophers theorized about the existence of a force that kept people alive and left them at death. They called this force *psyche,* or soul, and several said it resided in the brain. Some split the soul into three spirits. Humans and all living creatures took in life-giving essence from *pneuma,* or air. As pneuma moved through the body, it changed in ways that animated and strengthened its host. Digested food provided energy for the liver, where the pneuma became "natural spirit." This spirit traveled to the heart to become the "vital spirit." Then it traveled to the brain, where it transformed into the "animal spirit" that creates the conscious mind. Plato considered the soul that resides in the brain to be immortal, surviving the death of the body.

GALEN LOOKS INSIDE

Centuries later, Galen, a Roman physician who lived in the eastern Mediterranean in the second century of the Christian era, went beyond such mental exercises to test the brain for himself. He took a more hands-on approach and cut the sensory and motor fibers in pigs' brains to observe the results.

FAST FACT The idea that the mind survives the body's death appears quite ancient. Burial sites from 100,000 years ago reveal bodies interred with tools and food, possibly to help on journeys in the afterlife. Cave art possibly depicts spirit worlds.

Galen became the first to speculate that particular functions are carried out in specific parts of the brain. Furthermore, as a healer to wounded gladiators, Galen peered into holes rent in human bodies by the violent combat of the arena. He made rudimentary descriptions of the body's major organs and fleshed out the description of what he saw as the varieties of human spirit. The liver created desire and pleasure, he said. The heart gave rise to courage and the warmer passions. And the brain contained the rational soul.

Vital spirits swirling in the spaces of the brain carried the spark of human intelligence. He believed they navigated throughout the body via a network of hollow nerve fibers. The brain's instructions thus moved through tunnels like puffs of wind in pneumatic tubes.

Hippocrates, physician of ancient Greece, dissects a cow's head in a 16th-century woodcut.

Feelings, understanding, and consciousness arising in the physical structure of the brain? Unseen spirits causing the physical body to move? These were ideas that raised serious questions. If the qualities of thought that set humans apart from other animals—be it the sense of self, the mind, or the soul—resided in a physical organ, where, exactly, could such thoughts be found? And if thoughts and feelings had no substance, how could they act upon the physical matter of the human body?

MIND-BRAIN PROBLEM

Thus was born a conundrum that has sparked debate for many centuries. It's called the mind-body or mind-brain problem. Attempts to solve the problem had to await the rebirth of the Renaissance. Direct observation and systematic testing of hypotheses provided the keys.

The first direct, systematic observation of the human brain occurred in the 1300s when Italian medical schools began allowing human cadavers to be dissected. Authorizations came slowly at first, with one university permitting only one male and one female body to be cut up each year. With time, however, human autopsies became more commonplace.

Leonardo da Vinci (1452–1519) drew his extensive knowledge of anatomy from dissecting bodies. He fashioned a wax cast of an ox brain and assigned functions such as imagination, reason, and memory to its various parts. Without a way to test his hypothesis, however, he left room for disagreement

FAST FACT Are you a morning or evening person? Your brain is wired to prefer one or the other.

about what he observed. Critics said the part of the brain Leonardo assigned to the function of imagination was more likely related to sensation. As it was close to the sense organs, they said it must be the home of *sensus communis,* or common sense.

COGITO, ERGO SUM

More than a century later, mathematician and philosopher René Descartes aimed to ascertain more about the brain, and with greater clarity and certainty. The way to ascertain things with certainty, he believed, was to break them into their smallest parts and solve the pieces first.

Descartes began in the 1620s by addressing how humans know about the world. He wondered whether he could trust the reality he perceived with his senses. Such questions flourished in an age when Galileo and Copernicus were rewriting the laws that governed the movement of celestial bodies. Descartes knew that the rising and setting of the sun were functions of the Earth's rotation instead of the physical movement of the sun around the Earth, but his eyes tricked him into believing the sun actually rose in the east and set in the west. To get at the heart of how he could know something with certainty, Descartes sat inside a Dutch inn and pondered the nature of knowledge. He looked at the furniture and asked himself how he could know for certain it existed. The answer: he could not. All he could settle on was that his *perception* of the furniture existed. His consciousness, his awareness of the world, lay beyond the pale of any doubt. *Cogito, ergo sum:* I think, therefore I am, he said, and thus the ultimate reality of the world lies in the mind's perception of it. To Descartes, if a tree falls in a forest and there is nobody there to hear it, the lack of perception guides the answer as to whether it makes any sound.

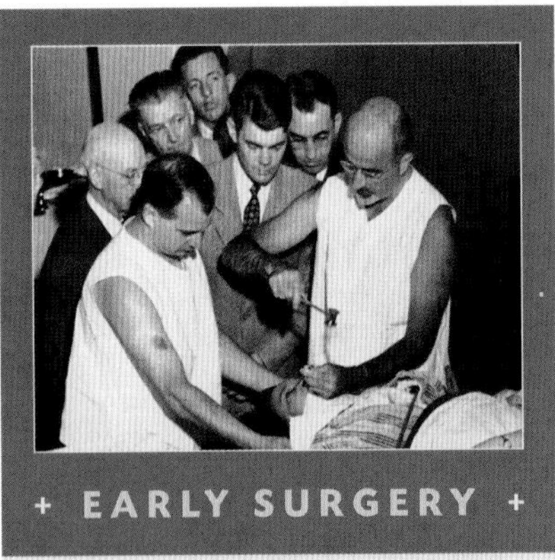

+ EARLY SURGERY +

+ A 17-century Jamestown colony skull shows signs of brain surgery.
+ William Macewen removed a tumor from a young woman's brain in 1879. She survived the surgery.
+ American physician Harvey Cushing (1869–1939) removed more than 2,000 brain tumors.
+ Portuguese doctor António Egas Moniz performed the first prefrontal lobotomies on humans in the 1930s. While the surgery, which cut key fibers in the frontal lobes, had the desired effect of calming agitated patients, it also drained them of emotion. Lobotomies now are considered radical procedures.

DESCARTES DISSECTS

Not content to just consider the function of the brain, Descartes began to physically examine brain and nerve specimens to gather more data. He bought the carcasses of slaughtered animals at the butcher shops of Amsterdam and dissected them to learn more, through his observation, about the brain, nerves, and body. "These are my books," he told visitors who asked to see his library.

Despite adopting first principles that demanded skepticism, Descartes took some leaps of faith as he examined brain and body. He considered nerves to be tubes that swelled and pulsated with living spirits, which pushed and pulled at muscle tissue. Nerves swollen with animal spirits could pull a foot back from a fire or turn the gaze from one object to another. Much of the action of movement was pure reflex, he said, carried out independent of will. (It's not

CROSS REFERENCE: See "Awareness," PAGE 178

hard to see where this idea arose: Push your fingertip into a candle flame and see whether the idea to remove it occurs before you yank it back to safety.) According to Descartes, mechanical operations of the body and brain, working like an elaborate clock, recorded images through the eyes, engraved memories in the mind, and moved the body through the coordination of nerves.

BRAIN & SOUL

Descartes saw no physical soul in his tours of the body. Instead, he conceived the soul as noncorporeal and thus above the mechanics that animated all flesh. Operating within the machine but not part of it, the soul oversaw humanity's consciousness, will, and all other attributes that separate mankind from the animals. Furthermore, he said, "There is only one soul in us, and that soul does not have in itself any diversity of parts."

Where, specifically, could that soul, or mind, reside within a person? Descartes sought his answer by going to his "books." Dissecting the brains of calves—even though they supposedly had no souls—Descartes settled on a tiny gland deep in the brain. The pineal gland appeared to reside in a central location where nerves and the ventricles, or spaces, of the brain converged. Thus, he thought it a perfect candidate for the role of

GHOST IN THE MACHINE

The turning of clock gears once provided a simple, mechanical metaphor for the brain.

OUR UNDERSTANDING of how the brain functions often is expressed in the language of metaphor. The brain is sometimes a computer, a phone bank, a black box. The choice of metaphor often builds upon the dominant technology of the day.

René Descartes, the 17th-century philosopher, likened the brain to the animated statues in Paris's Royal Gardens of Saint-Germain. Descartes described how the weight of a visitor's foot on particular garden tiles opened or closed hidden valves and redirected water flowing through a network of pipes. Streams of water flowing internally caused the statues, called automatons, to move. Descartes pictured the mind as an engineer who chose to open or close certain valves and redirect vital fluids through the brain's ventricles.

With the dawn of the industrial revolution, scientists turned to clockwork mechanics for metaphors. Philosopher Gilbert Ryle coined the phrase "ghost in the machine," a bodiless substance somehow throwing switches and moving axles and gears, in framing one of the popular theories about the mind.

Telephone metaphors arose in the 20th century but were not complex enough to describe the vast, organic circuitry of the brain. The function of brain circuitry and the importance of neuronal networking gave rise to metaphors including computers and the integrated complexity of the Web. But even the most sophisticated computer cannot rewrite its own programming or be aware of its own existence. The brain so far has eluded the perfect metaphor.

central actor in the drama of perception and action.

"Let us then conceive here that the soul [mind] has its principal seat in the little gland which exists in the middle of the brain," he wrote, "from whence it radiates forth through all the remainder of the body by means of the animal spirits, nerves, and even the blood, which, participating in the impressions of the spirits, can carry them by the arteries into all the members." Inside the pineal gland, at an infinitesimally small point, Descartes envisioned the mind orchestrating the actions of the body.

+ A CRUDE MAP +

SCIENTISTS CAN trace some of the earliest efforts to map the brain to Thomas Willis. In some of his experiments, he used live test subjects to illustrate his findings. In one particularly gruesome one, Willis tried to show that the brain's functions are localized by removing a dog's cerebellum, which he erroneously believed controlled heartbeat and respiration. No surprise—the dog died.

FAST FACT Thomas Willis was buried in Westminster Abbey. Poet Nathaniel Williams eulogized him as one who "knewst the wondrous art, / And order of each part, / In the whole lump, how every sense / Contributes to the health's defense."

This dualism separating the mind and brain has been thoroughly challenged by modern science. The mind cannot exist without the brain; damage to the brain results in compromises to the mind. Nevertheless, the view espoused by Descartes still colors our view of ourselves to this day. Neurologists treat disorders of the brain. Psychiatrists and psychologists treat disorders of the mind. Only now, as neuroscience begins to tease out the biological processes at the root of emotional and behavioral illnesses, are the mind and brain coming together again.

BIRTH OF NEUROSCIENCE

Today's scholars of the human brain and mind owe a great debt to Thomas Willis. Working in England in the middle of the 17th century, he meticulously observed and cataloged the anatomy of the human brain through dissection. Working in a medieval house at Oxford known as Beam Hall, Willis—a short man with a mop of hair an observer once described as "like a dark red pigge"—cut open cadaver skulls to observe and examine the brains and nervous tissue inside. He snipped the nerves that held fast to the nose and eyes. Then he flipped the brain to gently remove the membranes clustered around the nerves, veins, and arteries at its base. Finally, he held up the brain and described it for his audience of natural philosophers, doctors, and the merely curious who had assembled to watch the spectacle.

Watching carefully, Willis's assistant sketched the brain as he saw it laid bare. That artist, who illustrated Willis's 1664 book *Cerebri Anatome (Anatomy of the Brain),* was none other than Christopher Wren, who went on to design St. Paul's Cathedral in London. Wren's careful drawings of the human

BREAKTHROUGH

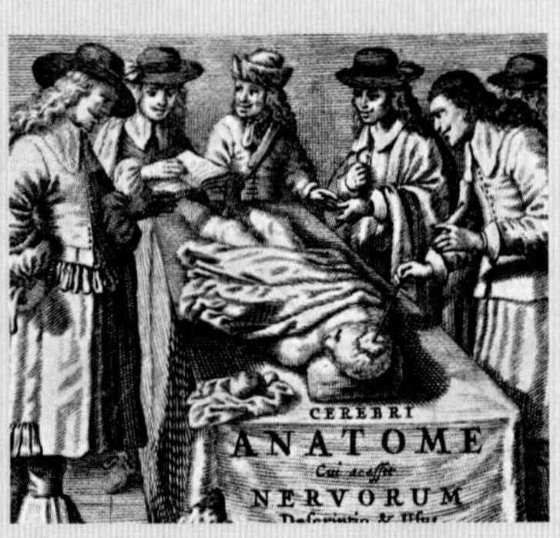

"I ADDICTED MYSELF to the opening of heads," wrote Thomas Willis, the founder of neurology. Willis (1621–1675) found direct examination of the human brain so much more enlightening than the course of study that had dominated medicine for 2,000 years: reading the works of Aristotle and Galen.

Willis dreamed of nothing less than to "unlock the secret places of man's mind." He performed countless autopsies during his practice as a doctor in Oxford, England. He knew that the classic descriptions of the brain didn't match what he saw with his own eyes. Thus, he set about removing and dissecting brains.

brain reproduce with nearly photographic clarity the contours and divisions easily recognized by modern medical students.

Wren developed a revolutionary method of inserting chemicals into the blood vessels of animals to better highlight the networks between them. Working with Willis, Wren injected india ink mixed with a hardening agent into vessels entering the brain. The ink made the vessels stand out like rivers and their tributaries drawn on a map.

A BETTER ASSESSMENT

Willis examined the complicated flesh of the brain and discarded the notion that its key functions lay in its ventricles, or spaces, where the ancients had conceived of spirits flowing and animating flesh. Instead, he correctly settled on the substance of the brain itself as the location of all the action.

Willis fancied fanciful language. He likened the brain's two main hemispheres to a pair of military towers, stronger for their reliance on each other. He also compared two masses that shared an artery to two provinces bordering a river. But in the significance of his observations, Willis, who was a founder of the Royal Society, stayed true to science. He argued that in the brain's convoluted folds and wrinkles, all memories, ideas, and passions found a home. All had a physical basis in the brain, he said. His studies became the first scientific investigation of the brain and nervous system. He called his work *neurologie.*

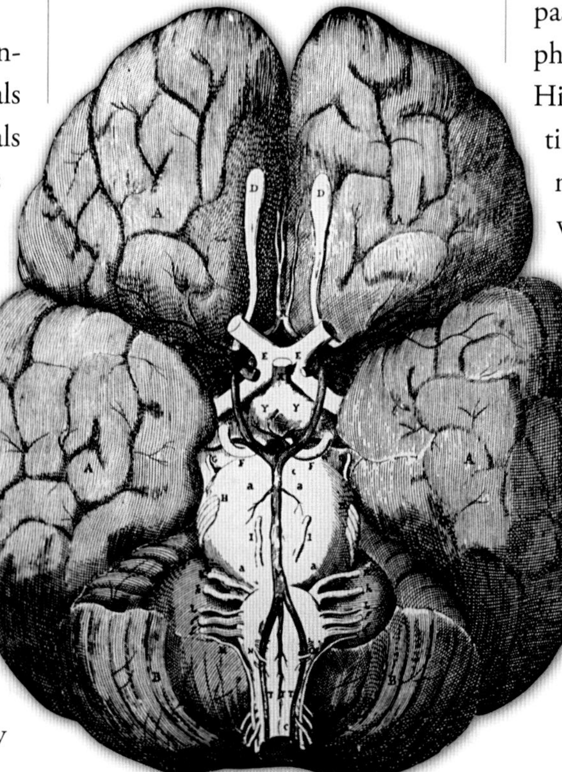

Christopher Wren's 1664 drawing traces the brain's blood supply.

Willis's dissections were crude by modern standards. Yet neuroscientists continue to follow the methods of Willis and Descartes: Look at the brain and the nervous system. Examine their parts. Trace the workings of the small bits and try to assemble them into a greater whole.

How far down the rabbit hole can the process go? Today's neuroscientists are examining not just molecules, but also the atoms—and subatomic particles—that the universe of brain chemistry comprises. Like peeling an onion, each layer takes the researcher deeper and deeper, closer to the heart of the matter.

BREAKTHROUGH

He examined the cerebellum, cerebral hemispheres, medulla oblongata, and other distinct parts. He tried to show how damage to particular areas of the brain might correspond with symptoms of diseases observed before death.

Willis experimented on a dog to demonstrate that blood reached the brain even if all but one artery were tied off. He got the idea from a human autopsy. The man had complained of headaches, but they went away and he lived for years. After the man's death, Willis's autopsy revealed that one carotid artery had become clogged, while the other had grown larger than normal. Willis guessed that the initial blockage of one artery had caused the headaches, and the enlargement of the other had made them disappear. Willis and his experiments with a dog set him on a path any modern scientist would have recognized: observe, hypothesize, and test.

NERVE CELLS

THE FUNDAMENTAL units of the brain, too small to see in Willis's time, are two types of nerve cells. One type, the neuroglia (or glial—"glue"—cells), has the rather pedestrian task of supporting the nervous system. Neuroglia play a role in guiding neurons toward making connections, promoting neuron health, insulating neuronal processes, and otherwise influencing neuronal functioning and, thus, information processing in the brain. Glial cells continue to divide over the course of a lifetime and fill in spaces in the brain. Glial cells come in six varieties, with some playing a key role in physical health by attacking invading microbes.

The human brain has about 100 billion neurons and about 50 trillion neuroglia.

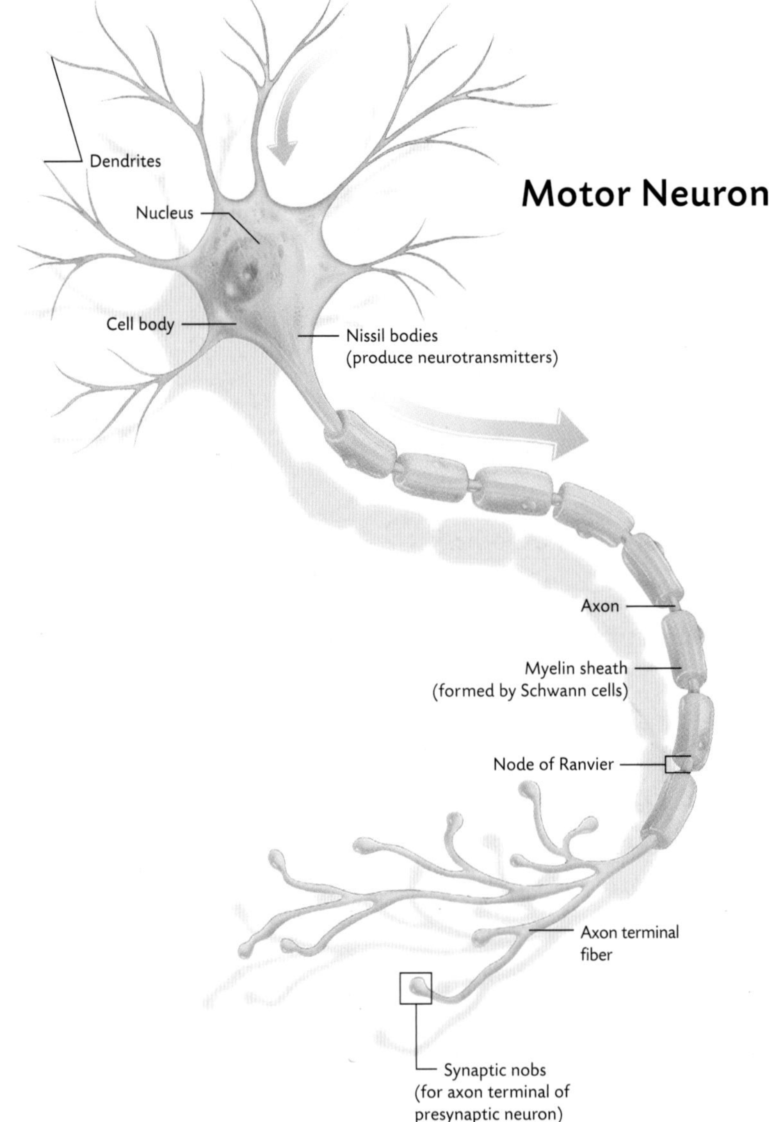

NEURONS

The other type of cell in the brain is the nerve cell, or neuron. In the late 1800s, a Spanish neuroscientist, Santiago Ramón y Cajal, used a special solution containing silver to stain nerve cells and examine them under a microscope in great detail. Ramón y Cajal's method worked on only about one in a hundred cells. Nevertheless, he was able to observe enough of the silver-encrusted neurons to describe them in vivid detail. The nerve cell was the "aristocrat among the structures of the body," he said, "with its giant arms stretched out like the tentacles of an octopus to the provinces on the frontier of the outside world, to watch for the constant struggles of physical and chemical forces."

Seen en masse in the outer regions of the human brain, neurons appear gray to the naked eye. Hence, scientists exploring the brain described neurons as gray matter. When Agatha Christie's fictional detective Hercule Poirot brags of the detective work of his "little gray cells," he is praising his neurons.

CROSS REFERENCE: See "Messengers," PAGE 42

ANATOMY OF A NEURON

Each neuron has a main cell body. Like all cells, the neuron contains a nucleus and an exterior membrane, which sometimes receives electrochemical messages from other neurons. Chains of neurons send messages from the body to the brain: "Here is pain, in the left wrist." "Here is the odor of soup." "Here is a stony surface beneath the feet." Chains also send messages from brain to body: "Shake your hand." "Eat." "Take a step."

Each neuron has an array of branching fibers called dendrites that extend outward toward other neurons. Dendrites expand the surface area of the neuron, increasing its sensitivity to its neighboring neurons. While some neurons have only a few dendrites, others have hundreds. They act as receptors for signals traveling from other neurons, carrying information toward the main body of the nerve cell.

Each neuron also contains one electrically sensitive fiber called an axon extending from one end of the cell body. Axons may be as short as a fraction of an inch or as long as several feet, as is the case with axons extending from the spine to the toes. At the axon's terminal end, as many as 10,000 branches spread out toward the dendrites of other neurons. Every branch terminates in a knoblike projection, like the business end of a paper match. These bulbs are called axon

COMPETITIVE DISCOVERIES

Santiago Ramón y Cajal, in a 1906 portrait, documented the existence of synapses.

OFTEN THE spirited competition between two great minds can yield amazing discoveries. Such was the case between Spanish neuroscientist Santiago Ramón y Cajal (1852–1934) and his Italian contemporary, Camillo Golgi (1843–1926), who shared the Nobel Prize in physiology or medicine in 1906. Ramón y Cajal was recognized for his deduction on the anatomy of a neuron; Golgi, for the staining process that made that deduction possible. Like most scientists at the time, Golgi held that neurons operate as one continuous, tangled network. Nerve cells must be fused, he said, to pass electrical impulses. Ramón y Cajal, however, envisioned chemical codes traveling across a synaptic gap between a single axon and the dendrites of the next cell. In 1887, Ramón y Cajal learned of Golgi's staining technique and realized its superiority. He modified it, finding it worked well with thicker sections of nervous tissue. Bird samples and tissue from younger animals were best, he surmised, because their axons lacked the protein sheath that obscures most nerve fibers. When impregnated with silver nitrate and viewed by microscope, these nerve cells jumped out as inky strokes on a yellowish background. *La reazione nera*—"the black reaction," as Golgi called it—illuminated the infinitesimal as well as the road toward Ramón y Cajal's revelation.

+ TYPES OF GLIAL CELLS +	
TYPE	**FUNCTION**
Astrocytes	Most abundant type. Support neurons' connections to capillaries, guide neuron migration and synapse formation. Control chemical environment around neurons.
Microglia	Monitor health of neurons. Protect neurons under attack and "clean up" dead ones.
Ependymal cells	Line the central cavities of brain and spinal cord. Help circulate cerebrospinal fluid filling those spaces.
Oligodendrocytes	Form insulating myelin sheaths around the axons of neurons in central nervous system.
Schwann cells	Form myelin sheaths around the axons of cells in the peripheral nervous system
Satellite cells	Surround nerve cell bodies in peripheral nervous system. Regulate chemical environment around neurons.

terminals, synaptic knobs, and *boutons,* or buttons.

Around the length of most axons lies a special wrapping of fatty tissue called a myelin sheath. The sheath is formed by two kinds of glial cells, called Schwann cells in the peripheral nervous system and oligodendrocytes in the central nervous system. The wrap is not continuous; small gaps called nodes of Ranvier separate the cylinders of fatty tissue that surround the axons. The axon's encompassing myelin acts as insulation, speeding the transmission of information in the form of nerve impulses moving at 9 to 400 feet per second.

When an electrical impulse reaches an axon terminal, it communicates across a tiny gap, called a synapse, separating it from the dendrite of another neuron. A few can connect directly with tissues of the skeletal muscles and glands, allowing direct communication.

Neurons differ in shape and complexity. Most, in particular the vast majority of those in the brain, are multipolar—they have one axon and a multitude of dendrites. The rest of the neurons are bipolar or unipolar. The former can be found in the retina, where neurons have a single dendrite. The latter, found in the peripheral nervous system, have a single extension from the main cell body that divides, like the cap of the letter "T," into branches for an axon and dendrites.

NEURONS AT WORK

Neurons serve different functions. Motor neurons carry impulses to activate glands and muscles. Sensory neurons send impulses from the skin and other body parts to the central nervous system. Interneurons, residing in the brain and spinal cord, integrate the signals and are crucial in making decisions. Thus, neurons allow for information from the body to reach the brain, be processed, and sometimes result in responses.

Some liken the neuron to an old-fashioned, landline telephone. The body of the neuron compares to the body of the phone, where signals are processed. The telephone receiver compares to the dendrites and their ability to gather information. And the axon compares to a telephone line, sending information processed in the phone body along an electrically conductive wire. It has the potential to pass information along to any other phone on the planet.

+ NEW CIRCUITS +

IF NEURONAL CIRCUITRY rewires itself in response to stimulation, do the brains of teens raised on the Internet and high-tech gadgets differ from those of older generations? The answer most likely is yes.

UCLA psychiatrist Gary Small believes tech-savvy children strengthen synaptic connections for electronic communication while their circuitry for a face-to-face world, such as reading body language, fades. Meanwhile, late adopters of technology lag in their ability to master new communication media.

MAKING CONNECTIONS

The human brain contains in the neighborhood of 100 billion neurons. Each neuron reaches out toward others with an array of dendrites and axon terminals. Each is capable of communicating with any other and, in the process, forging thousands of synaptic connections through the thickets of dendrites and axon terminals. All told, the brain has hundreds of trillions of synapses. No computer can match the human brain for its complexity and its potential for creative thought.

Communication occurs where two neurons come together. Camillo Golgi, a contemporary of Ramón y Cajal's, believed that neurons physically touched each other, forming a continuous network of neural fibers. Ramón y Cajal disagreed. In his sketches, he painstakingly drew neurons whose dendrites invariably terminated at a tiny gap that prevented them from touching other neurons. His drawings did not lie.

In the synaptic cleft, a neuron communicates with its neighbors by issuing electrochemical commands that may be strictly localized or extend the length of the longest chains of axons.

PLASTICITY

Neurons are not physically bound to each other like so many lengths of pipe, so they have the flexibility to make, break, and remake relationships with other neurons. The ability to reshape neural interactions in the brain is referred to as plasticity. The brain's ability to rewire itself helps it stay sharp.

As the brain ages, it loses individual neurons, but it retains its power to form new connections that increase the mind's complexity. In short, if new educational experiences challenge the brain to form new synaptic connections, its neurons will do more with less.

Experimental data with laboratory animals demonstrate the principle of "use it or lose it." When lab animals are placed in an environment with challenging toys, their brains develop a far greater number of neuronal connections than those raised in a dull environment. The brains of animals from stimulating environments will even weigh more because of the greater number of synapses.

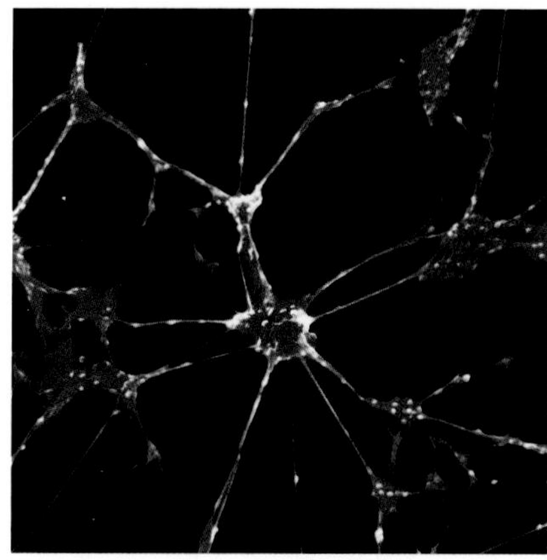

CONNECTIONS

Axons and dendrites spread in ghostly webs from the bodies of nerve cells.

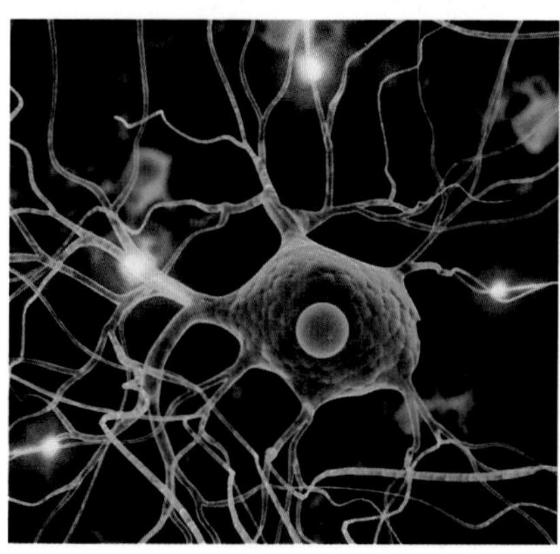

ACTIONS

Nerve cells flash with electrochemical activity in a three-dimensional rendering.

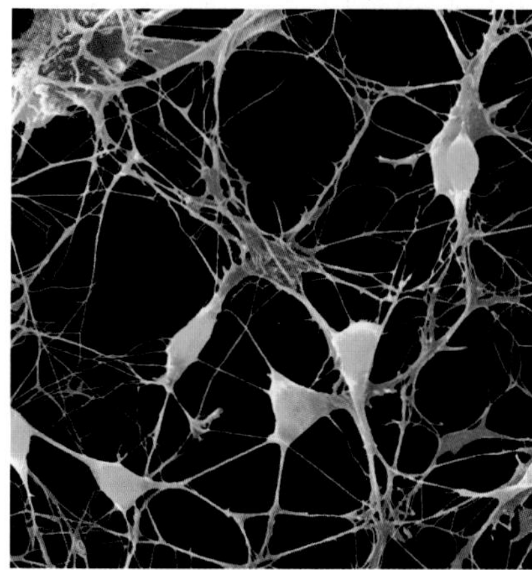

GROWTH & SUPPORT

Cortical neurons and a supporting neuroglial cell, top left, grow in a culture.

FAST FACT

The number of synapses may be as high as one thousand trillion, or the number 1 followed by 15 zeroes.

COMMUNICATIONS

Tim Berners-Lee, a creator of the World Wide Web, likens the brain's complexity to the nearly infinite capacity for Web sites to connect to each other. "A piece of information is really defined only by what it's related to," he said. "The structure is everything. There are billions of

Communicating with another cell, neurotransmitters journey across a synapse.

neurons in our brains, but what are neurons? Just cells. The brain has no knowledge until connections are made between neurons. All that we know, all that we are, comes from the way our neurons are connected."

Transmissions between neurons take place in two stages. The first is electrical. An electrical discharge travels the length of an axon. When it reaches the axon terminal that abuts the synaptic space, it sets the second stage in motion. This button, like the rest of the nerve cell, has an outer wall called a membrane. Its envelope contains a solution of messenger chemicals. These electrically charged chemicals move in the solution, constantly poised to respond to an impulse and exit through small openings of the membrane and into the synapse. When an electrical impulse arrives from the axon, if it is of sufficient strength it trips a trigger that releases one of the messenger chemicals, called a neurotransmitter, from storage in the button.

NEUROTRANSMITTERS

The neurotransmitting chemical then enters the synapse. Like a ferryboat crossing a small stream, the neurotransmitter traverses the synaptic cleft and attempts to link up with the dendritic membrane of a receptor cell. The journey across the synapse takes only a thousandth of a second. The receptor cell's surface

BREAKTHROUGH

WAKING IN THE middle of the night on the eve of Easter, 1921, German-born pharmacologist Otto Loewi (1873–1961) recalled an inspiring dream that gave him an idea for an experiment that would shatter scientists' conception of neural communication.

Most turn-of-the-century brain scientists believed nerves sent impulses via electric waves, firing sparks across the synaptic gap, neuron to neuron. In this way, they thought, motor intentions born in the cerebral cortex could be transmitted to receptor muscles and organs throughout the body. Only a handful of scientists—most notably Loewi and his English counterpart, Henry Dale—argued that chemical neurotransmitters are released at the synapse. An accelerant, noradrenaline, causes the heart to beat more quickly, Dale said. An inhibitor, acetylcholine, induces the opposite. Yet Dale was unable to extract

+ NEUROTRANSMITTERS +		
NEUROTRANSMITTER	**LOCATION**	**FUNCTION**
Acetylcholine	Parts of the nervous system associated with motion, including the brain's motor cortex.	Makes muscles contract. Also plays a key role in attention, memory, and sleep.
Dopamine	Brain and the peripheral nervous system	Important for body motion and reward experiences, including pleasure. Sufferers of Parkinson's disease lack normal levels of dopamine.
Endorphins	Brain, pituitary gland, and spinal cord	Powerful, natural opiates, endorphins block pain.
Gamma-aminobutyric Acid (GABA)	Retina, spinal cord, hypothalamus, and cerebellum	The most common inhibitory neurotransmitter, it quiets rather than excites neurons. Exists in as many as a third of all synapses.
Glutamate	Brain and spinal cord	Crucial for learning and memory. At small doses, glutamate excites cells to higher states of activity. At larger doses, however, glutamate kills neurons.
Norepinephrine	Brain and the peripheral nervous system	Regulates moods, blood pressure, heartbeat, and arousal.
Serotonin	Brain stem, cerebellum, pineal gland, and spinal cord	Crucial for proper sleep and appetite. Linked to depression and anxiety.

contains specially shaped docking sites, so particular neurotransmitters can dock only at the appropriate places, just as a key needs exactly the right shape to fit into a lock. The neurotransmitter either excites the receptor cell into action or dampens it into inaction. Once the receptor cell has been stimulated by the neurotransmitting chemical, the communication reverts to an electrical signal. It travels the length of the new cell until it reaches the synapse of another receptor cell, and starts the process all over

BREAKTHROUGH

either chemical organically, and lacking proof, his case remained dormant.

Then, as Loewi recalled, a fateful frog experiment flashed to him in a dream, and he dashed to his laboratory. He began with two frogs' hearts. Stimulating the vagus nerve of one to slow its beating, he applied a residual solution from this donor to a second heart, from which he'd severed the vagus nerve. The second heart immediately slowed, as if discouraged by an unseen force. Loewi's hypothesis was correct: A neurotransmitter (acetylcholine) had slowed the first heart, leaving a trace fluid—enough to slow the second, isolated heart.

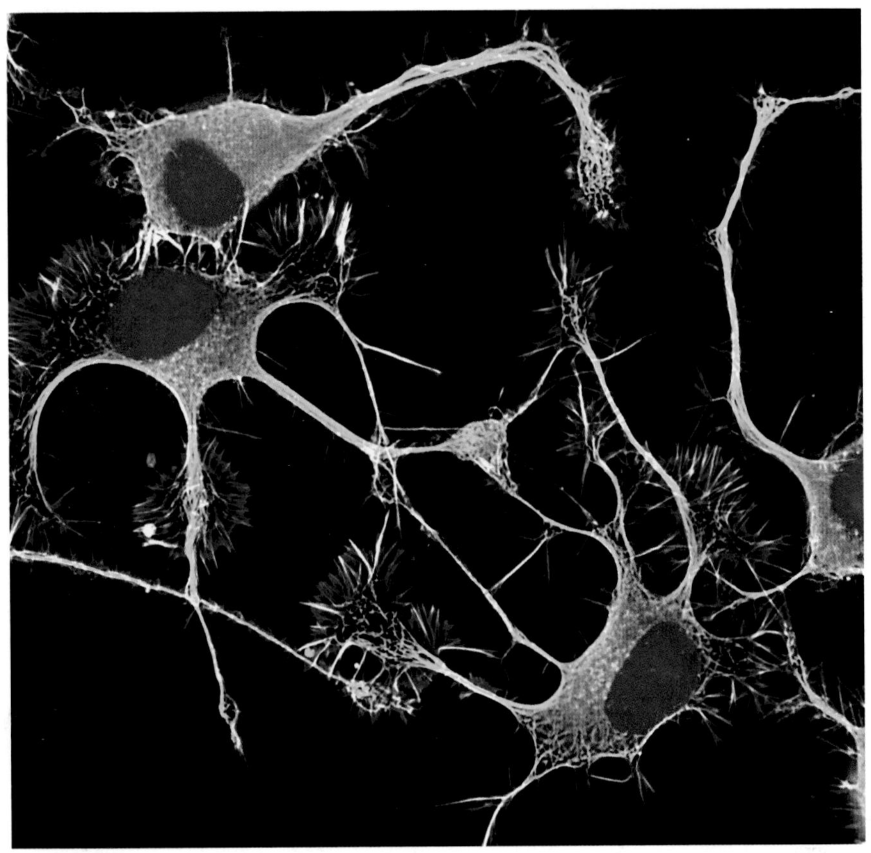

Precursors to axons and dendrites, in yellow and blue, respond to nerve growth stimulation.

again. After they have done their job in the synaptic space between nerve cells, neurotransmitting chemicals are reabsorbed by the transmitting neuron and prepared for rerelease (a process known as reuptake) or broken down and metabolized by enzymes in the synaptic space. It sounds like a lot of work, but neurons can repeat the electrochemical firing process up to a thousand times a second.

FAST FACT The brain devotes huge amounts of neural circuitry to the hands, lips, and tongue.

Dozens of neurotransmitters have been identified, and more discoveries are expected. Certain neurotransmitters make muscles contract, help regulate sleep, and block pain. Research into the role of neurotransmitters in mental and physical health is constantly expanding, and neurotransmitter disorders have been linked to Parkinson's disease, depression, Alzheimer's disease, schizophrenia, and a host of other illnesses.

LIFE SPAN

Amazingly, the cells that perform the complicated ballet of electrochemical transmission can live more than a hundred years, but they do not get replaced like most other body cells. Except for the hippocampus and the olfactory bulb, where new neurons have been shown to grow from stem cells, the neurons a person has at birth are all he or she will ever have. During the busiest times of neuron generation in the developing brain of a fetus, about a quarter million neurons are created every minute. They start from precursor cells and then migrate and differentiate.

When a neuron in the central nervous system dies or its long fibers are cut, it does not regenerate. Medical science currently has no cure for catastrophic nerve injuries of the spinal cord, and once a major communication line to or from the brain has been cut, it cannot be repaired. But new research with neural stem cells suggests neurons may yet be coaxed into regeneration.

+ REEVE'S RESEARCH +

RESEARCH INTO HOW TO regenerate nerve tissue after injuries like transections, a complete severing of the spinal cord, owes a great deal to the late actor Christopher Reeve. In 1995, Reeve shattered a cervical vertebra in a horseback riding accident and became paralyzed from the neck down, a condition known as quadriplegia. The injury was not quite a transection—he eventually regained some sensation—but nevertheless proved devastating. His public appearances in a wheelchair until his 2004 death drew attention to spinal injuries and ultimately raised millions of dollars to help seek a cure for nerve damage.

CROSS REFERENCE: *See "Living Longer,"* PAGE 290

GLOSSARY

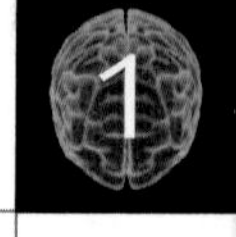

AMYGDALA. An almond-shaped section of the forebrain. This component of the limbic system plays a central role in response to fear and terror.

ARACHNOID. The weblike middle layer of the three membranes surrounding the brain and spinal cord.

AXON. The electrically sensitive fiber of a neuron responsible for the transmission of information away from the nerve cell.

AXON TERMINALS. The knoblike endings of the terminal branches of an axon.

BLOOD-BRAIN BARRIER. Membranes in the brain's blood vessels with limited permeability that inhibit the transfer of many substances from the blood into brain tissues.

CENTRAL FISSURE. Groove separating the frontal lobe from the parietal lobe.

CEREBROSPINAL FLUID. Plasmalike fluid that cushions and protects the brain and spinal cord and provides nourishment to brain tissues.

CRANIUM. The fused bones that encases and protects the brain and the organs for equilibrium and hearing; skull.

DENDRITES. Branching fibers of axons that act as receptors of information. These receive messages from other neurons and deliver them to the main body of the nerve cell.

DURA MATER. The tough outermost layer of the three membranes surrounding the brain and spinal cord, located directly beneath the cranium.

ENDORPHINS. Proteins that reduce the perception of pain.

FISSURES. The deepest inward folds or grooves of the brain.

GYRI (SING., GYRUS). The outward, elevated folds of the cerebral cortex.

HYPOTHALAMUS. Brain region located directly above the brain stem. The center of emotional response. Regulates body temperature, hunger, thirst, and sleeping.

INTERNEURONS. Neurons confined to the brain and spinal cord that integrate information between motor and sensory neurons.

LONGITUDINAL SULCUS. A deep groove that separates the cerebral hemispheres.

MEDULLA OBLONGATA. Part of the brain stem, it connects the spinal cord to higher brain centers. Controls heartbeat and respiration.

MOTOR NEURONS. Neurons carrying impulses away from the central nervous system to activate muscles and glands.

MYELIN SHEATH. An insulating and protective wrapping of fatty tissue that surrounds axons and increases the speed of the transmission of nerve impulses.

NEUROGLIA. Glial cells; these brain cells insulate, guide, and protect neurons.

NEURON. Nerve cell; a central nervous system cell that generates and transmits information from nerve impulses.

NOCICEPTOR. Pain receptor that responds to potentially harmful stimuli.

NODES OF RANVIER. Regular gaps in the myelin sheath occurring along the coated axons.

PIA MATER. The innermost cerebral membrane, this thin layer of connective tissue clings to every dip and curve of the cerebral cortex.

PLASTICITY. The brain's ability to reshape neural interactions.

PONS. Part of the brain stem that serves as a bridge between the medulla and midbrain and aids the medulla in respiratory regulation.

PRECENTRAL GYRUS. Area of the cerebral cortex containing the primary motor complex and thus responsible for body movement. Located on the frontal lobe of each hemisphere.

PREFRONTAL CORTEX. Brain region located in the anterior frontal lobe. Responsible for reasoning, planning, judgment, empathy, abstract ideas, and conscience.

SENSORY NEURONS. Neurons that send impulses from the skin and parts of the body to the central nervous system.

SULCI (SING., SULCUS). Inward folds of the cerebral cortex, more shallow than fissures.

SYLVIAN FISSURE. Groove separating the parietal lobe from the temporal lobe.

SYNAPSE. Tiny gap between the axon terminals of two neurons through which communication occurs.

TRANSECT. To completely sever.

ANATOMY DIFFERENT PARTS, DIFFERENT RESPONSIBILITIES

THE FIRST STEP to a better understanding of the brain is getting acquainted with its parts. From the protective structures on the outside to the hardworking parts on the inside—knowing where each structure is and how it interacts with the world gives greater insight into brain function and the problems that may arise.

PROTECTION

To take a tour of the human brain, begin with the crown of the skull, a collection of 22 bones that house the brain and protect it from harm. Except for the mandible (or jawbone), all of these bones are fused together and immovable. The topmost and rearmost bony parts form the cranium, the brain's tough, protective shell.

Inside, three membranes present themselves to provide more layers of protection. Immediately underneath the skull is the dura mater,

The eight bones that form the cranium shield the brain from injury.

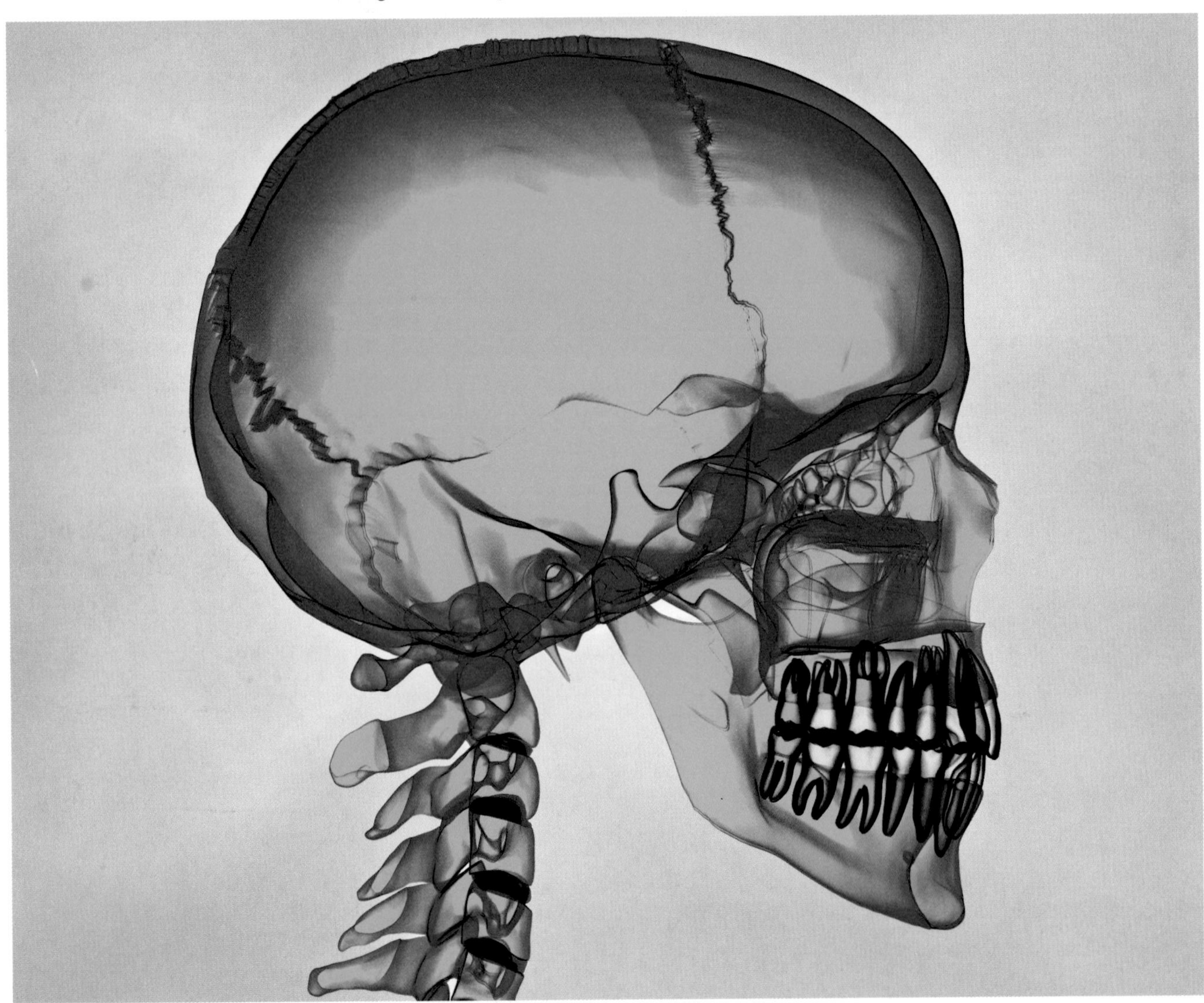

Latin for "hard mother." The next layer, the arachnoid, overlays the brain's network of crevasses. Early observers likened it to the spun lace of a spider, giving it a name that means "cobweb." The lowest of the three membranes, the pia mater ("tender mother"), is filled with tiny blood vessels. It embraces the brain surface like a mother cradling a child in her arms; every dip and rise in the brain matter is form-fitted by the pia. The ridges are called gyri, which means "twisters," while its grooves are sulci, or furrows.

Layers of coverings combine to cushion, protect, and support the brain.

BRAIN CUSHION

Flowing between the arachnoid and pia membranes is the brain's cerebrospinal fluid. This liquid bathes the brain's gyri and sulci, including the deepest grooves, which are known as fissures. Fluid-filled ventricles—the hollows that some philosophers such as Thomas Aquinas considered home to the mind—curve deep into the brain and connect to the spinal cord's central canal. Cerebrospinal fluid cushions the brain, provides nourishment for tissues, and perhaps acts as an internal channel of chemical communication.

FAST FACT

Poet Lord Byron's brain weighed 79 ounces, well above the average human brain's weight of 48 ounces.

PROTECTION

The body has evolved formidable defenses to protect its most vital organ. While capillaries in other parts of the body allow cells to absorb harmful substances from the blood, the brain has the so-called blood-brain barrier with only limited permeability. Thick, tight membranes in the brain's blood vessels screen out many substances in the bloodstream. Crucial chemicals such as oxygen and glucose can cross into the brain, as well as a few harmful ones, such as alcohol and nicotine. Frustratingly, many beneficial chemical compounds, such as drugs designed to attack tumors, are turned back.

FOUR DIVISIONS

Moving inward, we come to the organ itself. The brain may appear to be a uniform mass of folded, pink tissue. But a closer look reveals different lobes, regions, structures, and parts that all help regulate body functions, interpret information from the body, and react to stimuli.

The brain has four main parts: the cerebrum, diencephalon, cerebellum, and brain stem.

+ THE BARD +

SHAKESPEARE WEIGHS IN on the human brain in his plays:

+ "Tell me where is fancy bred, Or in the heart, or in the head?"—*The Merchant of Venice*

+ "The brain may devise laws for the blood, but a hot temper leaps o'er a cold decree."—*The Merchant of Venice*

+ "Her beauty and her brain go not together."—*Cymbeline*

+ "He has not so much brain as ear-wax." —*Troilus and Cressida*

The Brain

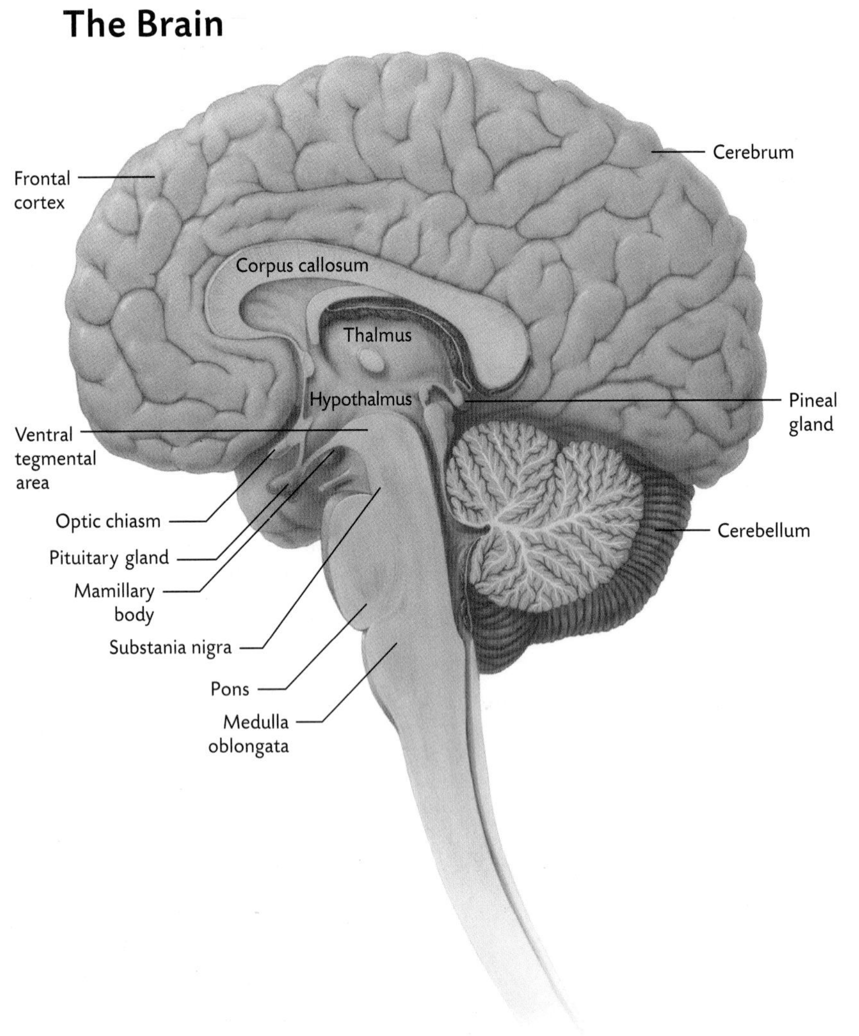

CEREBRUM

This largest, topmost layer of the brain is the cerebrum. It's what most people visualize when they use their brains to picture their brains. The external layer is called the cerebral cortex. Its outer portion is gray from the presence of billions of nerve cell bodies, while the inner portion is white from the tangle of axons coated in their myelin sheaths.

In the cerebral cortex lies the core of information processing that separates humans from other animals, including reason, language, and creative thought. *Homo sapiens* has more of its brain in the cerebral cortex—approximately 76 percent—than any other animal. (Chimpanzees rank second at 72 percent, while dolphins have only 60 percent.)

FAST FACT In 1999, scientists discovered that Albert Einstein's inferior parietal lobe, associated with mathematical and spatial reasoning, was 15 percent wider than that of an average brain.

FISSURES & HEMISPHERES

The cerebrum is divided into parts by deep fissures. The largest of the brain's fissures is immediately evident to the naked eye. Down the center of the cerebrum, separating it into left and right hemispheres, is the longitudinal fissure. The left and right halves of the cerebrum appear to be nearly mirror images of each other.

While they look alike, the two halves perform and control very different functions. The left hemisphere long has been considered the dominant half because of its role in processing language, but the right hemisphere is gaining new attention for its role in emotions and spatial cognition, as well as the integrative function that helps bring bits of information together to create a rich image of the world.

Connecting the two hemispheres are bands of nerve fibers that allow information to be passed back and forth between the two halves of the brain. The largest bundle, containing about 200 million nerve fibers, is the corpus callosum.

Two divides known as the Sylvian fissure and central sulcus lie on the outside edges of the hemispheres. Their locations serve as boundaries on a map, dividing

CROSS REFERENCE: See "In Harmony," PAGE 34

the hemispheres further into four lobes. The frontal lobe lies forward of the central fissure. Between the Sylvian and central fissures are two lobes that merge together, the parietal followed by the occipital. Behind the Sylvian fissure is the temporal lobe.

THE FRONTAL LOBE

A portion of the frontal lobe of each hemisphere called the precentral gyrus controls the body's movements. Oddly, each hemisphere moves the opposite side of the body, as if the brain's wiring somehow became crossed. Hence, the movements of the right hand and right foot, as well as the rightward gaze of both eyes, are governed by the left side of the brain. This phenomenon has been observed for centuries. Hippocrates noted that a sword injury to one side of the head impaired movement on the body's opposite side. And while observing combat wounds during the Prusso-Danish War of 1864, German doctor Gustav Theodor Fritsch noted that if he touched the cerebral cortex as he dressed a head wound, the patient twitched on the opposite side of his body. If one hemisphere's precentral gyrus is destroyed—during a stroke, for instance—paralysis will result in half the body.

In front of the precentral gyrus lie the premotor cortex and the prefrontal fibers. The former organizes the body's complex physical movements, whereas the latter inhibit actions. Inhibition is useful in a variety of social settings, such as preventing shouting in a quiet movie theater.

Crossword puzzles and other challenges keep the brain sharp.

STAYING SHARP

THE BRAIN NEEDS regular exercise if its neurons are to remain sharp. Repetition of newly learned tasks helps make those new connections stronger. Without stimulation, dendrites recede and the brain settles into simpler patterns of operation. Neurologist Robert Friedland has shown that posing new challenges to the brain can help in the defense against Alzheimer's disease.

Perhaps not surprisingly, "Use it or lose it" appears to be true not only of mental exercise but also of physical stimulation of the brain. The brain is like other organs and works better when the body is healthy. Exercising the body regularly appears to help ward off Alzheimer's disease, as do reducing body weight, lowering blood pressure, and eating a more healthful diet. General exercise that builds up cardiovascular endurance improves blood flow to the brain. A healthy heart usually is linked to a healthy brain, especially in the brain's "executive function," which is crucial to a slew of mental tasks.

A combination of physical exercise and mental gymnastics protects the brain against deterioration with age. To spur on the brain to make new neuronal connections and protect the ones it has, there are a number of activities to try, such as:

- ✓ Learning a new language.
- ✓ Listening to classical music.
- ✓ Solving mental puzzles and games, like crossword puzzles and Sudoku.
- ✓ Eating a healthful diet.
- ✓ Walking, jogging, or cycling regularly to promote cardiovascular health.
- ✓ Maintaining a healthy weight.

PARIETAL LOBE & TEMPORAL LOBE

In the parietal lobe lies the somatosensory cortex, which takes in stimulations of touch and other sensations. While lower parts of the brain register pain and pressure, the sensory cortex helps localize such feelings. Damage to the sensory cortex may result in confusion about which part of the body may be registering pain.

The temporal lobe is home to the functions of hearing and appreciation of music and to some aspects of memory. Self-experience also resides in this lobe. Electrical stimulation of the temporal lobe may dredge up intense feelings from the memory—the experience of reliving the past, known as déjà vu—or do just the opposite, causing familiar people and objects to become unrecognizable.

At its base, the temporal lobe connects with the limbic system, a series of brain structures also known as the animal brain. This system allows humans to experience intense emotions such as anger and fear as well as react to these feelings.

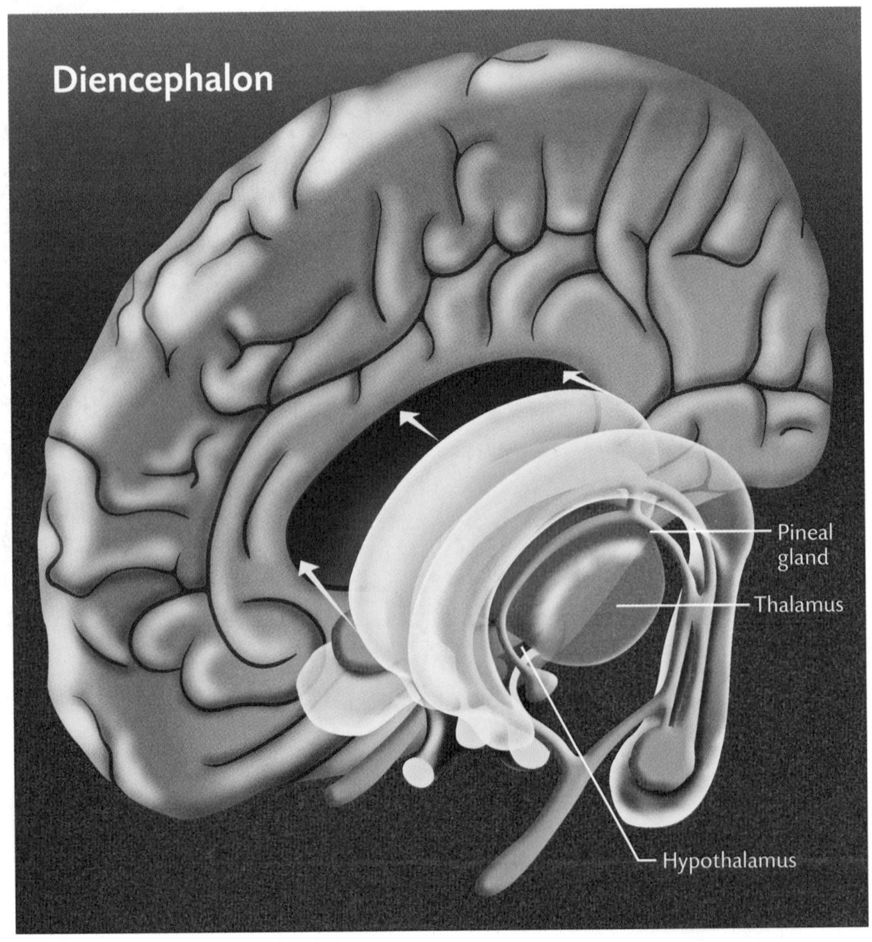

Elements of the diencephalon link the left and right hemispheres.

OCCIPITAL LOBE

Behind the temporal lobe, near the rear of the head, lies the brain's visual center in the occipital lobe. Far from the eyeballs, which take in visual information, this portion of the cerebral cortex processes electrical impulses that begin with light waves striking the retina. Wounds to the back of the head injuring the visual cortex can sometimes cause blindness.

DIENCEPHALON

In the center of the brain, between the cerebrum's two hemispheres, lies the diencephalon. It consists largely of three important structures: the thalamus, hypothalamus, and epithalamus.

The thalamus acts as a relay for sensory information on its way to the cerebrum and is crucial to memory and emotions.

The tiny hypothalamus exerts control over the autonomic nervous system and performs other functions, including regulating body temperature.

The epithalamus includes the pineal gland, which drew Descartes's attentions. Instead of housing the soul, scientists now know it helps to regulate the body's rhythms of sleeping and wakefulness.

CEREBELLUM

At the back and bottom of the skull rests the cerebellum. Like the cerebrum, it too is divided into

FAST FACT Misunderstanding of the work of neuroscientist Roger Sperry in the 1970s fed the notion that everyone is either "left brained" or "right brained." Although each hemisphere has special functions, the two halves work closely together in a healthy mind. Humans are whole brained.

halves and deeply fissured. Its role is to coordinate movement and balance. Precise physical activities that must be practiced to be performed well—hitting a golf ball, doing gymnastics, picking a pattern of notes on the strings of a guitar—are processed in the cerebellum. The cerebellum also is known to play a role in emotion and action.

MEDULLA OBLONGATA

Where the brain meets the spinal cord is the brain stem. The spinal cord, the central route of nerve cells connecting brain and body, terminates in a 1.2-inch extension into the lower brain known as the medulla oblongata, home to motor and sensory nerves. Here is where the nerves from the body's left and right sides cross each other on their way toward the cerebrum. Basic body functions such as heartbeat and respiration are controlled in the medulla.

Above the medulla lie the pons and midbrain. Pons means "bridge," and that's what it does—it acts as a bridge between the medulla and other brain regions. The midbrain links the pons to the diencephalon and controls reflexes of the ear and eye, such as the jolt the body experiences when startled.

FUELING THE BRAIN

Blood pumped from the heart pushes upward into the brain through two main sets of blood vessels, the internal carotid and vertebral arteries. Spiderwebs of smaller vessels, like distributary waterways at a river's mouth, send blood into every region of the brain.

The brain uses oxygen in a hurry. While the brain weighs only about three pounds, a mere fraction of body weight, it burns 20 percent of the body's oxygen and glucose. Most of that energy is mere upkeep, keeping the brain on the razor-sharp edge of action by maintaining the electric fields of the membranes surrounding the synaptic clefts. Actually thinking adds very little to the demand for energy—a fact that is somewhat counterintuitive for anyone who has ever struggled with a particularly difficult math problem or foreign language translation.

To get fuel to hungry brain cells, the body relies on the constant circulation of glucose. It's a kind of sugar that circulates via the bloodstream. Neurons can't stockpile glucose like coins in a bank, so they require a ready supply of this source of chemical energy. Neurons use the fuel of glucose to manufacture and transport molecules of neurotransmitters and enzymes. They also use plenty of energy— half of the brain's total, in fact—to transmit electrochemical signals from cell to cell. The body obtains glucose from starches and sugars in the daily diet. Good sources include grain, fruits, and vegetables. During periods of intense concentration, glucose levels decline in brain regions associated with memory and learning. Such a decline can cause a feeling of fatigue in the body and the brain.

+ AGED BRAINS +

AN OLD BRAIN can be an amazingly healthy and creative one. Consider:

+ Ben Franklin left public service at age 82.
+ Mary Baker Eddy founded *The Christian Science Monitor* at age 86.
+ Robert Frost published his last collection of poems at age 88.
+ George Bernard Shaw was still writing plays at age 94.
+ Grandma Moses received a painting commission at age 99.

LOOKING INSIDE

SEEING THE BRAIN AT WORK

ONCE THE brain's true purpose was ascertained, scientists began finding new ways to observe it and its functions. Starting with noninvasive methods, like IQ tests, they tried to learn more about the living brain and measure how it worked. These intelligence tests painted a picture of how the brain collected information, processed it, and then made conclusions.

Peering inside a living brain was virtually impossible—most of what scientists knew about the brain's anatomy was based on autopsies. But in the late 19th century, the invention of the x-ray made it possible to take a look inside the skull. In the 20th century, new scanning methods came along and gave greater insight into how the living brain works.

OUTSIDE LOOKING IN

Scientists have long dreamed of examining how the brain works

CT scans open windows into the brain's interior structure.

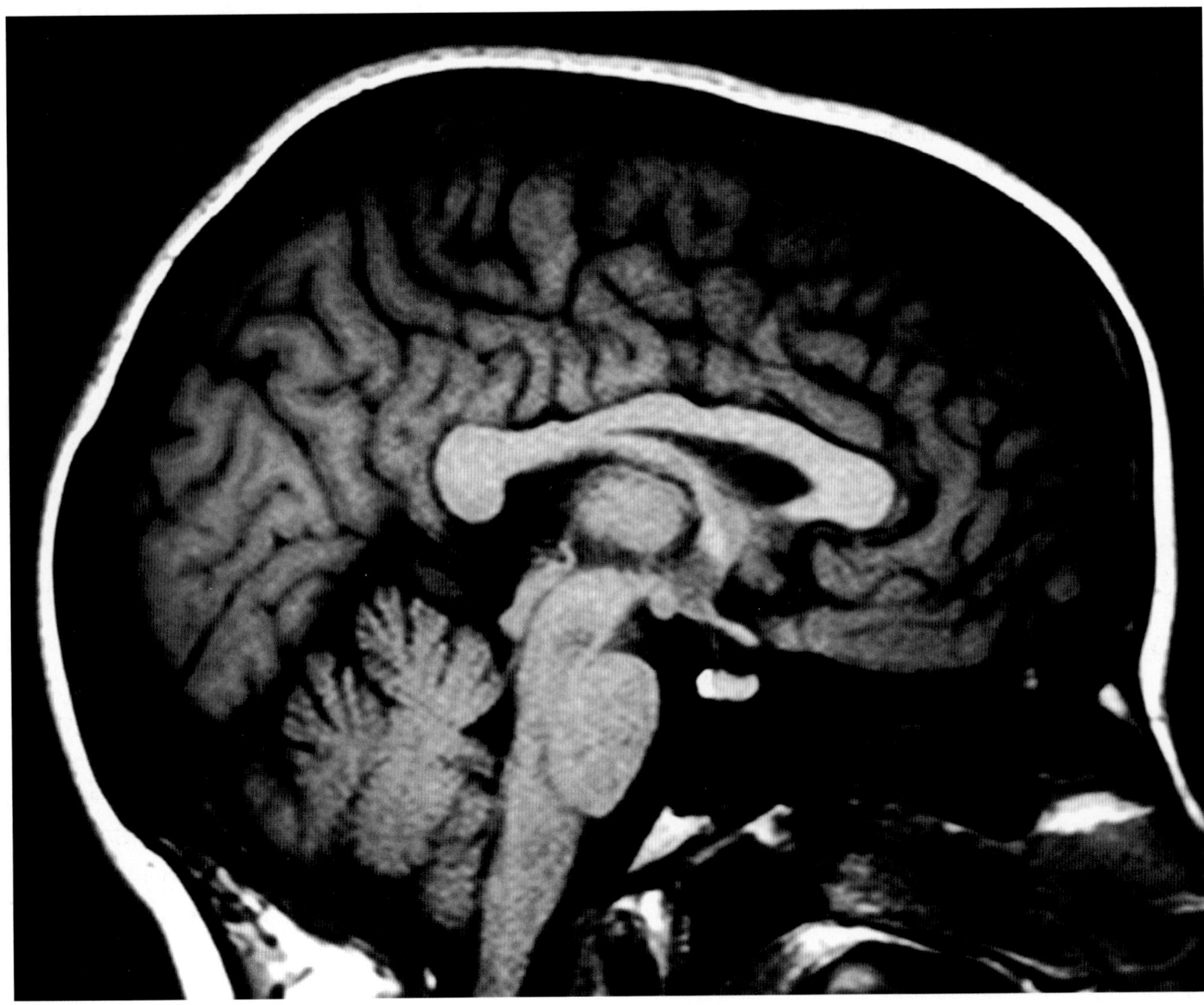

within a living body. The problem, though, was figuring out how to get inside the head without causing injury or even death. Doctors treating wounds from wars and accidents have been able to get glimpses of living brain tissue, but aside from poking or prodding, have had little to do with experimental observation.

Some early noninvasive attempts included phrenology, the pseudoscience developed in the early 19th century that measured the bumps on the outside of the skull as a means of analyzing the mental powers and characteristics. They stemmed from the theories of a German doctor, Franz Joseph Gall, who argued in the late 18th century that the separate faculties of the brain must manifest themselves in the shape of the overlying bone. Phrenology's popularity peaked between the 1820s and the 1840s but soon waned as the century progressed.

FAST FACT Overall, at least half of all cases of dementia—formerly known as senility—can be traced to Alzheimer's disease.

Toward the end of the 19th century, a new method of probing the hidden workings of the brain arose, again in central Europe. Wilhelm Wundt, known as the founder of experimental psychology, created a laboratory in the mid-1870s in Leipzig to perform research into

TESTING INTELLIGENCE

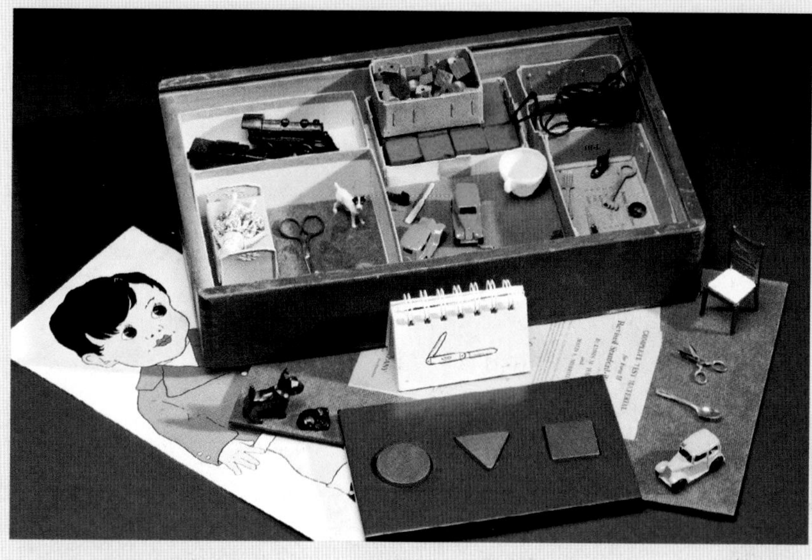

A 1937 Stanford-Binet intelligence test includes miniatures and printed matter.

ALFRED BINET (1857-1911) made the first serious effort to chart intelligence. In 1905, France commissioned him to create a test to identify students whose intelligence was below average. Binet and his doctoral student, Théodore Simon, devised a series of tasks for children. They then tested how well children of various ages performed the tasks, which gradually increased in complexity. Their work led them to create a scale of normal mental functioning. Binet's intelligence scores compared a child's mental abilities with those of his or her peer group. The test has been updated many times.

During World War II, the American government gave Army recruits intelligence tests to screen them for war work. Plenty of other groups have been given IQ tests since then, all over the world. If you look only at their scores, you might think humans are getting smarter all the time. New Zealand political scientist James R. Flynn observed that standardized intelligence test scores from 20 countries historically have kept rising by about three points a decade. The reason isn't entirely clear, but it's possible that improvements in nutrition, coupled with the more stimulating environments in which children are raised, contribute to greater neuronal complexity.

Today, scientists still wrestle not only with what intelligence is, but also how it can be measured. Harvard University's Howard Gardner believes at least seven types of intelligence exist, from the mathematical to the athletic.

WHAT IS INTELLIGENCE?

PERHAPS NO scientific book of the past half century stirred up as much controversy as *The Bell Curve: Intelligence and Class Structure in American Life*. The 1994 book, by Richard J. Herrnstein and Charles Murray, begins simply: "That the word *intelligence* describes something real and that it varies from person to person is as universal and ancient as any understanding about the state of being human." From there, the authors delve into definitions of intelligence and how it can serve as a good predictor for success in life.

Then they argue that different levels of intelligence lead to social outcomes, instead of the other way around—a person of low intelligence is more likely to end up a criminal or unemployed, for instance—and that intelligence levels have an observable correlation to biology.

Following the track linking genetics to intelligence, the authors make claims linking racial differences to intelligence, and thus the positive and negative social outcomes that define modern life. If a group of people can't change their biology, goes this hypothesis, they cannot change their social outcomes.

Does the brain's biology determine intelligence, and thus lock humans in to paths toward success or failure? It's a potent question.

DEFINING INTELLIGENCE

Part of the problem lies in the definition of intelligence. Neuroscientists don't agree on what the word means. Nor do they agree on what intelligence tests are actually measuring. Tests don't measure motivation, persistence, social skills, and a host of other attributes of a life that's well lived. Some say, only half facetiously, that IQ tests measure only one's ability to perform well on IQ tests.

Studies of identical twins have shown that certain regions of the brain are highly inheritable, affecting overall intelligence.

Neurologist Richard Restak likes to deliberately cloud the issue during his lectures by showing students images of two PET scans. Each reveals the level of brain activity of a student doing a problem in a Raven's Colored Progressive Matrices test, which aims to measure "fluid intelligence," or the ability to solve an unfamiliar kind of problem. In one scan, the image is illuminated in red and orange, representing an increase in brain activity. In the other, the cool shades of blue and green represent a less intense level of brain function. When Restak asks the students to guess which of the two students scored higher on the Raven's test, and thus (one assumes) possesses superior intelligence, the students invariably pick the brain lighted up like a Christmas tree. Instead, the student with the less active PET scan posted a higher Raven's score. The explanation: The brain that finds a problem easy to solve doesn't have to work as hard.

TYPES OF SMARTS

There are several aspects of intelligence. Most are related, but historically not all have tested what they set out to test. For example, some early IQ tests measured knowledge of facts, which actually is a function of education and memory rather than the ability to reason. In general, however, a person's performance on a test of fluid intelligence is a good predictor of performance on a wide range of mental exercises. For example, increased fluid intelligence correlates to a high level of "working memory"—one's ability to remember information temporarily—which can range from remembering where you parked your car to which words or number combinations you tried and rejected in doing a crossword puzzle or Sudoku. People with powerful working memories are more focused in solving problems.

Scientists use the term "g-factor" when discussing the general measure of mental ability, found in vocabulary size, mechanical reasoning, and arithmetical computations. They relate it to the properties of efficient neural functioning, rather than the value of knowledge in its own right. The prefrontal cortex, right behind the forehead, is the most likely home for much of the neural processes associated with one's g-factor abilities. When it's damaged, a person suffers a variety of impairments to abstract reasoning, and it lights up during brain scans taken during a variety of intelligence tests.

"You have less frontal development than I should have expected," says the evil Professor James Moriarty when he first lays eyes on Sherlock Holmes in a story by Arthur Conan Doyle. As scientists have discovered, the size of the prefrontal cortex in healthy brains generally correlates to fluid intelligence. (Perhaps Moriarty subscribed to the theory of phrenology and believed cortex size correlated to the bulging of a forehead. It's not so.)

Psychologist John Raven devised the Raven's Colored Progressive Matrices Test in 1938, a non-verbal test of intelligence in children.

But the size of a cortex doesn't mean, QED, that biology causes intelligence the same way gravity causes an apple to fall. Identical twins vary in their performance on IQ tests. In some cases, one twin develops schizophrenia or some other disorder, and the other does not. Furthermore, when identical twins are separated at birth and raised separately in similar environments, they show only a 72 percent correlation in intelligence.

FAMILY INFLUENCE

At best, genetics accounts for only a substantial fraction of intelligence. Perhaps heredity sets an upper limit for intelligence (through the potential ability to make neuronal connections), which then becomes subject to other forces. An environment with plenty of books and challenging toys plays a key role in increasing aspects of a child's intelligence but so does willingness to exercise the brain. Political scientist James R. Flynn noted that IQ scores have dramatically increased over the past several decades in many countries. He attributes the so-called Flynn effect to increases in modern humans' greater ability to solve abstract problems, possibly from living in a more intellectually stimulating world.

The brain's ability to rewire neuronal networks no matter how old the nerve cells provides the means to improve mental function. Instead of looking at family or ancestral heritage and deciding it determines mental performance, humans can set about learning new skills and tasks. Challenging the brain may not raise the score on a particular IQ test, but it will help the brain to perform better.

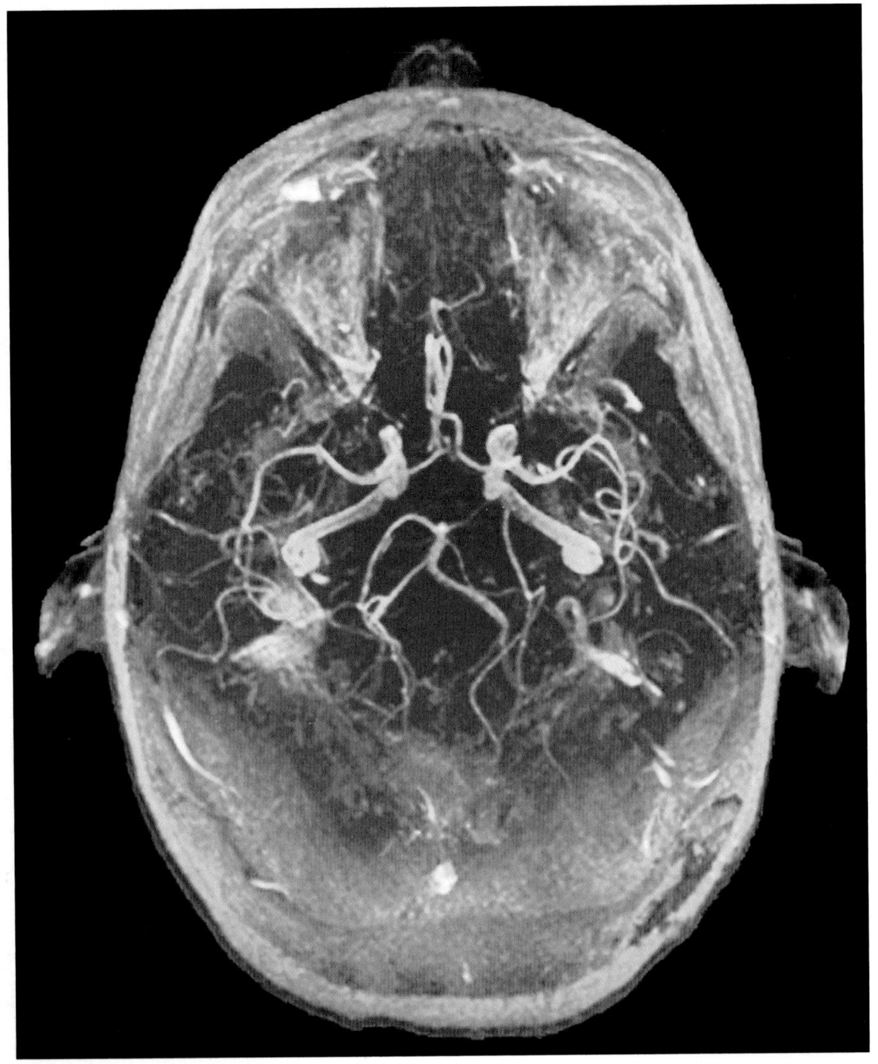

An angio-MRI of a 27-year-old woman reveals the arteries that provide oxygen to her brain.

psychology. The word derives from the Greek *psyche,* meaning "mind" or "soul." Wundt considered his research a way to get at the workings of the mind, which many still considered to be separate from the tissue of the brain.

In particular, Wundt aimed to examine the elements that made up consciousness and explain how they worked together to create the mind. Wundt concentrated on stimulus-response experiments, as he considered sensation the contact point between the external, physical world and the inner, psychological world. He recorded when and how sensations entered consciousness, including such mundane facts as whether one musical tone sounded higher or lower than another one did.

A contemporary of Wundt's, the American William James, also took up psychology as a tool to probe the mind. In his famous 1890 textbook *The Principles of Psychology,* James described processes including the sense of self, memory, movement, and sensation.

Assessing the brain's performance through intelligence testing was another way science attempted to access the living brain. In the 1900s, French psychologist Alfred Binet created the first IQ test as a way to measure intelligence. That test, designed to see which French schoolchildren needed special assistance, became the genesis of all IQ tests that followed.

FAST FACT
Your brain uses about 12 watts of power—a fraction of the energy of a household lightbulb.

Meanwhile, in Austria, Sigmund Freud (1856–1939), the founder of the psychoanalytic school of psychology, turned his interest in neurology into the study of the workings of the brain and the ways in which they affect behavior. He predicted, correctly, that someday the study of the physical workings of the brain would dovetail with his observations about unconscious drives.

FIRST GLIMPSE

The first technology to peer into the brain was the x-ray, invented by Wilhelm Röntgen (1845–1923) in 1895. The German scientist discovered a form of radiation that could penetrate the body; the rays were absorbed by dense bones, which then appeared as shadows on film.

When applied to the brain, simple x-rays, harnessed to make photographic images of bone, permitted doctors to make a basic examination of the structure of the head. However, x-rays give only a two-dimensional view, and show relatively little of the soft tissues of organs. As the human brain is a three-dimensional object, whatever appeared in a 2-D image usually was murky and confusing. Often, structures lying in different planes of the brain overlapped each other, making analysis difficult.

A BETTER LOOK

Scientists first peered at real-time brain functions in 1929 with the invention of the electroencephalogram, or EEG. Electrodes fitted to the scalp record electrical activity within the brain as neurons discharge. Unusual brainwave activity registered on an EEG may indicate brain disorders. This technique records electrical activity in real time.

More recently, scientists have employed a variety of tools to get a more detailed and localized look at structure and action inside the brain.

COMPUTERIZED VISIONS

Computerized axial tomograms, or CT scans, have substantially improved the ability of x-rays to probe the secrets of the brain. A patient receiving a CT scan lies inside a doughnut-shaped array of sensitive detectors while a movable x-ray emitter rotates around the brain. Computers convert the images into a three-dimensional

+ GREETING YOURSELF +

SWEDISH SCIENTISTS in 2008 created the illusion of shaking hands with yourself. They had volunteers and a mannequin wear virtual reality goggles. Images in the volunteers' goggles came from the dummy. Most test subjects felt the weird sensation of the dummy's point of view when shaking their own hands.

+ BRAIN IMAGING +

TYPE	FULL NAME	DESCRIPTION
EEG	Electroencephalogram	Electrodes on the scalp record brain waves. Unusual activity may indicate brain disorders.
CT scan or CAT scan	Computerized Axial Tomography	Series of x-rays of the head are taken from many different directions. Useful for quickly assessing brain injuries.
MRI	Magnetic Resonance Imaging	Yields a more detailed, three-dimensional image. Allows precise mapping of the physical shape of the brain.
fMRI	Functional Magnetic Resonance Imaging	Uses changes in blood flow to generate images that show brain activity and performance.
MEG	Magnetoencephalograph	Magnetic sensors placed on the skull reveal neural activity with little interference from other structures. Used to locate tumors and to determine functions of brain parts.
PET	Positron-Emission Tomography	Radioactive isotopes injected into the blood are then tracked by a computer through the brain. Reveals blood flow, oxygen levels, and glucose metabolism.

image of the brain. Slices of the interior—the word *tomos* is Greek for "section"—can be teased from the data and shown on a screen to give doctors a narrow look at particular points in the brain. For example, a CT scan might reveal a tumor located deep inside the tissue of a living brain, far too deep to be visible during routine exploratory surgery.

BRAIN MAPPING

Magnetic resonance imaging (MRI) gives a more detailed 3-D picture than a CT scan. An MRI relies on an intense magnetic field generated in a cylinder that surrounds the patient. It allows precise mapping of the physical shape of the brain. Its magnetic field is so powerful that it causes some of the atoms inside the brain to jerk into alignment. Then a series of radio waves from the MRI scanner bounce off the affected atoms and push them slightly out of line. When the energy from the radio signals is turned off, the atoms move back into their magnetic alignment, emitting telltale energy patterns along the way.

Computers read these minuscule bits of energy and assemble images of cross-sections of the brain. Slices can be placed atop each other, like the layers of a cake, to represent the entire brain in three dimensions, or they can be examined individually, providing a closer look at localized phenomena. Comparisons of MRI scans of a single brain over time can show its growth—or reveal its deterioration.

In addition to mere structure, an MRI can also capture a snapshot of thought. A variation called a functional MRI, or fMRI, builds upon the fact that a blood cell's magnetic properties change according to how much oxygen it contains. Receptor cells use oxygen as they take in signals from surrounding cells; burning oxygen causes cells to require more oxygen-rich blood. As blood surges toward neurons where synapses are firing with thought, emotion, or other impulses, the oxygen they carry gives off a traceable signature of radio waves. Different thoughts light up different areas of the brain in an MRI. The processes of peaking, reading, appreciating humor and music, and recognizing faces illuminate various groups of neurons. MRI techniques thus help localize areas associated with certain brain functions.

SEEING THOUGHTS

Magnetoencephalography (MEG) also relies on magnetism to examine the brain. In this case, it's the body's ambient magnetic fields, not those generated by an external machine, that form the basis of brain imaging. These magnetic fields are extremely weak—perhaps only a billionth of the power that causes a compass needle to point toward the north magnetic pole. Yet, when read by sensors placed on the skull, MEG scans reveal the electrical currents created by neural discharges. The resolution is as fine as a thousandth of a second and as small as a cubic centimeter. The

FAST FACT Addictive drugs work by mimicking neurotransmitters or altering their work. Brain scans reveal physical changes in the synaptic activity of a drug user. The drug known as Ecstasy, for example, can permanently damage neurons that produce serotonin.

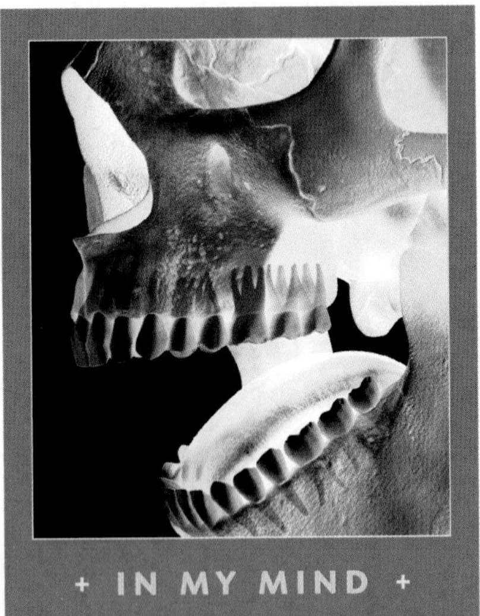

+ IN MY MIND +

WHEN THE DENTIST asked British philosopher Bertrand Russell where he felt pain, Russell replied, with humor and honesty, "In my mind, of course." Russell knew the brain uses the senses to collect data about the world and construct a version of "reality." Whether that world actually exists independent of the mind makes little difference to the sufferer of a toothache—the pain hurts just the same. In fact, some philosophers, such as George Berkeley (1685–1753), have questioned whether "reality" exists.

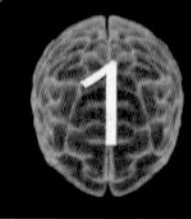

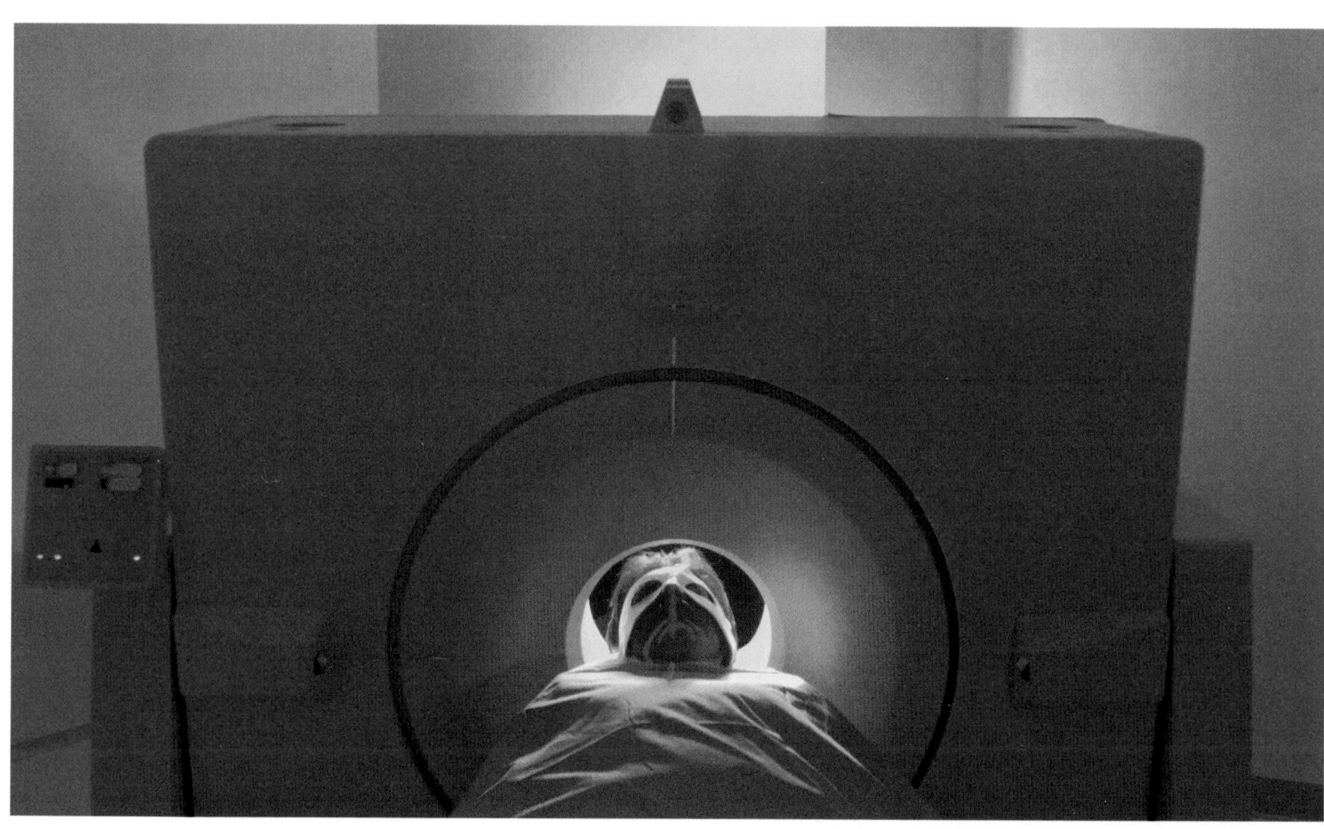

A patient receives a PET scan to pinpoint regions of the brain that are most active.

MEG scan and EEG are the only observational techniques capable of anything approaching real-time revelations. When a patient thinks a specific thought, it shows up, in progress, on an MEG.

Mental functions also can be localized with a technique called positron-emission tomography, or PET. A radioactive isotope is injected into a patient. Because all radioactive atoms decay into stable atoms at a known rate, the decay of the isotope, which is usually paired with glucose, is recorded and turned into images with computer programs. Like MRI and CT scans, PET scans let observers localize activity inside the brain.

The array of brain-imaging techniques serves like the variety of hammers, saws, and other tools in a mechanic's toolbox. A scientist observing the brain chooses the right tool based on what kind of information is needed. A CT or MRI scan would be the choice if a doctor suspects the growth of a tumor or physical damage to part of the cerebrum. A PET scan might be the appropriate choice for investigation of deficiencies associated with language or reason. And lack of oxygen use in stroke-damaged sections of a brain would call for a functional MRI.

True to the rational and observational methods of Descartes and Willis, science has made great strides in describing how the brain's parts, both large and small, function. But understanding any organ that is "wider than the sky" is not as easy as toting up small pieces of information. The brain is an integrated unit, with its complexity arising from the synergy created by the simultaneous functioning of its billions of neurons and trillions of synapses in nonlinear ways. Science has learned much about movement, sensations, emotions, and the sense of self. Yet much is yet to be gleaned about the most complicated object in the universe. There will always be more to learn about the brain.

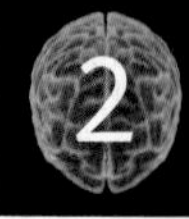

CHAPTER TWO

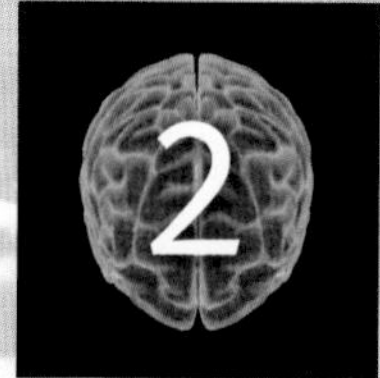

THE NERVOUS SYSTEM

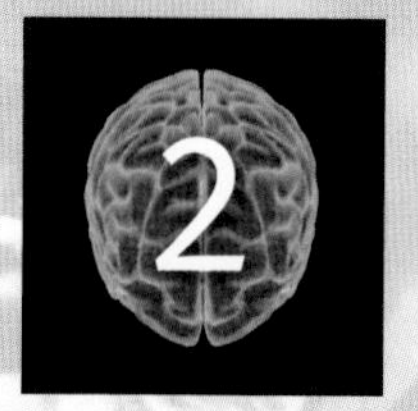

WHETHER IT BE a surprise, a startle, or a scare, how the brain reacts to a situation is determined by the information that is gathered by the nervous system. Through this vast interconnected network, the brain is able to collect data, interpret them, and then react to them in a matter of milliseconds—governing such things as how fast our heart races, how hard we laugh, or how loud we scream. Every reaction, thought, action, and emotion is regulated by the nervous system, which excels at communication and control.

The shock of an ice-cold victory celebration causes a full-body startle reaction.

IN HARMONY

HOW THE NERVOUS SYSTEM RUNS THE BODY

Just as the conductor of an orchestra directs the flow and tempo of music, so the brain controls the flow and tempo of the body.

THINK OF THE brain as a symphony orchestra. When everything goes right, the brain remains in constant communication with the entire body at all times. Sometimes, as when musicians are warming up or the mind's attention is unfocused, the signals are muted or lack direction. But when the conductor walks to the podium and taps the baton, all snap to attention.

Then, with the downsweep of the maestro's arms, everyone springs into action. Each musician, like every nerve that registers and transmits information, watches for instructions. Upon recognizing the conductor's intent, each carries out orders to speed up or slow down, emphasize or downplay a particular action, or otherwise fine-tune the adjustments that create music out of a hundred different sounds—or the thoughts of the brain into physical action.

The conductor, like the brain's executive function, also is watching

FAST FACT Cells in your brain, as in all tissues, have their own genetic code made up of just four nucleotide bases. They're usually referred to by their first letters: G, C, T, and A, for guanine, cytosine, thymine, and adenine. Out of these letters come the combinations that make you unique.

CROSS REFERENCE: See "Nerve Cells," PAGE 10

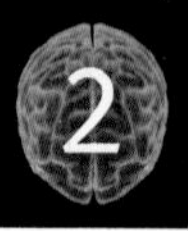

for incoming signals. Each musician's performance makes an impression upon the maestro, who processes the information and calls for any necessary changes. At the same time, the brass section perhaps may be reacting to the percussion without any intervention by the conductor, just as some reflexes travel only from a nerve in the leg to the spinal cord and back again.

As the musicians play together, their individual contributions unite in harmonious song. Thus, the brain has its many functions that, when added together, lead not only to consciousness, but also to overall health.

MANY PARTS

Much of what goes into making music takes place without thought. Professional musicians don't stop to ask themselves, How do I play a C major chord? Instead, their actions have become automatic. Likewise, some learned actions are so routinely processed that they pass out of the conscious thoughts of the cortex and are pushed deeper into the rote performance of the cerebellum.

The similarities continue. The noise of some instruments may be drowned out by the trumpets and drums, but those sounds are still there, just as the brain's control of breathing and heartbeat continues regardless of whether they register on the mind. The conductor may step down from the podium and

The Nervous System

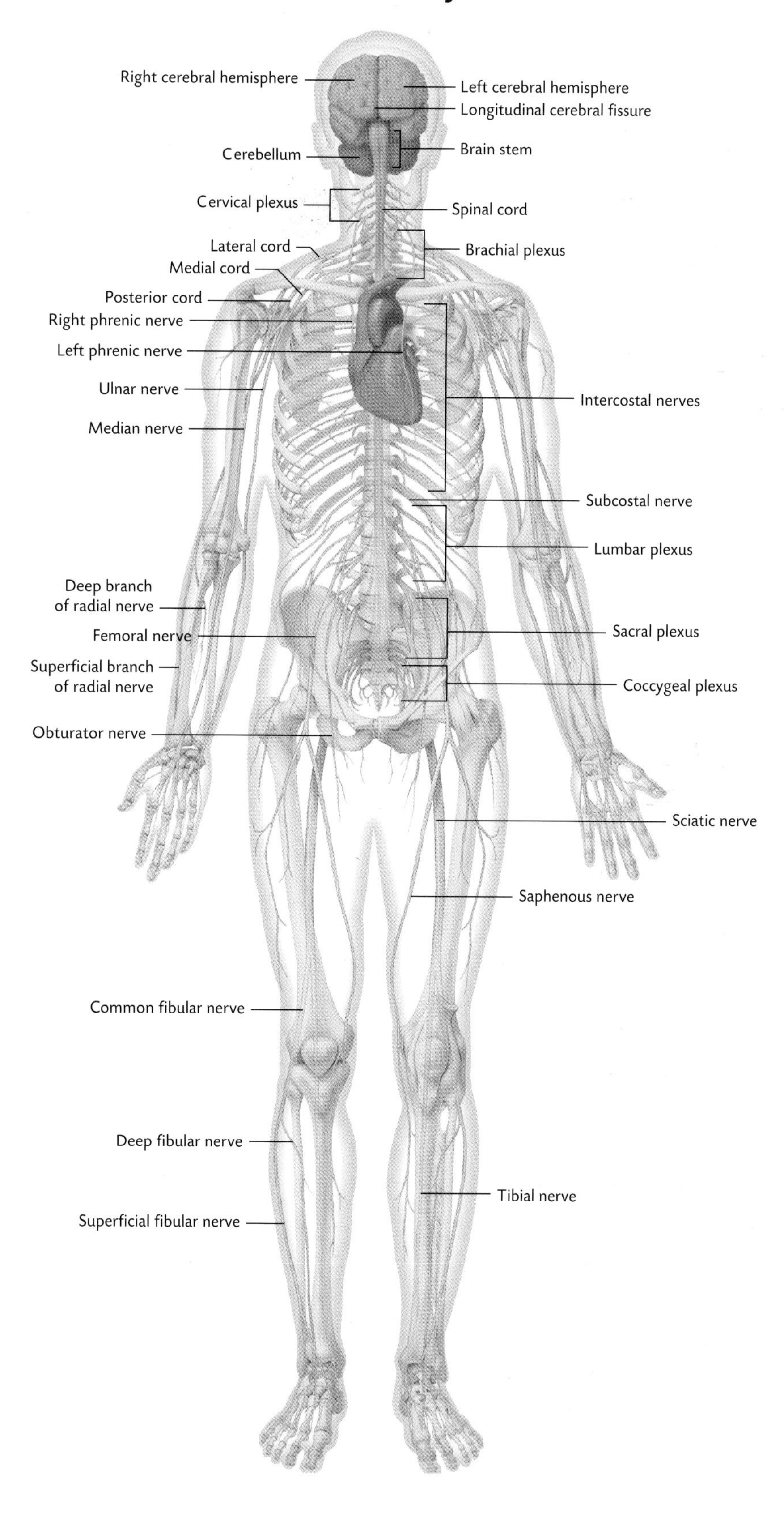

Right cerebral hemisphere
Left cerebral hemisphere
Longitudinal cerebral fissure
Cerebellum
Brain stem
Cervical plexus
Spinal cord
Lateral cord
Brachial plexus
Medial cord
Posterior cord
Right phrenic nerve
Left phrenic nerve
Ulnar nerve
Intercostal nerves
Median nerve
Subcostal nerve
Lumbar plexus
Deep branch of radial nerve
Sacral plexus
Femoral nerve
Superficial branch of radial nerve
Coccygeal plexus
Obturator nerve
Sciatic nerve
Saphenous nerve
Common fibular nerve
Deep fibular nerve
Tibial nerve
Superficial fibular nerve

PHRENOLOGY

As a schoolboy of nine, Franz Joseph Gall (1758–1828) was intrigued by a classmate with large, protruding eyes and a knack for rote memorization. The student's appearance and skills made a lasting impression, one that years later Gall would trace to his theory of cortical localization. All the best memorizers, the German anatomist recalled, seemed to share these bulging, "ox-like" eyes. So it followed, Gall concluded, that the function of verbal memory is governed by the frontal lobe of the cerebral cortex. The better the memory, the larger the lobe, and hence the jutting eyes.

Though he did not coin the term—and shuddered at its usage—Gall would become a leading exponent of phrenology, the pseudoscience of interpreting personal characteristics and mental abilities from cranial knobs and knots.

In interviewing hundreds of personalities across the continent and amassing a collection of some 600 skulls—not the interviewees', fortunately—he determined the human brain to house 27 faculties. Each, he said, is controlled by different areas of the brain.

Among those faculties we share with animals, Gall included "reproductive instinct," "pride," and "destructiveness, carnivorous instinct, or tendency to murder." Unique to humans were "poetic talent," "religious sentiment," and "wisdom."

Determining each faculty's cortical coordinates was simple enough. A large percentage of pickpockets, for example, had a sizable bulge on the side of the head. This area, Gall assumed, was the location of a faculty he called "desire to possess things." The logic of Gall's classification system had made it widely appealing by the 1830s.

An ivory phrenological head maps skull lumps for pseudo-scientific analysis.

Phrenology has since been lumped with the likes of astrology, palm reading, and graphology (handwriting analysis). Yet Gall unwittingly contributed to true science. His theory of cortical localization would prompt future neuroscientists to rethink their concept of the brain, paving the way for groundbreaking discoveries at the turn of the century.

lower his arms; the brain rests and the body falls asleep. Or the pianist may have injured an arm and play badly or not at all, just as the signals to or from the brain may fail, and the body consequently suffers.

HEAD & BODY

The human body has been shaped through cephalization, an evolutionary force that concentrates nervous and sensory tissue at one end of the body. Animals undergoing this process enjoy advantages in natural selection. When vision, hearing, smell, and other faculties work with a nearby brain, they provide a rich picture of the world. Specifically, having a head improves efficiency in locating food and avoiding predators.

A narrow gap between brain and sensory organs, such as eyes, creates the shortest pathways for information to move back and forth between the two. That reduces reaction time. Imagine the alternative: if you had organs of vision in your toes, it would take a moment longer for any images they register to reach a brain at the other end of your body, and another moment or two for the brain to send them feedback. That's a long delay when the eyes detect a potential threat. There's not typically a lot of variation from one head to another.

Each brain lies encased within a hard, bony skull, a series of 22 fused bones that protect it. Inside

Divisions of the Nervous System

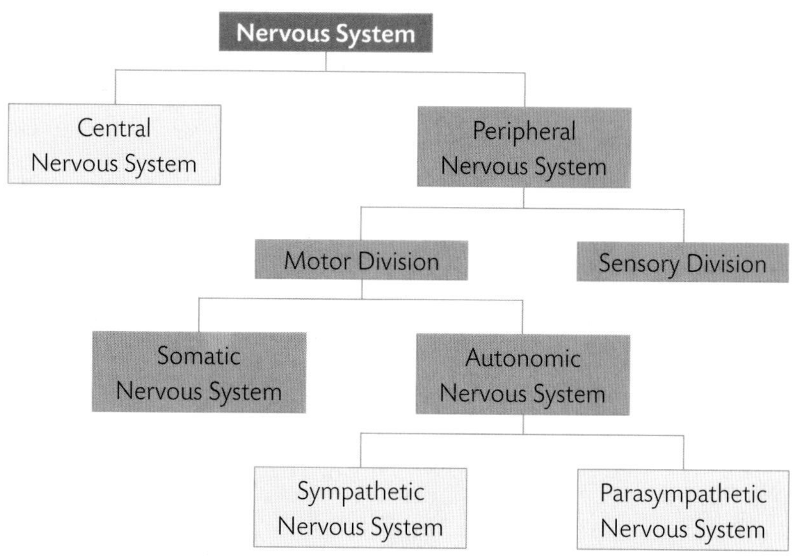

Each division is responsible for the collection of and response to different stimuli.

the skull is a series of protective membranes called meninges that cover the brain tissue and blood vessels, and a shock-absorbing liquid called cerebrospinal fluid. The average man's brain weighs about 3.5 pounds; the average woman's, 3.2. Taken as a pure ratio between brain size and body mass, that's not a significant difference.

Like a captain on the bridge of a ship, the brain issues commands atop the spinal cord, which also lies within protective membranes, a column of bones called vertebrae, and cerebrospinal fluid. The brain communicates with most of the body through nerves that pass through the thumbwide bundle of the spinal cord inside the vertebrae, and branch out in 31 pairs of spinal nerves, each serving its own region. A few nerves, such as those that serve the face, connect directly to the brain.

ORGANIZATION

The brain's internal organization makes performances like the orchestra's possible. Resembling Russian dolls that nest one inside another, the systems of the brain are organized with greater or lesser degrees of scale, but with the same principles. At the brain's behavioral level—the largest doll—humans carry out actions originating in the cortex. These behaviors include speech and written language.

At the next level, the microscopic, behavioral activity is processed by the sum of electrochemical signals pinging among the brain's billions of neurons. At a still smaller, molecular, level, behavior is influenced by the neurotransmitters that pass information across the synaptic clefts that separate individual neurons.

For communication to occur through, say, language, every level has to operate in harmony and virtually simultaneously. Electrochemical processes must pass information from neuron to neuron; neural pathways must interact; and those interactions must come together to create speech.

CENTRAL & PERIPHERAL

The very concept of the brain's whole being greater than its parts continues throughout the nervous system. The body contains only one nervous system, but for purposes of study it often is divided into parts, each of which has smaller and smaller divisions.

The nervous system's two biggest parts are the central nervous system and the peripheral nervous system. The former consists of the brain and the spinal cord. It interprets sensations and issues commands in the form of motor responses, which are based on current sensations, reflexes, and experiences. The peripheral consists mainly of the axons that branch out of the brain and spinal cord, carrying nerve impulses to and fro. Spinal nerves send impulses to and from the spinal cord, while cranial nerves do the same for the brain itself. All cranial nerves terminate in the head and neck

Thanks to evolution's hard wiring for survival, gazelles in Botswana react without thinking when a lioness attacks.

except for the vagus nerve, which extends into the chest and abdomen. Cranial nerves in the head include those that interact with eyes, ears, nose, and tongue.

DIVISIONS

The peripheral nervous system has two key parts. The sensory division is sometimes called afferent, for the Latin for "carrying toward." It sends signals from sensory receptors all over the body toward the central nervous system. Sensors in the skin, muscles, and joints are called somatic ("body") afferent fibers, while those from the internal organs are called visceral afferent fibers.

The other part, the motor or efferent division, sends signals from the central nervous system to the muscles and glands. As these signals cause, or "effect," changes, they create the motor responses that make the body move. Most nerve cells act as two-way streets, sending signals back and forth between the brain and extremities. Purely afferent or efferent cells are rare.

The motor division also is divided into parts. The somatic nervous system sends signals from the central nervous system to the skeletal muscles. As it is usually under conscious control, this is sometimes called the voluntary nervous system. The other part is the autonomic nervous system,

FAST FACT Handwriting analysts claim forward-sloping handwriting indicates an outgoing personality, while a backward slope signifies bashfulness. This discounts or ignores contradictory evidence, such as noted extrovert Bill Clinton's backward slant.

CROSS REFERENCE: See "Nerve Cells," PAGE 10

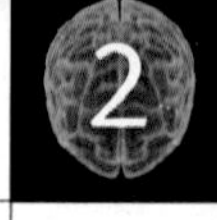

which comprises visceral motor fibers that automatically activate the heart, digestive tract, and other body functions.

THE CEREBRAL CORTEX

Seven-tenths of the volume of the human nervous system lies in the cerebral cortex. Given that the human cortex is many times larger than that of any other creature, scientists are convinced its huge size is the main source of what sets humans apart from the animals. Creativity, emotion, perception, language, imagination—all have strong connections to the workings of the cortex.

Beginning in the late 19th century, researchers began cataloging variations in the thickness and structure of the cerebral cortex. Korbinian Brodmann, a German neuroscientist, created a numbered map of the cortex in 1906, based on the organizational architecture of the cells that he observed after staining them. He numbered 52 sites in the brain, now called Brodmann areas. While the significance of these areas has been widely debated, further investigation has linked some of the sites to particular functions of the brain. PET scans and functional MRI scans have linked specific motor and sensory functions to specific cortical areas called domains. Brodmann areas 1, 2, and 3, for example, reside right behind the central sulcus and are closely linked to the primary somatosensory cortex, while Brodmann areas 41, 42, and 43 are associated with hearing.

The map is not a precise atlas with domains neatly separated by boundary lines, the way countries are separated by political divisions inked on paper. Many functions such as language and memory overlap domains and may in fact be scattered throughout much of the brain.

Nor is the map an indicator of destiny, as other scientists would find. In the early 19th century, Franz Joseph Gall made his own maps of the brain and skull, but they proved faulty. He examined the bumps on the head and drew erroneous conclusions about the functions of the underlying portions of the brain. Physical variations in the size and shape of the head have nothing to do with the workings of the brain power beneath. Damage to a particular Brodmann area, however, may manifest itself in predictable ways, such as language deficiencies resulting from lesions in areas 44 and 45.

+ GRAPHOLOGY +

IS IT POSSIBLE to have handwriting like a serial killer's? Does a physician's scrawl indicate a love for humanity? Much like the phrenologists who thought a bumpy skull could reveal insights into the human psyche, so do today's graphologists, or handwriting experts, believe that penmanship can tell us a great deal about who we are. Handwriting analysts have succeeded more than phrenologists in selling their pseudoscience. Witness the TV ads in 2008 that analyzed car buyers' signatures. Proponents claim that because the brain controls psychological traits and muscles that produce handwriting, they must be linked. No causal link has been found. Graphologists lack scientific rigor, often analyzing the writing of people with known traits–kind of like shooting an arrow at a barn, then drawing a bull's-eye around it.

THE AUTONOMIC NERVOUS SYSTEM

Much of what the brain does takes place beyond our ability to sense it—or appreciate it. In the midbrain's pons and medulla lie the centers that regulate the vital, everyday functions of life. Think about it: How fortunate you are that you don't have to concentrate in order to breathe, or make your heart pump blood.

The first rule of the living brain is to go on living. Thus, these crucial areas of the midbrain, called the autonomic ("involuntary") nervous system, are not easily overruled by the higher functions of the cortex. While it's possible to hold your breath while underwater or throwing a tantrum, the midbrain will eventually overrule the efforts of the cortex and force the lungs to inhale. However, some drugs, such as tranquilizers and stimulants, can affect the autonomic nervous system, altering things like the heart

STAYING SHARP

OVERREACTING TO stress hurts the brain through chronic exposure to hormones like cortisol. Stress can worsen psychiatric disorders and damage the hippocampus, impairing the ability to store memories and to learn new things. To lessen the impact, lower the cortisol through stress management. Some proven methods:

✓ Slow your thoughts through meditation, deep-breathing exercises, and yoga.

✓ Maintain ties with family and close friends. Strong social connections foster a sense of well-being.

✓ Laugh. Laughter increases oxygen intake and the release of endorphins, the feel-good neurotransmitters.

✓ Stay rested. Being overtired can raise levels of stress hormones, but sleep can lower them.

✓ Exercise. Physical activity lowers cortisol levels.

rate and blood pressure for good or ill.

TWO BRANCHES

Like day and night, the autonomic nervous system has two equally important halves. They are reciprocal and complementary. The daylight side of wakefulness and work is called the sympathetic branch. It works when the body's sense of self-preservation, developed over eons of evolution, calls for energy. In extreme cases, the sympathetic branch triggers the so-called fight or flight response. When a threat looms, the body prepares to meet it or quickly escape from it. Blood pressure and heartbeat skyrocket, breathing speeds up, and in a multitude of other ways the midbrain signals to the body to prepare itself for action.

The parasympathetic branch is the calmer, quieter side of the nervous system. It's responsible for the so-called relaxation response. The midbrain signals to the body to lower breathing rate, heartbeat, and blood pressure. As a result, the brain promotes and recognizes a feeling of well-being.

Modern pharmacology can bring about a similar result, but much of the self-help books of the past few decades have focused on meditation and other forms of stress management to stimulate the parasympathetic branch while soothing the sympathetic.

SHOCK TO THE SYSTEMS

When you're startled, the two branches work together, regulating the body without any conscious thought needing to be involved. Thanks to these automatic responses, the brain's cortex is allowed to remain free to do other things—process sensory information, register emotion, pursue rational thoughts, and initiate voluntary movements. This can happen because the parasympathetic nervous system briefly lowers the heart rate, breathing, and other functions. That gives the cortex time to do its job, assessing any possible threats from the external world. Within a flash, the sympathetic nervous system sends signals to release neurotransmitters that put the body on full alert to prepare for the next step.

Meanwhile, the cortex uses the data it has collected to make a decision on an appropriate response to the startling stimulus. If the cortex perceives a real threat—a tiger on the loose from the zoo, for example—the brain automatically sends signals straight to the hypothalamus. The hypothalamus then releases a stress hormone known as CRF. It increases anxiety, puts the senses on extreme alert, and orders the release of the stress hormones cortisol and epinephrine (adrenaline) from the adrenal glands.

Next, the hypothalamus also signals to the pituitary gland to release hormones into the bloodstream that energize all of the body's organs. Thanks to all this interaction and coordination, a person is now primed to run from the tiger, climb a tree, or fight back if necessary.

The tiny hypothalamus, less than one percent of the brain, is rich in neural connections and receptors for hormones, and it strongly influences the pituitary gland. Damage to the hypothalamus weakens the immune system and its response to viruses and germs. Conversely, electrical stimulation boosts immunity.

CROSS REFERENCE: See "Dark Emotions," PAGE 214

GLOSSARY

ABSENCE SEIZURES. Mild epileptic seizures in which consciousness is lost and facial muscles twitch briefly. Most occur in young children and disappear by age ten. Formerly known as petit mal seizures.

AUTONOMIC NERVOUS SYSTEM. Involuntary nervous system. Consists of visceral motor fibers that activate the heart, digestive tract, and glands.

BRODMANN AREAS. Fifty-two sites located and mapped on the cerebral cortex by neuroscientist Korbinian Brodmann in 1906. Damage to a particular area manifests itself in a distinct, predictable way.

CENTRAL NERVOUS SYSTEM. The brain and spinal cord. This control hub integrates incoming sensory information and issues motor responses.

CEREBRAL CORTEX. The outermost layer of the brain, responsible for creativity, planning, language, and perception.

CHEMORECEPTORS. Neural receptors that respond to the presence of chemicals.

CORTICOTROPIN RELEASING HORMONE. Released by the hypothalamus, increasing anxiety, putting the body on alert, and causing adrenaline and the stress hormone cortisol to be released.

CRANIAL NERVE. Carries information to and from the brain. Except for the vagus nerve, all terminate in the head and neck.

DIABETES INSIPIDUS. A condition caused by a lack of sufficient amounts of antidiuretic hormone (ADH) in the body. Patients experience extreme thirst and frequent urination.

DIABETES MELLITUS. A condition caused by a lack of insulin, resulting in heavy blood sugar loss through urination.

DOMAINS. Specific cortical areas that have been linked to specific motor and sensory functions by PET scans and fMRIs.

ENKEPHALINS. Natural pain suppressants that inhibit the discharge of pain-exciting neurotransmitters.

FIGHT OR FLIGHT RESPONSE. Triggered by the sympathetic branch of the automatic nervous system, the brain's response to flee from or defend itself against a potential or perceived threat.

G-FACTOR. Short for "general intelligence factor." A psychological measure of cognitive mental abilities.

HOMEOSTASIS. A state of equilibrium referring to the body's ability to remain internally stable while external environments vary.

HYPERALGESIA. An increased sensitivity to pain.

MECHANORECEPTORS. These receptors create nerve impulses when their shape is deformed by a mechanical force such as pressure or touch.

PARALLEL PROCESSING. The transmission of information through the body whereby one neuron excites multiple others and several paths are utilized at once.

PARASYMPATHETIC BRANCH. This branch of the autonomic nervous system is responsible for relaxing the body. It lowers heart rate and blood pressure and reduces breathing rate.

PERIPHERAL NERVOUS SYSTEM. The nerves that branch out of the brain and spinal cord.

PHOTORECEPTORS. Neurological receptors that react to light energy.

PHRENOLOGY. Pseudoscience popular in the 19th century believing that personal characteristics and mental abilities can be derived from knots and knobs of the skull.

REFLEX. An automatic and uncontrolled reaction to stimuli.

SERIAL PROCESSING. Transmission of information along a direct chain of neurons; a single neuron is excited at a time.

SOMATIC NERVOUS SYSTEM. Voluntary nervous system. Sends signals from the central nervous system to the skeletal muscles. Usually under conscious control.

SPINAL NERVES. The nerves branching out from and relaying information to and from the spinal cord.

SYMPATHETIC BRANCH. A branch of the autonomic nervous system that puts the body on alert and supplies it with energy in response to fear or excitement.

THERMORECEPTORS. Neuroreceptors that register changes in temperature.

TONIC-CLONIC SEIZURES. The most severe epileptic seizures, often causing loss of bowel and bladder control, tongue biting, and strong convulsions. Formerly known as grand mal seizures.

WORKING MEMORY. Where information is stored temporarily in the brain. Also called short-term memory.

MESSENGERS

RELAYING INFORMATION TO & FROM THE BRAIN

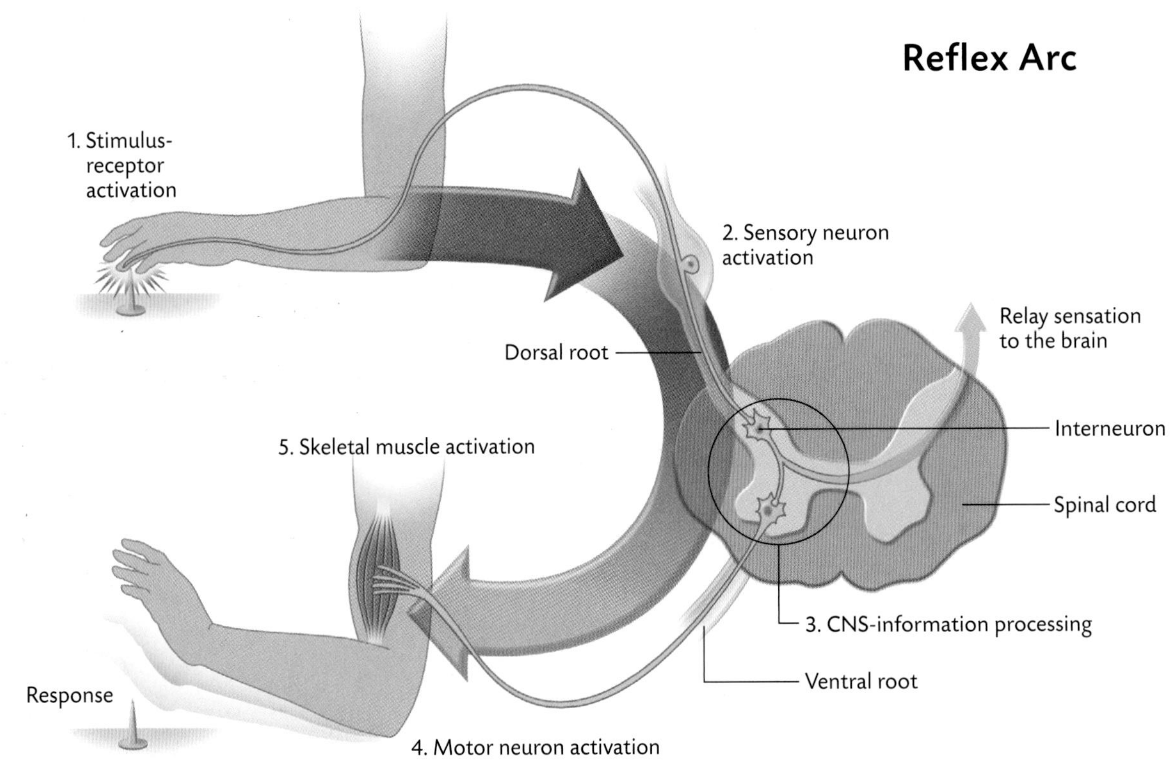

Reflexes are almost instantaneous. They provide protective, involuntary reactions to a stimulus.

THE COMPLEXITY of the brain and how it collects data and reacts to them lies in the very integration of its many neurons. Neural integration not only results in the interplay of sensations associated with motor activity but also influences the ways humans remember, think, and create. In the central nervous system, neurons form organizations called neuronal pools that process information brought in from either the peripheral nervous system or the neighboring neuronal pools.

SERIAL PROCESSING

Sometimes, one neuron excites only one other neuron, which excites only one other neuron, and so on, like a single row of toppling dominoes. The result of such "serial processing" is a clear-cut response. You can see a good example when the doctor taps your knee with his hammer, and the reflex action makes you jerk your leg. The links in the chain, called a reflex arc, must include a receptor responding to an external stimulus, a sensory neuron to carry the information to the central nervous system, an integration center in the spinal cord, a motor neuron to carry a return signal, and a muscle or gland to react.

Indigestion can hurt your chest. Packed spinal nerves sometimes confuse paths of pain signals.

PARALLEL PROCESSING

Other times, sensory information branches into many pathways. A single neuron may excite several others, like one domino setting a dozen rows in motion. This causes "parallel processing" of information

CROSS REFERENCE: See "Silent Running," PAGE 146

+ ITCH & SCRATCH +

NERVE ENDINGS sensitive to the sensation of itching proved hard to find. Not until 1997 were these receptors isolated in the skin; their extreme thinness helped hide them from prying eyes. The sensation—the itch—and its response—the scratch—still remain mysterious for neuroscientists. In 2008, findings showed that there are different kinds of itches, which activate different neural pathways. The relief of a scratch depends on the type of itch. Insight into how an itch works can help neuroscientists understand how to control it—and other sensations, like pain.

as circuits diverge and converge in the central nervous system. Each neural circuit delivers different information at the same time.

For example, seeing a kitten may remind you of the cat you raised as a child; the scar on your hand that you got when you bathed your kitty the first (and possibly last) time; the subtle hints your daughter has made in the last few days that she would like to own a pet; or the pleasant purring a happy kitten makes when you gently stroke it. Or all of these associations may appear in quick succession. Each response to the stimulus—"kitten"—is unique, not only among every human, but also from instance to instance in a single brain, thanks to the addition of new experiences and environments.

Parallel processing creates complexity several orders of magnitude above serial processing. For instance, when you see a driver's license, you quickly recognize it as such because your brain's neuronal circuits are assimilating various inputs from it at the same time. The shape of the license, its colors, the photograph of a face on one side, the identifying information about the card's owner, the state's name and artwork, and perhaps the fact that you saw it being removed from a wallet—all pass along through a variety of parallel circuits to allow a bartender to quickly say, "You're underage," or a traffic officer to remark, "You need to renew that next month." In contrast, it takes a much longer time for a computer using serial processing to analyze the object and declare what it is. Its circuits are not as efficient as the brain's systems.

RECEPTION

A healthy brain needs a constant stream of incoming information. Picture what happens without it: When volunteers enter a sensory deprivation tank—a body-temperature pool of water in which they are forced to go without sights, sounds, smells, tastes, and skin sensations—they begin to hallucinate; their brain creates stimuli to stay occupied. Insanity awaits those whose brain starves for external stimulation. Conversely, a healthy body needs the brain to send it signals. Deprived of adequate motion because of nerve damage or a sedentary lifestyle, for example, once strong muscles of the body will quickly atrophy.

Sensory receptors come in five types. The mechanoreceptors create nerve impulses when their physical

Touching a devil's club thorn stimulates pressure-sensitive mechanoreceptors and, possibly, pain-sensitive nociceptors in the fingertips.

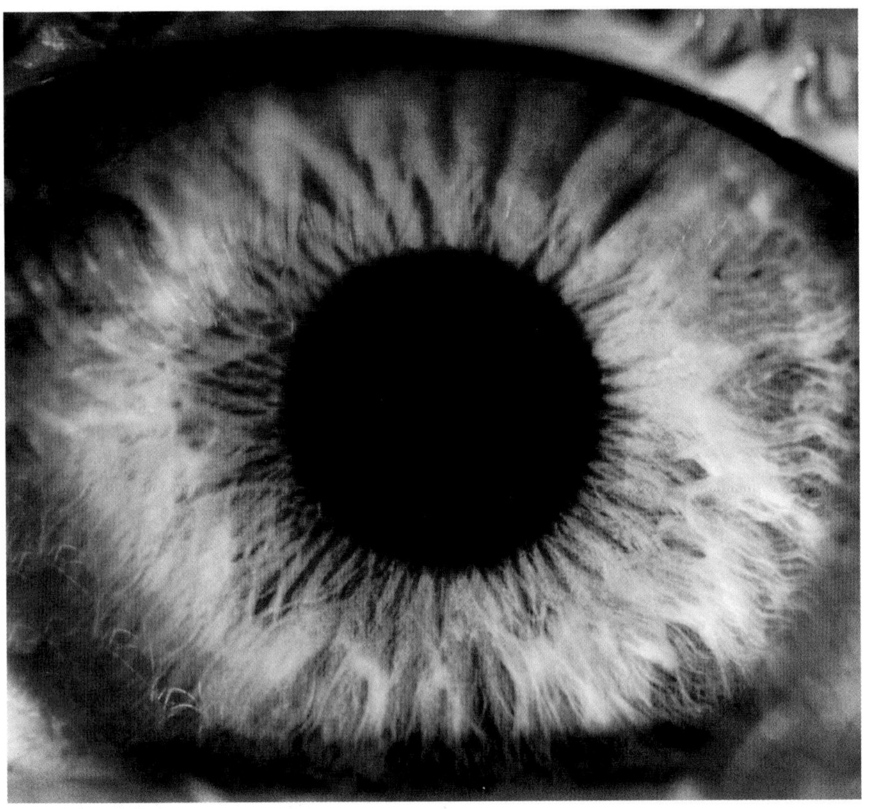

Photoreceptors in the eye begin the neural circuitry that registers sensations of visible light.

shape changes in response to external force, such as pressure or touch.

Photoreceptors respond to light. Curiously, not all photoreceptors exist in the eyes; some are found in the skin. Scientists at Cornell University and at White Plains, New York, found they could combat jet lag and insomnia by shining lights on the back side of the sufferer's knees. Thermoreceptors register heat and cold. Chemoreceptors register the presence of chemicals, such as the sugars in an orange when you bite into it.

And last are the nociceptors, which respond to external stimuli that have the potential to create, or do create, pain. The body needs to process painful feelings in order to warn it of possible larger dangers that pose threats to life and limb.

Nociceptors are able to act in concert with other sensory receptors. For example, the warmth of a fire on a wickedly cold day feels good on the feet because it stimulates thermoreceptors in the skin. If the toes get too close to the flames, however, extreme heat activates the nociceptors and the sensation changes from pleasure to pain.

PAIN GATEWAY

The nervous system does have natural responses that can ease minor pains, like the sting of a scrape or ache of a bump. When you were a child and trying to learn to roller-skate, perhaps you once fell and skinned your knee. To stop your tears, Mama may have given you a kiss, rubbed the area around the injured flesh, cleaned up the wound, and given you a bandage to show off to your friends. Miraculously, you felt better.

Turns out it was no miracle. Mama really did know best. According to research published in

+ SENSORY RECEPTORS +

NAME	DESCRIPTION
Mechano-receptors	Create nerve impulses in response to external forces such as: pressure, touch, itch, stretch, and vibration
Photo-receptors	Respond to light. Found in the eyes as well as parts of the skin
Thermo-receptors	Register temperature changes
Chemo-receptors	Register the presence of chemicals, as in smell, taste, and changes in blood chemistry
Nociceptors	Respond to potentially damaging and/or painful stimuli

CROSS REFERENCE: See "Perception," PAGE 100

the 1960s about the so-called gate control theory of pain, stimulation of the injured skin through rubbing temporarily overwhelms the brain. These tactile sensations send a second set of sensations along the bundles of nerve fibers whose neighbors are already sending pain signals to the brain. As the brain doesn't have the ability to entirely focus on multiple tactile sensations at once, the second set of sensations (the mother's touch) lowers the perceived intensity of the first set (the skinned knee). The gateway to pain closes a bit. Researchers call this competitive inhibition.

FAST FACT About 100,000 genes interact to create a human being. Perhaps 30,000 are specific to the brain.

Rubbing also results in the release of natural painkillers that act like opiates. They interact with receptors in the synapses of the amygdala and hypothalamus. Those collections of neurons, in turn, send signals via the medulla and spinal cord to offset the afferent pain signals from the nociceptors. The result: a decrease in the transmission of pain sensations. That's great for a skinned knee. But what if the pain is more acute, or even life-threatening?

BIG PAIN

It turns out, the brain has automatic defenses cued up for a quick response to more serious pain. The perception of pain warns the brain of actual or potential tissue damage. The brain's recognition of pain sets in motion actions to reduce or remove it, and thus the threat.

Most pain receptors consist of the bare ends of sensory nerves embedded throughout all body tissues, except the brain, whose cells cannot experience sensation. These nociceptors react to any "noxious" stimulation, anything that damages the body's cells.

Damage makes the cells release chemicals that activate neurotransmitter receptors (substance P is the transmitter for pain) and send pain signals via the peripheral nervous system to the central nervous system, where it may take a while to be felt. Pain doesn't reach the brain instantly because of the distance the signal must travel; in a tall man, injury to the toe may take two seconds to register in the brain.

In the skin, muscles, and joints, cell damage is likely to cause relatively brief and sharp pains. That's because nerve cells in the spinal cord release natural pain suppressants known as enkephalins, which inhibit the discharge of more pain-exciting neurotransmitters and keep the sensation short. As a result, sharp pains usually fade into dull aches.

Deeper cell damage is more likely to create burns and aches that last longer. The difference lies in the kinds of nerve fibers that transmit the pain signals, and how quickly that information travels.

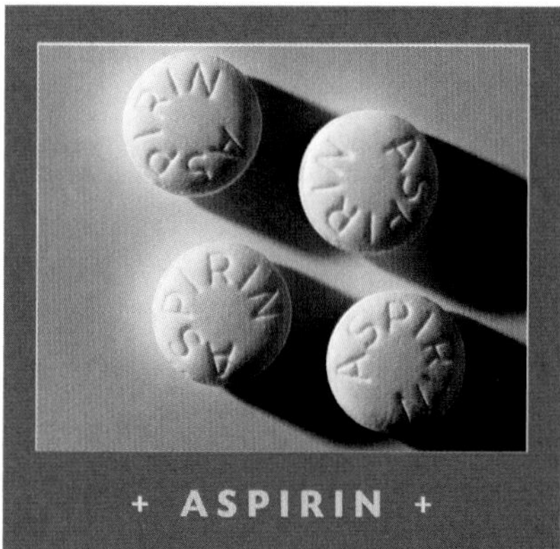

+ ASPIRIN +

HIPPOCRATES, the founder of modern medicine, knew that chewing willow bark alleviated pain. Thousands of years later, scientists discovered why: The bark contains salicylic acid. When cells are damaged, they release an enzyme called cyclooxygenase-2. That chemical in turn produces prostaglandin, which signals to the brain that part of the body is in pain. Prostaglandin also causes the injured flesh to swell and become inflamed. Salicylic acid binds to cyclooxygenase-2, blocking the creation of prostaglandin. Less prostaglandin means fewer pain signals reaching the brain, and less inflammation of the cells around the injury.

Damage to the internal organs, or viscera, usually results in dull aches, burning sensations, and gnawing pain. As the pathways for the visceral and somatic nerves of organs and body converge in the spinal cord, the brain sometimes gets confused and assigns visceral pains to other parts of the body that are not actually injured. A heart attack, for example, may seem to cause shooting arm pains.

PATHWAYS

Pain signals take two tracks on their way to the brain. The express line,

FAST FACT

Many Hollywood movies depict characters with unusual states of mind. While some include pure fancy—machines wiping the memory clean, for example—others are more or less accurate. Among the latter: *Awakenings, A Beautiful Mind,* and *Memento.*

like a nonstop train between cities, sends signals through the spinal cord and connects directly to the thalamus. While some pain signals are diverted along the way, those that reach the thalamus are relayed to the cerebral cortex, where they quickly get analyzed.

When you cut your finger while slicing an onion, the quick pathway of pain activates the cortex to figure out how much pain you feel and where you feel it. The brain's quick recognition of the danger may stop you from bringing down the knife blade again and slicing your finger a second time.

The other, slower pathway travels through slow, narrow nerve fibers with frequent synaptic connections, lumbering like a commuter train that stops at every little burg. These sensations register in the brain stem and hypothalamus, as well as in other deep brain regions, before a portion of them reach the thalamus. Effects include longer-lasting aches as well as emotional reactions to pain, such as the sheepishness of realizing you injured yourself through either clumsiness or negligence (or both). These slow-action pains include the unremitting discomfort of chronic diseases such as cancer.

GRAY MATTER

But not all pain sensations terminate in the thalamus. Many halt at a portion of the brain stem known as the mesencephalic central gray matter. It's a tiny spot that is difficult to locate. But as a convergence zone for pain impulses, this area is highly sensitive. When lab animals have their mesencephalic gray matter stimulated by electricity, they can be operated on without painkillers. Yet they maintain their sensitivity to touch, heat, and other sensations in the pain-affected body parts.

DEGREES OF PAIN

Similar pains don't always register with the same intensity. Although nearly all humans—besides the very few who lack the ability to feel pain—recognize extreme heat or a deep cut as painful, they can react differently. Some tolerate pain more easily, whereas others feel it more intensely. Physical, cultural, and psychological variables may also influence a person's individual degree of pain tolerance.

Cultural and psychological influences on an individual's tolerance of pain are more ethereal and hard to measure than physiological influences. During World War II, British soldiers injured in the brutal fighting at Anzio, Italy, in 1943 routinely refused morphine to kill their pain, while civilians who suffered far less serious wounds demanded it to ease their pain. The surgeon who noted the difference came to the conclusion that certain kinds of pain could be a matter of mind, not of the body.

Long-term, intense pain can create a different perception in the

WHAT CAN GO WRONG

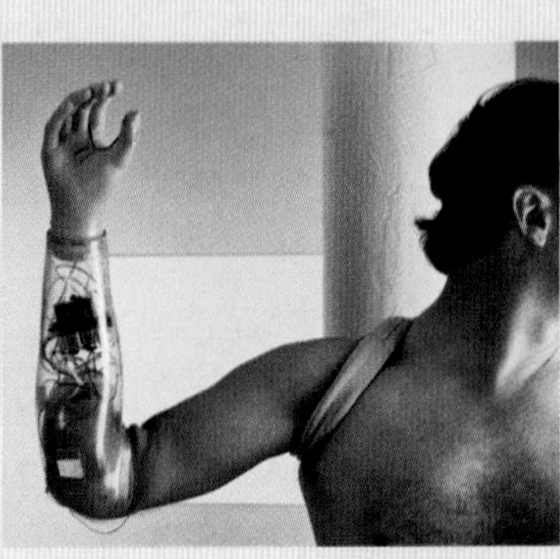

CAPTAIN AHAB asked his ship's carpenter for a special bit of work in the novel *Moby-Dick.* Ahab, who had lost a leg to the teeth of a white whale, hoped a replacement limb might expunge the feeling of "another leg in the same identical place with . . . my lost leg." "Phantom" limbs, such as Ahab's lost leg, have been reported since ancient times. American neurologist Silas Weir Mitchell cataloged many varieties in the Civil War. About 70 percent of phantom limbs proved excruciatingly and chronically painful. How could a missing leg create the illusion of existence, or even pain? The answer lies in the brain.

Ritual mortification of the flesh at the Hindu festival of Thaipusam in Malaysia demonstrates the power of brain over pain.

brain. This chronic sensation may confuse the central nervous system and result in hyperalgesia, or pain amplification. Such pain registers on the same kind of synaptic receptors that are activated during certain kinds of learning. Under the worst-case scenarios, the chronic pain causes the spinal cord to "learn" hyperalgesia, and pain's sensitivity increases. Examples include the lingering pain of phantom limbs—the sensation of pain from an amputated arm or leg.

GOOD FEELINGS

Pleasure also has its centers in the brain. A Tulane University

WHAT CAN GO WRONG

Neural networks that process stimuli from a limb remain primed to respond to signals even after it's gone. Random signals may get misinterpreted as tingling, itching, pain, or some other sensation. Neuroscientist Vilayanur Ramachandran found he could create sensations in phantom limbs by applying pressure to various skin surfaces. His conclusion: The cerebral cortex relocated sensation pathways associated with the old limb. These pathways may always have existed in a weak state, but loss of the limb amplified them. Unfortunately, neural networks that continue to recognize "pain" signals from a missing limb become more strongly primed to repeat the mistake. Treatments for phantom pain range from drug therapy to acupuncture and deep brain stimulation. Newer treatments, using mirrors or virtual reality goggles, trick the brain into thinking it can control the amputated limb.

neurologist stumbled across one such center in the 1950s when he tried to electrically stimulate the brains of schizophrenics to break them out of their passivity. His patients told him their implanted electrodes created pleasant sensations. The neurologist, Robert G. Heath, seized upon the results, focused his attention on the brain's pleasure centers, and published the 1964 book *The Role of Pleasure in Behavior*.

Together with the discovery of pain centers in the brain, research on the physical causes of the sense of pleasure seemed to prove the ancient wisdom that humans seek to act in ways that bring them pleasure and reduce or avoid pain. New paths of investigation have led to innovative treatments for addiction, which is a form of behavior based on compulsive forms of pleasure seeking. PET scans reveal how drugs such as cocaine and heroin activate the brain's pleasure centers. Cocaine, for example, blocks a neuron's reuptake mechanism, which causes dopamine to linger in the synaptic cleft.

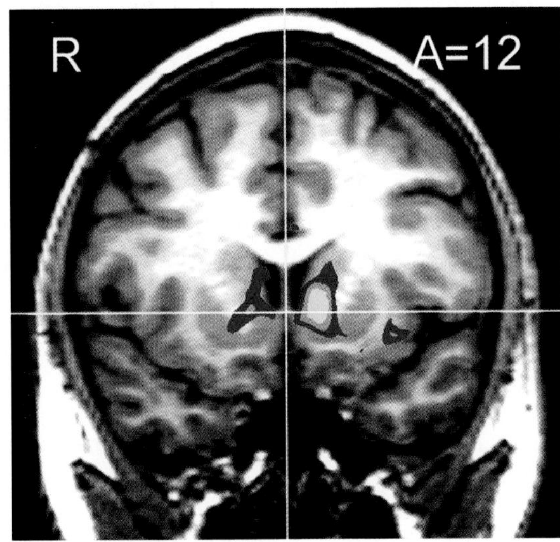

FIRST THE PLEASURE

A desirable purchase activates nucleus accumbens.

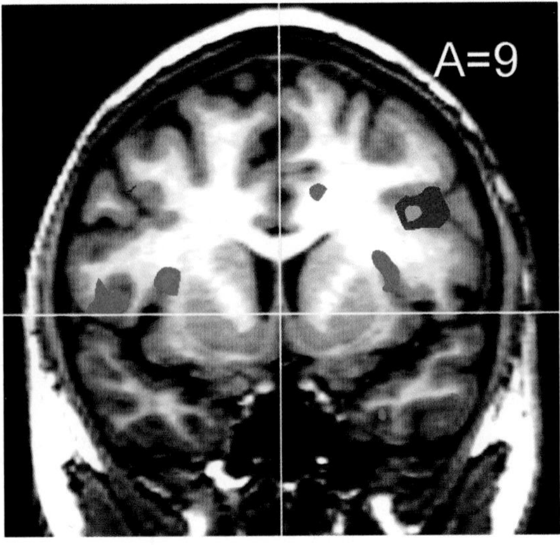

THEN THE PAIN

Insula reacts to a shopper learning item's high price.

PLEASURE CENTERS

Joy, happiness, pleasure—whatever you want to call the positive feelings that bring rewarding sensations and make life worth living—arise from the sensations of security, warmth, and social well-being combined with an awareness of the rightness of such feelings. A healthy brain recognizes the conditions that give rise to pleasure and responds to them appropriately. An unhealthy brain, or one that has learned negative behaviors such as addiction, can miss out on experiencing life's joys. Both are primarily a matter of chemistry.

The sensation of pleasure registers in several brain regions, including significant centers in the hypothalamus and nucleus accumbens, which lies below a portion of the basal ganglia linked to movement. All such pleasure centers rely on the chemical work performed by endorphins and neurotransmitters, particularly dopamine, to create and sustain a happy mood. Experiments with rats have demonstrated the key role of dopamine. In the 1950s, scientists wired rats' brains so that when they pressed a bar, they received a mild electric shock to the hypothalamus. This stimulation registered as pleasure; the rats would rather press the bar than eat. However, in later experiments, rats wired for self-stimulation first received injections of drugs that block the receptors where dopamine normally binds, denying its pleasure-giving action. The rats no longer felt a pleasant reward from pressing a lever to stimulate their brain, and they stopped doing so. When humans take a similar dopamine-lowering medication, often in order to ward off hallucinations and other psychotic behavior, the

CROSS REFERENCE: See "Good Feelings," PAGE 220

FAST FACT When the skin warms, the sympathetic division of the autonomic nervous system dilates blood vessels near the surface and activates the sweat glands. When body temperature cools, the autonomic nervous system narrows surface vessels to send blood to deeper, more vital organs.

drug's success comes at a price. Delusions may leave, but so do joy and motivation. Conversely, drugs like amphetamines that increase the activity of dopamine in the brain lower the threshold for the perception of pleasure. Too much of a drug-induced pleasant sensation, however, can lead to addiction and manic moods.

"The greatest pleasure of life is love," said the Greek playwright Euripides nearly 2,500 years ago. Like other forms of pleasure, love is processed by brain chemistry, specifically by heightened levels of neurotransmitters in the pleasure centers. MRI scans of the brain relate the feeling of lust to estrogen and androgens; attraction—more emotional than physical—appears to be associated with serotonin and dopamine. The brain chemistry that supports long-term relationships such as lifelong commitment has been harder to pin down.

Playing key roles in the sensation of pleasure are oxytocin, endorphins, and phenylethylamine, or PEA, sometimes called the love drug. These chemicals help foster the "high" felt in the first stages of love, as well as the euphoria sometimes reported by long-distance runners. Even a small pleasure, such as finding your lost car keys, begins with a tiny rise of these and similar neurotransmitters in the brain's pleasure centers.

WORKING TOGETHER

How do all of these systems—central and peripheral, somatic and autonomic—and receptors work together in the symphony of the brain? From simple actions to complex ones, these systems must work in concert.

Consider the "simple" act of catching a ball. It's an amazingly complex process that requires some basic anatomical structures and neural circuitry before it can be attempted. Obviously, most animals cannot toss an object. Nearly all lack hands with fingers and opposable thumbs, as well as the dexterity that has developed in human beings, across millennia of evolution, through the growth of increasingly complex neural circuits in the cerebellum and cerebral cortex. Thanks to evolution providing the basic tools of manual dexterity and the expansion of specialized brain functions such as those children develop when learning how to throw a ball, adults have basic skills ready to be activated when a ball comes their way.

SEEING THE BALL

The simplified version goes like this. When someone throws you a ball, photoreceptors in your eyes register the action and send it along afferent nerve fibers to specific portions of the frontal lobes of the cerebral cortex. Parallel processing of various sensations—including the motion of the pitching arm, the path of the ball as it travels through the air, and its speed—occurs within milliseconds. The cortex registers the perception "The ball has been thrown" and works with the cerebellum to calculate its likely point of arrival.

If it's thrown particularly hard, say, and right at your head, the

+ WHAT IS PLEASURE? +

"OUR ENTIRE psychical activity is bent upon procuring pleasure and avoiding pain," Sigmund Freud said in 1920. More than a century earlier, British philosopher Jeremy Bentham had a similar idea: What humans seek to do is maximize pleasure and minimize pain.

But what is pleasure? Bentham equated it with happiness. Freud named things (especially sex) that make us feel good. It's not an abstract argument for neurochemists. So-called recreational drugs affect the centers of the brain that register pleasure. How ironic that Freud championed cocaine as a treatment for neural disorders.

Catching a baseball requires a complex chain of actions in the sensory and skeletal muscle nerves, cerebrum, cerebellum, and basal ganglia.

autonomic nervous system registers the action as a possible threat, sends out efferent signals that release a chemical soup of neurotransmitters, and may prompt you to duck. But if the ball arrives as an ordinary pitch you've experienced a thousand times, the motor areas of the cortex, which control voluntary movement, work with the cerebellum and basal ganglia to move your gloved hand to the right place for the catch.

HERE COMES THE BALL

The cerebellum, at the rear and bottom of the brain, is a key brain area for practiced, complex motor skills. It maintains the body's balance during the catch and coordinates with the portions of the cerebral cortex that involve thinking. You may realize, "Here comes the ball," but little thinking is involved in moving your hand to make the catch if you've practiced that motion. Instead, the cerebellum moves

WHAT CAN GO WRONG

LOU GEHRIG, the "Iron Horse," played in 2,130 consecutive games for the New York Yankees from 1925 to 1939. In May of his final year as a Yankee, when his batting average dipped to an uncharacteristic .143 and he began feeling inexplicably weak and sluggish, he took himself out of the lineup. He told the manager he thought the club would do better if someone else replaced him at first base. Two months later, Gehrig knew the reason for his sluggishness. Doctors at the Mayo Clinic diagnosed him as suffering from a degenerative disease of nerve cells in the brain and spinal cord. Two years after that, he was dead.

CROSS REFERENCE: See "Shared Roles," PAGE 152

FAST FACT

Luigi Galvani discovered in the 18th century that nerves use electricity. It was an accident. An aide touched a frog nerve with a scalpel, and its legs contracted. Galvani substituted electric sparks and got the same effect. His name lives as a verb: when sparked into action, we are "galvanized."

the body smoothly and quickly in response to the cortex's analysis of the sensory stimuli. The movement occurs because somatic motor neurons were prompted to release the neurotransmitter acetylcholine at their synapses in the skeletal muscle fibers. Acetylcholine always excites action rather than suppressing it. Once acetylcholine's effect reaches a threshold, the fibers of the muscles in the arms and legs contract, moving the hand into position to make the catch. Continuing sensory input from the eyes creates a feedback loop of information between the brain and the hand. The brain continues to make fine motor adjustments as the ball comes near.

A GOOD CATCH

When the ball hits the glove, mechanoreceptors in the hand register the arrival as pressure. Those in the ear, attuned to the vibrations of sound waves, record the *thwack* of ball hitting leather. (And if the leather in the palm is too thin, cell damage in the hand may release noxious chemicals such as prostaglandins, which set off a chemical chain reaction ending with nociceptors initiating pain signals to the brain.) The cortex processes the new sensory stimuli, perceives that the ball has arrived, and sets in motion, with the cerebellum's help, the voluntary muscle contractions that squeeze the gloved fingers.

Another way to think of the integration of brain functions, in a metaphor of psychiatry professor John J. Ratey's, is to picture a house. Some functions exist on only one floor—the furnace kicks on automatically in the basement when the thermostat tells it to—but others require communication among all the floors. The basement has the brain stem and spinal cord, which automatically oversee reflexes and respiration. The first floor houses basal ganglia and the cerebellum, which oversee the basement and communicate information to the upper floors. The second floor has centers of increasing control over the nervous system such as the motor and premotor cortex. The top floor is home to the prefrontal cortex, decision-maker of the brain. The top floor's decisions get communicated downward, receiving feedback as they are carried out.

+ MANUAL DEXTERITY +

EVOLUTION HAS selected for the development of eye-hand coordination in human beings. As humanity's ancestors swung from branches, they refined their performance by figuring out how to grasp one limb after another. Later, as they stood on two feet, they freed their hands for manipulating objects. Manual dexterity improved through brain-hand feedback, leading to the creation of tools and other developments that aided survival. Today, the hand is so closely integrated to the neural circuitry of the brain that neurologist Richard Restak suggests it is best thought of as an extension of the brain.

WHAT CAN GO WRONG

His disease was amyotrophic lateral sclerosis. Such was the sudden drama of his situation that the illness claiming his life is sometimes called Lou Gehrig's disease. This devastating disease gradually destroys motor neurons. As motor nerves lose their ability to send signals that move muscles, the muscles atrophy. Those afflicted lose their ability to speak and swallow, and eventually even to breathe. Researchers hypothesize that the motor neurons are killed by an attack of the sufferer's own immune system, the production of too much of the neurotransmitter glutamate, or both. In making his farewell to 62,000 fans at Yankee Stadium on July 4, 1939, Gehrig called himself "the luckiest man on the face of the Earth." For his performance on the field, as well as his demeanor while facing a final opponent he could not defeat, Gehrig is remembered as a "Gibraltar in cleats."

DELICATE BALANCE

THE BRAIN'S EQUILIBRIUM

Buddhists in Java engage in meditation, which has been found to decrease stress and anxiety and promote calm feelings.

THANKS TO THE autonomic nervous system, the human body pretty much takes care of itself without conscious effort. The weather changes but core temperature is maintained, food gets digested, cycles of sleeping and waking follow upon one another, and the body's status remains fairly even from one day to the next. It's a system in a delicate balance, self-regulating in an attempt to keep the entire body stable and healthy.

HOMEOSTASIS

American physiologist Walter Cannon came up with the word *homeostasis* to refer to the body's ability to stay relatively stable while internal and external environments are changing. While homeostasis

WHAT CAN GO WRONG

ABOUT ONE in a hundred Americans older than age 65 suffer from Parkinson's disease, a neurological condition that mysteriously kills off cells in the brain. They include preacher Billy Graham and former Attorney General Janet Reno. Younger people, like actor Michael J. Fox, can also be stricken with the disease. Symptoms of the disease first appear with the onset of small tremors during voluntary movements. Over time, it becomes harder to initiate motion. Finally, muscles grow rigid, and even making the simplest movements takes extended time and effort. The condition is caused when cells in a region of the brain beneath the cortex that produces and stores the neurotransmitter dopamine die. This region, including the basal ganglia and an area called the substantia nigra (because it appears black in autopsies), plays a key role in coordinating movement.

literally means "unchanging," the body does indeed change when sensory receptors detect changes in the environment and automatically react, causing the release of appropriate neurotransmitters and hormones to help the body adapt to the world around it. The body then reacts to the changes, those alterations get fed back into the nervous system, and the process repeats itself.

This is known as dynamic equilibrium. It occurs when change after change keeps the body healthy. And it's complicated. Think of the body's constant need to adjust heartbeat and respiration, regulate temperature, as well as maintain the smooth functioning of neurons throughout it. Think of how distracting it might be if the brain didn't adjust to our environment on a regular basis; hearts would beat rapidly long after a moment of fear had passed; the body wouldn't adjust to changes in temperature. The unconscious efforts of the brain go by virtually undetected as the body goes about its business.

BALANCING ACT

Some feedback mechanisms suppress actions in the brain and body. Others excite them. Their delicate balance keeps the body between extremes. To have too much or too little of one can throw the system out of whack.

To take one example, the lack or overabundance of neurotransmitters such as dopamine causes health problems—Parkinson's disease in one case, schizophrenia in the other. Because the brain and body are so closely interrelated—you could think of the glands, organs, bones, muscles, and other parts of the body as functionally integrated appendages of the brain—damage to the brain and the rest of the nervous system can knock the body dangerously out of homeostasis.

FAST FACT Transplants of fetal neurons producing dopamine show promise as a Parkinson's treatment.

Physical damage to the brain is an obvious source of homeostatic imbalance. Shrapnel from an artillery shell, tumors and lesions that arise organically, and atrophy or death of neural groups in the brain reduce and sometimes destroy the brain's ability to monitor the body and respond to its needs. Headaches, seizures (and epilepsy in particular), diabetes, and Parkinson's disease are examples of the consequences of a body getting out of a healthful dynamic balance.

HEADACHES

In the waning days of the Civil War, Union general Ulysses S. Grant was suffering from a terrible headache. He stopped at a farmhouse in the rear of his army, which had been pressing the forces of Confederate general Robert E. Lee. "I spent the night in bathing my feet in hot water and mustard, and putting mustard plasters on my wrists and the back part of my neck, hoping to be cured by morning," Grant wrote in his journal on April 9, 1865.

Shortly afterward, Grant was visited by a messenger who carried a note saying Lee, who had refused to surrender the previous day, had

WHAT CAN GO WRONG

Treatments vary. Neurochemical treatments seek to replace the dopamine depleted by the death of the brain's dopamine-producing cells. Drugs like levodopa, also known as L-dopa, are able to pass through the blood-brain barrier. Once inside the brain, L-dopa is transformed into dopamine. It works only up to a point, and it can have side effects, including hallucinations. Furthermore, as the disease progresses, larger and larger doses are required to get the same benefits, with an increased risk of bad reactions. The drug interferes with other neurotransmitters, so large doses often have multiple reactions.

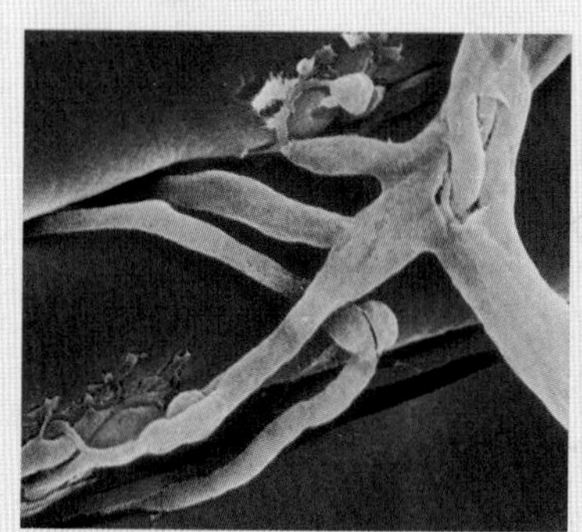

changed his mind and would be willing to meet to discuss a formal end of hostilities. "When the officer reached me," Grant said, "I was still suffering from the sick headache; but the instant I saw the contents of the note I was cured."

Grant probably suffered from a muscle-contraction, or "tension," headache. Typically, a tension headache begins when the neck, scalp, and face muscles are tensely held stiff for a long time. The most usual source is prolonged anxiety, a debilitating form of stress. Grant needed Lee to surrender; Lee's announcement of his plans took the worries, and the agony, away. "The pain in my head seemed to leave me the moment I got Lee's letter," Grant reportedly told an aide as he rode off to end the war.

Red indicates pain in a map of common headache sites, none of which is in the brain itself.

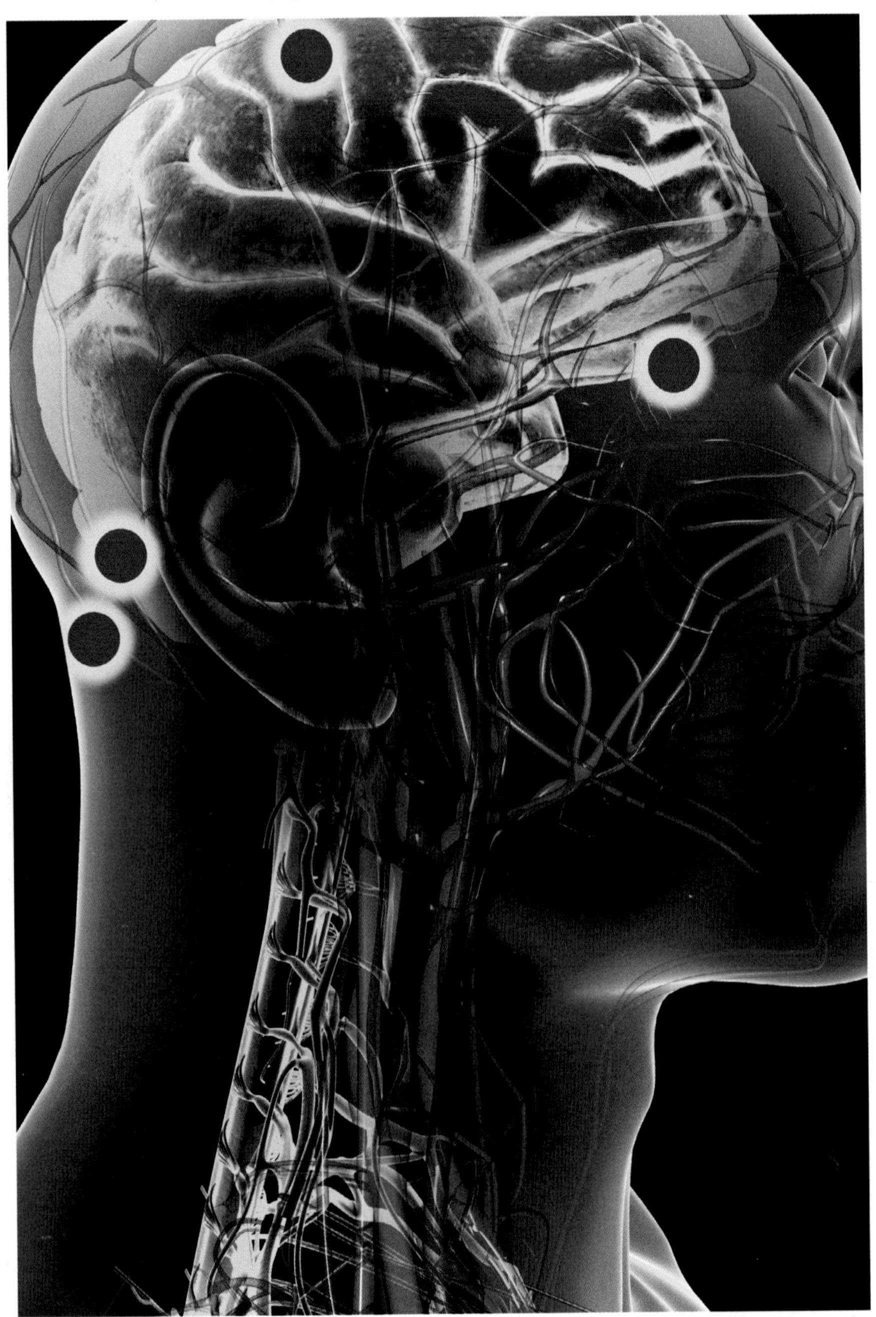

CATEGORIES

Even as it serves as an indicator that homeostasis is being disrupted, a headache is not a disease per se. Instead, it may a symptom of some other problem. It can manifest itself in response to irritation of blood vessels in the head, or to an injury or imbalance, or to inflammation of bodily tissues, to disorders related to stress—or to a host of other possible triggers. While it may feel as if the brain screams in pain, a headache can only occur outside the brain itself, which contains no pain receptors.

Headaches come in dozens of varieties. An easy way to categorize them is by the ways they cause pain. Muscle contractions such as Grant's are one of the most common sources, especially among those living with high levels of stress. Dilation of blood vessels is a second typical cause. When arteries expand in the head, they squeeze against surrounding tissues, producing viselike pressure and pain. Fever, migraines, drug reactions, changes in blood pressure, and carbon dioxide poisoning

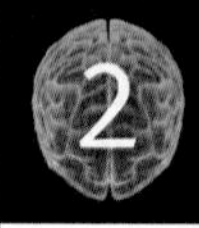

+ KINDS OF HEADACHES +

A headache may seem to punish your brain, but it's just not so. The brain has no pain-sensitive nerve fibers. It cannot "ache." However, the brain plays a role in headaches. Although the exact reasons for onset remain unclear, headaches occur when nerves of the head's muscles and blood vessels send pain signals to the brain.

KIND	SYMPTOMS	DESCRIPTION
Tension headaches	Constant ache or pressure around the head; strongest at temples or base of the neck. Caused by muscle contractions.	Most common form of headache for adults and adolescents; typically come and go for a prolonged period. Can be treated with over-the-counter drugs.
Migraine headaches	Last a few hours to a few days; typically return. Moderate to severe pain often associated with light sensitivity, odors, noises, and nausea.	Related to distension of larger blood vessels within the brain, along with activation of the trigeminal cranial nerve and its connections to the brain stem and upper spinal cord.
Cluster headaches	Rare form carries intense pain characterized by constant burning, piercing, or throbbing. Pain settles behind one eye.	"Cluster" name arises from attacks coming in patterns of one to three a day for weeks or months. Strikes more men than women.
Sinus headaches	Characterized by pain in cheeks, forehead, or bridge of nose; often accompanies nasal discharge and other symptoms of sinus problems.	Pain spikes when the head is moved suddenly. A doctor should diagnose the infection and may prescribe antibiotics to treat the root cause of the headache.
Rebound headaches	Occurs daily, usually in the morning. Persists throughout day with worsening pain as withdrawal from medication or caffeine increases.	Caused by overuse of pain medication and/or caffeine. Reducing intake may worsen the headaches at first but will eventually break the cycle.

can provoke dilation. Internal traction—an abnormal growth in the head, for example—is a third trigger. When a tumor presses against other tissues, or the brain itself begins to swell, the pressure causes pain. Inflammation is a fourth common source. Allergic reactions and infections such as meningitis can irritate pain-sensitive receptors in the head. Finally, headaches can occur without an obvious physical cause. These headaches are called psychogenic, meaning they arise in the psyche. They may spring from an emotional problem, as the sufferer converts emotional pain into real, physical symptoms.

The word *migraine* evolved from the Greek word *hemikrania,* meaning "half-skull."

Many of these disorders strike not next to the brain, but in the eyes, sinuses, and other facial organs and tissues. Cranial nerves intimately connect the face and neck muscles to the brain, so it is no wonder pain sensations can spread until they feel as if they overwhelm the entire head.

Treating chronic headaches requires a proper diagnosis. Given the wide range of headaches and their causes, as well as the possibility of triggers working in combination, medical treatment often relies on detective work. At least, however, the efficacy of treatment has advanced since humanity first tried to cure a headache. A thousand years ago, Arabs recommended applying hot irons to the head, while a French medical treatise written in Latin urged sufferers to mix the brain of a vulture with oil and shove it up the nose. Today, modern pharmaceuticals, relaxation techniques, and proper diet target dilation, tension, and other causes. One of the most effective pain relievers is common aspirin.

SEIZURES

Abnormal electrical activity in the brain produces seizures, which have a broad range of manifestations.

MIGRAINES

THE DAMAGE caused by headaches is eye-popping. About 45 million Americans suffer them regularly, and about half of the sufferers find the pain severe and sometimes disabling. The result: lost time from work, play, the day-to-day stuff of life. Counting only the 30 million who suffer migraine headaches—one of the 150 described categories of headaches—American victims lose 157 million work days each year.

ALL IN YOUR HEAD?

Victims often describe the pain as throbbing or pounding. Other related symptoms include sensitivity to light, sound, and odor. Some experience nausea, abdominal pain, or vomiting, and some sufferers report seeing auras or streaks of light shortly before the pain begins. Young victims may also complain of blurred vision, fever, dizziness, and upset stomach. A few children get migraines about once a month accompanied by vomiting; such headaches are sometimes referred to as abdominal migraines. About 5 percent of children younger than 15 report having had migraines, compared with 15 percent who experienced tension headaches.

ANATOMY OF A MIGRAINE

Headaches occur when nerve cells that are pain sensitive, for reasons that are still not clearly understood, begin sending pain signals to the brain. These nociceptor cells often act in response to stress, tension, hormonal changes, or the dilation of blood vessels.

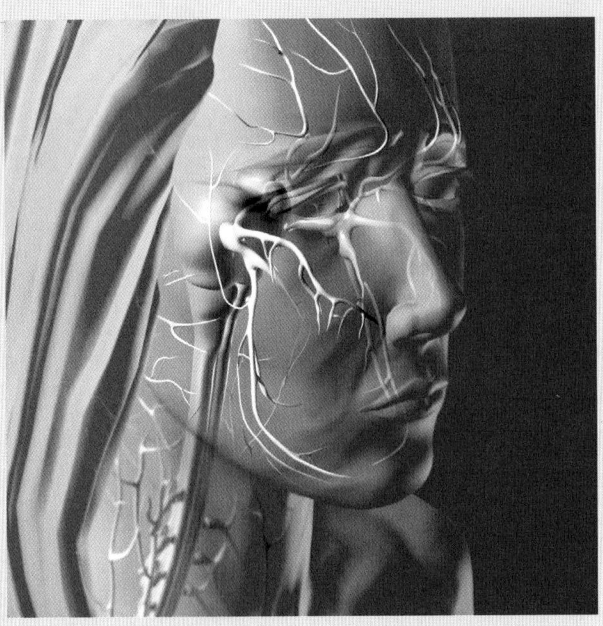

Pain from migraine headaches is typically located on only one side of the head, behind the eye.

Some researchers theorize that chronic headache sufferers lack normal levels of pain-blocking neurotransmitters called endorphins, a Greek word that means "the morphine within." This deficiency means that their pain signals are more likely to cause severe discomfort than those in people who have higher endorphin levels.

Migraines are particularly devastating because of their severity and recurrence. They begin with impulses in hyperactive nerve cells. These impulses tell blood vessels in the head to constrict, and then to dilate. The process releases serotonin, prostaglandins, and other chemicals that inflame nerve cells surrounding the blood vessels in the brain. Specifically targeted are the trigeminal cranial nerve and its connections to the upper spinal cord and brain stem. The result: pain.

Researchers long believed migraines arose from the narrowing and expanding of blood vessels on the surface of the brain; now, the most common theory traces migraines to hereditary abnormalities of the brain itself.

TRIGGERS

Emotional stress, anxiety, changes in weather, depression, lights, loud noises, alterations in sleep routines, and foods and beverages have been identified as migraine triggers. Stress causes the release of so-called fight or flight hormones, which can prompt changes in blood vessels that bring on headaches. Chemicals in foods and food additives also are commonly linked to the onset of

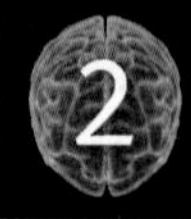

migraines. These include chocolate, aged cheese, red wine, yeast, monosodium glutamate, wheat, tea, meat containing sodium nitrates, coffee, oranges, milk, and corn syrup.

ALL IN THE FAMILY

Family genetics tell the tale: When one parent suffers migraines, his or her child has a 25 to 50 percent chance of befalling the same fate, but when both parents are victims, their child's risk rises to 70 percent.

Then there's the genetics of the two sexes. Boys who suffer from migraines tend to outgrow them by the time they get to high school. Girls, however, are more likely to see their migraines become more frequent after they enter puberty because of body changes related to female hormones. Teenage girls are three times more likely than teenage boys to suffer migraines.

Triggers for migraines range from foods including red wine and aged cheese to stress and changes in the weather.

PREVENTION

Eliminating or reducing severe stress and the intake of risky foods may help stop migraines from returning. In addition, physicians may prescribe regular doses of antidepressants, anti-seizure medicine, and cardiovascular drugs to ward off migraines. Migraines that are mild may respond to nonsteroidal anti-inflammatory drugs, or NSAIDs, such as ibuprofen or naproxen sodium. Stronger migraines may respond to drugs containing acetaminophen, aspirin, caffeine, or triptans, which mimic the neurotransmitter serotonin. Even when taken under a physician's care, these drugs can have unpleasant side effects. Doctors often advise patients to stop taking them as soon as possible.

TREATMENT

Biofeedback also has shown promise for some sufferers as a treatment, and particularly as an early response to the onset of a migraine. A headache sufferer who has learned biofeedback techniques monitors vital body signals and then attempts to control them through "mind over matter." Biofeedback methods often help reduce stress or anxiety; among headache sufferers, a common technique aims to alter the pulse or raise the temperature of one or both hands, diverting blood there and helping restore circulatory balance to the head. However, some researchers believe biofeedback works better with other forms of headaches than with migraines. Showing new promise in lowering the pain and frequency of migraines, according to a study in the journal *Headache*, is a mix of yoga, relaxation techniques, and breathing exercises.

RESEARCH

The link between body chemistry and migraines was underscored by a study associating migraines with depression. The study, conducted by the Henry Ford Health System, found that compared with people who are free from headaches, migraine sufferers are five times more likely to develop major depression. Likewise, people who began the study suffering from depression were three times as likely to develop migraines as those who weren't depressed. Researchers concluded that the two disorders are biologically related, perhaps through hormones or neurochemistry. If one disease exists, they say, doctors should be on the lookout for the other.

+ TYPES OF SEIZURES +	
NAME	**SYMPTOMS**
Tonic-clonic ("Grand mal")	Unconsciousness, convulsions, muscle rigidity
Absence ("Petit mal")	Brief loss of consciousness
Myoclonic	Sporadic jerking movements
Simple motor	Jerking, muscle rigidity, spasms, head-turning
Simple sensory	Unusual sensations affecting taste, smell, hearing, and touch
Complex	Impairment of awareness, repetitive movements like lip smacking, fidgeting, and pacing

Some are so minor that they may occur unnoticed, while others can cause violent spasms and convulsions. Victims may even lose consciousness. They can be a one-time event or occur frequently.

A number of things can cause seizures: Serious conditions like strokes, brain tumors, and severe head injuries can generate them, as well as other seemingly harmless things like bright, rapidly flashing lights and low blood sugar.

There are two general types of seizures: generalized and partial. Generalized seizures involve both sides of the brain from the beginning of an episode while partial seizures begin in specific regions of the brain and may spread to the entire brain. Generalized seizures have several subtypes, from tonic-clonic seizures (formerly known as grand mal) to absence seizures (also known as petit mal).

MAPPING SEIZURES

Seizures may occur in any part of the brain; their point of origin often can be mapped. Some occur as a result of lesions in specific domains. Nineteenth-century doctor John Hughlings Jackson, an aloof but meticulous researcher, posited that lesions would produce two effects. He based this belief on the idea that most of the neurotransmitters in the brain at any given moment inhibit action. A minority of neurons at any one time release neurotransmitters that bind to receptors. Others do nothing. Thus, Jackson said lesions would produce negative reactions because of the destruction of brain tissue. However, they also would have the opposite reaction of freeing other, healthy areas of the brain, which previously had been suppressed.

FAST FACT Four ions—sodium, potassium, calcium, and chloride—regulate electrical charges in synapses.

The minus and plus aspects of brain damage appeared to match the observed effects of a brain tumor in a teenage girl named Bhagawhandi in the 1970s. A neuroscientist who observed the girl diagnosed a malignant brain

BREAKTHROUGH

FIRST THEY felt hyperactive and frenzied. Then their body motions became more violent, and they would twitch and convulse. Finally, they fell into a deep trance. And there they remained, these sufferers of the disease encephalitis lethargica, until neuroscientist Oliver Sacks found them in the 1960s—40 years later. As depicted in the movie *Awakenings* (1990), Sacks gave them L-dopa, which the brain transforms into dopamine. The dopamine levels in the postencephalitic patients had been greatly diminished by their disease. The patients woke up from their stupor, and health seemed to be restored to them.

CROSS REFERENCE: See "Motion Sickness," PAGE 160

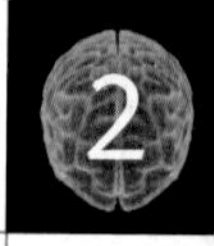

A photomicrograph of L-dopa, suggestive of an abstract painting, hints at the complex world of neurochemistry.

tumor. As the tumor grew to press on her temporal lobe and her brain started to swell, she suffered a series of seizures. They grew more frequent. However, whereas her initial seizures were intense grand mal convulsions, her new manifestations, localized in the temporal lobe, were weaker. She began experiencing dreamy states in which she saw visions of her home in India. Far from being unpleasant, they made her happy—"They take me back home," she said. She remained peaceful and lucid during her episodes. The seizures killed her in a few weeks, but doctors often noted the rapt expression on her face as

BREAKTHROUGH

The beauty of L-dopa lay in a seemingly simple but startling idea for treatment: If the neurons' ability to make dopamine had dramatically decreased, why not merely supplement the supply of the drug in the brain? Not only did L-dopa help the encephalitis lethargica patients, it also became a popular treatment for a far more common disease, Parkinson's disease, marked by muscle rigidity and loss of motor control.

Despite its ability to ease suffering, though, L-dopa is no "magic bullet," no magic cure. Sacks's patients began relapsing into their former patterns of tics and frenzies. Parkinson's sufferers also found that over time, L-dopa lost some of its power to help them. Still, the tangible results of L-dopa treatments have encouraged neuroscientists to seek the right combination of medications to restore balance to brain chemistry for a variety of illnesses.

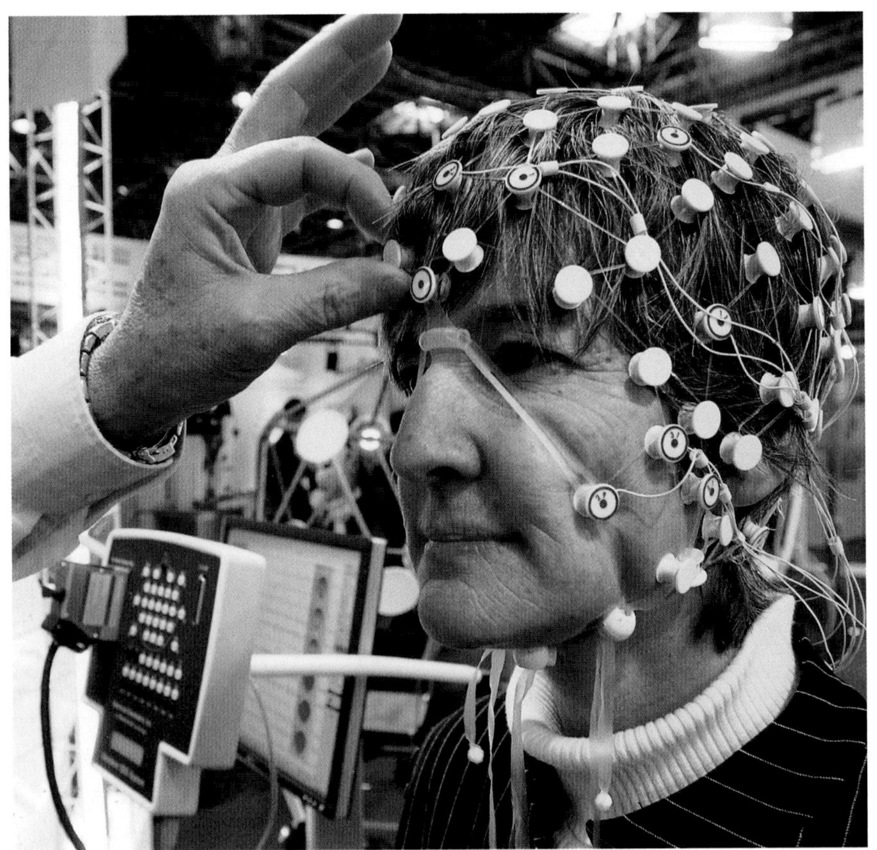

A woman wears an EEG sensor net that aids in epilepsy analysis.

she moved deeper into her visions. Only a few diseases of the central nervous system produce pleasure. Anything that pushes the brain out of homeostasis is more likely to bring pain and discomfort to the body.

EPILEPSY

On a summer day, storm clouds can suddenly gather and transform an afternoon of sunshine into a violent monster of rain, hail, lightning bolts, and the occasional twister. Sunlight and warmth get blotted out. So it is with the nervous system. The brain's higher functions, working in harmony with the body, promote consciousness and a sense of well-being. But because the brain functions through the medium of electrochemical reactions, the occasional storm knocks the brain out of balance.

Epilepsy is a flood of electrical discharges in groups of cranial neurons. While the brain suffers through its own electrical storms, no other signals get passed through. Those who suffer an attack may fall to the ground, black out, foam at the mouth, and jerk about uncontrollably. Epileptic seizures can last from a few seconds to a few minutes, and can vary widely in their ferocity.

TYPES OF EPILEPTIC SEIZURE

The mildest used to be called *petit mal,* French for "little illness." Now they're referred to as absence seizures. Sufferers, usually young children, lose consciousness for a few seconds, often staring blankly into space. They typically do not know what has happened to them. Such seizures usually go away by age ten.

Stronger, convulsive seizures are called tonic-clonic, which replaces the old term, *grand mal,* French for "big illness." Epileptics in the midst of a tonic-clonic seizure lose consciousness and may experience loss of bowel or bladder control, as well as muscle contractions so severe they have been known to break bones. After a few minutes, when a major seizure dissipates, the

+ DIVINE ILLNESS +

A NEUROSCIENCE JOURNAL article in 1997 listed religious figures thought to be linked with epilepsy because of recorded accounts that match its symptoms. The historical figures included:

+ Saint Paul, apostle and writer of much of the New Testament.
+ Joan of Arc, 15th-century saint and heroine of France.
+ Emanuel Swedenborg, 18th-century theologian.
+ Ann Lee, 18th-century leader of the "Shaking Quakers," or Shakers.
+ Joseph Smith, 19th-century founder of the Church of Jesus Christ of Latter-day Saints, commonly called the Mormon Church.

sufferer slowly regains awareness. Some tonic-clonic attacks give fair warning. Sensory hallucinations known as auras, including smells and bright lights, give the sufferer a chance to lie on the floor before the onset to avoid the potential injury of falling.

CAUSES & TREATMENTS

Epilepsy has a variety of causes. Some are genetic in origin and caused by an inherent problem in the brain. Typically, the disease strikes far more men than it does women. Other cases have their onset after physical injuries to the brain, such as strokes, fevers, tumors, or head wounds.

FAST FACT About the size of an almond, the small hypothalamus plays a big role in both the nervous and endocrine systems.

Treatment options include anticonvulsive drugs and vagus nerve stimulation. In the latter, stimulators are implanted in the chest to send regular pulses of electricity through the vagus nerve to the brain. These pulses aim to keep the brain's electrical activity from tipping from order to chaos.

New possibilities include the implantation of monitoring devices combined with electrical stimulators or drugs. The idea is to detect the subtle electrical changes that signal an oncoming epileptic seizure, then deliver a small shock or dose of medicine to ward off the attack before it strikes.

THE ENDOCRINE SYSTEM

The nervous system isn't the only method by which the brain controls the body and maintains homeostasis. The direct, electrochemical means by which the nervous system collects information from stimuli and then formulates responses is augmented by the endocrine system, which works with the nervous system to regulate the body's cells. The autonomic nervous system responds to changes in the body's dynamic balance by releasing electrochemical impulses to the body's endocrine organs. These include the testes and ovaries, pancreas, adrenal glands atop the kidneys, thymus and parathyroid glands, and three glands in the brain: the pineal, hypothalamus, and pituitary.

Endocrine glands respond to the nervous system's orders by releasing hormones into the bloodstream. Hormones (from the Greek for "to excite") bind to specific cell receptors and affect virtually every cell in the body. For example, instructions

Abnormal neuronal firing causes intense electrical activity of frontal lobe epilepsy.

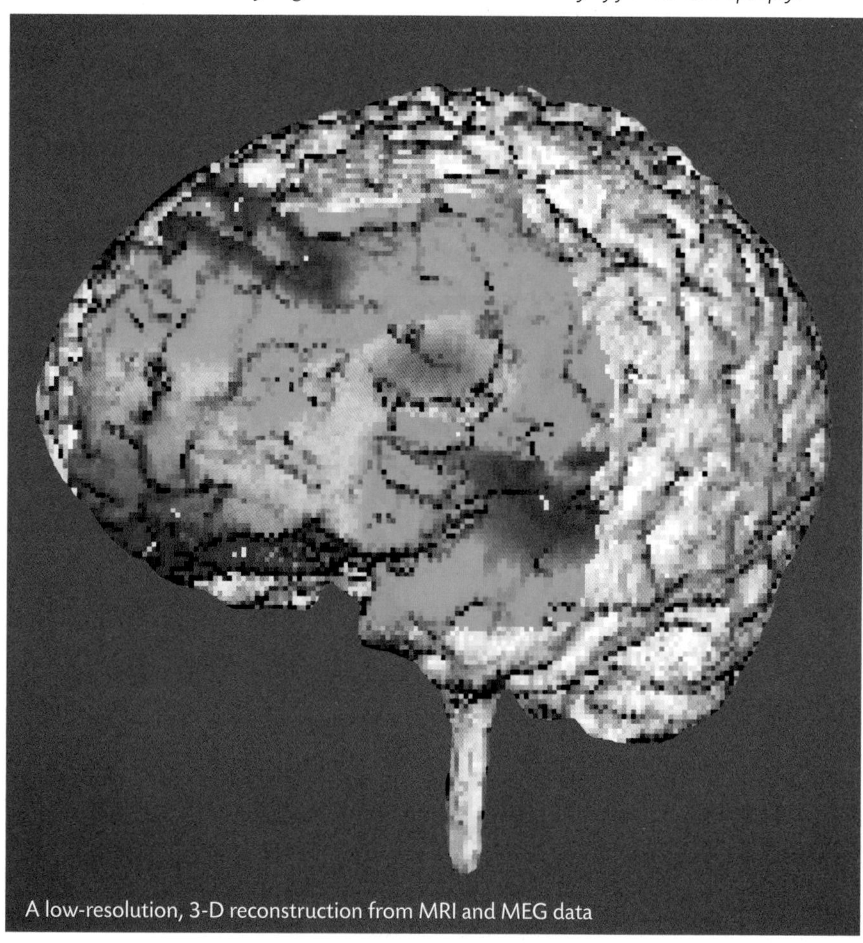

A low-resolution, 3-D reconstruction from MRI and MEG data

STAYING SHARP

FOR A HEALTHY BRAIN, good foods are a key part of optimizing your brain's performance. Here are some foods your brain will welcome:

✓ Fresh fruits and vegetables. These include blueberries, leafy vegetables, broccoli, and cauliflower. They contain high amounts of acetylcholine and useful vitamins. Certain vitamins, notably vitamins C and E, and beta-carotene, a precursor to vitamin A, act as antioxidants. They neutralize destructive molecules and atoms known as free radicals, which damage brain cells by stealing electrons from cellular molecules or atoms.

✓ Unsalted nuts. Their omega-3 fats help keep the brain and nervous system healthy. Neurons require fats in their myelin sheaths to function properly.

✓ Fish. It's a better source of proteins than high-fat meat, and it's another source of omega-3 fats.

✓ Chicken without the skin and lean meats. Protein in the meat helps build tissue and supply the amino acids that form neurotransmitters.

✓ Fruit juice. It's a natural source of beneficial vitamins, including antioxidants. Be sure to drink plenty of water, too, to keep your brain and body hydrated.

✓ Small amounts of alcohol, such as one glass of wine a day. This may increase blood flow to the brain and lower the risk of strokes.

✓ Small amounts of caffeine. It activates the cerebral cortex and helps release the neurotransmitter epinephrine.

✓ Pasta, cereal, and bread. They contain carbohydrates for energy as well as being rich in serotonin.

from the brain, given at the proper time, order the endrocrine glands to release the hormones responsible for sexual development to trigger puberty at adolescence. Other hormones maintain the body's balance of energy, keep the blood's supply of electrolytes in balance, and muster the immune defenses against infection. The nervous system and the endocrine system share a special relationship, as their functions can seem intricately intertwined.

DIABETES INSIPIDUS

When the nervous and endocrine systems get out of balance, the resulting dearth or overabundance of hormones can cause havoc. Consider just one hormone. The pituitary gland in the brain stores antidiuretic hormone (ADH), also called vasopressin, which is created by the hypothalamus. ADH helps regulate the body's water content through its ability to prevent the formation of urine, which contains water expelled by cells.

Neurons in the hypothalamus monitor the water content of the blood and call for the release or withholding of ADH when the blood contains too much or too little water. The dry mouth you experience on the morning of January 1 may be a result of too much partying the night before; excessive alcohol consumption suppresses the release of ADH, causing excessive urination—and thus dehydration and cotton mouth.

When the hypothalamus and pituitary fail to regularly create and release enough ADH, often through damage to the hypothalamus or the pituitary, the result is diabetes insipidus. Patients with this disorder urinate frequently and are constantly thirsty. Mild forms of diabetes insipidus can be treated simply: As long as the brain's ability to recognize thirst is undamaged, patients can compensate for dehydration by drinking plenty of water whenever they feel the need.

Blueberries are rich in acetylcholine and antioxidants, making them an excellent food for brain health.

DIABETES MELLITUS

Diabetes mellitus creates a lack of the hormone insulin, resulting in heavy losses of blood sugar through urination. Insulin arises in the pancreas, a gland that produces enzymes important for digestion. Insulin's influence is most apparent just after a meal, as it works to take glucose out of the bloodstream to use it for energy in the body's cells. Insulin also helps store fat and synthesize proteins.

Diabetes mellitus occurs when the pancreas doesn't produce enough insulin. Its lack leads to excess blood sugar levels, resulting in dehydration through urination, fatigue, weight loss, nausea, abdominal pain, as well as extreme thirst and hunger. The most common treatment is for the afflicted to test their blood sugar levels and inject themselves with insulin when needed. Accidental overdoses are the most common cause of hypoglycemia, which occurs when too much insulin in the bloodstream lowers blood sugar dangerously. Eating a piece of candy or sipping a glass of orange juice helps restore sugar levels.

Regular tests help diabetics monitor levels of glucose in the bloodstream.

CLASSIFICATIONS

Diabetes formerly was classified into "juvenile onset" and "adult onset" varieties because of the typical time frame for diagnoses—ages eight to twelve in children, and forty to sixty in adults. The classification system changed when doctors analyzed symptoms that did not match up well with ages. Patients whose body produced no insulin at all were reclassified as "insulin dependent," while those whose body made insufficient amounts became "non insulin dependent." The former now is called Type 1, and the latter Type 2.

Type 1 diabetes is commonly diagnosed in children, teens, and young adults. Symptoms usually come in a rush, shortly after the patient's immune system turns on itself and destroys the insulin-producing cells of the pancreas. Lack of insulin used to be a death sentence. Now patients survive with regular injections of insulin, either by syringe or an automatic pump and catheter.

Type 2 is the more common variety and can begin at any age. It usually starts because the body's liver, fat, and muscle cells fail to use insulin efficiently. That causes glucose levels to rise in the bloodstream. Feedback mechanisms in the peripheral nervous system detect the increase and trigger production and release of more insulin in the pancreas to offset the higher glucose levels and maintain homeostasis. However, the pancreas cannot keep up the extra production forever. Diet, exercise, weight loss, and medication are common methods of keeping Type 2 diabetes in check.

FAST FACT

Diabetes mellitus gets its modern name from the Greek for "overflow" (diabetes) and the Latin for "honey" (mellitus). Overflow is a reference to the symptom of frequent urination, and honey refers to the glucose that appears in the urine. Ancient physicians would diagnose the condition by tasting urine for sweetness.

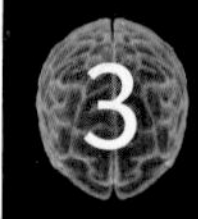

CHAPTER THREE

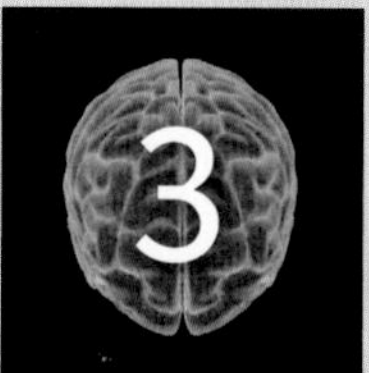

BRAIN DEVELOPMENT

FROM THE single-celled product of conception, the human animal grows into a complex, uniquely cognitive being. Evolution has built upon older, more primitive animal brain forms to lead humanity to emotion and rational thought. Over eons of time, neural circuitry has developed to promote—and continue to promote—individual and collective survival. That's because the human brain is "plastic," primed from an extremely young age to learn and change.

A six-month-old girl examines her reflection. From birth, humans appear to be drawn toward faces.

EVOLUTION

GROWTH & ADAPTATION OF THE HUMAN BRAIN

THE DEVELOPMENT of the human brain is written in millions of years of evolution, its story still unfolding.

Neurons began to emerge with the appearance of multicellular animals. The earliest neural connections formed primitive networks of cells in tiny life-forms swimming in primordial oceans. Today, such systems can still be found in simple life-forms such as jellyfish.

SIMPLE BRAINS

Animals with only the barest collection of neurons can function with surprising sophistication. The marine snail *Aplysia* has only about 2,000 neurons, yet it is capable of movement, reaction to touch, sensation, and all of the things that make a snail live like a snail. It even can learn despite lacking a true brain. *Aplysia's* neurons organize themselves into clumps called ganglia at various points on its tiny body, creating a maze of connections. These neural clumps can amplify or tamp down electrochemical signals as they pass from neuron to neuron; its neural connections can be strengthened or weakened just as in human brains. Scientists have found that when they shock *Aplysia's* tail, it reacts by reflex—its neural network contracts the affected flesh to pull it away from the source of the shock. However, things get interesting when the shock is preceded by a light touch against the snail's flesh. After a few repetitions, the lowly *Aplysia* has enough neural complexity to connect the two sensations: touch, followed by pain. In time, the light touch alone, with no electric shock afterward, is enough to make the snail recoil as if in pain.

FAST FACT An octopus's brain is dime size, but it can solve simple problems such as moving barriers to get food.

GROWING COMPLEXITY

If 2,000 neurons are sufficient for simple learning, imagine the explosion of complex behavior that accompanied the growth of neural complexity about 530 million years ago. Larger clumps of neurons in the diverse animal population that seemingly emerged overnight encouraged the flourishing of new animal species. The variety of new species could better react to, and survive, changes in their environments. Ocean life diversified into the ancestors of today's worms, mollusks, and crustaceans.

The forward tip of the neural cords in the first proto-vertebrates began swelling and folding to create primitive brains. Neural networks in those early brains began to diversify. Some connections began to specialize in vision. Some took on the function of hearing. Among the sharks, neural connections specializing in smell became hypersensitive, empowering them

BREAKTHROUGH

CHARLES DARWIN KNEW he had opened a tinderbox when he published *On the Origin of Species* in 1859. He laid out a theory of evolution through natural selection: Individuals that have a biological advantage are more likely to outlive their peers and pass their edge to offspring. A gazelle that is a bit faster than another may outrun the lion and breed fast children the next day. *Cuidado,* Darwin wrote in his notebook, using the Spanish for "careful." Taken to its logical conclusion, even humans fell under his theory—an idea Darwin downplayed at first because he knew it would be unpopular.

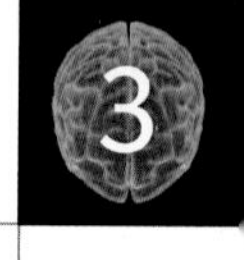

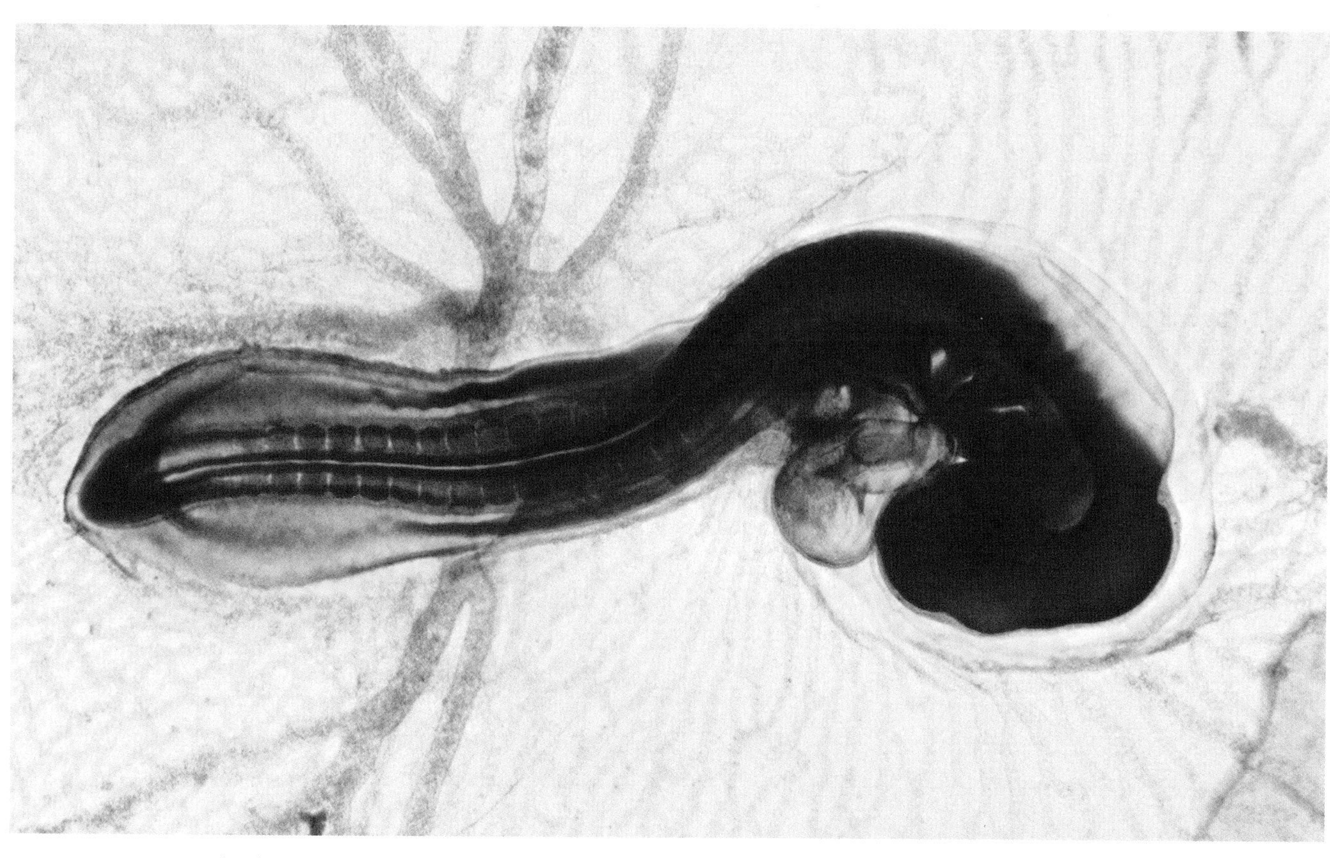

A developing spinal cord is already visible in a three-day-old chicken embryo developing inside its eggshell.

to detect blood in concentrations as small as 1 part per 25 million of water. That allowed them to smell bloody prey a third of a mile away (and, not coincidentally, strengthened their chances for survival in the constant interspecies combat of evolution).

As animals began crawling out of the ocean onto the shore, around 360 million years ago, their brain didn't begin anew. Instead, new experiences and new evolutionary developments were laid down atop their existing neural networks. Birds and reptiles added new levels of behavior, and new brain matter developed as well. Mammals put their own layers on top of their evolutionary predecessors. And finally, humans with their gigantic brain added the newest and most complex layers in the wrinkly pink walnut of the cerebral cortex.

BREAKTHROUGH

Darwin explicitly put humans in the crosshairs of his theory with the 1871 publication of *The Descent of Man.* Human bodies—and brains—evolved and continue to do so.

The human brain differs physically from those of other mammals in its size, complexity, and dominance of its cerebral cortex. Just like speed and strength, early advantages in the brain such as analytical power ("How can I trap that animal?") and capacity for speech ("How can I get others to help me trap that animal?") improved the odds of early humans' survival. Advantages spread to new generations and became common.

Networks of synapses constantly compete with each other, roughly like animal species fighting for limited food. Networks that get steady stimulation grow stronger, while others atrophy. Nobel laureate Gerald Edelman calls the process neural Darwinism.

EVOLUTIONARY CLUES

Some of humanity's evolutionary history can be observed in the development of a human fetus. As chicken and human embryos develop, for example, they experience a stage where they both have a tail, as well as arches and slits in their neck remarkably like the gill slits and arches found in fish. Thus, scientists in the late 20th century concluded that chickens and humans most likely shared a fishlike ancestor, based not only on visual evidence but also on DNA and fossil records. Not all ancestral characteristics become evident during fetal development, but enough similarities exist to suggest an evolutionary thread.

A few days after conception, a human embryo's cells begin to specialize. Some form a simple neural plate, which changes into a groove and then a tube. The huge cerebral cortex that distinguishes the human brain develops last, in the final months before birth, just as it evolved from humanity's simian ancestors two million years ago—relatively late on the evolutionary tree. Like an hour-long film compressed into a few seconds, the pageant of growth and diversity in the fetal brain roughly condenses a half billion years of animal evolution into nine months of flesh-and-blood transformation.

The common animal ancestors of humans and other animals are

PAVLOV'S DOGS

Ivan Pavlov observes one of the dogs he subjected to conditioned-behavior experiments.

AT FIRST, Russian physiologist Ivan Pavlov (1849–1936) wanted only to know the neural link between dinner and dog drool. To find out, he anesthetized his test subject and detached its salivary duct, lightly stitching this to the dog's outer cheek. Then, placing food in the dog's mouth, he could easily collect and calculate its salivary response. In this way he hoped to unlock the mysteries of the canine nervous system.

After repeated experiments, unfortunately, the dog seemed to catch on and began to salivate before the food had arrived. Clearly this was a problem. How could Pavlov understand salivary response to food in the mouth if the response occurred in the absence of food? Initially puzzled, Pavlov realized he'd stumbled upon something even more intriguing than his original objective. As environmental factors determine evolutionary adaptations within a species, he concluded, so too must external forces mold the behavior of an individual.

From a knee-jerk defense mechanism to the performing of Rachmaninoff, acquired reflexes are the building blocks of learning. And if dogs' brains were sophisticated enough to make such connections, imagine what human brains could do.

Pavlov soon discovered he could condition animals to respond to arbitrary stimuli. If a snack was repeatedly paired with buzzer, whistle, or A-minor triad on the piano—he rarely used that legendary bell—the dog would begin to salivate at sound alone. But a slight variation—B-flat minor, perhaps, or A minor in a different octave—triggered no response. The same held for shapes, clocks, shades of gray, melodic patterns, light, and rotating objects.

CROSS REFERENCE: See "Anatomy," PAGE 18

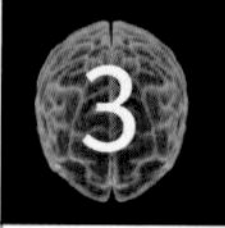

suggested by common elements of animal brains. The more complex structures of the late developers overlie the simpler forms of creatures that evolved earlier, and thus lower on the evolutionary tree.

THE THREE BRAINS

Neuroscientist Paul MacLean suggested in 1967 that the human brain functions as three separate "brains," each of which represents a stage in evolutionary development. He referred to the three-way unity as humanity's triune brain. Through evolution's penchant for preserving genetic code that proves useful for survival and discarding mutations that prove useless, MacLean suggested that human brains evolved by adding to successful brain structures of earlier vertebrates. Thus, both fish and dogs have brain structures in common with people. But instead of the evolutionary structures being uniformly mixed throughout the human brain, they nest one inside another like Russian dolls. The most primitive lies deepest in the brain, under more modern layers.

FAST FACT Charles Darwin observed that domesticated animals have thinner cortical layers than their wild cousins in the forest. Wild animals' exposure to a wider variety of environmental stimuli may create richer neural connections.

MacLean's first "brain" is the R-complex, which takes its name from its resemblance to the simple brains of reptiles. The R-complex formed from an extension of the upper brain stem. It's enough to keep a snake or a salamander alive as well as ensure the continuation of the species. The R-complex oversees sleeping and waking, breathing and heartbeat, temperature regulation, and automatic muscle movements. It also plays a crucial role in the processing of sensory signals from the peripheral nervous system. MacLean's experiments with a variety of animals demonstrated that the neural connections in the R-complex provide sufficient mental firepower for hunting, mating, establishing territory, and fighting. In other words, everything necessary for finding food, competing with other animals for survival, and passing along the genes of the dominant, strongest individuals. Humans may think of themselves as being far above turtles and alligators, but their brain shares the same mechanics for regulating basic body functions. Furthermore, whenever humans engage in a schoolyard scuffle or compete for the affections of another, they're exercising the reptilian cores of their brain.

+ THE TRIUNE BRAIN +

BRAIN LAYER	NICKNAME	LOCATION	FUNCTION
R-complex	Reptilian brain	Brain stem and cerebellum	Oversees tasks crucial to physical survival and regulation of the body: sleeping and waking, breathing and heartbeat, temperature regulation, and automatic muscle movements.
Limbic system	Paleomammalian system	Amygdala, hypothalamus, hippocampus	Houses primary centers of emotion. Plays important role in simple memory formation. Coordinates and refines movement.
Cerebral cortex	Neomammalian brain	Cerebrum	Responsible for language, including speech and writing. Handles problem solving, memory, and planning for the future. Controls voluntary movement. Processes sensory information.

Swinging through forest has been linked in theory to brain hemisphere specialization.

The second "brain" is the limbic, or paleomammalian, system. It's common to all mammals, including humans, but is lacking in reptiles. The limbic system coordinates and refines movement. It gives rise to emotions and simple memory, as well as the rudimentary social behaviors they make possible. When MacLean destroyed part of the limbic system in the brain of young mammals, their behavior regressed toward the reptilian. They stopped playing and exhibited weaker mother-offspring bonds. Humans who flush with anger when they get slapped across the face, or glow with happiness when kissed, are using their limbic systems. If they choose to ignore the slap or the kiss, however, they need to exercise the third and highest level of the brain.

The third "brain" is the cerebral cortex. Many mammals possess a cortex, but it is most highly developed in humans. It adds the benefits of problem solving and both long-term and complex working memory to the lower two "brains." The neomammalian brain, as MacLean dubbed it, gives humanity its capacity for language, culture, memory of the past, and anticipation of the future. It also makes humans the first species with empathy, the ability to see the world through the eyes of others.

"It is this new development that makes possible the insight required to plan for the needs of others as well as the self. . . . In creating for the first time a creature with a concern for all living things, nature accomplished a 180-degree turnabout from what had previously been a reptile-eat-reptile and dog-eat-dog world," MacLean said.

SPECIALIZATION BEGINS

As modern humans evolved from their hominid ancestors, their brain development continued with increasing specialization of regions and functions. One hypothesis suggests that the differences between the left and right hemispheres of the human brain can be traced to

CROSS REFERENCE: See "Language," PAGE 258

humans' simian ancestors swinging through trees. Grasping one limb after another requires the arms to act independently instead of in unison. Perhaps the ancestors of humans began emphasizing the use of one arm over another, encouraging greater neuronal development in the hemisphere that controlled action on that side of the body.

One of the most pronounced differences between brain hemispheres can be observed in dissection of cadavers. The brain region mainly responsible for speech, the planum temporale, is larger in the left hemisphere of two-thirds of human brains. The left-handed nature of language is evident across time and stage of life. Full-term fetuses exhibit larger, speech-related regions in the left hemisphere than in mirror locations on the right hemisphere. The same was true of Neanderthals, according to the telltale marks on the inside of their 50,000-year-old skulls made by contact with their gyri and sulci.

GENDER DIFFERENCES

The two sexes also experience differences in brain function. Men are more likely to be left-handed, dyslexic, hyperactive, and autistic. Women are more likely to suffer migraines and, on average, have weaker spatial functioning. Women, though, generally outperform men in the fine motor skills of their fingers, and they learn to speak their native language earlier and foreign languages more easily than men. The bottom line, however, is that if you were to look at two brains on a laboratory table—one from a man, and the other from a woman—you probably wouldn't be able to tell any difference.

In men, the third interstitial nucleus of the hypothalamus typically is twice as big as it is in women's brains. The hypothalamus is crucial to sexual behavior, as well as regulation of body temperature, eating, and drinking. Furthermore, women's and men's brains differ in response to orgasm. PET scans show less activity in a woman's prefrontal cortex and in a man's amygdala during sexual climax, while both sexes experience more neuronal firing in the cerebellum.

FAST FACT Neuroradiologist Majorie LeMay examined the Sylvian fissures of human skulls 30,000 to 300,000 years old. These fissures revealed an asymmetry that suggested dominant left hemispheres. Perhaps the asymmetry provides evidence of an ancient capacity for language, which favors the brain's left side.

+ GENDERED BRAIN +

THE SEXES DIFFER in cognitive ways. A big one involves spatial orientation. Men typically use mental maps, while women prefer landmarks. Men would likely give directions by saying, "Drive north 2.2 miles, turn east, and drive 1.5 miles," whereas women would more likely say, "Drive toward the mountains until you see the barn, turn right, and go to the pond." Small wonder that one sex may get frustrated giving directions to the other. Women take the prize for remembering objects' locations—*where are those keys?*—while men win at abstract spatial reasoning, such as mentally rotating objects. As a group, men have a wider dispersal of scores on some mental tests.

PREPROGRAMMING

Much human behavior arises from culture and environment. Some, however, appears to be prewired into the brain. The capacity for language appears to be so strongly encoded that children raised without exposure to any language will make up their own.

Communication is an evolutionarily favored social activity that helps humans compete with other animals for resources necessary for life. Similarly, the brain's ability to process and integrate visual stimuli exists almost immediately after birth. At only a few weeks old, an infant raises its arms to protect itself from the approach of an object. Sight, texture, and size appear to be aspects of object recognition that the brain is prewired to bring together for self-defense.

A NEW BRAIN

FROM THE WOMB TO CHILDHOOD

Brain Development in the Womb

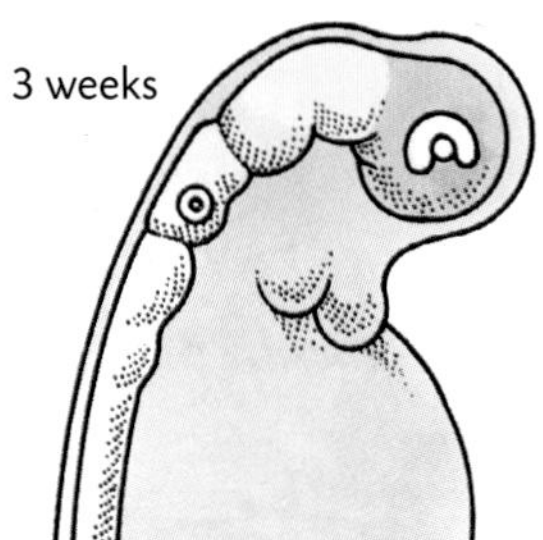

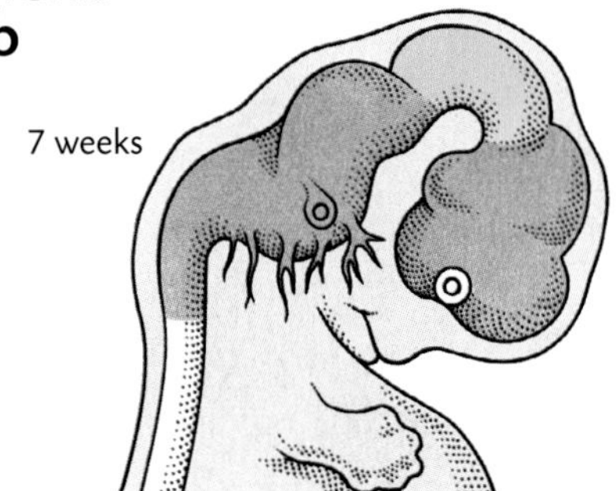

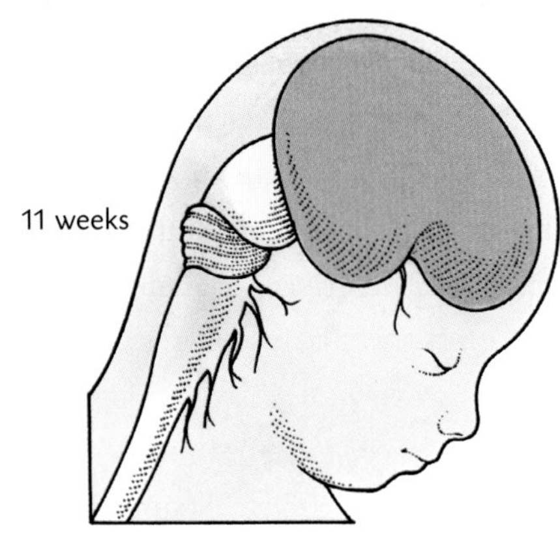

As an embryo develops into a fetus, the brain grows and differentiates rapidly.

WHEN SPERM meets egg, the merger of a father's and mother's DNA triggers the start of a new life. Encoded in the tens of thousands of genes that make up a human being are a substantial fraction that will create the brain and central nervous system. You won't find the child's personality, emotions, and ideas buried in the code; they arise instead as the brain develops and interacts with its environment after birth. Nevertheless, the explosion of cell development that begins with conception is the first step toward forming the brain and all of the hopes and dreams it will one day contain.

DIVISIONS & LAYERS

In its first phases of development, the fertilized egg, or zygote, undergoes a rapid series of divisions. One cell becomes two, two become four, four become eight, and so on until the exponential divisions create a tiny, hollow ball of hundreds of cells nearly uniform in design. Two weeks after conception, the sphere of cells, still dividing, takes the first step in the series of physical changes to construct a differentiated body and begin the process of becoming human.

First, a dent appears in the sphere. Cells move into the indentation, which folds under the surface of the sphere. The folding creates three layers of cells: an outer layer called the ectoderm, an inner layer called the endoderm, and a middle layer called the mesoderm. In the following weeks, these three layers grow into the tissues that give rise to the body's major systems: Endoderm becomes digestive tract; mesoderm creates muscles, skeleton, heart, and genitalia; and ectoderm forms brain, spine, nerves, and skin.

BUDDING BRAIN

The nascent brain makes its first appearance at about four weeks after conception, when a thin, spoon-shaped layer of cells called a neural plate emerges at the head end of the embryo. Major characteristics of the future brain already are in place just one month into fetal development. Hemispheres later will develop on either side of a groove down the center of the neural plate, creating the bilateral symmetry of the human brain.

As the fetus grows, the bowl of the spoon will become the brain itself, while its handle grows into the spinal cord. And as the neural plate folds to form a tube, swellings

FAST FACT Lots of gentle handling produced increased serotonin, a neurotransmitter that dampens aggression, in baby rats. Grown into adults, the rats lived longer and handled stress better.

in the original spoon shape become the forebrain, midbrain, and hindbrain. As they develop, they work together to form the major sections of the brain, from the cerebrum at the top of the head to the thalamus, hypothalamus, cerebellum, and spinal cord at the back and lower end.

NEURON MIGRATION

The most dynamic growth occurs in the cerebral cortex, the largest and outermost layer of the brain. During the first months of fetal development, when 250,000 new nerve cells are being created every minute, neurons begin to take on specialized tasks.

First, they inch their way from where they were formed by cell division to their permanent home in other regions of the brain. Most go toward the cortex, but some move into the cerebellum and other portions of the brain. This process, known as migration, is quite remarkable for the distance the neurons must travel as well as their ability to navigate surely along the tangled pathways of the developing brain. Millions of neurons migrate a distance equivalent to a person hiking from Los Angeles to Boston. Amazingly, they manage to arrive at Paul Revere's house, the U.S.S. *Constitution,* or Faneuil Hall without ever consulting a map.

Once the migrating neurons reach their destination, they develop axons and dendrites to

+ STAGES IN FETAL BRAIN DEVELOPMENT +

TRIMESTER	WEEK	EVENT
First	Conception to 14 days	Fertilized egg divides repeatedly, creating a ball of cells.
First	About day 14	Cells of embryo fold and turn to create three layers: mesoderm, endoderm, and ectoderm.
First	Around week three	Brain development emerges from the ectoderm as glial cells form and young neurons known as stem cells divide and create neuroblasts, or primitive nerve cells, at a rate of a quarter million every minute. Neurons begin migrating and forming connections.
First	Around week four	The ectoderm grows thicker and forms a spoon-shaped neural plate. The plate folds on itself to create a neural groove, dividing the plate into hemispheres. The groove closes to make the hollow neural tube.
First	Weeks four through eight	The neural tube develops with the explosive growth of neurons and glial cells, giving rise to the spinal cord and brain. By week eight, the developing brain forms three major regions—forebrain, midbrain, and hindbrain—seen in adults.
First	Week 11	The hemispheres of the cerebrum are evident in the developing forebrain. So too are the midbrain and, in the hindbrain, the cerebellum, pons, and medulla.
Second	Five months	Cerebral hemispheres enlarge to dwarf other brain components, but they continue to appear smooth on the surface.
Third	Six through nine months	Cerebrum folds continuously, creating gyri and sulci that create the mature brain's characteristic look of a wrinkly pink walnut.

FAST FACT You can't clone a brain. And even if you could, it wouldn't turn out like the original. Sensitivity to initial conditions in the womb coupled with differences in environment after birth would significantly alter development despite the identical genetic code.

reach out and make connections with other neurons. Like roads that connect to create a grid for traffic, neurons set up systems of communication that link all parts of the brain. Some pathways receive huge amounts of sensory traffic and become the equivalent of information highways. Others turn into dead ends or decay into crumbling blacktop from lack of use.

UNDERSTANDING MIGRATION

The brain reacts with extreme sensitivity to anything that influences neuronal migration. Only a few decades ago, neuroscientists believed that each neuron had its own special, predetermined location when it set out on its trek across the brain. Now, researchers have found that neurons take on different characteristics *because* of their journey and their destination. To take just one example, neurons that process oral communication are not inherently preprogrammed to be speech neurons. Instead, they become speech neurons by migrating to the areas of the brain associated with language.

This discovery prompted new understanding of a wide variety of brain disorders. If something interferes with neurons migrating to their intended destinations—and not overshooting or undershooting their targets—the results can be powerful. Such disorders as autism, schizophrenia, dyslexia, and epilepsy have been at least partly linked with abnormalities in neuronal migration.

Fetal alcohol syndrome has also been linked to problems in migration. The brain's hypersensitivity to toxins that impede migration underscores the warnings given to expectant mothers to avoid exposing a developing baby to alcohol, tobacco smoke, drugs, or other chemicals that may interfere with healthy brain development.

SUPPORT & SURVIVAL

Migrating neurons are helped along by glial cells. They support and nourish the neurons on their journeys. Some help regulate the neurons' metabolism, and others coat the nerve cells' axons with myelin, a fatty substance that provides electrical insulation and thus controls the speed of communication along neural networks.

Although the brain of a fetus at about eight months after conception weighs only a pound, or about a third of an adult's, it contains twice as many neurons. Chemical signals called trophic factors influence how individual neurons connect to each other, but the survival of those connections depends on repeated communication across the synapses.

The brain cannot possibly sustain biochemical reactions across all of its neural connections, and so the weakest connections begin to die, through a process known as

WHAT CAN GO WRONG

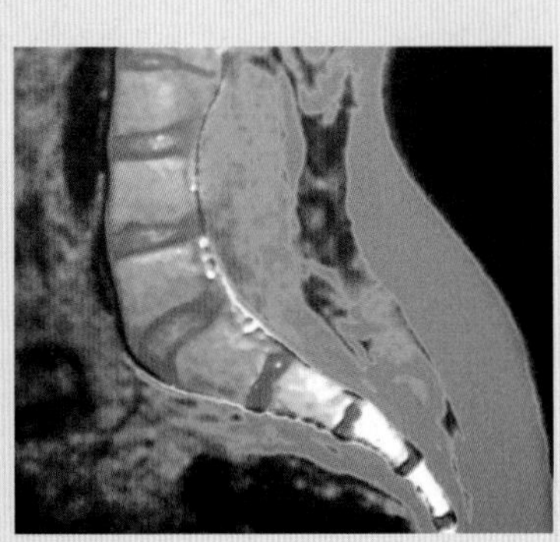

FIRST DESCRIBED 4,000 years ago, spina bifida is a malformation of the fetal spinal column that has been linked to a diet deficient in folate, a B vitamin, in pregnant women.

From the Latin for "spine split in two," the birth defect occurs in 1 to 2 births per 1,000. One or more vertebrae, particularly in the small of the back, don't grow the bony projections called vertebral arches that point away from the center of the body. Often a cyst bulges outward from the spine, encompassing spinal tissues, cerebrospinal fluid and even parts of the cord itself. Large cysts likely signal severe neurological

CROSS REFERENCE: See "Nerve Cells," PAGE 10

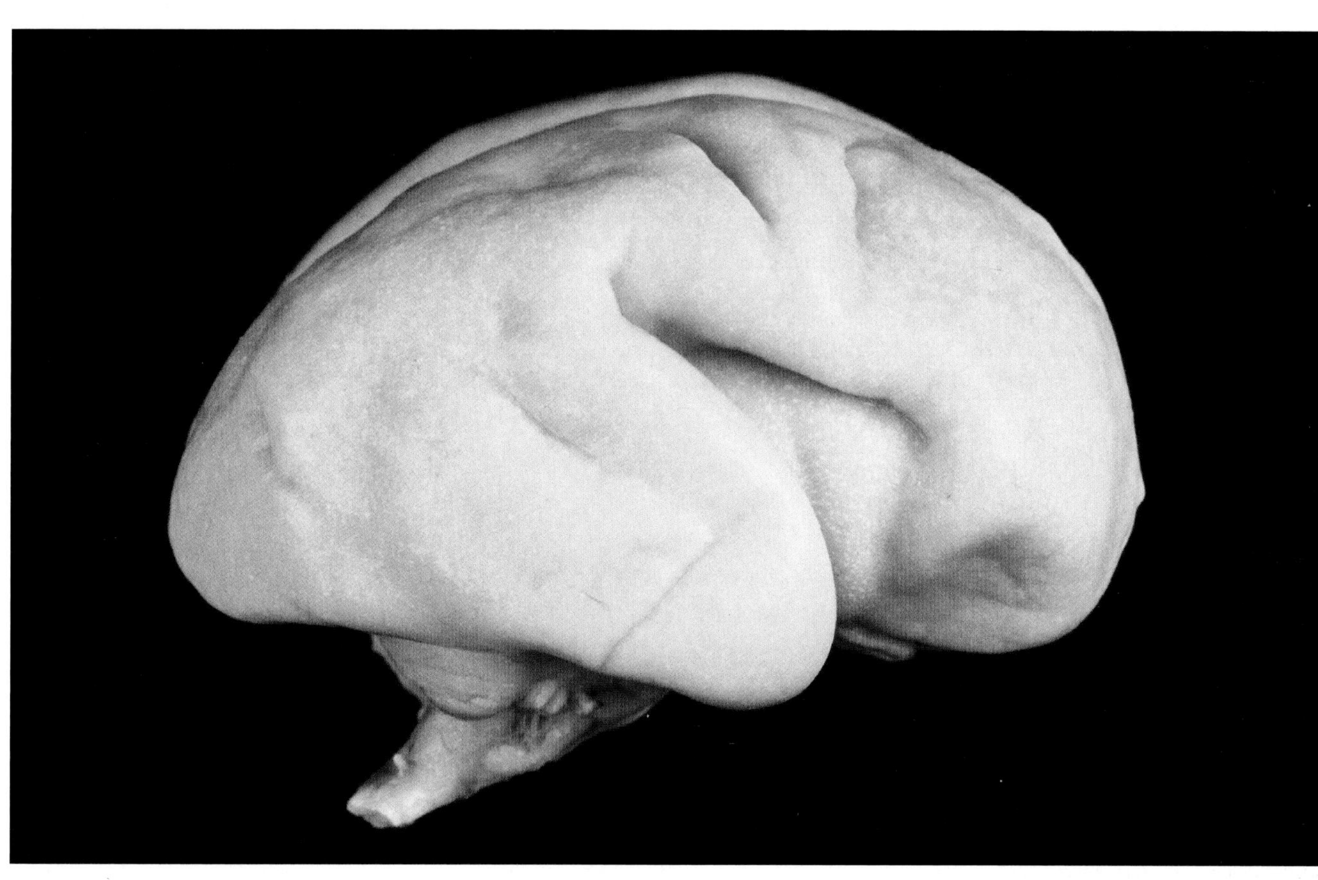

A fetal brain at 24 weeks, with spinal cord at left, has yet to develop characteristic cerebral folding.

pruning. In the last stages of fetal development in the womb, about half of all neurons die. The loss is normal; it eliminates many of the connections that are weak or improper for efficient brain function, leaving behind the strongest and fittest neurons.

NEWBORN NEURONS

As a baby emerges from the womb, brain development expands to include processing responses to the baby's new experiences—sights, sounds, smells, actions, sensations, and emotions. Networks of neurons, primed to receive new stimuli, compete for survival. It's a random battle at first, but soon becomes more organized as environmental stimuli strengthen some connections while others wither. If the baby is exposed to a broad vocabulary and a wide range of music, the connections for language and

impairment; a portion of the body's central nervous system, designed to be safely protected from the outside world behind walls of tissue and bone, lies exposed. When the spinal cord is so compromised as to lose function, the infant may suffer paralysis of the legs and bladder, as well as bowel incontinence.

As a preventative measure, since 1998 all bread, pasta, and flour produced in America contains supplemental amounts of folate. The vitamin, found in green, leafy vegetables, helps the body grow new cells, but how its lack can trigger the disorder remains unclear. Genetics play a role, as the highest incidence rates occur among the citizens of Ireland and Wales as well as their immigrant descendants.

Surgery often can close openings over the exposed portion of a spine and reconstruct misshapen vertebrae, but many impairments remain for a lifetime.

A HEALTHY START

THE WISDOM linking a pregnant mother's health to that of her baby is as old as the Bible. In Judges 13:7, an angel appears to the wife of Manoah and proclaims, "Behold, thou shalt conceive and bear a son; and now drink no wine or strong drink, neither eat any unclean thing." Manoah's wife did as she was told and received her reward: A son not just healthy, but exceedingly strong. His name? Samson.

Not all women are fortunate enough to receive prenatal advice from an angel, but luckily thousands of years of science have led to a set of recommendations that pregnant women can follow to increase the chances of their baby's having a strong, healthy brain.

Getting plenty of exercise is important to both the mother and her developing baby.

EAT WELL

The first, and easiest, thing a mother-to-be can do is to eat for two: This doesn't mean doubling up on servings—it means remembering that the vitamins and minerals from a well-balanced diet not only nourish mom's brain and body but the brain and body of her developing baby. Pregnant women need proper amounts of folic acid, vitamin B_{12} (crucial to the functioning of the central nervous system), fatty acids, iron, and other nutrients. She should consult her obstetrician about taking prenatal vitamins, which contain many of these substances and fill in any nutritional gaps in her diet.

Good nutrition is vital for healthy brain development. Lack of nutrients at crucial moments in fetal brain development leads to a drop or even a halt in the creation of neurons. Babies born after suffering malnutrition often display a smaller brain and have cognitive disabilities. Lack of folic acid (found abundantly in bread, beans, pasta, spinach, and orange juice) raises the chances of a child being born with spina bifida. On the other hand, too much of a good thing can be bad. Overabundance of certain vitamins, including A and D, can cause toxic reactions in the fetal brain. The best advice for a mother-to-be is to consult her doctor about the best diet for her, one with lots of fresh fruits, leafy green vegetables, legumes, whole grains, and lean meats.

AVOID ALCOHOL

To decrease the chances of neurological defects, moms-to-be should also avoid many substances that can harm an unborn child's brain, such as alcohol. In 1899, William Sullivan, a doctor who studied babies born in an English women's prison, discovered much higher rates of stillbirths among mothers who drank heavily. He suspected a link between alcohol and fetal health when he noted that mothers who gave birth to babies with severe birth defects in the outside world had healthy babies in prison, where they were denied alcohol.

It would take more than seven decades before researchers at the University of Washington cataloged the recurring patterns of birth defects as fetal alcohol syndrome.

When pregnant women drink heavily, their children are at high risk of having a malformed heart and limbs, a smaller brain, reading and math disabilities, hyperactivity, depression, and distinctive facial abnormalities. Mental retardation also is possible. Unfortunately, alcohol's most devastating impact on a developing fetus occurs early in the pregnancy, when the mother may not even know she is carrying a child. And small amounts in the first trimester cause more damage than greater alcohol consumption later on, apparently because of alcohol's impact on the migration of developing neurons in the fetal brain. Normally, neurons stop their travels when they reach their intended destinations. The presence of alcohol makes them overshoot and die.

JUST SAY NO

Other substances harmful to adults are even more so to a developing fetus, whose brain is especially sensitive to its chemical environment. Tobacco, illegal drugs such as cocaine, and environmental toxins, all of which do some level of harm to an adult's body, deliver hammer blows to a developing fetus and can even cause harmful impacts on sperm cells, so men should consider their levels of exposure before trying to start a family. Sperm live for about three months. To minimize the chances of their sperm being adversely affected by alcohol, tobacco, drugs, and toxins, fathers-to-be should avoid exposure to such harmful substances for 90 days.

For pregnant women, tobacco smoke is the most common environmental hazard to a fetus. Nicotine in tobacco causes blood vessels to constrict; an affected fetus gets less blood, and its heart rate decreases. Furthermore, nicotine becomes more concentrated in the fetus's body than in that of the mother. Like alcohol, nicotine is believed to interfere with neuronal migration, connection, and development. Spontaneous abortion rates nearly double for mothers who smoke. Babies carried to term are more likely to be mentally retarded and have congenital abnormalities.

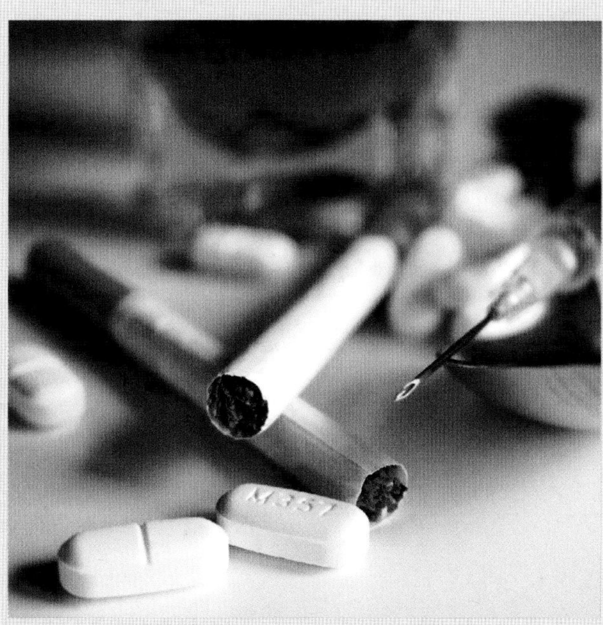

Drugs taken by pregnant women can cause abnormalities in the developing fetus.

SEEK OUT HIDDEN RISKS

Toxins harmful to a fetus range from obvious hazards such as the poisons in pesticides to common and seemingly harmless substances such as vitamin A, which in high concentrations (such as in acne medication) harms a fetus's brain. Lead particles, many over-the-counter and prescription medicines, x-rays, and some cancer drugs also poison a developing brain.

The jury is out on the possible impact of antidepressants. A pregnant woman's use of Prozac, a common prescription-only treatment for depression, so far has been shown to have no impact on her child's behavior, language, or intellectual abilities. Yet children of mothers who took Prozac during pregnancy have been found to have a higher rate of minor congenital abnormalities, such as more wrinkles in the palms of their hands. Because the effects of medication on mother and child are complex and in many cases not fully understood, women who are pregnant or plan to become pregnant should consult with their doctors about their use of prescription drugs.

sound recognition grow stronger. If the baby is kept in an environment lacking in toys and visual stimulation, the baby's analytical powers may be slow to develop.

ESTABLISHING NETWORKS

Defects in infants' eyes illustrate the sensitivity of a newborn's brain and the competing neural networks. When a child is born with a cataract in one eye, that eye is deprived of normal vision, and the portion of the brain that processes information from that eye suffers lack of stimulation. The baby's one normally functioning eye begins to process all visual information.

The "use it or lose it" principle starts to work—with a vengeance. Neural connections develop for the good eye but fail to do so for the eye with the cataract. Unless the cataract is removed shortly after birth, the child will remain blind in that eye. Even if the cataract is removed later, the brain has lost its one chance to develop the neural circuitry to process visual signals from the eye; the eyeball may appear healthy, but it cannot communicate with the brain.

If surgery removes the cataract in time, the strong, already existing neural connections of the stronger eye give it a favored place in brain development. In order to make both eyes work with the same acuity, doctors often patch the stronger eye for a few hours every day. That way, for extended periods, all of the neural development for vision is processed via the weaker eye. Its brain circuitry grows stronger by not having to compete all the time with the good eye.

The process of establishing and strengthening connections in the brain to process vision underscores the fact that certain periods are absolutely critical to proper functional development. While the brain retains a measure of plasticity among existing networks, it also seldom offers a second chance for *establishing* those networks at an early age. In other words, the brain cannot expand and reconnect a neural network that doesn't exist—or one that exists, like a dead-end road, without functional traffic.

NEURAL DARWINISM

Some scientists argue that as the brain incorporates new experiences and makes new connections

FAST FACT

A hemispherectomy surgically removes one hemisphere of a damaged brain. Infants who undergo one generally grow up just fine. Their neural networks are so plastic they rewire themselves. After childhood, however, hemispheres become so specialized that removal makes full recovery extremely difficult.

+ NEWBORN SIGHT +

WE CAN'T KNOW for certain what the world looks like to a newborn; babies don't answer interviewers' questions. However, scientists who study the makeup of newborns' eyes and test for whether babies will gaze at objects believe that for the first months of life, children lack the ability to see fine lines and a full spectrum of colors. The world probably looks like a blurred, faded photograph as seen through a cardboard tube.

Newborns appear to be hardwired for looking at faces. Shortly after birth, infants will look at faces longer than they will look at any other object.

BREAKTHROUGH

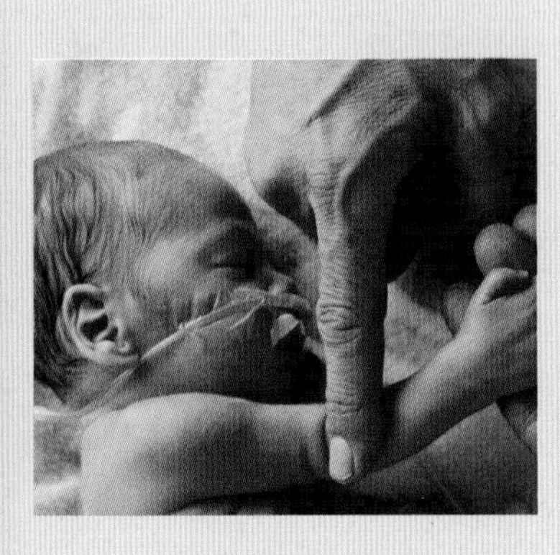

PREMATURE births pose special challenges to the brain. The child emerges from the womb before its neural networks have been established and have gone through initial stages of pruning. Much of the brain development must occur in the buzzing confusion of the world rather than a calm womb, which psychologist Sigmund Freud called the baby's stimulus barrier. Development of the preemie's brain occurs without the nutrients and protection of the uterine environment. In addition to difficulties involving regulation of body temperature, digestion of food, and weakened breathing, many preemies suffer brain

CROSS REFERENCE: See "Sights & Sounds," PAGE 106

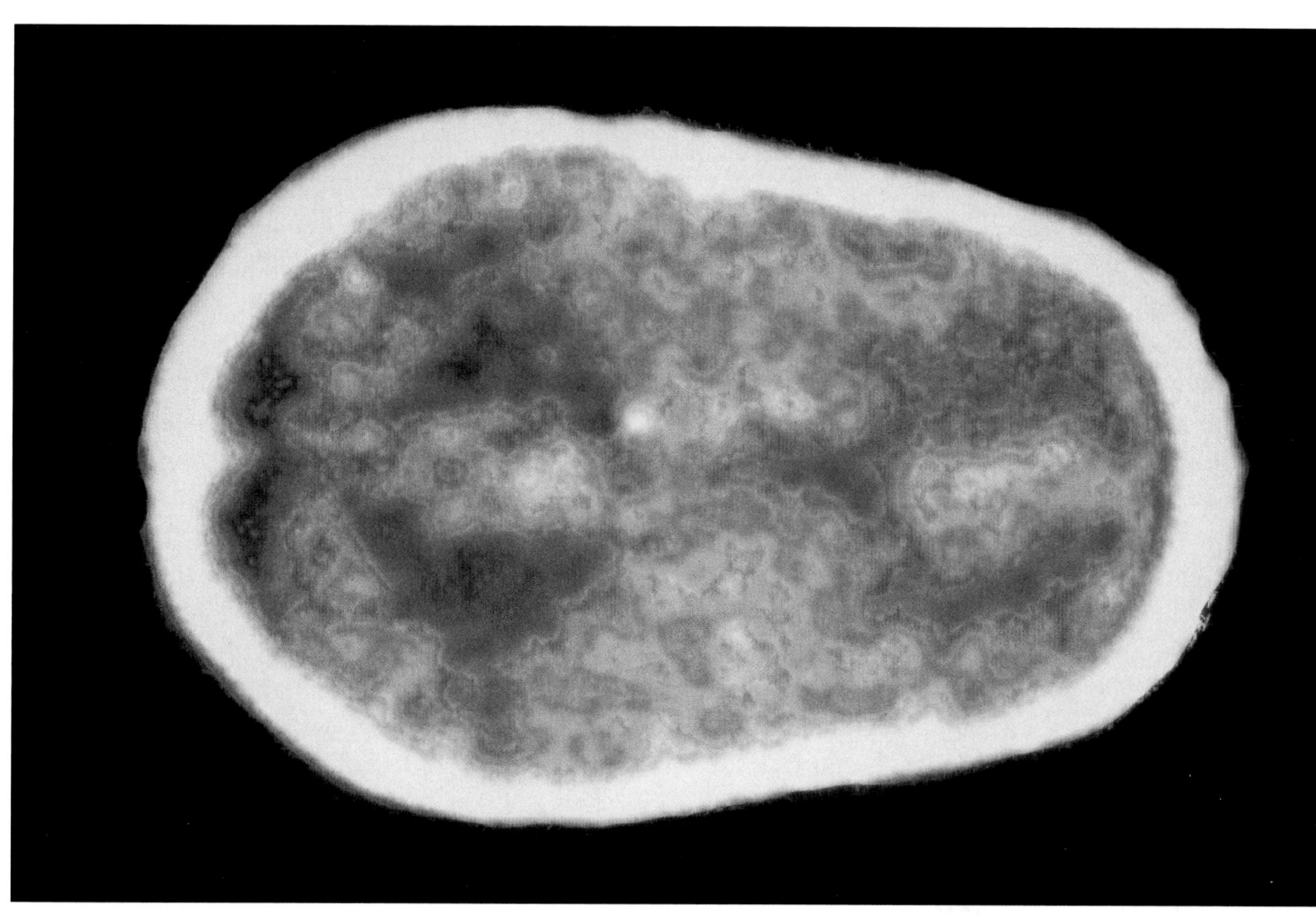

A newborn's brain (seen above in an MRI) is ready to begin making, remaking, and pruning neural connections by the million.

among neurons, it expresses a form of evolution through the competition of its various neural networks. Nobel Prize-winning neuroscientist Gerald Edelman suggests that the brain's many networks vie against each other in "neural Darwinism."

While genes determine how the brain begins to grow in an embryo, the brain's extreme complexity and plasticity make it nearly impossible to predict how it will develop in response to a particular stimulus. The complexity of the brain makes it like the weather. Short-term weather forecasts are possible with some degree of confidence, but long-range forecasts become more and more difficult because of the interaction of so many variables. The so-called butterfly effect, which

BREAKTHROUGH

hemorrhages. Babies who survive amid the chaos of lights and sounds in a hospital nursery may have their brain overstimulated and may develop problems such as attention disorders and learning disabilities later in life.

Brigham and Women's Hospital in Boston has attempted to re-create the conditions of the womb in its neonatal intensive care unit. A preemie's brain reacts with extreme sensitivity to light and loud noises, so the hospital keeps its NICU dark and quiet. Babies get plenty of skin-to-skin contact, to mimic the touch of the womb. They feed on demand. And they're allowed some freedom of movement, as they would experience inside the womb, rather than being swaddled tightly. The result: These babies leave the hospital earlier than those raised in a standard intensive care unit and have an accelerated developmental curve compared with other preemies.

was discovered during computer-generated weather simulations in the 1960s, posits that under the right conditions, the flapping of a butterfly's wings in China can be magnified until it causes a tornado in Texas. As expressed in the brain, a small change in biochemistry under sensitive conditions may have a tremendous impact on the brain's future development.

Consider how neural Darwinism finds expression in the early stages of fetal brain growth. Neurons forming from stem cells move through the brain, guided by basic genetic coding. Genes determine how the neurons connect, axon to dendrite, to create the foundation and basic architecture of the brain. However, the precise chemical environment surrounding the newly formed neurons strongly influences how far they migrate and which neighboring neurons they link with. Exposure to substances in the womb, such as alcohol, can disrupt neuronal migration, but there is no guarantee that exposure will or won't lead to fetal alcohol syndrome. The unpredictability of the complex system that is the human brain makes such precise calculations impossible.

FAST FACT Babies don't learn to walk until about a year after birth, but they are born with the neural program already hardwired.

As people grow older, they take in new experiences. There may be changes in climate, social networks, formal education, and career. To get on in life, people have to adapt to change. Successful adaptation is a matter of rewiring the brain by creating new neuronal connections. Links that promote

Toys and a mentally stimulating environment help a baby's brain grow complex neural connections.

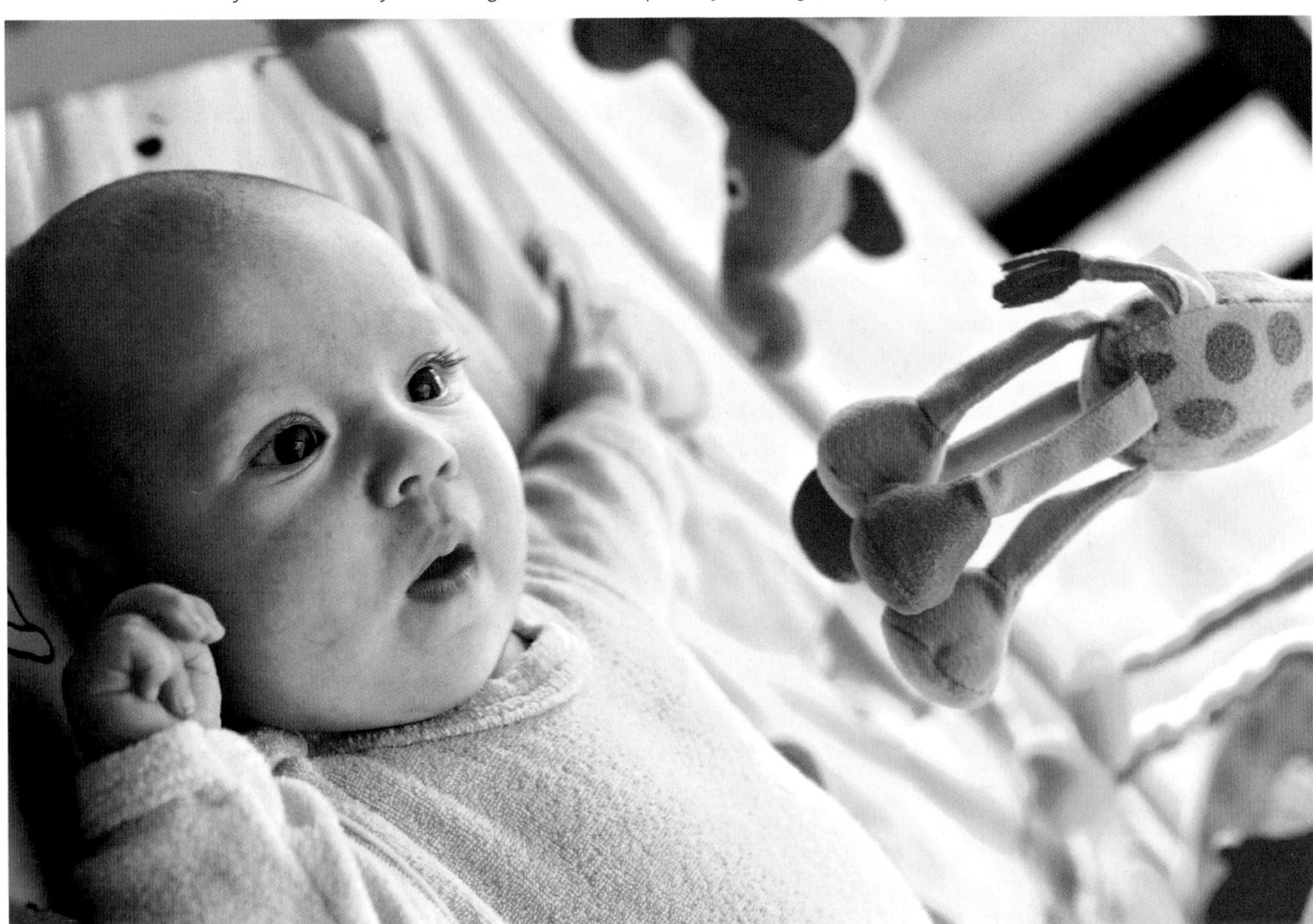

survival and well-being grow stronger. Those that lose their usefulness grow weaker. In a process that resembles natural selection, they lose the competition to stronger neural networks, and they die.

Neural Darwinism provides a new perspective on the brain's plasticity: As neural networks compete, those that function best get stronger. Changes in the environment encourage changes in the brain by giving new neural networks a chance to flourish. Such evolution of a single brain continues over an entire lifetime.

CHANGES IN PLASTICITY

By the time a baby is three or four months of age, its behavior provides clues to its having reached new milestones in brain development. At that age, individual infants differ widely in their reaction to events and in their patterns of brain activity as measured in EEG scans.

A pattern of responses known as behavioral inhibition, which includes shyness and fear when exposed to new people and experiences, occurs in one in five healthy four-month-olds. Their brains show higher levels of electrical activity in the right frontal lobes. Likewise, older babies who cry upon being separated from their mother have more activity in the prefrontal cortex of their right hemisphere than do children who

ALBERT & THE RAT

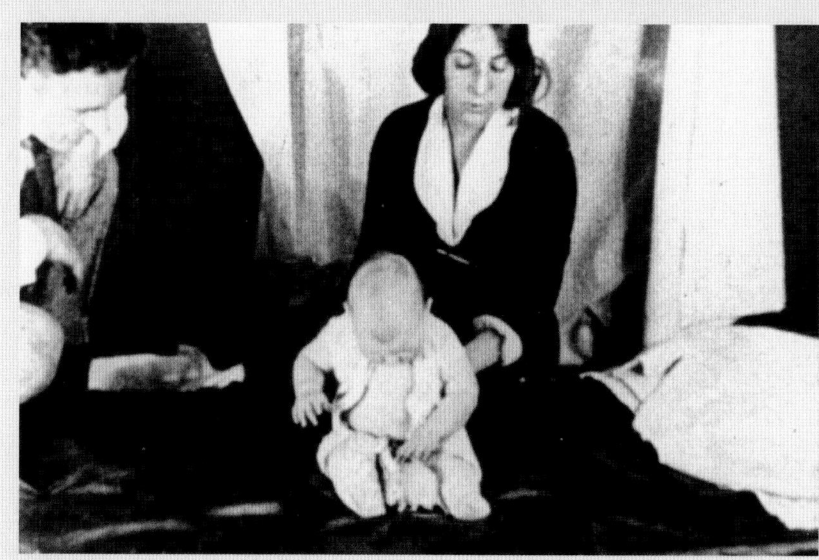

An 11-month-old called Little Albert plays his part in a famous behaviorist experiment.

IN A 1913 manifesto, John B. Watson introduced the term behaviorism, which, he wrote, eliminated the "dividing line between man and brute" in asserting that emotions are determined not by DNA but by external stimuli. Watson built on Ivan Pavlov's foundation of conditioned stimulus response. Foreshadowing the 1932 publication of Aldous Huxley's novel *Brave New World,* Watson theorized that "man and brute" alike can be made to order. He guaranteed, for instance, to rear any of 12 random infants to take on the occupation of his choosing. Yet Watson is remembered most, perhaps, for instilling in an infant boy an irrational fear of all things white and furry.

In 1919, Watson began to work with 11-month-old Little Albert, conditioning him to fear a white rat. To begin with, Albert liked his pet, trying to touch and even hold it. Watson believed this reflected a curiosity innate in all children. Later, a new stimulus was introduced: When Albert reached for the rat, Watson banged a metal bar with a carpenter's hammer. Albert fell face-forward on the mattress, whimpering. The rat was shown repeatedly, with gong and without, until Little Albert's congenital fear of loud noises was transferred to the rat. This phobia, Watson later learned, applied also to white rabbits, dogs, a fur coat, and even a Santa Claus mask. Presumably, Watson wrote, Albert could eventually become unconditioned, but the boy was adopted before further experiments could be performed.

+ Rs & Ls +

JAPANESE WHO BEGIN studying the English language as adults struggle with the sound of the letters *R* and *L*. It's not the tongue that's to blame—it's the brain. Newborns can distinguish all phonemes, or language sounds. Between six months and one year of age, however, children lose the ability to process previously unheard language sounds. Their loss is called phoneme contraction. Since the Japanese language slurs *R* and *L* phonemes, adults who are exposed to the separate sounds in English for the first time cannot hear, or articulate, the difference. It's the same for English speakers learning Japanese. They can learn the words, but it's too late for the neuronal circuits to get the sounds exactly right.

remain calm when mom disappears from sight.

LEARNING LANGUAGE

The enhancement and pruning of neural networks occurs most apparently as the baby begins to develop language. Spoken languages can sound very different from each other. In all, human languages produce about 200 different spoken sounds, called phonemes. Spoken English contains just over one-sixth of those possible sounds.

Brain scans of newborns reveal that in the first few months of life, their brain recognizes the subtle differences in phonemes other than those spoken at home. Japanese infants easily recognize the difference between the sounds made by the letters *R* and *L*. However, as the Japanese language has no sound like the letter *L*, adults raised speaking Japanese lose their ability to distinguish it from the letter *R*. Similarly, English speakers learning Spanish as adults struggle to separate the subtle sounds of the letters *B* and *P* in spoken Spanish.

But babies are able to tell such differences. That's why it's far easier to learn a variety of languages as a child. However, as infant brains focus on processing the auditory signals of their native languages, starting at about age 11 months they lose their ability to differentiate some nonnative phonemes. Children and adults who learn new languages after having undergone "phoneme contraction" speak with an accent.

YOUTH & PLASTICITY

The younger the brain, the more plastic it is. Young brains have the ability to learn and adapt with great ease. Young brains even have the potential to rewire themselves, and are thus able to overcome severe trauma, such as damage to an entire hemisphere.

For example, a seven-year-old named Michael in upstate New York began suffering as many as 300 or 400 seizures each day. Doctors said the only way to stop the crippling seizures was to remove the site of their origin in the brain—the left hemisphere. Its removal resulted in the right side of Michael's body initially being paralyzed.

But thanks to plasticity, the right hemisphere of Michael's

A Japanese-language keyboard suggests some of the potential complexity of learning language.

CROSS REFERENCE: See "Language," PAGE 258

brain reorganized itself to take on the tasks that had formerly been performed by the left. He slowly regained the use of his right leg and arm as his brain recruited neural networks for new motor skills. Speech returned more slowly, however, as the right hemisphere is not organized as efficiently as the left side for processing and articulating words.

At 14, Michael was speaking in simple sentences and racing mini stock cars.

FAST FACT Adults use only half the glucose in their brains compared with brains of children between ages 3 and 10.

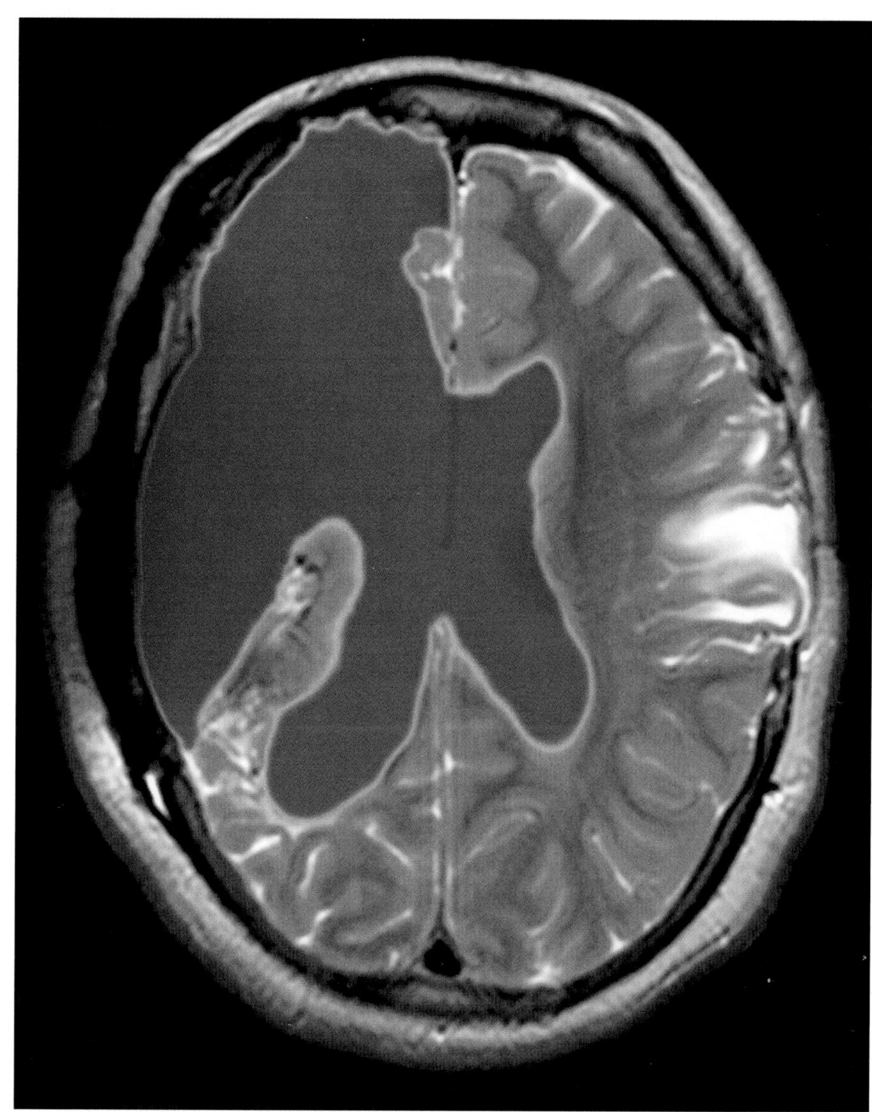

Cerebrospinal fluid (red) fills space in an MRI of a near-total hemispherectomy to ease seizures.

SPECIALIZATION

As brains age, they lose a large measure of plasticity. Things become harder to learn, and recovery from brain damage gets progressively more difficult. But there is a trade-off that benefits the maturing brain: specialization. It creates a more complicated, and therefore more sophisticated, brain.

Neurologist Richard Restak likens the benefits of specialization of neural connections to the construction of a house. Imagine, he says, if your house were built by people who specialized in particular construction tasks. Plumbers connect the water and gas pipes, carpenters put up the walls, electricians run the wires, and carpet layers install the floor coverings. Such a house should look good and be completely functional.

Now, he says, imagine the same house if it were built by people who had special skills but didn't use them on their assigned tasks. The plumbers do the carpentry, and the electricians shingle the roof. While each set of specialists is familiar with the others' work, your house nevertheless probably would look like a dog's breakfast—the doors don't close right, the walls don't meet at 90-degree angles, there's no water pressure in the shower, and the paint is uneven and splotchy. It's the same with the brain. Specialization in the various neural systems makes each perform at a high level. When many systems work at their best, the result may be an Einstein or a Mozart. Or it may be an ordinary person working at peak performance.

WAVES OF CHANGES AN ADOLESCENT BRAIN

As CHILDREN become teens, the brain continues to develop, grow, and change along with the rest of their body. Brain specialization becomes more apparent at adolescence. The adolescent brain can imagine and ponder at a much higher level than a child's brain.

Whereas a child tends to dwell on immediate sensory experiences, adolescents gradually gain the capacity to imagine a variety of futures. They start analyzing problems instead of merely reacting to them. They develop a preference for wrestling with issues instead of other people. In short, they begin to grasp abstract concepts and test them in their imagination.

They also develop these qualities at different rates, and some not at all. Plenty of adults still haven't developed these specialized brain functions. Even among adolescents whose neural networks begin to make the transition, however, their brain still cannot be considered mature. That of adolescents may lack the nuanced reactions of the brain of most mature adults, and they thus may strike others as indecisive, moody, and rebellious. Often idealists, adolescents argue and overreact, or react inappropriately, when challenged. Yet while the adolescent brain begins to experience the push and pull of adulthood, it has not left behind all the ways of childhood. In this twilight zone between impulsiveness and control, between plasticity and specialization, the adolescent brain represents possibilities in transit to realities.

The brain of six-year-old boys is about equally developed verbally to that of five-year-old girls.

DEVELOPMENT RATES

As the child grows into an adolescent, different regions of the brain develop at different rates. For example, the regions that govern reflexes and the ability to process new information are as mature in a teenager's brain as in that of an adult. Teenagers learn facts quickly but often forget them just as easily. Their brain still is in its peak learning time, like that of children, but not yet fully developed, like that of adults. A mature brain makes strong connections, starting at the back of the head and working its way toward the front, as myelin gets laid down over neural networks to make them conduct signals more quickly. The last place to be fully covered in myelin is the prefrontal cortex in the adult brain. Thus, while the adult brain may lack the learning power of a teenager's brain, it typically makes up for it with faster, sharper judgment in the frontal lobes.

GENDER DIFFERENCES

Boys' and girls' brains mature at different rates. Girls typically are a year or two ahead of boys in the

WHAT CAN GO WRONG

YOUNG ADULTS do strange things—and not just among humans. Among a variety of mammalian species, animals advancing from youth to maturity lose a measure of impulse control, take more risks, put greater emphasis on social activities, and derive greater joy from doing new things.

Between ages six and sixteen, the body grows to resemble an adult's but at age six, the brain is already at 90 percent of its adult volume. As the brain adds that last 10 percent, it slowly improves its capacity for emotional self-control, memory, and ability to focus attention and forecast behavior. Some of the last

CROSS REFERENCE: See "Learning," PAGE 236

timing of their myelination. The brain in boys finishes laying down myelin when they are in their late teens and early 20s, while that of girls completes the job in the mid to late teenage years. Some observers have said that might be a good reason to teach academic subjects earlier to teenage girls than to teenage boys. Others, however, say that because brains vary not only between the sexes but dramatically among individuals, the best instruction ideally would be tailored to each student.

Child's Changing Neuron Network

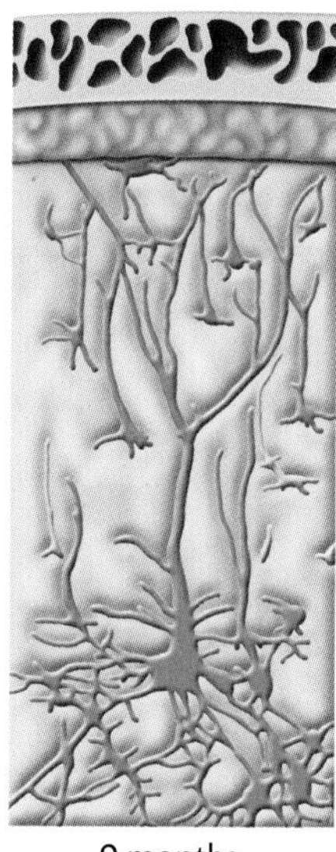
9 months

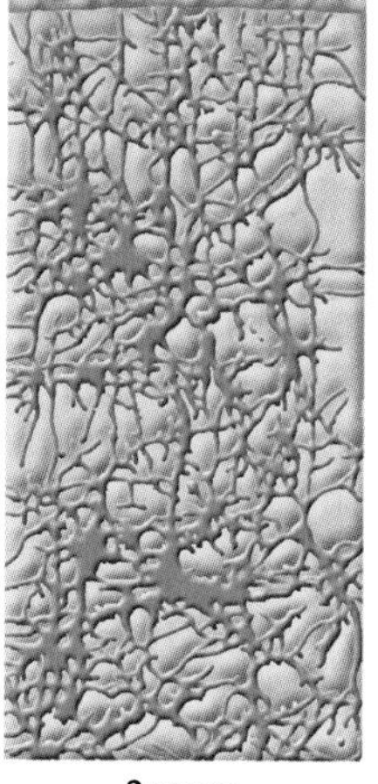
2 years

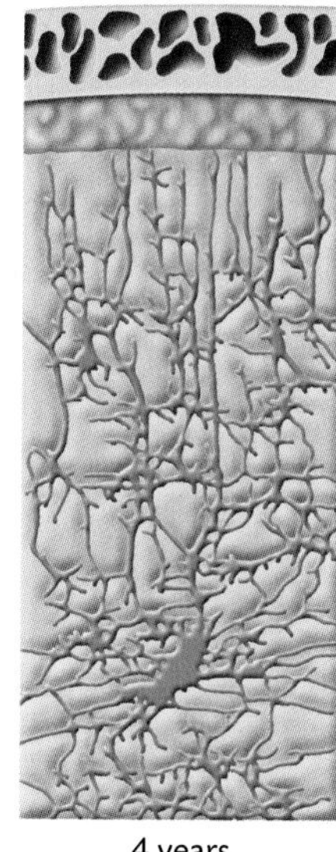
4 years

Growth and pruning of a child's neurons at nine months, two years, and four years.

GAINING CONTROL

Control of physical movements is nearly complete among adolescents. A teenager may have an adult's command of the physical skills of basketball or tennis. However, the regions of the brain that control emotions may still be under development in the teenager's brain. Small wonder, then, that a young athlete may be able to hit a golf ball 250 yards yet struggle to avoid an emotional eruption when it slices into the woods.

During adolescence, neurons create connections that may last a lifetime. But they do so in an environment where the hormones of puberty change the cellular structure of brain and body. The result is a volatile time of brain development that matches the chaos of the body's adolescent years.

Take just the gray matter—the neurons—as an example. Neuroscientists at the National Institute of Mental Health (NIMH) in Bethesda, Maryland, found surprising changes in preadolescent and adolescent brains when they performed a series of MRI scans in the 1990s and first decade of the 21st century. While white matter stayed nearly constant over the age range, the volume of gray matter underwent two shifts—increasing toward the end of childhood, and then decreasing with the onset of adolescence. The researchers believe the brain undergoes a wave

WHAT CAN GO WRONG

complex neural systems to develop form in the prefrontal cortex, which plays a key role in making plans and deciding what's right and wrong. The teenage brain rewards them for behaviors that provide immediate gratification, without the warnings that a mature brain would provide about long-term consequences.

Scientists examining behavior and the brain have focused their attention on neurotransmitters and the sensitivity of dopamine receptors. In early adolescence, the dopamine balance in the prefrontal cortex appears to reward new kinds of behaviors. Later, however, the system shifts to favor rewarding the comforts of familiar actions and surroundings.

During the time from childhood to adulthood, the cortex is sensitive to stress. Depression, anxiety, and other psychiatric disorders often appear as a result of the volcanic brain chemistry of the teenager.

The brain of teens can master learned movements such as basketball skills but lag in control of emotion-laden behavior.

of neuronal connections around age 11 in girls and 12 in boys, followed by a wave of pruning. It's as if a plant sent out too many roots, and then let die the ones that failed to find water.

Jay Giedd of NIMH described the adolescent brain as follows: "There's an enormous potential for change through the teen years. And this is great because those years are a time when choices have to be made and skills acquired as adolescents learn how to adapt to their environment. During the process, adolescents must learn to gain control over their sexual and aggressive impulses, adapt their behavior to the reasonable expectations of parents and teachers, accept authority, and generally get along with others."

ATTAINING MATURITY

The frontal lobes, last to receive their myelin coating, play a crucial role in choosing proper behavior, inhibiting inappropriate behaviors and selecting the best actions for meeting goals. If not fully myelinated, a teenager's brain may not be well equipped to make good choices. Teenagers most likely react to events with more of an emotional charge because their limbic system, including amygdala, is more fully formed than their frontal lobes.

The prefrontal cortex, the part of the frontal lobe closest to the forehead, takes a long time to mature. It first expresses function with the ability to pay attention, which develops a few weeks after birth as infants focus on and anticipate events around them. By a year old, a child can move an object out of the way to get to a toy, which demonstrates a choice of action as

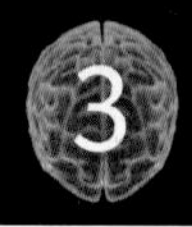

the means to an end. At the same time, the child begins to use language and symbols to represent the world, a crucial step in the formation of memories. Self-control, an important executive function, develops over the next year or two. Attention and alertness increase through age six, and major cortical development occurs between ages seven and fifteen. Afterward, pruning sharpens the prefrontal cortex as the youth acquires greater control over behavior. Until the youth's prefrontal cortex is able to make good decisions, parents and teachers help regulate behavior with their more highly developed frontal lobes.

WINDOWS OF VULNERABILITY

In the meantime, the adolescent brain is vulnerable. The prefrontal cortex's inhibitions—its ability to avoid choosing risky behaviors—aren't fully developed at the age when many youths are most tempted to engage in such acts. Unprotected sex, drug use, smoking, and drinking—all are temptations with well-advertised consequences, yet the adolescent who encounters them lacks the adult's ability to weigh long-term actions and say no. For a variety of reasons, adolescents who give in to peer pressure and decide to take drugs or engage in other risky behaviors are unlikely to be doing so because of weak moral character. Most never exhibited such risky behavior as children, arguing against a deeply rooted character flaw. Instead, many turn to risky behaviors as a result of mental disorders including attention deficit hyperactivity disorder (ADHD) and depression. Many of their family members likely experienced similar attractions and addictions, suggesting a genetic component to heightened risk of dangerous behaviors. Taken together, these clues point toward addictions—whether to physical substances or to the thrills of new sensations—as a brain disorder linked to neurotransmitter functions.

ATTENTION DEFICITS

Much scientific research has historically focused on the brain of infants as they begin to develop their mental abilities, and on diseases that begin to sap the brainpower of elderly. Only recently has the adolescent and teenage brain gotten its proper due. Neuroscientists are probing not only the sensitive stages where the brain makes a healthy move from childhood to adulthood but also the potential snags that can upset the transition.

+ BOY VS. GIRL +

NEARLY EVERY psychiatric disorder differs between girls and boys. Only eating disorders appear to be more common in girls, who also suffer more migraines because of hormonal changes. Boys are more likely to have autism, ADHD, Tourette's syndrome, dyslexia, and a host of other complications. Researchers note that females have larger basal ganglia than males and wonder whether that difference influences women's greater protection against some mental disorders. Basal ganglia assist the frontal lobes in performing their executive function. Could it be that greater influence of the basal ganglia provides some protection against certain learning disabilities?

FAST FACT Gray matter is thickest in girls at age 11, in boys at 12 years of age.

ADHD can occur in children when the prefrontal lobes of the brain haven't developed enough for an adolescent to exercise self-control. About 4 percent to 5 percent of children share this disorder, which becomes increasingly important as children are forced to assimilate information rapidly to keep up in an increasingly technological world.

"As the number of sensations increase, the time which we have for reacting to and digesting them becomes less . . . the rhythm of our life becomes quicker, the wave lengths of our mental life grow shorter," wrote historian James Thurlow Adams. "Such a life tends to become a mere search for more and more exciting sensations, undermining yet more our power of concentration in thought. Relief from fatigue and ennui is sought in mere excitation of our nerves, as in speeding cars and emotional movies." Adams wrote those words in 1931, but his observations about more sensations arriving at the brain with more and more rapidity could have been penned yesterday and applied to our current understanding of modern adolescent confusion.

Adolescents grow up in an environment that places increasing demands to do multiple tasks at the same time. Their brain attempts to adapt by rapidly shifting attention from one thing to another. While many manage the problem and still function well in a learning environment, others find it impossible to maintain their attention long enough for significant learning to occur.

SIGNS OF ADHD

Children with ADHD typically squirm and cannot sit still, have trouble playing quietly, have difficulty managing their time, talk all the time, fail to focus on the details of their school lessons, and are easily distracted. The disorder usually appears before age seven, but it lingers in most cases into the teenage years and about half the time into adulthood.

FAST FACT Adolescents who take medication for ADHD cut their risk for subsequent alcohol use by two-thirds.

In the brain of a child with ADHD, the lack of development of the prefrontal cortex causes the brain to process information in other, less efficient regions. Some neuroscientists hypothesize that a fully functioning prefrontal cortex may act as a damper, preventing the rapid choice of an inappropriate action before it can happen.

Lacking the ability to focus on lessons, children with ADHD often fall behind in school and may experience a decline in their grades. As a result, their self-image may suffer, and they become more at risk for substance abuse. Some may try to self-medicate by turning to drugs to help them focus or relieve their depression or anxiety.

CAUSES & TREATMENTS

Research points to a lack of efficient dopamine use in the brain of those with ADHD. Medications such as Ritalin, Cylert, and Dexedrine increase the amount of dopamine and other neurotransmitters, either through introducing more of them into the synapses or inhibiting their reuptake so existing neurotransmitters linger longer between axon and dendrite.

The disorder appears to have a genetic component. Parents of many children who have attention deficit hyperactivity disorder often have the symptoms of adult ADHD, or upon being interviewed recall exhibiting the telltale behaviors in their own adolescence, before the disorder was widely diagnosed. However, there also appears to be a wide range of ADHD cases without any genetic history, suggesting that the disorder also may be introduced by environmental factors.

Beeping and buzzing electronics, multimedia available at the click of a button or a mouse, and

BREAKTHROUGH

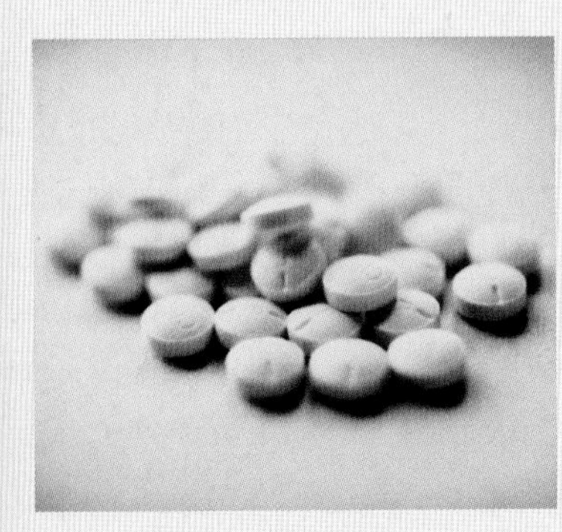

RITALIN IS THE best known medication for treating attention deficit hyperactivity disorder. Apparently it works by binding to dopamine receptors in the brain. Dopamine molecules cannot link up with their docking sites because Ritalin got there first, so they linger in the spaces between neurons. Doctors first prescribed the drug in the 1950s as a treatment for narcolepsy, an illness in which patients suddenly fall asleep during the day. A rise in available dopamine stimulates the brain like high-octane fuel in an engine, thus warding off sleep. A decade later doctors began prescribing Ritalin for attention disor-

CROSS REFERENCE: See "Nerve Cells," PAGE 10

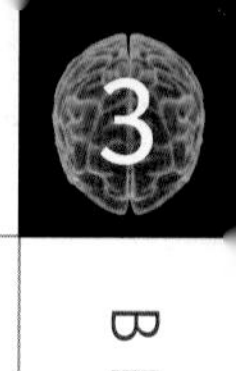

An aggressive four-year-old boy screams. Most psychiatric disorders are more common in boys than girls.

instant communication—and gratification—have created a learning environment for adolescents far different from the text-and-voice world of the 19th century. The result may help explain the large number of ADHD diagnoses. The disorder can be thought of as an addiction to the present, according to psychiatry professor John J. Ratey. Children with ADHD organize their tasks to turn first toward those that offer the swiftest gratification. They fail to break away from the moment to ponder or evaluate the long- and short-term consequences of their actions.

"With so many distracted people running around, we could be becoming the first society with Attention Deficit Disorder," said a

BREAKTHROUGH

ders because the increased stimulation it provided appeared to help patients focus their thoughts.

The action of Ritalin in the prefrontal lobes and an underlying region known as the striatum acts to brake sudden impulses. Properly administered, Ritalin helps people with ADHD to slow their thoughts, consider their actions, and live more productive lives. But it can be overused and abused. Pressure from parents, from teachers, and from society to improve children's behavior and boost their grades may be behind the phenomenon of perhaps 10 percent of American children being prescribed Ritalin at one time or another. And that's not counting the many college students who buy the drug illicitly online and use it to help them pull all-nighters as they study for tests—they say it helps them concentrate without the buzz-and-crash effects of caffeine.

cyberspace analyst writing in *Wired* magazine. ADHD, he said, might be the "official brain syndrome of the information age." It also may be so prevalent a response to the high-tech world that it deserves classification not as a disorder but rather as a particular type of brain organization.

SCHIZOPHRENIA

As the prefrontal cortex completes its final wiring, about one percent of adolescents and young adults develop the brain disease schizophrenia. People with the disease typically hear voices, experience other delusions and hallucinations, and have emotional disturbances.

+ MADNESS +

UNTIL THE 19th century, observers viewed mental illness as a form of madness. They burned the afflicted at the stake or locked them up. Doctors typically considered mental illnesses incurable and lumped them together under the common label dementia praecox (or early insanity). Not so Swiss psychiatrist Eugen Bleuler. In 1911 he divided mental illnesses into groups that shared symptoms and dubbed a common group of abnormalities schizophrenia, for the splitting of emotion and reason in a patient's mind. Bleuler noted that some patients in a Zurich mental hospital got better on their own, sometimes spontaneously. Unlike his contemporaries, he believed schizophrenia could be treated as a disease, a tenet that's commonly accepted in the medical community today.

FAST FACT

Stress is believed to contribute to the severity of schizophrenia. An environment that lowers stress, such as a supportive family, can help delay the disease or lessen its impact.

Neuroscientists believe that the foundation for the disease is laid in infancy, as neurons migrate throughout the brain.

The complex process creates the possibility of faulty connections. Somehow, in a process not fully understood, the defective neuronal networks lie dormant until activation during adolescence or early adulthood.

The schizophrenic experiences spontaneous stimulation of sensory areas of the brain. Neurons wired for the sensation of sound discharge on their own, like gas-soaked rags igniting spontaneously in a hot, dark garage. In the absence of sights and sounds, the schizophrenic's brain creates a powerful illusion of reality.

RESEARCH & TREATMENT

Research for treatment and a future cure focuses on imbalances of dopamine and other neurotransmitters in the brain's neural networks. In particular, the excitatory neurotransmitter glutamate is believed to play a significant role in schizophrenic episodes. Imbalances damage brain cells, and the longer the disorder goes untreated, the more brain cells are harmed or destroyed. CAT scans of schizophrenics reveal decreased brain volume in the medial temporal lobe, and increased size in the ventricles, possibly as a result of changes in volume of gray matter. Such scans also reveal the right and left hemispheres to be the same size, as opposed to the right side being slightly larger in unaffected brains. As the brain usually develops its slight asymmetry during fetal development, the more rigid symmetry of the schizophrenic's brain is believed to have begun forming in the uterus, evidence of a further link to the fetal origins of schizophrenia.

The disease's appearance in adolescence appears to follow physical and mental triggers that occur during and after puberty. Being a teenager is a very stressful time. The body goes through significant anatomical changes in response to the release of hormones and other biochemicals. Add on the stresses of the developing brain from school, social relationships, self-image, and the battle between emotions and reason, and it's easy to understand how schizophrenia might be awakened from its sleep. "The frontal lobe is fighting to adapt to the environment, to deal with all these instinctual urges," said Daniel Weinberger, a psychiatrist at the National Institute of

Mental Health. "Indeed, it's difficult enough for people with a normal frontal lobe to make it through adolescence. But we believe that patients with schizophrenia don't have normal frontal lobes. We believe they didn't develop normally from early in life."

AGE DIFFERENCES

Basic research is beginning to fill in the gaps of knowledge between the more commonly studied young and old brains to reveal more about the changes experienced by the brain as it grows. According to neurologist Frances E. Jensen at Children's Hospital in Boston, the brain of teens and young adults forms an exciting new frontier. "We kind of needed the two ends of life to sort of anchor us so then we could move in and understand there's a huge difference from early life to late life and from early life to adult," she said. "The early adult's brain development does not finish until sometimes 23, 24, 25, so there's a whole story there that's probably yet to be mined."

Adding significance to such brain research are the rites of adulthood available to those whose brains have not yet fully matured. Americans can legally drink at 21, vote at 18, and marry sometime in their teenage years. Yet many apparently lack the fully developed capacity to completely weigh the potential consequences of their actions.

JOHN NASH'S LONG JOURNEY

John Forbes Nash, Jr., during an economic colloquium in Beijing in 2005

ECONOMIST and mathematician John Forbes Nash, Jr., started hearing things when he was young. He became convinced that aliens were communicating through the *New York Times*, and he rambled through Europe in a vague quest of becoming a refugee. At Princeton University he scribbled late into the night on blackboards, earning him the nickname "The Phantom."

Yet he learned to ignore the voices. He got tired of delusional, irrational thinking, he said. "I think people become mentally ill when they're somehow not too happy—not just after you've won the lottery you go crazy. It's when you don't win the lottery."

Schizophrenia typically strikes men in late adolescence and women in their early 20s. Symptoms include delusions, hallucinations, disorganized speech, and chronic fear.

Delusions and fears led Nash to believe he played a role in a great game among the superpowers. He viewed others as supporters or opponents and his hospitalization for schizophrenia as a coup for the bad guys. He received a variety of treatments, eventually returning to the academic world.

Nash went on to win science's version of the biggest prize when he shared a Nobel in economics in 1994. The actor Russell Crowe portrayed him in the 2001 movie *A Beautiful Mind*. Nash is philosophical about schizophrenia and his struggles. The stigma of mental illness will disappear only when the disease does the same, he said.

MATURITY

FINDING BALANCE AS THE BRAIN AGES

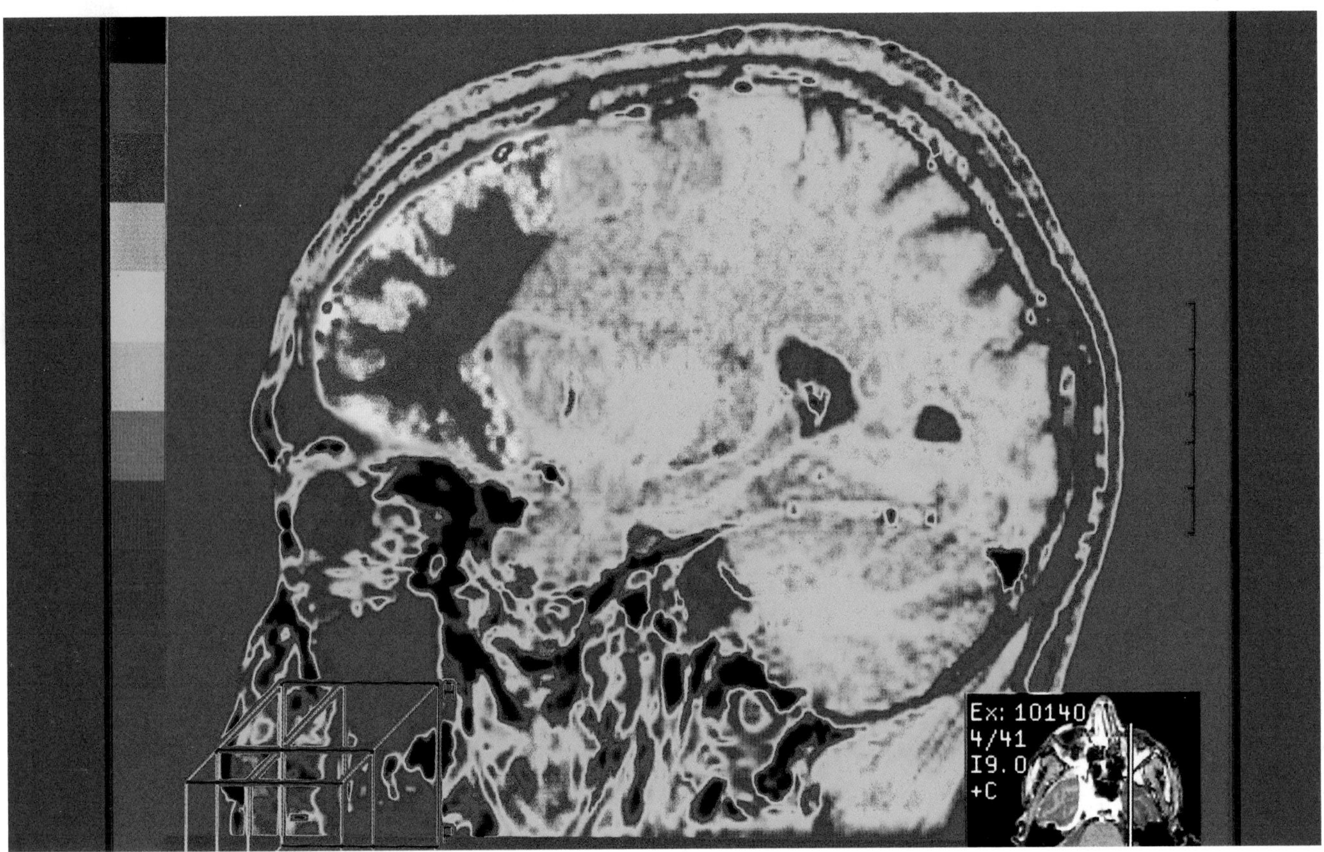

An MRI shows emotional activity in the frontal lobe of the human brain.

After adolescence, the brain transitions into adulthood. A fully functioning prefrontal cortex is the highest expression of the human brain. The balance of reason and emotion marks the emergence of the adult brain. Only in functional adulthood does society expect a person to have the ability to imagine the consequences of possible actions, understand their possible emotional impact, and make reasoned choices. For this reason, children and the mentally retarded are not held to the same legal standards in criminal cases as are clear-thinking adults. In secular society, convicted juveniles go to detention centers, while adults go to jail. Similarly, the Catholic Church insists that youths must reach an age of reason before they can be held responsible for the choices they make.

REGISTERING EMOTIONS

The adult brain associates external stimuli with a broad range of emotions. The news of a long-absent relative coming home to live for a while creates mental images of the future event. Depending on the relationship with the relative, these images may be paired with emotions such as love, anxiety, anger, depression, joy, or relief.

Mapping of electrical activity in the adult brain suggests that each emotion has its own particular neural links, and it activates, as well as turns off, different regions. PET scans reveal activity in widespread regions of the brain as it cycles through emotions. To take just

CROSS REFERENCE: See "Emotions," PAGE 206

FAST FACT Where can an old brain find new challenges? Leonardo da Vinci said: "Stop . . . and look into the stains of walls, or ashes of a fire, or clouds, or mud or like places, in which, if you consider them well, you may find really marvelous ideas."

one example, the emotion of love activates regions deep in the brain: the caudate nucleus, putamen, and insula; the anterior cingulate and cerebellum; and the hippocampus in both hemispheres.

Meanwhile, other regions get deactivated by the emotion of love. These regions, located primarily in the right hemisphere, lie on the surface of the brain, except for the deep-seated amygdala. The amygdala plays a crucial role in response to fear and terror and is increasingly active among unhappy people; other regions suppressed during the emotional ecstasy of love include those linked to depression, anxiety, and sadness. Love makes life's highs higher while tamping down the lows. No wonder it delivers such a heady cocktail of sweetness.

EMOTIONAL APPRAISALS

The mature brain has the ability not only to register emotions but also to reappraise them before taking action. Adults, unlike children and adolescents, can mentally detach themselves from the emotions they experience, label them, and place them in context—in short, realize what they feel, why they feel it, and how they choose to react to it.

It's a useful skill in many situations, particularly when faced with stress or a profound emotional disturbance. Reappraising a situation can make you feel better or worse about it. Adults who crash the car on an ice-covered road initially may feel anger or embarrassment. However, after a few moments' reflection, they can choose to think about how much worse the accident could have been. Nobody was injured, the car was insured, and as soon as the tow truck arrives, they can get on with the day. An adolescent brain, with a less developed prefrontal cortex, may dwell on the negatives and not see the other sides of the issue. That's why, to a teenager, a romantic breakup may indeed seem to be the end of the world.

Researchers at the University of California at Los Angeles reported in 2005 that adults can decrease their emotional responses if they examine them with an outside observer's detachment. Neuroscientist Golnaz Tabibnia found activity in the amygdala signifying an emotional response when test subjects looked at pictures of angry faces. As soon as the subjects thought to themselves, That's anger, they activated the linguistic regions of their prefrontal cortices, and the emotional activity in their amygdala decreased. Labeling the emotion reduces its impact, Tabibnia said.

EMOTIONAL CONTROL

Further research indicated a difference between adults and children

STAYING SHARP

CHILDREN RAISED in a home where two languages are spoken grow up fluent in both and don't have an accent. Adults, on the other hand, often struggle to pick up a second language, and even when they succeed they don't sound like native speakers.

The difference lies in the greater plasticity of the child's brain. Young children recognize a greater range of language sounds than an adult. They pick up vocabulary and syntax more easily. And they process languages more efficiently, activating smaller regions of their brain than do adult learners, who draw on more widespread cortical regions when communicating in their nonnative tongues.

Although the brain is particularly sensitive to learning languages at a young age, it's never too late to benefit from the mental gymnastics of wrestling with a new tongue. Adding a second language improves cognitive skills and memory, as well as exposing the learner to new ideas. Studies in Britain in 2004 revealed that those who spoke a second language had denser gray matter in their left inferior parietal cortex. Age even offers certain advantages to learning a second language: The mature learner already knows something about grammar and has a wide set of skills for learning, including literacy skills and memory aids.

in their ability to lessen emotional responses. Mario Beauregard of the University of Montreal demonstrated the importance of age in emotional control with a two-part experiment. In the first part, he performed fMRIs on women between ages twenty and thirty, and on girls between ages eight and ten. While being brain scanned, both groups watched clips from motion pictures designed to induce sadness. Both groups' brain activated the usual regions associated with sadness, including the ventrolateral prefrontal cortex and the anterior temporal pole.

In the follow-up test, Beauregard continued the fMRIs and asked both groups to suppress their feelings of sadness. This time, the scans differed. Both groups' brain activated its prefrontal cortex and anterior cingulate, but the girls also showed activity in their hypothalamus, a region associated with intense emotion. The women had no such reactions. Beauregard's conclusion: The women had a fully developed prefrontal cortex, allowing them greater control over their emotions. The girls lacked such development and therefore the associated self-control.

Like a brakeman on a train, a mature prefrontal cortex in a healthy adult functions as an emotional modulator. It can release to express emotions at appropriate times, such as joy at births and

A MATURE VISION

Primitive-style artist Anna "Grandma" Moses took up painting in her late 70s.

SHE WAS BORN on a New York farm before the Civil War, when high technology meant steam railroad engines and telegraph lines. By the time of her death, America had sent rockets into space. In between those events, Anna Mary Robertson "Grandma" Moses learned to adapt, not only with the changing times but also with her own talents. She took advantage of her mature brain's plasticity and focus, becoming an artist whose skills flowered at an age when many are happy simply to retire and take it easy.

Grandma Moses (1860-1961) spent much of the first seven decades of her 101 years doing farmwork and creating embroidery. When she passed age 76 and arthritis made holding a needle too painful for delicate work, Grandma Moses took up painting. She had her first one-woman show in 1940 and attracted a broad audience with her delightful, primitive country scenes. Even more impressive, she painted many of her historical images from memory. Among her more than 3,600 paintings are ones that hang in the White House and Smithsonian Institution. Typical scenes depict happy children and farmworkers, rural vistas, and the charm of the changing seasons—in short, works that, like Grandma Moses, defy the passage of time.

CROSS REFERENCE: See "Mature Minds," PAGE 268

weddings, and conversely it can dampen emotions at inappropriate times, such as sexual arousal or rage at an office party.

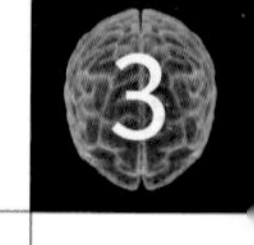

"The ability to modulate emotions is at the heart of the human experience; a defect in this ability may have disastrous social and emotional consequences," Beauregard said.

THE AGING BRAIN

As the adult brain ages, it continues to experience pruning while still being open to new opportunities for making new connections among neurons. Unless compromised by disease, the adult brain remains plastic and grows more efficient. Adults remain capable of learning new tasks—grandmothers replacing their handwritten letters with email, for example—and finding fulfillment through creative and intellectual pursuits.

The aging of the brain begins in the early 20s and continues steadily onward. Variations in human brains lead some people to a healthy intellectual life into their ninth and tenth decades, while others suffer from degenerative disorders such as Alzheimer's disease at a much earlier age. No doubt the individual's biological predisposition toward healthy or unhealthy aging plays a role, but the brain's lingering plasticity affords the opportunity to make the most of one's gray matter no matter how old or young. Keeping the brain active and challenged is the single most important factor in maintaining a brain at its peak possible performance level.

CHANGES IN PERFORMANCE

Nevertheless, some decline in brain function is inevitable as we age. The first system to show its age is memory. Brains begin requiring more time to learn and store information in late middle age.

At the same time, the prefrontal cortex experiences a drop in its ability to hold information in so-called working memory. Dr. Restak likens working memory to the desktop of a computer. It's where information is kept ready at hand for immediate use, such as when each new sentence you read in a romance novel builds on what you've read immediately before. Or, it's what you use when you enter a grocery store and check off items from your mental list of things to buy.

As both long-term and working memory decline, the brain takes longer to file information for later use, longer to retrieve it when it's needed, and longer to make decisions. People in their late 60s or older typically find it harder to filter out the "noise" of distractions and concentrate on tasks at hand. An overabundance of information, all too common in a heavily wired world of instant communication, may overload the elderly person's mental desktop in ways that a younger mind could more easily handle.

Deterioration of the frontal lobes' ability to maintain sufficient working memory explains why elderly drivers often struggle with traffic that they easily negotiated in their youth. As you drive, your prefrontal cortex constantly manipulates information arriving through the peripheral nervous system. Speed, direction, information on roadside signs, weather conditions, and constant feedback on the position of vehicles and pedestrians must get processed simultaneously as you change lanes, keep an eye on the cars around you, and search for your exit. A young adult's brain handles such variables with little difficulty. An older driver's brain, however, may get overwhelmed. As a result, grandpa may prefer to drive the old highway rather than the eight-lane interstate.

YOUNG & OLD

Brain scans reveal the different ways young and old brains process information. When young and

FAST FACT

Certain brain activities need not decrease as we grow older. The frontal lobes of a young adult and a healthy 75-year-old glow with equal brightness in a PET scan when taking the same memory test in an experiment.

> **+ 10 PERCENT MYTH +**
>
> IT'S A MYTH that humans use only a tenth of their brain. Perhaps this error can be traced to author Norman Vincent Peale, who wrote (without attribution) that people use 10 to 20 percent of their mental capacity. Or perhaps the source lies with the observation that humans sometimes recover from extreme brain damage. Mathematically, it's impossible to figure a percentage of active or inactive neurons in a human brain. Brain scans don't keep a tally of neurons as they fire, and nobody knows exactly how many nerve cells exist in a healthy adult brain. So the numerator and denominator of any fraction representing the activity of neurons in the brain are mere guesses.

old test subjects look at pictures, the young brains experience most activity in the right hemisphere, in a region known as the right visual cortex. Older brains activate the visual cortices of the right and left hemispheres about equally. When asked to observe a picture and hold the image in their mind, young people activate their frontal cortex more than older adults, whose brain lights up more diffusely in the temporal and parietal lobes.

Failure to remember names is an example of decreased memory performance of the aging brain. Although it's common for people of all ages to struggle with names from time to time, the problem becomes pronounced among the elderly. Theories about the increased difficulty of storing and retrieving names from long-term memory focus not only on the general decline of such memory function of the aged, but also on the lack of context that would more easily call names forth. A person's name usually has no connection with how he or she looks, dresses, or talks. Thus, there's no associative link to aid memory retrieval, as there would be for, say, a woman named Rose who has pink hair. Without such a link, names have no ready place for storage in memory. Futhermore, elderly brains with their overtasked working memory may be more prone to distraction during introductions. If you hear someone's name for the first time while simultaneously thinking about other things, it's likely you'll have more trouble remembering it.

MENTAL FITNESS

There is good news, however. When an elderly brain gets regular mental exercise and remains free of disorders such as dementia, it maintains capacities for abstract and analytical thinking, expression, and other higher functions. If memory remains intact, vocabulary and knowledge of the world expand with time, and communication can become more sophisticated. The storehouse of wisdom accumulated in a well-aged brain becomes a treasure of experience built up over a lifetime. An elderly brain may not react as quickly as a youthful one, but it can be just as complex, or even more so.

Frontal lobe deterioration, affecting working memory, impairs the driving ability of the elderly.

CROSS REFERENCE: See "Brain Changes," PAGE 278

GLOSSARY

ADRENAL GLANDS. Produce hormone and regulate metabolism and blood flow and volume. Play an active role in the fight or flight response.

ATTENTION DEFICIT HYPERACTIVITY DISORDER (ADHD). A common disorder in which individuals have difficulty concentrating and multitasking and are easily distracted. Affects 4 to 5 percent of children and may continue into adolescence and adulthood.

BEHAVIORAL INHIBITION. A group of responses that develop in some children faced with new people or situations, who may express fear or shyness, and may cry or seek comfort in a familiar person.

BEHAVIORISM. The science of predicting and controlling behavior through use of stimuli and conditioned responses.

ECTODERM. Outermost layer of cells of a developing embryo that becomes the skin and nervous system.

ENDODERM. Innermost layer of cells of a developing embryo that becomes the digestive tract.

EPINEPHRINE. Also called adrenaline. Primary hormone produced by the adrenal medulla. This works with norepinephrine to put body systems on alert.

FETAL ALCOHOL SYNDROME. A range of developmental disorders and birth defects caused by excessive drinking during pregnancy. Children may be born with abnormal features and malformed organs and are at risk of mental retardation.

GLUTAMATE. Excitatory neurotransmitter prevalent in the central nervous system.

HEMISPHERECTOMY. The surgical removal of one of the hemispheres of the brain.

MESODERM. The middle layer of cells in a developing embryo. This forms the muscles, skeleton, heart, and genitalia.

NEURAL DARWINISM. A term coined by Nobel laureate Gerald Edelman, this describes the process in which neurons that receive constant simulation grow, and those that do not atrophy.

NEURAL GROOVE. In a developing embryo, the second stage of brain and spinal cord development, this occurs as the neural plate begins to fold inward.

NEURAL PLATE. Formed during the third week of embryonic development as the ectoderm thickens. The first stage of brain and spinal cord formation.

NEURAL TUBE. Formed as the neural groove fuses together. Occurs by day 22 in embryonic development.

NEUROBLASTS. Primitive nerve cells.

NEUROLEPTICS. Antipsychotics, a class of drugs generally used to treat schizophrenia and other psychotic disorders.

NEURONAL MIGRATION. Process where new neurons created in the prefrontal cortex relocate to other parts of the brain and assume new tasks. This occurs during the final months of fetal development.

NOREPINEPHRINE. A neurotransmitter and adrenal medullary hormone, this works with epinephrine to activate the sympathetic branch of the autonomic nervous system. Also involved in arousal and the regulation of sleep and mood.

PHONEME. The smallest sound element in a language that can be altered to change the meaning of a word. These have no meaning on their own.

PHONEME CONTRACTION. The loss of the ability to hear and differentiate all language sounds. Occurs between the age of six months and one year.

PLANUM TEMPORALE. Region of the brain associated with speech and sign language comprehension. Larger in the left hemisphere in two-thirds of the population.

PRUNING. A natural process of the brain where weak neural connections die off. Occurs on a large scale during fetal development and during teenage years, and on a lesser scale through adulthood.

SCHIZOPHRENIA. A chronic neurological disease of distorted thoughts and perceptions. This affects both men and women and usually surfaces during adolescence or young adulthood. Symptoms include delusions and hallucinations.

TRIUNE BRAIN. A 1967 theory on the evolution of the brain suggested by Paul MacLean in which there are three separate areas of the brain representing evolutionary development.

TROPHIC FACTORS. Proteins that promote the survival, function, and growth of neurons, and that are responsible for the correct wiring of neurons during brain development.

ZYGOTE. A fertilized egg.

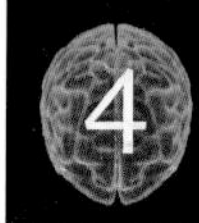

CHAPTER FOUR

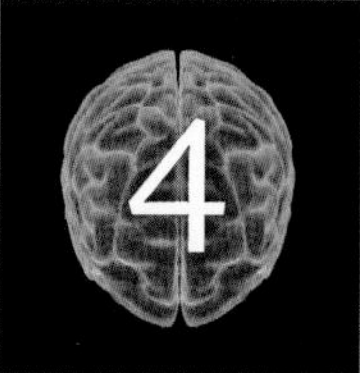

THE SENSES

WE EXPERIENCE the world through sensations. Pictures in our head, complete with sounds, smells, and other data gathered through the senses and supplemented by uniquely human cognition, create an internal universe to match the external one. The brain's ability to gather and decode information beyond the body is crucial not only to survival but also to the emotions, feelings, and complex social behavior that can give life richness and meaning.

Travelers in a Spanish train station gawk at the illusion of a performer who appears attached to a wall.

PERCEPTION

HOW THE BRAIN EXPERIENCES THE OUTSIDE WORLD

LOCKED UP safely within the skull, the brain experiences the outside world through the senses, five major ways we gather information about our environment. Seeing, hearing, tasting, smelling, and feeling are how human beings collect information about the world. These data are relayed to the brain, which uses them to formulate ideas and opinions, assess situations, generate reactions, and then store what it has learned as memories. Information that enters the brain through the senses powerfully influences thoughts, emotions, and personality. Put another way, what you see and hear—and taste, smell, and touch, for that matter—has much to do with who you are and what you think about the world.

THE SENSING PROCESS

Sensory receptors are specialized neurons that react to environmental changes, known as stimuli when they register on the nervous system. Sensation is the awareness of the stimulus, such as the knowledge of music coming from your stereo. Perception is an interpretation of what the stimulus means, such as that the music is a little bit too loud or that the song is a favorite. Both occur in the brain, after information is carried there by both the peripheral and central nervous system.

Alice's Parlor, an optical illusion room at the Virginia Science Museum, makes visitors question what they see.

CROSS REFERENCE: See "Messengers," PAGE 42

Science once perceived sensation much like philosopher and mathematician René Descartes's mechanistic view of the brain, based on the observation that stepping on stones in Paris's Royal Gardens triggered the rush of water through pipes and caused statues to move. In this view, sights and sounds and other stimuli hit the brain and, by some invisible process like water passing through unseen channels, automatically led to perception and action.

RICKIE'S BRAIN

It's much more complicated than that, as the experience of a girl named Rickie demonstrated. Cornell University researcher Frederic Flach in his 1990 book *Rickie* related the case of a girl who began perceiving the world in an unusual way when she was three years old. As she stood with her father and looked through a picture window at a stand of trees, she began trembling and yelling, "The trees are coming into the house! They're all coming in here!" Her father dismissed the incident as merely part of a child's vivid imagination. He was wrong.

> **FAST FACT** Architect Buckminster Fuller wore earplugs and special glasses to block out sensory stimuli and free his mind.

As she grew, Rickie had trouble learning in school and particularly

SEEING THINGS

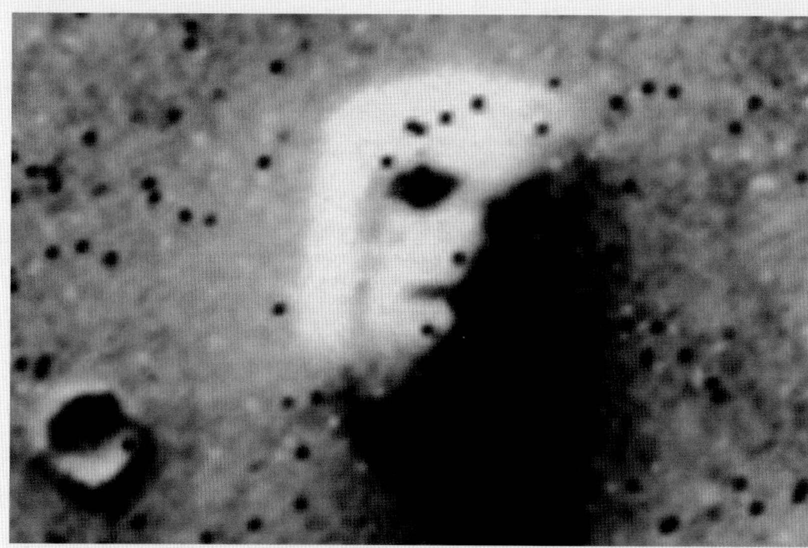

A "face" on Mars, caused by shadows, in a 1976 image from NASA's Viking 1 orbiter.

IT WAS AS IF THE longtime mayor of Rosemont, Illinois, didn't want to say goodbye. Two months after Donald E. Stephens died in 2007, residents spied his face amid the peeling bark of a 50-foot sycamore tree near a health club—a tree that Stephens twice had saved from being cut down. So many people jammed around the sycamore that police erected a crowd-control barricade. It's highly unlikely—and impossible to prove—that Stephens's ghost created his image in the bark. A more rational explanation is called pareidolia, the human capacity to see significance in random patterns. It provides a scientific reason for such visions as a giant human face spotted in 1976 on the surface of Mars, as well as an image of the Virgin Mary in the blackened bread of a grilled cheese sandwich that sold for $28,000 during an online auction in 2004.

"We're hardwired to pay attention to faces," said Stewart Elliott Guthrie, a cultural anthropologist who wrote the book *Faces in the Clouds: A New Theory of Religion.* The brain's predilection to sift sensory stimuli for meaning underlies the interpretation of Rorschach tests. Swiss psychiatrist Hermann Rorschach created the set of ten inkblots in 1921 as a means to examine the mind. Patients interpret splotches of black on white cards; descriptions of what they see are assumed to arise from underlying personality issues. While the cards are still used today to assess mental health, detractors say they yield no more than other tests, and that doctors often read too much into them.

struggled with reading. Sometimes as she looked at things, the world disappeared except for the object on which she focused her attention. Even then, she had to concentrate to keep that object sharply within her field of vision.

Tests revealed nothing physically wrong with Rickie's eyes. Doctors mistakenly diagnosed her with a variety of psychological disorders. Some even suggested a lobotomy to resolve Rickie's way of interpreting the world.

FINDING AN ANSWER

Rickie's problem began to become clear when a doctor performed a series of vision tests on the girl. The doctor asked Rickie to focus her eyes on an object, then look away. "When you look at something, how long does the image stay," he asked Rickie. "Does it stay or does it disappear, vanish?"

"It stays. I mean, I can make it stay," she answered. As the doctor asked follow-up questions, it became apparent that when Rickie focused her vision on something, her brain soon began to shut out visual stimuli from surrounding objects. She had to work harder and harder to keep the object of her attention in sight as her vision narrowed into a tunnel, and then the tunnel collapsed.

It's hard to describe what Rickie saw, but some likened it to looking into a room through a door's keyhole. The anomaly of her visual processing erased depth perception and had led to her vision of the moving trees when she was a little girl.

She had thought everyone experienced the world that way, and thus never considered telling others about the "normal" way she saw things. It would have been like a fish remarking on the amazing qualities of water to other fish.

The doctor fitted her with a special pair of glasses that allowed her to focus for longer periods of time without the images breaking apart. After months of wearing them and working through visual exercises to realign her perception of the visual world, the neural networks of her brain began to change. After six months, Rickie saw the world like most everyone else.

FAST FACT British arts professor Alexander Wallace Rimington applied for a patent in 1893 for a "colour-organ," a device that matched colors to musical tones. He based his work on Sir Isaac Newton's idea that colors and sounds are both vibrations.

+ WHAT GORILLA? +

WHEN ATTENTION is narrowly focused, people can miss some big changes in their surroundings. Psychologists call this change blindness. In a famous test, Harvard researchers showed a video clip of two teams, dressed in white and black, tossing and dribbling a basketball. Viewers were told to count the white team's passes. Halfway through the clip, someone wearing a gorilla costume walked through. Most of the test subjects never even saw the gorilla.

SENSES SHAPE THE WORLD

The lesson taught by Rickie's experience is the crucial role of sensory perception in how the brain creates a simulacrum of the "real" world based on input from the five senses. Vision is nothing more than the creation of symbols in our head that represent what exists outside the body. Similarly, hearing decodes auditory stimuli in ways that give them meaning, and we react to that meaning—when we hear a car backfire and we think "gunshot," for example.

Touch, smell, and taste interact with molecules outside the body to create their own sensations in the brain, leading to everything from pleasure to disgust. In each instance, we react to the symbols, not to the reality upon which they are based.

Rickie's case underscores how distortions in perception create distortions in the complex webs of neural networks in the cerebral

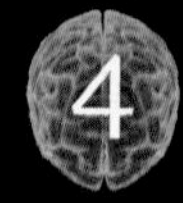

Moviegoers wearing 3-D glasses with special blue and red lenses react to the illusion of objects coming at them.

cortex that synthesize information. Eventually, they may affect the emotional centers in the limbic system and the decision-making functions of the prefrontal lobes. As a result, some problems diagnosed as psychological may be strictly perceptual.

This is not to say that one set of perceptions is "right" and another "wrong," only that some are more common than others. Since we react to the pictures in our head instead of to the world itself, who is to say that one set of images is the only correct one? The brain plays the crucial role in determining what it is reality.

Poet and mystery writer Edgar Allan Poe could have been talking about the role of perception in defining the world when he wrote, "All that we see or seem is but a dream within a dream."

PATTERNS & STEREOTYPES

As the brain collects information from the outside world, it becomes accustomed to patterns of perception. For instance, when you see a cocker spaniel, your brain processes its shape, the length and quality of its hair, the yip of its bark, the smell of its fur, the colors and texture of its coat, its height and length, and other details. The brain then reaches the conclusion that you are seeing a dog of a particular breed. If you've seen enough cocker spaniels, your brain makes this connection more quickly than if you were seeing the breed for the first time. Such rapid processing of sensory information has its uses. Walter Lippmann, a 20th-century journalist, observed that political propaganda works most effectively when it plays upon preexisting attitudes,

This person's silhouette registers in the brain as a man because it fits a masculine stereotype.

formed over time by the repetition of sensual experience. The brain categorizes new mental images according to what it has already experienced previously. Images of the world get filtered through a personal lens, making a crowded city street a symphony of light and sound for one viewer, and a dismal garbage dump for another. Both categorize the same observations by different experiences. Lippmann called the process stereotyping, a reference to the process of producing printing plates from molds—the plate matches the preexisting form.

SIMPLIFYING THE WORLD

And it's not just propaganda that relies on the volume of preexisting perceptions. Lippmann saw stereotyping as crucial to everyday understanding of the world. If everyone saw every detail of life afresh at each moment, they would be overwhelmed by the constant mass of sensory information.

FAST FACT Some autistic people process sensory input more slowly than others and can easily become overstimulated.

The brain often simplifies perceptions to get through the day; stereotypes lead to expectations about the world. They allow humans to anticipate actions and reactions and to prepare for them in ways to minimize harm and maximize

CROSS REFERENCE: See "Learning," PAGE 236

pleasure. It is the stereotype of a snarling, barking dog that the brain recognizes as a threat. "We do not first see, and then define," Lippmann wrote, "we define first, and then see."

Potential danger arises when the pictures in our head don't match the world outside—when we jump to conclusions based on fitting a deficient amount of new perceptions to our stereotypes. Perhaps that dog is a wolf, and not a cocker spaniel. Perhaps that person rolling on the ground is suffering a seizure, and not dancing. Perhaps that clear liquid in the glass is alcohol, and not water.

EXPOSURE TO THE NEW

When neurons are repeatedly exposed to a particular stimulus, they self-organize to recognize and then respond quickly. A dog can recognize a bell as a call to dinner or the stirring of its master's legs under the bedcovers as an indication that it's time to get up and go for a morning walk. An infant hears her mother's voice over and over and soon is able to pick it out from all others. A trombonist listens to a recording of his favorite symphony and discern the trombone's notes from the rest of the brass section. The smell of percolating coffee practically galvanizes a slugabed before he even takes the first sip. The cerebral cortex learns these experiences, then pushes them into the lower regions of the brain so they can be automatically processed without serious cognitive effort.

New experiences add to the brain's neural connections, build up patterns, and refine how we see the world. Just as muscles expand and grow stronger with repeated exercise, neural networks become more efficient in their responses to external stimuli. Recording new information and new experiences keeps the brain fit and sharp. The process of creating and reacting to forms of expectations, or Lippmann's stereotypes, can easily involve distortion.

STAYING SHARP

"SEEING IS ALREADY a creative operation, one that demands an effort," artist Henri Matisse wrote. Matisse's instruction anticipated modern neuroscience. While a century ago researchers considered the brain's processing of vision and other senses to be a somewhat passive synthesis of information, they now understand a subtle difference. Visual experience is a synthesis of *separate* streams of sensation. These include color, motion, the outlines of forms, and so on. Each aspect gets processed separately and simultaneously, then synthesized.

Seeing art—really *seeing* art—challenges and expands not only the brain's sensory networks but also its creative powers and memory.

Neurologist Richard Restak suggests the following ideas:

✓Buy a bonsai and get to know its every branch and leaf through mental exercise.

✓First, put the miniature tree on the floor and look closely at it from above. Memorize its shape and patterns of branches. Close your eyes and envision it with your memory. Then open your eyes and check for accuracy. Repeat, with the tree at eye level.

✓Visually zoom in, like a camera lens, to see the exact number of branches. See the number of leaves on a single branch. Close your eyes and re-create the exact shape in your mind. Keep your thoughts focused on images, not words.

✓As you improve your ability to "see" the bonsai's details, with eyes open or shut, you'll be developing your powers of perception and their associated neural networks.

SIGHTS & SOUNDS HOW VISION & HEARING WORK

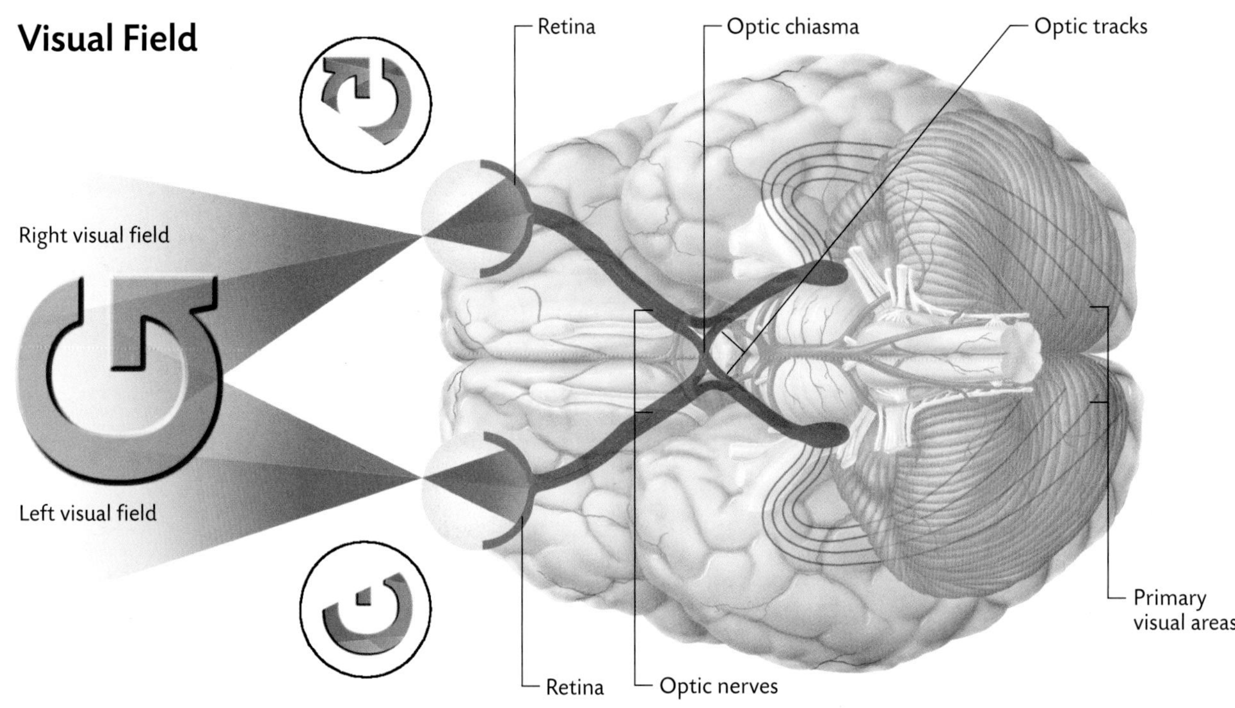

Each eye takes in a slice of the visual world, which gets processed in the opposite hemisphere before integration into a coherent image.

PEOPLE ARE often asked the hypothetical question "Which sense would you rather lose: vision or hearing?" It's a tough choice as both senses are critical to how we perceive and interpret the world around us. The important roles they play in feeding information to the brain cannot be overestimated. Both the eyes and ears are processing data constantly, from a near-infinite number of sources.

COMPLEXITY OF VISION

Consider your sense of sight. As you read this sentence, the visual networks of your brain are taking in more than 100 million bits of information. Your eyes constantly flit from place to place, usually never landing longer than a split second on any one word. Peripheral vision outside the dimensions of this page is a blur of color and shape; only a tiny region in the center of the eye called the fovea contains enough photoreceptors to see with great sharpness.

You may think you see the entire world as a sharp and seamless whole, but your retinas are segregating information into various categories, such as color, shape, and line, and permit acuity only on a small spot in the center of your field of vision. This screening process keeps the brain from getting overwhelmed by too much visual stimulation. Instead, it collects what the brain needs to create a useful image of the world as you shift the focus of your attention.

A MILLION LITTLE PIECES

As you perceive attributes of an object, whether it is a word on a page or a car going by on the street, your brain synthesizes the streams of information, matches them to images stored in memory, and makes the connection—and you recognize what you see. Because you reach these conclusions based on sketchy information, your brain fills in the blanks of perception. For example, each eye has a blind spot at its connection point with the optic nerve, a point where there are no photoreceptors. The overlap of binocular vision fills in that gap. You're probably not aware of the blind spot with either monocular or binocular vision until someone else tests for it by moving an object slowly across your field of vision.

Such perceptive synthesis fills in the gaps, for example, when you spot a deer on the other side of a latticed fence. After you spot it nibbling in the garden, it startles and runs away: You are actually seeing a hundred bits of deer through the multiple gaps of the fence, but your brain integrates them into a whole animal.

For most people, the brain relies on information from two eyes. As the brain develops in a young child, it learns to process information from both eyes into one coherent image. The evolutionary benefit of this development is depth perception. If you close your left eye, then open it and close your right, you will note that the difference between the two images is far greater for objects close to your face than far away. Judging distances is useful in fine tactile work, such as threading a needle, and in avoiding potential threats.

CONES & RODS

Vision begins with light of wavelengths between 400 and 700 nanometers striking the retina at the back of the eyeball. Four types of photoreceptive cells in the retina react to different wavelengths and intensities of light. Three of these neurons are varieties of cones, which react to wavelengths of bright light associated with green, red, and blue. As the intensity of the color grows stronger, these neurons ratchet up the strength of the electrochemical signals that eventually wend their way to the visual center located in the occipital lobes at the back of the brain.

+ BEAUTY OR HAG? +

OPTICAL ILLUSIONS occur because brains decode stimuli to create meanings that don't match reality. When we recognize the illusion, our brain usually compensates to create new meaning. But it's not always so simple. In illusions such as the one shown above, the brain decodes the image as a young girl in a fur or an old woman. As the brain cannot construct contradictory meanings at the same time, perception switches between the two.

Neural networks create other colors by mixing the sensation of the three primary colors of light (red, yellow, and blue) in varying intensities. But this mixing doesn't match the combination of colors of paint. If you blend red and green paint, you get brown. However, mixing red and green wavelengths of light creates yellow, which you can demonstrate by affixing red and green filters to the front of two flashlights and then overlapping the beams. If the cones have an impaired or absent ability to register all hues of the visual spectrum, the result is color blindness.

FAST FACT The average human eye contains 91 million rods but just 4.5 million cones.

The fourth type of light-sensitive neurons are called rods. They register light when its intensity is low, as on a moonless night, but do not add to the mix of primary colors from the cones. Rods rely on the extremely light-sensitive chemical rhodopsin, which contains vitamin A. It's concentrated in spinach, fruit, sweet potatoes, and carrots, which explains why eating carrots (just as your mother told you!) does indeed improve your vision.

VISION MECHANICS

Rods and cones work with other neurons. Some compare the relative

FAST FACT In the 1970s, Svyatoslav Fyodorov, a pioneering Russian eye surgeon, developed the first surgical procedure to correct eyesight. The surgery, called radial keratotomy, involved making small cuts in the cornea, the clear protective tissue over the eye. As the cuts healed, the cornea contracted, and vision improved.

brightness of two points of vision that are next to each other, which helps define the edge of an object. Brightness varies by day and night, making objects look different in different light and sometimes tricking the brain into wrong patterns of recognition. For example, a coat hanging on a rack in a darkened room may be interpreted as a threatening human figure.

Some neurons register the distance to an object based on visual cues, including stereoscopic vision. The bricks in a cobblestone street provide just such a cue; as you watch the street recede into the distance, bricks of uniform size appear smaller.

Other neurons register motion, an important artifact of evolution that allowed our ancestors to quickly recognize potential predators. Interestingly, various kinds of motion register in different neural networks. Straight-line motion, such as the fall of an apple from a tree on a windless day, activates one neural network. Spirals, such as debris in a tornado or the vortex of a whirlpool, register on another network. And expansion, such as the flyaway lines of stars when a starship jumps to light speed in the *Star Wars* movies, registers on a third.

SIGHT & THE BRAIN

How does it work? Imagine strolling down a hallway in the Louvre and suddenly seeing Leonardo's painting of the "Mona Lisa." Light waves of sufficient intensity cause synapses in the visual neurons to fire. That sends electrochemical signals via the optic tract into the lateral geniculate body of the thalamus and the superior colliculus. The latter, among other things, adjusts the head and eyes to maximize visual input. The former acts as a way station for visual signals on their way to the visual cortex in the occipital lobes. In the visual cortex lies an area known as V1, which redistributes the electrochemical information to at least 30 neural networks for further processing of visual attributes including color, shape, and texture.

Each lobe, in the right or left hemisphere, receives half the visual information. These regions integrate the two images and finally forward a unified single image to the frontal cortex for analysis. It is there that awareness occurs: Only then does the brain realize it is looking at one of the world's most famous paintings.

RECOGNITION

The revelation that different neurons respond selectively to different kinds of visual stimulation led to experiments in which single neurons were wired with electrodes to record what made them fire. Scientists once joked about finding the "grandmother" neuron that fired upon seeing and recognizing your lovable granny.

WHAT CAN GO WRONG

MACULAR degeneration, a disorder affecting the center of the retina, ranks as the leading cause of severe vision loss among Americans older than 55. It affects nearly two million. The disease harms the macula, a thin film of tissue about a fifth of an inch in diameter, situated on the inside back wall of the eyeball. Normally, proteins in the macula's light-sensitive rod and cone cells slowly degrade as they react to light and get shed as waste. For unknown reasons, macular degeneration interferes with the elimination of waste, causing a blurry buildup of yellow, fatlike deposits. As the neurons of the macula are

CROSS REFERENCE: See "A Memory Forms," PAGE 246

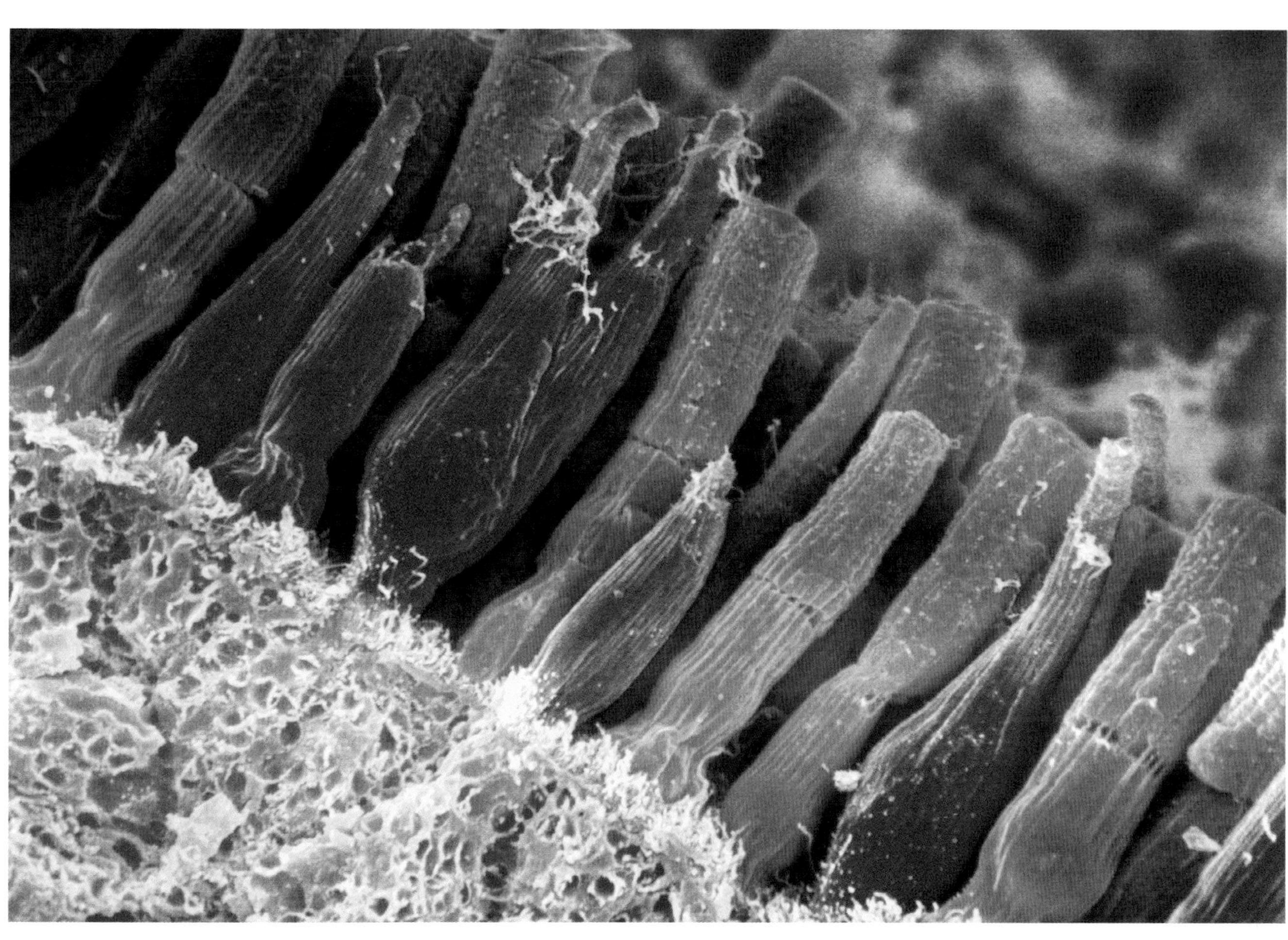

Rods, rendered in green by a colorized scanning electron micrograph, and pink cones detect and pass light signals to the optic nerve.

Recognizing your grandmother relies on more than one cell, however. The brain has learned to pay particular attention as it processes visual information about faces. Differences in the width of someone's eyes, tilt of a nose, or color of hair get synthesized into recognition of one person instead of another. Even tiny differences, as between twins, are noted and become the fundamental basis of identification. Flexible neural networks process varieties of information that may change. It is these networks that allow you to recognize your grandmother, even though since the last time you saw her, she lost some

WHAT CAN GO WRONG

responsible for sharp detail in the center of the field of vision, their degradation interferes with reading, driving, recognizing faces, and working with close eye-hand coordination. The disease strikes more often among women, whites, the elderly, the overweight, and smokers. No treatment can reverse the disease, but between 2005 and 2009, neuroscientists using MRIs demonstrated that the visual cortices of macular degeneration patients remap themselves in response to it. Patients who lose sharp central vision begin to focus with other portions of their visual field. Neurons in the cortex that lose signals from the retina's center begin to respond to stimuli from other parts of the retina, thanks to plasticity. If the neurons originally wired to receive information from the center of the retina remain active, a hypothetical treatment could one day reconnect the pathways for sharp central vision.

weight and is sporting a different hairdo and a new wardrobe.

Recognition depends on decoding information, and every object has its own characteristics. "It is no good painting a picture of a fork if it looks as if it is made of India rubber," said artist Anthony Green. "A fork is a hard thing, and it has a certain shine to it, which gives it its personality. It has got three or four prongs on it. It is forky. This applies to ears, noses, mouths, whiskers, whatever you want."

Through experience, the brain comes to recognize the essence of objects, which even the most sophisticated machines still struggle to do. The essence of objects allows their recognition even when important details are stripped away. That's why editorial cartoons of famous people—a handful of lines to represent an entire human face—remain recognizable. The artist has seized on the basic elements of the face and screened out the noise.

HOW THE BLIND SEE

Given the importance of vision, how do blind people "see" the world when their eyes are not collecting data? They certainly rely more than sighted people on their other senses. Hearing takes on greater importance as the detection and analysis of sounds help them to navigate through the world, such as listening for echoes from a tapping cane to detect objects around them. They also tend to have excellent memories, especially for

FAST FACT

German physiologist Adolf Fick developed the first wearable contact lens in 1887. It was made of blown glass.

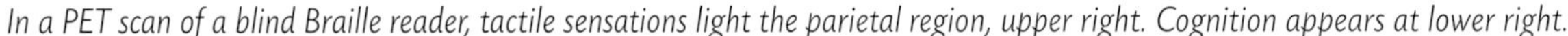

In a PET scan of a blind Braille reader, tactile sensations light the parietal region, upper right. Cognition appears at lower right.

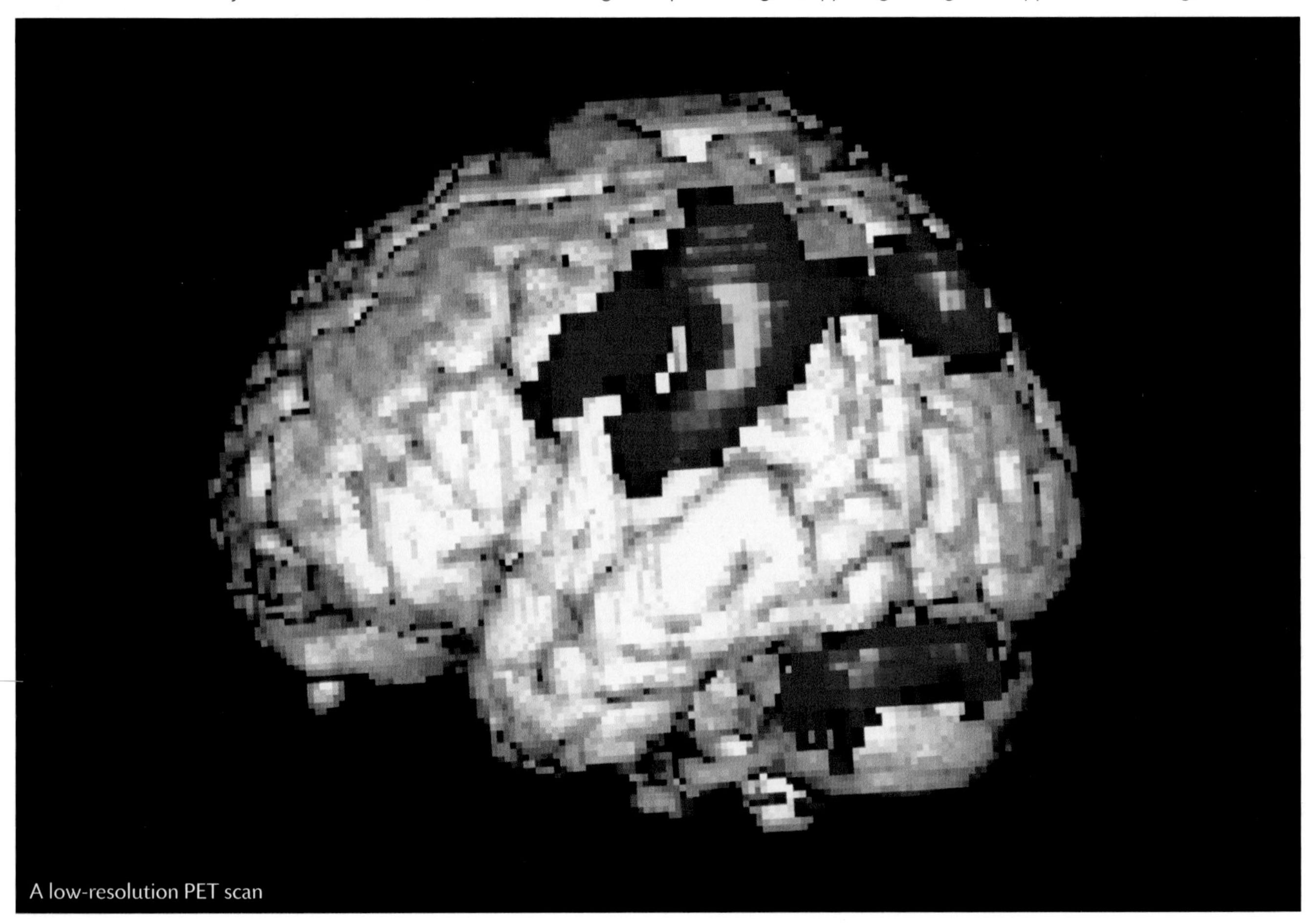

A low-resolution PET scan

language and spatial arrangements. Blind bards of pre-literate Europe memorized thousands of lines of stories and poems. Then, as now, blind people also call upon their spatial memories to map the orientation of objects around them. Since they can't see the location of a desk or a coffee cup on top of it, they must remember those spatial facts or face bumps and burns.

But the brain of the blind does adapt in response to the loss of sight. Blind people apparently make new use of neural networks that process visual information among the brain of people who can see.

For example, when blind people remember verbal information, they typically call upon neural pathways in the primary visual cortex. When researchers generate a magnetic field near a blind person's visual cortex, its electrical disruptions interfere with the ability to choose verbs. The field interferes with a sighted person's vision but doesn't affect language.

HEARING & LISTENING

Next to sight, hearing is one of the senses often ranked as most important. The brain collects vital information about the environment through the noises and sounds created around it. A creaky floorboard lets you know someone is sneaking up on you, while a leaky faucet can tell you to call a plumber. A favorite song summons you to get up and dance.

The main tools of sound collection are our ears. Inside the ear lie groups of mechanoreceptors that perform different functions. The first registers the existence of sound, including its pitch and intensity. The second acts independently to continually monitor and adjust movements of the head and the body in response to movement and orientation to gravity.

+ IS IT MAGIC? +

MAGICIANS DON'T NEED black arts. They need to understand the brain. It's crucial to many a stage act to misdirect attention. Sudden movements, sounds, introduction of new objects—anything can divert sensory focus from the nitty-gritty of the trick. The viewer's brain, not registering the true cause and effect, fills in the blanks of perception, and—voila!—"magic." "The principle of misdirection plays such an important role in magic that one might say that Magic is misdirection and misdirection is Magic," said magician/journalist Jean Hugard.

SOUNDS IN THE BRAIN

Sound is created by disturbances of pressure in a conducting medium, such as air. The brain registers sound when pressure stimulates the auditory region of the temporal lobe of the cerebral cortex. To get to that region, pressure waves must transfer their energy through the air to membranes, fluids, and bones in the ear, and on to receptor cells of the so-called spiral organ (organ of Corti) in the inner ear.

It's a long and somewhat complicated chain of vibrations. First, sounds strike the tympanic membrane and make it vibrate at the same frequency as the incoming sound waves. The louder the sound, the more the membrane moves back and forth. The tympanic membrane transfers its energy to the middle ear, where it gets concentrated and magnified.

The increased pressure sets waves in motion in the fluids of the cochlear canals of the inner ear. These waves move through other structures of the ear until they impact the fibers of the basilar membrane. Long fibers resonate in response to low-pitched sounds, while shorter fibers resonate to sounds of higher pitch.

All of this processing of varieties of sound occurs before they reach the organ of Corti, which rests atop the basilar membrane. This special organ consists of supporting cells and hearing receptors called cochlear hair cells. The hair cells and associated sensory neurons bend in response to particular

RAVEL & APHASIA

In a cruel twist, composer Maurice Ravel's brain lost its ability to appreciate music.

AFTER 14 MINUTES of invariant repetition, composer Maurice Ravel's *Bolero* stridently shifts from C to E major, then back again, as the delicately forged framework of rhythmic ostinato and mounting melody at last falls apart. In 1928, the year of the piece's composition, the mental equilibrium of its 53-year-old composer had started its own march toward collapse. Toward the end of his life, Ravel suffered from a debilitating combination of apraxia—the inability to perform coordinated movements—and aphasia combined with amusia, neurodegenerative conditions that affect the ability to express and understand language or music, respectively. Aphasia sets in after stroke, disease, or head injury—anything that results in damage to the brain's dominant (in most cases, left) lobe, where speech is formed and processed. Since the ability to comprehend music is housed in both hemispheres, including Broca's area on the left side and regions of the right hemisphere devoted to sound quality, some patients understand neither speech nor song, while others suffer only from aphasia. Ravel suffered from aphasia and amusia. Had modern music-speech therapists been available, they couldn't have helped him. As the diseases progressed, Ravel could hardly differentiate notes on a keyboard. Attempts to relearn the alphabet failed. By December 1937, his left lobe had atrophied to the point of collapse, and he died. "I haven't said anything yet, and I have still so much to say," he once lamented.

vibrations of the basilar membrane. As the cochlear hair cells react, they initiate electrochemical signals that get sent to the brainstem via the auditory nerve. This nerve contains only about 25,000 fibers, far fewer than the number of nerve bundles associated with vision. Therefore, the fibers must act efficiently. Even in the absence of sound, they remain primed to carry information.

PROCESSING SOUNDS

When the auditory stimuli reach the brain stem, neural networks sort them by tone and by quality. The brain stem simplifies comprehension by eliminating those echoes that are commonly created by vibrations bouncing off walls, ceilings, and floors. If a sound is new or strange, though—a potential threat, for example—the brain stem lets it through. The brain stem also begins the processing of phonemes to initiate the comprehension of speech.

Auditory impulses then are routed to the midbrain's superior colliculus. Once there the sensations of sound get synthesized with those of other senses to begin creating a unified experience of the world—such as hearing a boom and smelling gunpowder when witnessing the flash of a musket at a pioneer celebration.

Auditory sensations then rise through the thalamus to reach the

FAST FACT

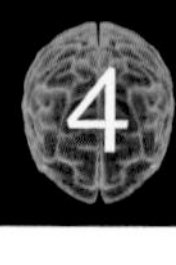

Sometimes the ears hear sound when there is none. Chronic ringing or clicking amid silence is called tinnitus. The loss of neurons in the auditory system creates a sensory void that nearby neurons fill, creating signals interpreted as noise.

primary auditory complex, where they interact other neural networks that link sound to memory, other senses, and awareness. The auditory complex has vast numbers of neurons to process different sound frequencies.

To appreciate the complexity of neural integration, consider that watching someone's lips move without speaking aloud activates networks in the auditory complex, while a facial movement unrelated to speech does not. At another level of integration, the brain's emotional centers can add meaning to music, sparking a bright cheerfulness for orchestral works in a major chord or sadness for works in minor chords.

REACTIONS

Neural processing of auditory information is not evenly divided between the brain's hemispheres. The left hemisphere decodes musical rhythms better than the right, while the right specializes in the quality, or timbre, of sound. The left side also processes fast sounds, such as specific consonant combinations of spoken language, better than the right.

Evolution of the brain has left the fight or flight response to threatening sounds in the deep portions of the brain. When there's no time to think about whether a sound represents danger, the thalamus handles sounds quickly and efficiently.

The upper portion of the brain, including the auditory complex in the temporal cortex, takes longer to process and react. Mixes of frequencies create patterns that the cortex compares with its library of experience and then acts upon. An incredible range of sounds, from a baby's cry to a whale's song, set

How We Hear

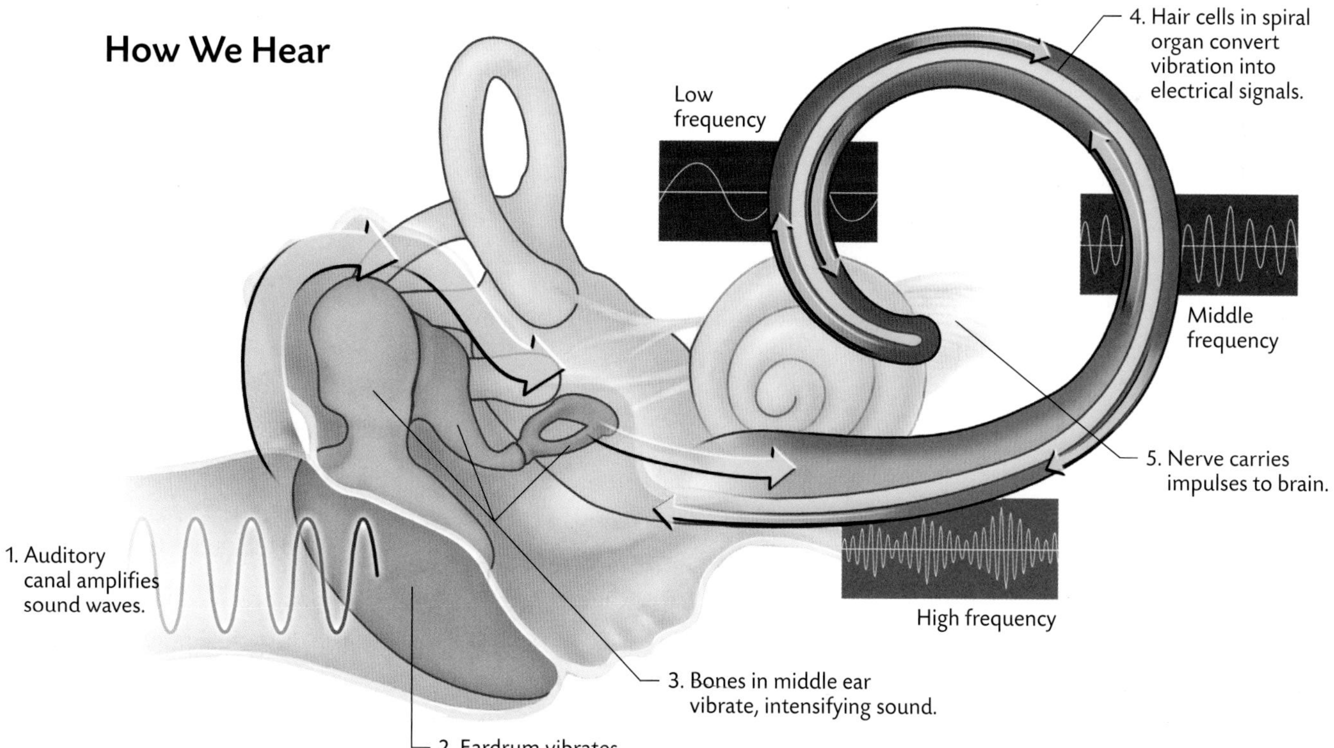

Sound waves are converted in the ear to nerve impulses that are carried to the brain's auditory center.

FAST FACT Appreciation of music often inspires creativity in other forms of expression. Dissident Russian writer Alexander Solzhenitsyn listened continually to Beethoven's Ninth Symphony while writing a long treatise against the abuses of communist censorship in 1967.

groups of neurons firing in the cortex. Blending the neural stimuli creates fine shades of difference, such as distinguishing the hard "th" of the word *that,* versus the soft "th" of *thought.*

FINE-TUNING

Most of the sensory neurons in the inner ear send information *to* the brain, but some cells in the outer ring of cochlear hairs receive information *from* the brain. The brain uses this feedback circuit to protect the ears. When the brain detects incoming sounds as dangerously loud, it sends signals to the inner ear to dampen sensations by spreading their energy over a broader portion of the basilar membrane. Sounds that are too loud can cause the loss of cochlear hair cells, a permanent condition that ends with some measure of deafness.

When the range of incoming sounds includes more than one pitch or volume, more than one group of hair cells on the basilar membrane vibrates in resonance. Two or more sounds get processed by the cortex. As a result, a music lover listening to a symphony performing Beethoven's *Eroica* Symphony can pick out individual instruments or sections within the wall of music.

Like the brain's system of visual processing, which separates and then synthesizes attributes of sight, the auditory region of the temporal cortex contains neurons that perform small, specific functions. Some neurons fire at the start of tone, and others fire at the end. Some fire quite easily, while others resist firing.

In general, louder sounds and noises release more neurotransmitters, more often, causing more neural networks to activate and the brain to interpret the heightened action as greater volume. Neurons in the brain stem interpret differences in the intensity and timing with which sound waves strike the two ears, which helps the brain locate the source of a sound.

Hearing aids work on the principle of amplifying sound waves that reach the ear. Early versions of these devices caused some confusion because they amplified everything without any discrimination. Modern refinements adjust the balance of sound to noise and help the impaired focus only on certain sounds, so they are able to block out background noise in favor of more important sounds.

The same thing happens naturally at a party. When you're attending a gathering at which lots of people are talking, your brain registers all of the sounds. You can follow your own conversation, however, because the brain classifies some sounds as significant and others as noise.

+ TUNING IN & OUT +

IT'S CALLED THE cocktail party problem: At a loud wingding, your friend says, "Hey, do *(whomp! bang!)* want *(boom! bash!)* or not?" Try as you might, you can't make out the words. You might have a problem resulting from high-frequency hearing loss—or perhaps a brain that's trying to keep from getting swamped by sound. Brains overwhelmed by sound tell their auditory-processing neurons to reduce input, allowing greater understanding. Young brains have a pretty good "dimmer switch," but it grows weaker with age. Researchers are looking at degradation of cells in the dimmer switch circuitry and loss of vibration-sensitive hairs in the ear.

ON BALANCE

In addition to the role that they play in hearing, the ears contain structures that are vital in keeping the body upright and even, a property that is better known as balance. The lower brain stems of all vertebrates also function to maintain balance and an awareness of spatial orientation.

The jellylike otolithic membrane, containing tiny calcium carbonate otoliths (Greek for "ear stones"), detects gravity. The membrane works with semicircular canals that detect rotation and forms the

CROSS REFERENCE: See "Silent Running," PAGE 146

When sound volume increases, more neural networks become active, prompting sensations of greater intensity.

brain's vestibular organ. It registers changes in the body's spatial orientation, such as when the body tilts one way or another, during such activities as swinging a golf club or a tennis racket. This organ works with vision and the motor complex to maintain balance.

The vestibular organ's performance in the presence of gravity worried space scientists as they considered the potential impact of long periods of weightlessness in space. Their fears were overblown. Astronauts got used to not knowing which way was "up." It even became fun to spin and turn without restraint. However, when astronauts returned to Earth they required a period of adjustment to regain their ability to balance under normal gravity.

Balance is an amazing state of body equilibrium. It keeps printed type steady in your field of vision if you move your head from side to side. Unpleasant disorders result when balance breaks down. Motion sickness, common at sea, occurs when the brain receives conflicting sensory information from eyes and the vestibular system. Passengers in a ship's cabin see the room as stable, but their otoliths insist their center of gravity is shifting beneath them, suggesting movement. Standing on the deck, so the motion of the sea registers on both eyes and ears, lessens the effect.

About two million Americans seek medical treatment per year for vestibular balance disorders. Treatments vary with the causes. Often, doctors will treat symptoms with medication, such as drugs to alleviate vertigo. Physical therapy may also be added to target the root of the problem.

SMELL & TASTE

COOPERATIVE SENSES

SMELL AND TASTE are usually thought of as separate senses, but they have much in common with each other. They analyze molecules entering the body from the outside world. They screen out harmful threats. And they work together to maximize appreciation of food and drink, two of life's great pleasures.

A DIRECT SENSE

Compared with the sense of smell, the other senses take a long, somewhat roundabout road to the brain. Smell, the most ancient of senses, takes a more direct path. Taste, touch, hearing, and a portion of vision send their electrochemical signals to the brain via the brain stem, which then relays them to the thalamus and on up to the cerebral cortex. Sensation of smell goes straight into amydgala and olfactory cortex, both parts of the limbic system, without stopping at the thalamus along the way.

Smell also is hardwired to the brain's emotional centers. When you smell something, the sensation rushes, practically unfiltered, into the frontal lobes. There's an evolutionary reason for the hotline. If you prepare to eat food that's spoiled, or ingest a noxious chemical, the nose acts as a screening system. Animals without a strong sense of smell faced a greater likelihood of dying before reproducing than those that had a keen nose.

As the amygdala directly influences the sympathetic branch of the nervous system as well as the nurturing bonds of family, smells can trigger a rise in heartbeat and blood pressure or bring on a feeling of calm and well-being. The latter forms the basis of aromatherapy.

The direct wiring into the emotional and memory centers of the brain also empowers odors to trigger powerful recall of events from the distant past. The limbic system, which processes smells, contains pleasure centers, and plays a crucial role in emotional response. How a smell is interpreted can be highly individual. The mixture of gasoline and cow manure, two odors generally considered unpleasant, may dredge up memories of happy summers spent with cousins on a Kansas farm.

FAST FACT Nasal cavities contain pain receptors, so some smells, like hot chilies or ammonia, actually trigger pain.

RELATIONSHIP TO TASTE

Smell also plays a crucial role in taste, which is why these two senses frequently are studied in tandem. Much of the enjoyment of the flavor of a meal is actually chemical information that is released as odor from plate or bowl or wafting upward as food is crushed by being chewed. Without the sense of smell, the world seems like a silent black-and-white movie compared with a noisy color cartoon.

BREAKTHROUGH

THE HUMAN NOSE can distinguish among thousands of scents, but how can subtle differences be expressed in words? German psychologist Hans Henning (1885–1946) developed an "odor prism" in 1916 in an attempt to chart the differences among smells. He recruited several volunteers to sniff more than 400 odorants and, after stripping each to its "bare sensory quality," classify them. Henning took the participants' responses and tried to chart them, without success, until he hit upon the form of a prism. He claimed that all odors could be located on the surface of a prism whose four-sided base consisted

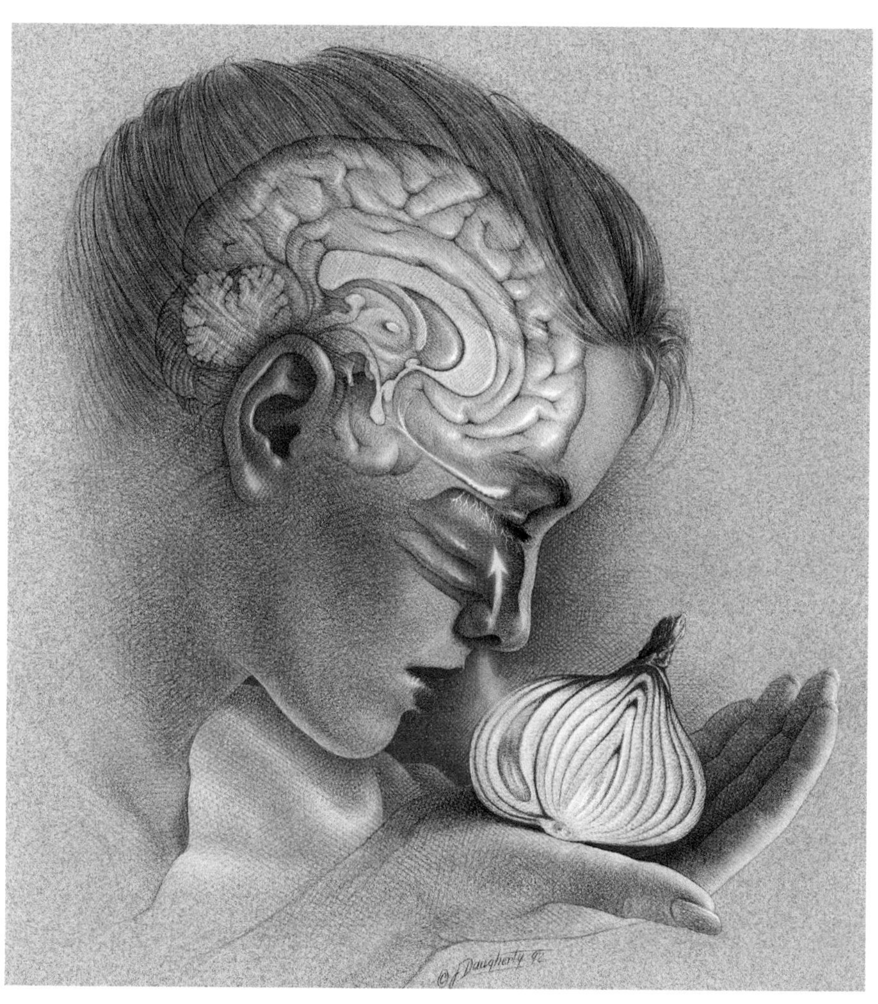

Smell circuitry includes receptor cells in the nose, olfactory bulb, cerebrum, and limbic system.

THE OLDEST SENSE

Smell is believed to be the oldest of the senses because layers of olfactory neural networks were the first to emerge atop ancient animals' primitive nerve cords. As a matter of fact, the limbic system is believed to have grown out of the primitive brain's need to interpret smells as possible sources of food and to emit chemical messengers called pheromones, which send sexual and social information to other animals.

Some pheromones act like magic bullets on insects and fish, initiating everything from sex to combat. Humans, though more highly evolved, are not immune. They release pheromones in all body fluids. Women who share living space tend to have their menstrual cycle begin within a day of each other, thanks to the presence of pheromones. Research even showed they didn't need shared living quarters—merely to have one woman's sweat rubbed on the upper lip of another woman sometimes triggered menstrual synchronicity.

As the brain evolved, visual, auditory, and other sensory networks became more important, and their portion of the brain expanded. Meanwhile, the day-to-day importance of smell has remained important but has not grown to keep pace with the neural networks of the other senses. The cerebral cortex contains relatively few neurons

BREAKTHROUGH

of fragrant, ethereal, spicy, and resinous odors, and whose peak ranged from putrid to burned smells. Henning tried to apply chemistry to the system, claiming, for example, that fragrant smells originated from adjacent corners of the ring of the benzene molecule. His system has been challenged on several grounds, including its simplicity. Some smells seem to be charted inside the prism instead of on its surface, and problems with what researchers call inter-coder reliability—in this case, the degree to which two or more people agree on how to categorize a smell—have undermined attempts at objective measurement. Furthermore, many biological smells are not by any means "pure." They may incorporate snippets of tens or hundreds of different odors in varying ratios, defying simple classification. Still, Henning prepared the way for scientific classification of smell that continues to undergo refinement.

PROUST: TASTE & MEMORY

Madeleine cakes sparked concrete memories of childhood for author Marcel Proust.

NOTHING BRINGS memories alive like a familiar odor. Marcel Proust, author of *Remembrance of Things Past,* found as an adult that the aromas of a cup of tea and a madeleine transported him to his childhood home, where his aunt gave him the same treat on Sundays. "The entire town, with its people, and houses, gardens, church, and surroundings taking shape and solidity, sprang into being from my cup of tea," he wrote.

Smell powerfully evokes memories because it is the only sense hardwired into the brain's limbic center. Other senses pass through intermediary circuitry. The scent of a perfume worn by a special person many years ago may evoke powerful emotional memories. For this reason, some real estate agents suggest that would-be sellers fill their home with the scent of flowers or freshly baked bread or cookies to promote warm emotional responses among potential buyers.

Neurologist Richard Restak recommends exploring pleasant scents with friends or family members as a means of enhancing emotional memory. His daughter Jennifer suggested an aromatic exercise in which three or four friends gather around a table where they have brought their favorite scents, such as sandalwood, leather, and fresh cookies. Each person samples the scents, describes which had the most pleasant effect and why, and shares memories associated with the smell by talking or writing about feelings the smells evoked.

associated with smell. Thus, it is difficult to imagine a smell.

The brain recognizes some smells, such as rotting food, at birth. Others must be cultivated, such as the delicate odors of some perfumes. In all, the olfactory network can recognize about 10,000 smells.

DETECTION OF SMELLS

The detection of smell begins when molecules from the surface of an object are released into the air and reach the nose. When air enters the nose, it carries molecules across the olfactory epithelium, a yellow blanket of sensory cells on the roof of the nasal cavity. The epithelium contains millions of sensory cells shaped like little bowling pins. Chemicals that reach the olfactory epithelium must dissolve in its mucous sheath and then bind with protein receptors if they are to be detected. When the neurons' action potential is reached by a minimum concentration of molecules, they fire to signal the presence of odor to the olfactory bulbs at the far end of the olfactory tract.

Complex neural structures called glomeruli activate for specific odors. Just as neurons in the visual and auditory systems identify individual characteristics of sight and sound, glomeruli have specialized functions. They're not as simple as taste, which recognizes five basic flavors, but they're not so complex that each odor has its unique

CROSS REFERENCE: See "A Memory Forms," PAGE 246

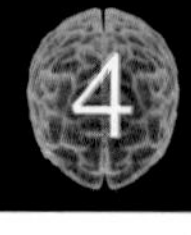

FAST FACT In 1908 Tokyo researcher Kikunae Ikeda discovered the fifth flavor while studying the taste of stock made from kelp. He isolated glutamate, or glutamic acid, as the source and christened its complex, savory flavor umami.

receptor. Perhaps one thousand or so "smell genes" encode receptor proteins, each of which responds to a variety of odors.

Meanwhile, each odor binds to a variety of receptor genes. Some receptors are so sensitive that they fire in the presence of only a few molecules—an infinitesimal fraction of the ten billion billion molecules in a cubic centimeter of air. When the glomeruli activate, their signals get amplified and refined by neurons called mitral cells and then sent via the olfactory tract to the limbic system.

The human nose isn't nearly as efficient as a dog's. Fido's olfactory receptors cover a much larger area inside the nose and process smells at least a thousand times better than a human olfactory system. The human nose is poorly designed, with a relatively small olfactory epithelium located beyond a sharp turn inside the nasal cavity. To get a good whiff of molecules suspended in air often requires sniffing or deeply breathing.

Like Pavlov's dogs, humans respond to external stimuli, like appetizing smells—by salivating and stimulating the digestive tract.

POWER OF TASTE

Remember the last time you had a really nasty cold—itchy eyes, sneezing, fever, and a nose clogged with mucus? Your dinner probably lost most of its flavor. The stuffiness in

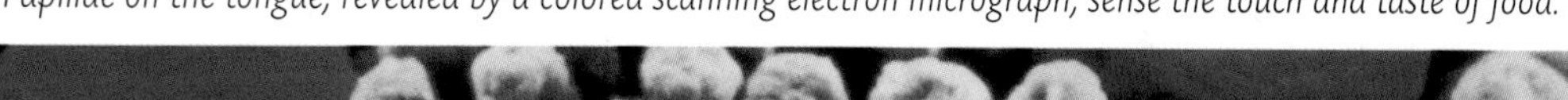
Papillae on the tongue, revealed by a colored scanning electron micrograph, sense the touch and taste of food.

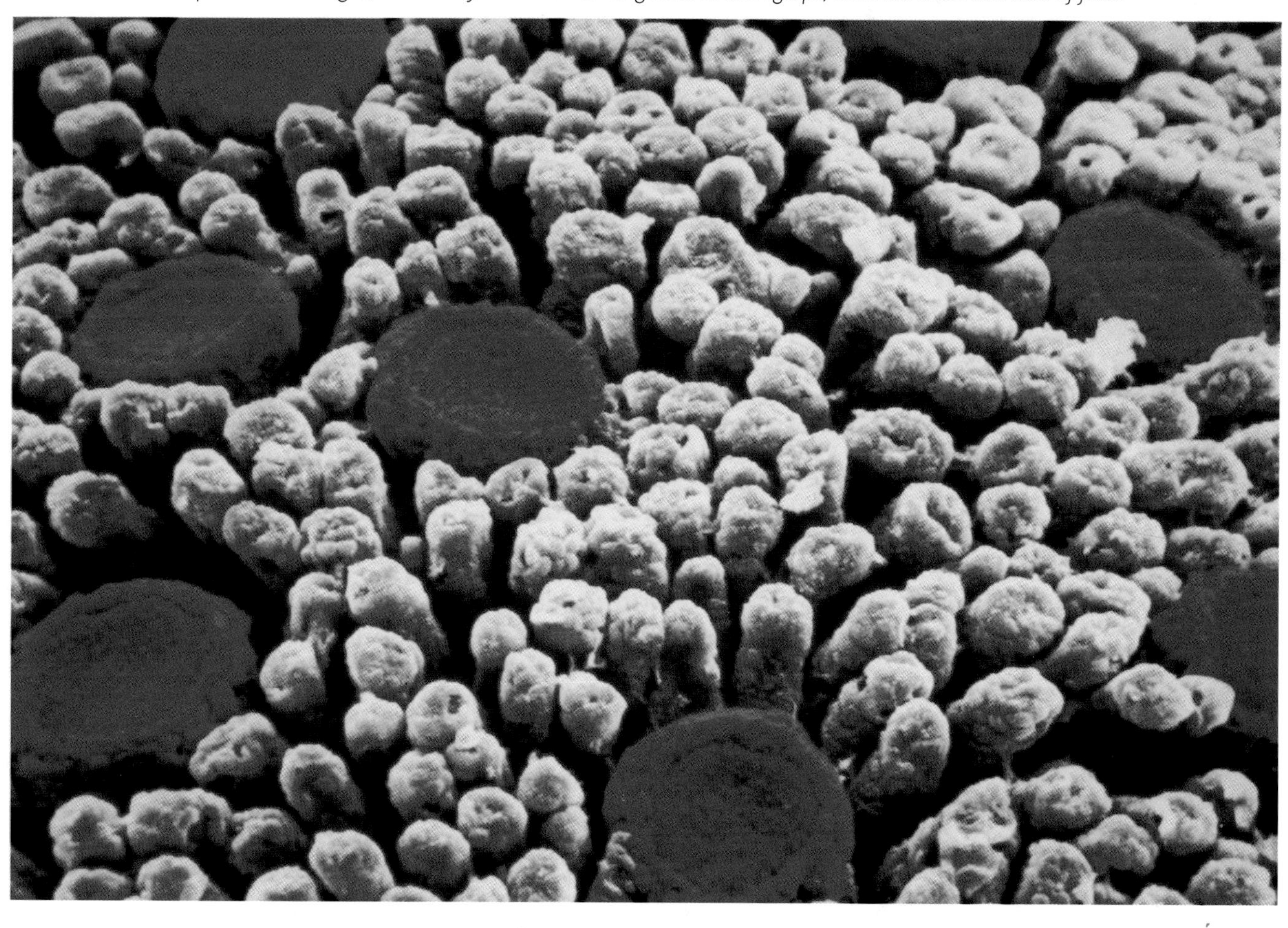

your nose probably prevented the aromas of your food from reaching the olfactory receptors inside your nasal passages.

About three-quarters of what the brain perceives as taste actually enters our perception through the nose. So, during the time the nose is blocked, a fraction of a particular flavor enters through the taste buds of the tongue, along with the perception of the food's texture and temperature.

The English word *taste* comes from the Latin *taxare,* meaning "to touch" or "to judge." Tasting has an ancient association with verification. "O taste and see the goodness of the Lord," says Psalm 34. Tasting, like smelling, allows close interaction with the environment. Chemicals that make up our surroundings get closely analyzed as they touch the tongue before being taken inside the body. The body recognizes potential threats in food by their unpleasant taste and smell, causing protective reflexes such as choking and vomiting.

TASTE BUDS

Humans have about 10,000 taste buds, most of which cover the tongue in a blanket of peglike bumps. Others reside on the soft palate, the insides of the cheeks, pharynx, and epiglottis. Each bud contains 50 to 100 epithelial cells made up of gustatory and basal subgroups. The gustatory cells project hairs through a taste pore to the surface of the epithelium, where they are bathed in saliva and serve as the receptors for gustatory cells. Because cells in the taste buds constantly experience friction and heat from chewing food, they often suffer damage. Fortunately, the last time you burned your tongue on hot mozzarella didn't forever ruin your taste for Sicilian pizza. Basal cells in the taste buds constantly divide and differentiate to create new gustatory cells, replacing the surface cells every week or so.

+ HOT PEPPERS +

CAPSAICIN PUTS the *pep* in peppers. This tasteless chemical irritates tongue receptors and registers as a spicy burn. Those same receptors register pure heat—in other words, they recognize two kinds of "hot." When receptors detect capsaicin, they relay both signals to neural processing. The brain, tricked into sensing body heat, lowers skin temperature through sweat. Capsaicin prompts the release of endorphins, which encourage good feelings. This helps explain the love of spicy food.

FIVE FLAVORS

Receptors on the tongue differentiate among five basic tastes: sweet, sour, salty, bitter, and umami. Each taste can be mapped to a general region on the tongue. Divisions of taste receptors on a map of the tongue are only approximations. All tastes can be elicited to some degree from any surface that contains taste buds. Furthermore, most buds react to two or more tastes, accounting for the broad spectrum of distinct flavors.

Taste buds at the front of the tongue react most strongly to sweet sensations, caused by exposure to sugars, saccharine, alcohol, and some amino acids. That's why the most effective way to enjoy a high-sugar lollipop is to lick it with the tip of the tongue.

+ THE FIVE FLAVORS +

TASTE	REGION OF TONGUE	FOODS
Bitter	Back of tongue	Food with alkaloids: coffee, citrus peel, unsweetened chocolate
Salt	On the sides of the front of the tongue	Foods with sodium ions: anchovy, salted popcorn
Sour	On the sides in the middle of the tongue	Foods with acids: lemon, grapefruit
Sweet	The tip of the tongue	Foods with sugars: candy, ripe fruit
Umami	Concentrated on the pharynx	Beef, lamb, Parmesan cheese, soy sauce, fish sauce

Sour tastes, like lemon juice, arise from exposure to acids, which contain hydrogen ions. Like salty tastes, which come from metal ions, sour flavors are best experienced on the sides of the tongue. Bitter tastes, perceived in alkaloids such as caffeine and nicotine, create the strongest reaction on the back of the tongue. A fifth taste, unidentified until the 20th century, is umami. It arises from the amino acids glutamate and aspartate and is linked to the characteristic tastes of beef and cheese. Umami receptors are concentrated on the pharynx.

HOW TASTE WORKS

When you take a sip of orange juice, the chemicals that give it its flavor mix with saliva, contact the taste pores of the taste buds, and touch the gustatory hairs. The chemicals of taste, called tastants, bind to receptors on the gustatory cells. If the orange juice hasn't been watered down, the chemical flavors of the orange, including its acids and sugars, cause the receptors to release neurotransmitters that bind to sensory dendrites and begin a reaction along three neural pathways to the brain—the facial nerve, glossopharyngeal nerve, and vagus nerve—to recognize the orange flavor.

These nerve fibers forward taste sensations to the solitary nucleus of the medulla, and from there to the thalamus and gustatory cortex. Gustatory fibers also connect to the hypothalamus and limbic system. There, in regions associated with emotions, the brain forms its appreciation of the flavor. Without the connections to the lower portions of the brain, the orange juice would still have its flavor but it wouldn't elicit that *mmm* reaction.

Sensitivity of the receptors varies with the flavor. Bitterness carries an extremely low threshold for recognition, an evolutionary artifact in which quick recognition of spoiled food can prevent poisoning.

Taste Regions

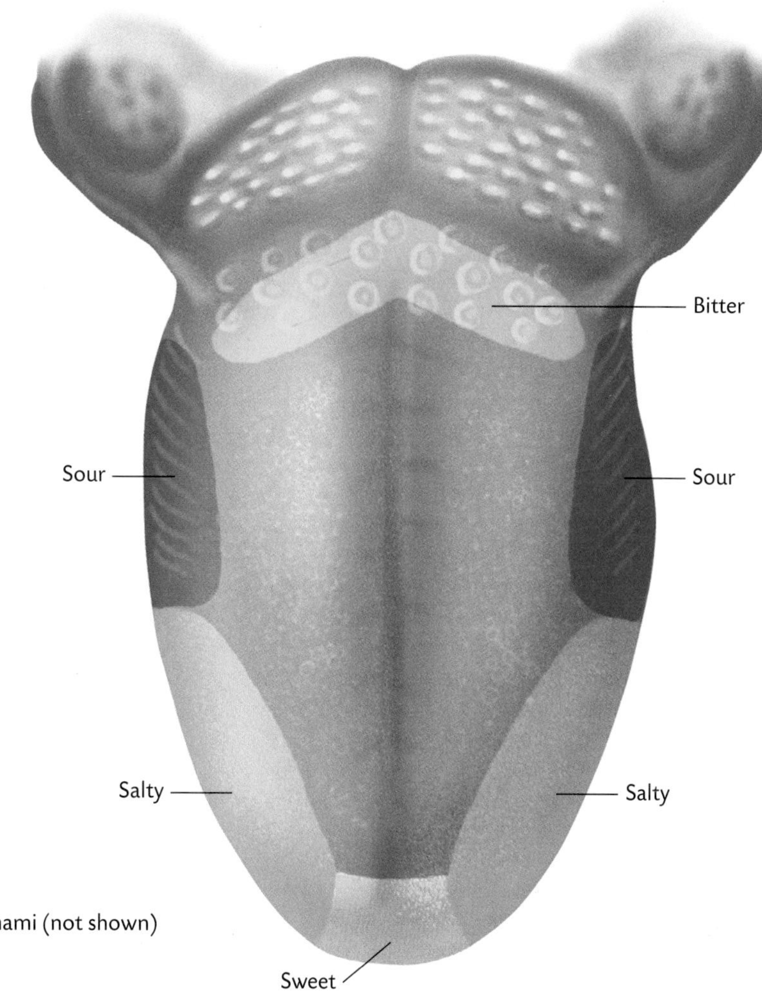

While found all over the tongue, taste receptors are concentrated in different regions.

FAST FACT A cold beer's low temperature dulls your taste buds, making it harder to taste. As the beer warms up, your gustatory receptors register its actual flavor. Really good beer tastes just fine at or slightly below room temperature, when your taste buds function normally.

TOUCH

REGISTERING A MULTITUDE OF SENSATIONS

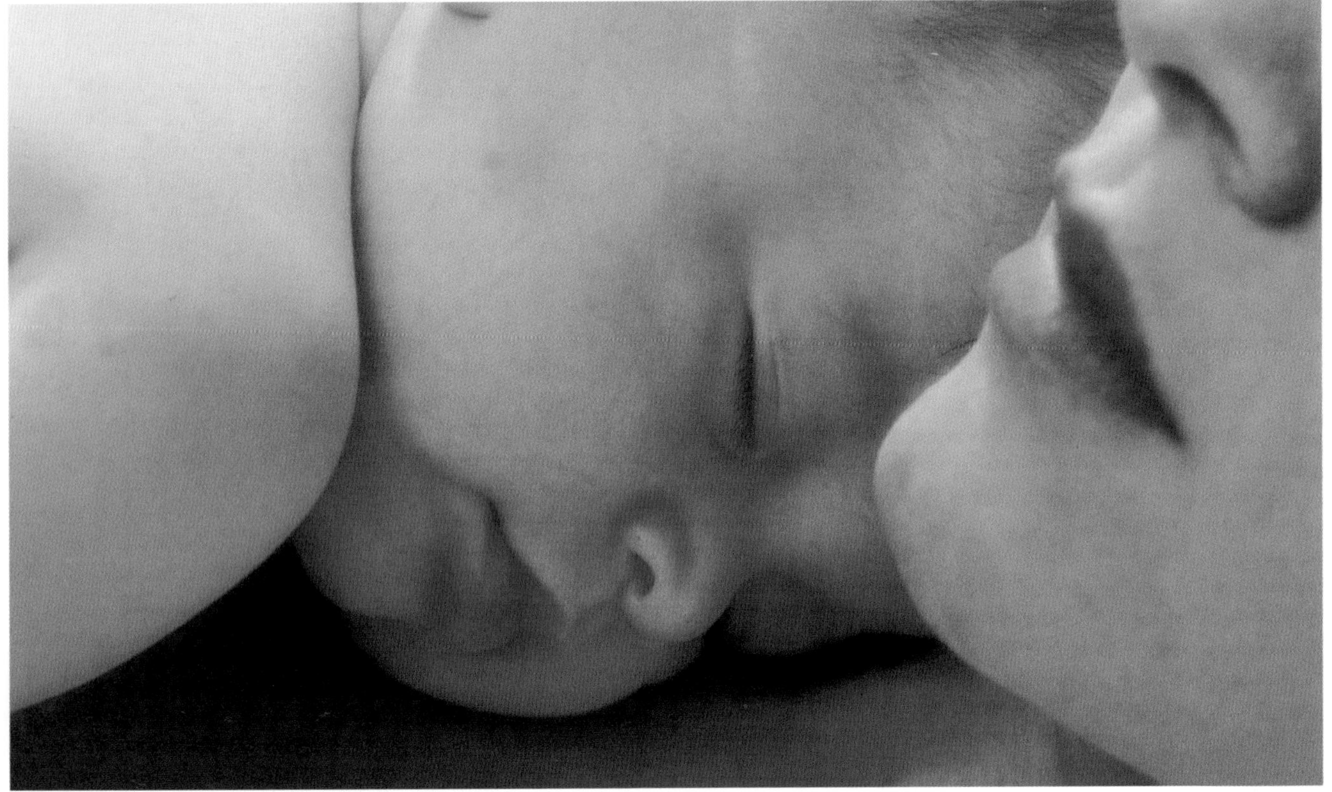

A mother holds her sleeping baby. Newborns require physical contact for proper development.

THINK THE SENSE of touch isn't as important as, say, vision or hearing? Think again. Among newborns, it can mean the difference between life and death.

Researchers believe human contact plays a crucial role in a child's development. Its lack can have serious consequences. "The easiest and quickest way to induce depression and alienation in an infant or child is not to touch it, hold it, or carry it on your body," said developmental psychologist James W. Prescott. He suggests that society's high levels of violence stem from insufficient mother-child bonding through touch.

A SOFT TOUCH

In a famous set of experiments in the 1960s, psychologist Harry Harlow tested infant rhesus monkeys. The monkeys had the choice of accepting one of two mothers. The first was draped in terry cloth but had no food; the second had a body of bare wire and a baby bottle filled with milk. The monkeys preferred to cling to the terry cloth mothers, which offered a reassuring contact, something apparently more important than food. Affection and emotional connection, given in small measure by the soft touch of the terry cloth, is crucial to the mother-child relationship, Harlow said.

He later went on to do studies that showed the complete psychological deterioration of monkeys that were kept in social isolation.

IMPORTANCE OF TOUCH

Unfortunately, similar experiments, whether planned or spontaneous, suggests the same is true for humans. In the 13th century,

CROSS REFERENCE: See "A New Brain," PAGE 72

FAST FACT

Haptic touch refers to manipulation of objects to sense tactile qualities like texture and hardness. If you manipulate a penny and a dime in your pocket, you can tell them apart. But if the coins are on a tabletop and your eyes are closed, you won't learn enough to tell them apart.

Frederick II of the Holy Roman Empire set out to find the original language of humanity. He ordered that a group of newborns be raised in the absence of language. In addition, their nurses, who fed them, were forbidden to touch them. All of the babies died before they could talk. Similarly, children found in Romanian orphanages after the fall of communism there in 1989 suffered from lack of touch. They were packed into cribs and fed from bottles propped over their cribs. They had high stress levels and low mental development, acting only about half their age.

THE FIRST SENSE

Touch develops before all other senses. Babies begin experiencing the sense in the uterus and are born with their neural wiring for touch significantly further developed than networks for vision and hearing. Rooting, a reflex that causes a baby to turn its head when its face or cheek is touched, helps locate the mother's nipple when it's feeding time. Infants also reflexively grip a finger that touches their hands. As a newborn reaches out to touch its environment, it develops its cerebral cortex. Touch literally relates to the initial stages of intelligence.

SENSATIONS

Touch receptors feel a variety of sensations. Pressure, heat, vibration, pain, et cetera, register on specialized receptors in the skin and organs. Receptors are unevenly scattered all over the skin. One of the sparsest concentrations lies in the middle of the back, while the highest concentrations exist in the fingertips, followed by the face. That's why you gain the most tactile information by running your fingertips over a surface, and why a kiss sets off a multitude of neural fireworks. The brain allocates space for tactile analysis based not on the size of a body part, but on how many receptors it contains. Thus, the number of neurons devoted to analyzing physical sensations against the skin of the face, a relatively small area, is larger than the number allocated for several other body parts combined.

Receptors in the skin send volumes of information to the central nervous system, to the benefit of the brain and body. Imagine if it were not so. If you walked barefoot onto the broken glass of a pop bottle or across the fiery sands of the desert, you would have no sensation of pain or burns. Your mind

In Harry Harlow's experiment, infant monkeys prefer a cloth mother to a wire one.

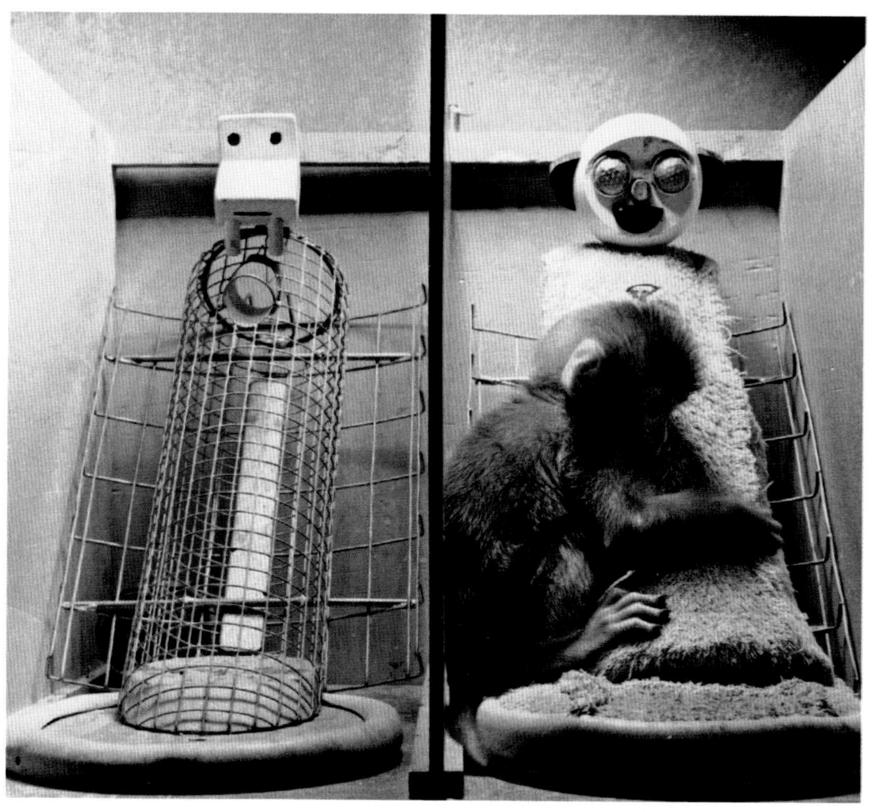

REDUCING STRESS

STRESS HURTS the body. Stress hurts the brain. Stress can even kill. If you think that's the bad news, wait for the kicker: Stress levels keep rising.

The faster pace of life in the 21st century, the constant sound and fury of the information age, and the lack of slow, quiet time keep raising stress to new heights. About two-thirds of Americans surveyed say their stress levels are too high, and further report that they've got more stress than a few years ago. And it's not just adults. The suicide rate of U.S. adolescents has increased fourfold since 1950.

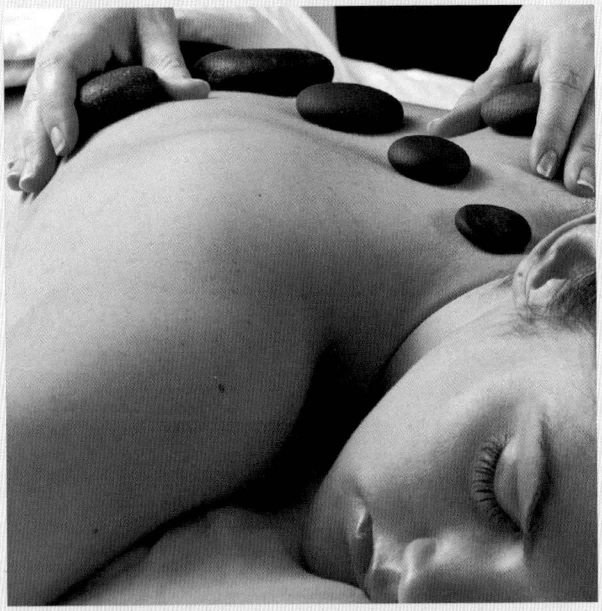

STRESSFUL SYMPTOMS

Symptoms of too much stress include a constant sense of being mentally tired, a lack of focus, difficulty concentrating, and trouble making decisions. Stress increases the number of neurons in the hypothalamus that produce a hormone known as corticotropin releasing factor, or CRF. That hormone stimulates the release of another hormone, called ACTH, that releases a group of stress hormones called glucocorticoids from the adrenal glands. Glucocorticoids perform an essential role when they help mobilize the body for fight or flight in an emergency. Too much of them, however, destroys neurons by greatly reducing their ability to store energy.

One specific site that is affected by the die-off is the hippocampus, which initiates the encoding of memories. In addition, chronic stress can trigger a chain of biochemical events that deposit fat in the coronary arteries.

Stress is a part of the fabric of life. A scale of the most stressful events, developed by two doctors, Thomas Holmes and Richard Rahe, includes joyful events like marriage and gaining a new family member, such as the birth of a child, on the list of top ten stressors. Not surprisingly, the death of the people closest to us ranks as the most stressful event, followed by divorce and marital separation. Death of a wife or husband and the start of a serious bout of depression have been linked with higher rates of cancer.

Spa treatments, such as massage or the application of hot stones to the back, can help relieve dangerous stress.

CONTROL ISSUES

Stress also relates to social status and control. Among societies of baboons on the Serengeti in East Africa, the animals with the highest social rank have the least stress and the lowest levels of glucocorticoids. The primates at the lowest levels of the social ladder suffer from chronic stress and its attendant maladies: high blood pressure, increased cholesterol levels, and weaker immune systems. Moving to the jungles of America's biggest corporations, a 1974 study of mortality rates by the Metropolitan Life Insurance Company found that men in the top rung of executive positions in Fortune 500 companies had substantially lower mortality rates than their peer group. One key reason for the difference appears to be the degree of control people exercise over their life. Both animals and humans are wired to prefer to choose when they experience stress and to have some meaningful response to it. Evidence comes from an experiment at the University of Colorado. Two groups of rats received electric shocks. One group could avoid the shocks by rotating a wheel. The other group had no recourse but to suffer through them. Without even the slimmest sense of control over the shocks, the latter group entered a state of "learned helplessness," resulting in a noticeably weaker immune system. Among humans, who have stresses the rats and baboons never dreamed of, the same holds true. No action creates more stress than facing a physically or emotionally painful experience without any measure of control over it.

POWER OF RELAXATION

Now the good news. Stress can be managed, within limits. The body and brain can be induced to relax and lessen the impact of stress by techniques that lower blood pressure and slow the heartbeat. Oxygen use declines, as does the expulsion of carbon dioxide. Muscles relax and the brain enters a calm state. Religious thinkers and meditation practitioners have known about the benefits of relaxation techniques for centuries. Witness the Zen practice of breath control as well as the 14th-century Christian practice of eliminating physical activity and blocking out distracting thoughts to approach a closer spiritual state.

You don't have to be a religious mystic to practice relaxation. If you can change the stressful situation, do so. If you can't, then change your attitude, and do whatever you can to compensate.

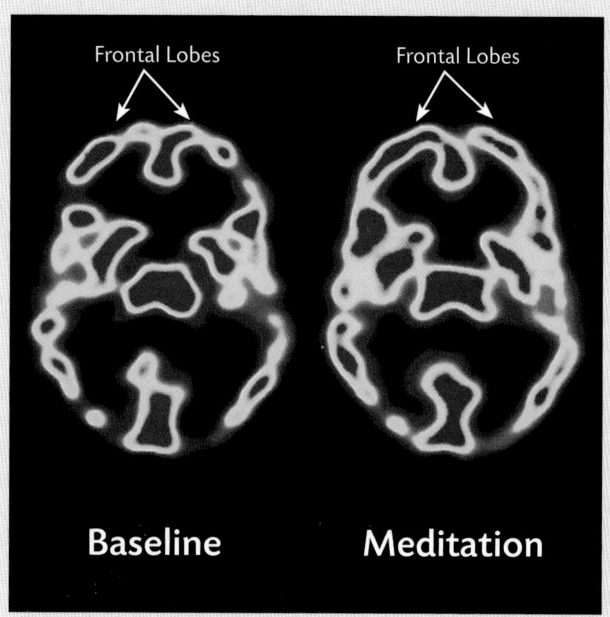

During meditation, the front part of the brain, which focuses attention, is more active (increased red activity).

STRESS FIGHTERS

To fight back when stress sets in, there a number of things to try. First, take control of your breathing. Shift from being a "chest breather" to a "stomach breather," like a newborn baby. Try lying on your back and putting a book on your stomach. With your stomach muscles relaxed, inhale deeply to make the book rise. Such deep breathing gets air into all parts of the lungs and maximizes the uptake of oxygen. Try inhaling slowly for five seconds and exhaling slowly for the next five seconds.

Next, unclutter your mind. Then slow it down. Try reading a classic bit of literature while you simultaneously listen to a recorded version. You'll find it difficult to skip ahead. The slow pace of reading aloud will bring your mind in line with the storytelling cadences of a calmer century.

A touching tactic: Get a massage from someone who knows how to give one. Massage has been shown to reduce stress and anxiety in depressed children, cancer victims, and people trying to quit smoking. Its manipulation of muscles and soft tissues, resulting in a relaxation response, even boosts the immune system.

And last, think positively before stressful situations. For instance, if you encounter stress from public performances, such as speaking to a large group, avoid dwelling on potential negative consequences. They tend to become self-fulfilling prophecies. If you can replace the negative images and your anxiety with a focus on positive outcomes, you will have prepared the brain for optimal performance.

Motor and Sensory Areas

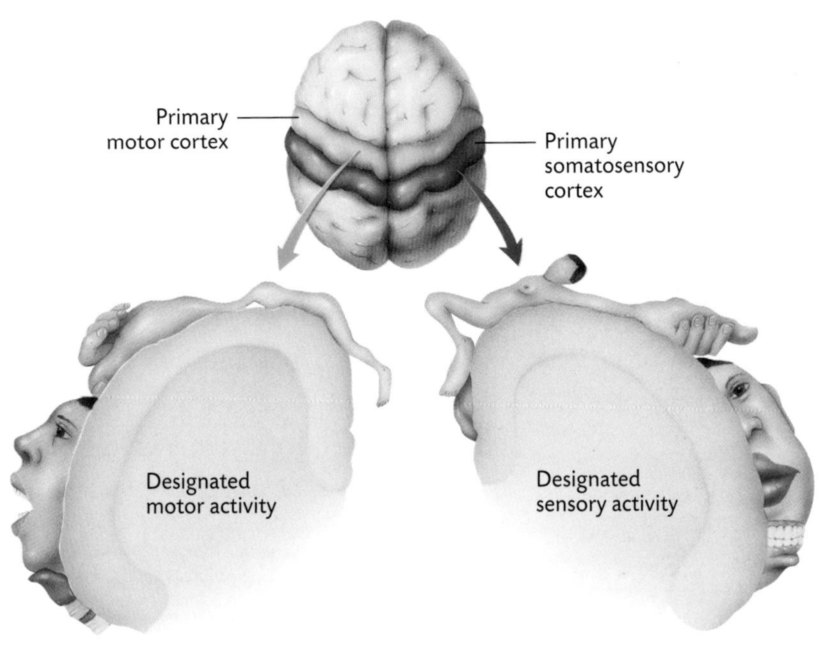

The primary motor cortex (blue) controls voluntary movement; the somatosensory cortex (purple) touch. The larger the representative body part on each, the more cortical tissue is involved in its control.

would have no information about the potential danger to the body from blood loss or infection, and the need to seek first aid. Your body also would have no way to sense a dangerous rise or drop in environmental temperature. When it gets cold, your blood loses heat. Receptor cells detect the loss and signal the brain to constrict the body's blood vessels to conserve warmth. Tiny muscles, surrounding the base of each hair, pull the hairs upright to trap an insulating layer of air next to the skin. A rise in temperature causes receptors to register the heat and signal for blood vessels in the skin to dilate in response. Either extreme can be deadly—as any high school chemistry student can tell you, heat speeds chemical reactions, which can kill cells, and cold slows such reactions and may halt cellular activity.

BENEFITS OF TOUCH

Given the importance of touch and the abundance of receptors in the skin, it's not surprising that touch can have therapeutic value for people of any age. Full-body massage has been shown to ease the symptoms of diabetes and hyperactivity and improve the immunity of HIV-positive patients. Massage can relieve the pain of migraines, help asthmatics breathe, and increase the mental focus of children with attention deficit disorder. The rubbing action of massage stimulates neural networks that cause the brain to lower the levels of the stress-related hormones cortisol and epinephrine.

Touch also communicates at a basic level, more profoundly than words. A gentle caress says I love you better than words.

COMPUTERS AND TOUCH

Touch is also being used in amazing new technologies. For instance, imagine a delicate operation being performed on a patient hospitalized in southern Africa. Now imagine the surgeon's scalpel being manipulated from a hospital in the United States.

It's not a far-fetched idea. In 2008, Ralph Hollis of the Robotics Institute at Carnegie Mellon

+ SELF-TICKLING +

EVEN IF YOU'RE the most ticklish person in the world, it is impossible to tickle yourself. Your brain keeps track of your body's motions through proprioceptors and anticipates the resulting sensations. That's why you can't do it: You know where you're wiggling your fingers and what it will feel like. The cerebellum distinguishes your fingers from someone else's but can't predict the outsider's actions, which leads to the tickling sensation.

University in Pittsburgh demonstrated a computer interface that allows users to experience a highly sophisticated sense of touch when manipulating objects in a virtual, three-dimensional world. The texture, shape, movement, and hardness of the object register on the hands of the user, making possible a computer-to-computer connection that could extend a person's tactile sensations to the other side of the planet.

FAST FACT In social situations, the French touch each other about 200 times in a half hour. Americans do so twice.

Hollis's so-called haptic interface—haptics being the science of touch—relies on the use of powerful magnets that cause a metal handle to levitate inside a device that looks like an upside-down umbrella connected to a computer. The user grasps the handle to manipulate objects on the screen of the computer.

Magnetic repulsion causes the handle to float freely. As the user moves the handle to control objects on the screen, the magnetic system creates tactile feedback, such as the resistance of a solid. While other virtual systems exist, the benefit of magnetic flotation lies in the sense of direct connection between the skin of the fingers and hands and the immediate feedback created via the handle.

ACUPUNCTURE

ACUPUNCTURE, the insertion of needles into the body for pain relief and therapeutic purposes, is an ancient technique that seems to work for many medical patients but has defied a clear scientific explanation. It relies on centuries of observation and refinement in Chinese medicine.

Although the practice has been documented for about 2,000 years, some historians believe it had its first crude flowering in the Stone Age. Early practitioners may have used stone knives and other sharp tools to puncture and drain abscesses. These tools were known by the character *Bian,* and the modern term *Bi,* representing a painful disease, probably came from the use of Bian stones in ancient medicine. Needles of stone and pottery, and later of metal, replaced sharp stone tools, giving rise to the modern practice of slipping thin needles into highly specific points on the body to bring relief from a variety of complaints. The Chinese refer to this practice as *Chen,* which means "to prick with a needle." The term acupuncture was coined by Willem Ten Rhyne, a Dutch physician who visited Japan in the 17th century.

Principal acupuncture points, called Xue, *lie along lines of vital energy.*

Chinese philosophy holds that acupuncture acts on the chi, a form of bodily energy. Needles supposedly block or rebalance the flow of energy when they are inserted at key points along lines called meridians. Western science cannot document the existence of chi, and consequently has difficulty explaining acupuncture's success rate. Perhaps, for some patients it has a placebo effect–the brain expects the body to feel better, and the expectation is enough to bring relief.

The practice first received extensive news coverage in the West after *New York Times* reporter James Reston, preparing for President Richard Nixon's groundbreaking trip to China in 1971, wrote about doctors successfully using acupuncture to treat his severe abdominal pain. California licensed its first acupuncturists as primary care providers in 1978. Today, thousands of American physicians practice acupuncture.

INTEGRATION BRINGING THE SENSES TOGETHER

Drawing on a multitude of senses—vision, touch, and balance—a gymnast executes a handspring on the balance beam.

THE SENSES all work together to create a complete rendering of the world around the body. Each one provides its own dimension, allowing the brain to collect and interpret the most information it can about a situation, a place, and a time. As discrete as each of the senses is, they all rely on common neural networks and receptors to sense and respond to a whole variety of stimuli.

SENSORY RECEPTORS

Vision, smell, hearing, taste, and the touch of skin rely on a class of sensory neurons called exteroceptors. That's a fancy way of saying they react to events happening outside the body. These receptors register touch, pain, pressure, and temperature in the skin, as well as taste, vision, smell, and hearing.

A second set of neurons are interoceptors, which lie inside the body and react to stimuli such as chemical changes in the blood, hunger, thirst, and the stretching of tissues. A third set, called

BREAKTHROUGH

EXPOSED TO new situations, the brain works to integrate them with experience. New sensations expand the brain's associated neural networks. That's plasticity: Neural connections spring up from interaction with touch, sight, sound, smell, and taste. How this occurs was suggested in the 1980s by neuroscientist Michael Merzenich at the University of California, San Francisco. Images he obtained of brains processing new information provided a window onto the physical changes of plasticity. Merzenich put electrodes into the cerebral cortex of six squirrel monkeys around the neural networks associated with fingers. He then put a cup containing a banana-flavored food pellet in a cup outside each monkey's cage. When the monkeys mastered retrieving the pellet from the cup, he made the cup narrower. Merzenich went through four cups, each narrower than the one

CROSS REFERENCE: See "Nerve Cells," PAGE 10

proprioceptors (from *propria,* Latin for "one's own"), react to internal stimuli but are confined to skeletal muscles, tendons, ligaments, joints, and possibly the equilibrium region of the inner ear. They register sensations that the brain interprets as location and movement of the body.

Without proprioceptors, it would be impossible to maintain balance while walking or to smash an overhead volley in tennis without falling over. Diseases that affect the proprioceptors and their associated regions of the brain can create the impression of being cut off from one's own body. Patients so affected report the odd symptom of looking at their hands and feet and not feeling as if they belong to their body.

FAST FACT Classical Indian philosophy holds that all human senses emerged from a single unity.

WORKING TOGETHER

While neural networks are attuned toward registering perceptions of light, sound, and so on, they often work together. Perception functions particularly well when objects are appreciated by a variety of senses. A person's walk, for example, registers in the brain as visual stimulation as you watch the moving body, as well as auditory stimulation from the clacking of the walker's heels.

An examination of a handful of coins registers not only as the sight of the images on "heads" and "tails," but also as the tactile stimulation of the hard, round, and smooth or ridged edge. A rich sensual image emerges when many senses combine to provide an appreciation of art and beauty.

+ TYPES OF NEURAL RECEPTORS +

TYPE OF RECEPTOR	FUNCTION
Exteroceptors	Sensory neurons that register external sensations like touch, pain, pressure, and temperature
Interoceptors	React to internal changes, such as changes in blood chemistry, hunger, and thirst
Proprioceptors	Register the sensations associated with movement and location. Helps maintain balance

REWIRING THE SENSES

Sensory stimuli are associated with specific regions of the brain, but these areas apparently are not predetermined genetically. This was demonstrated by a series of classic experiments on plasticity between 1990 and 2000. Neurophysiologist Mriganka Sur of the Massachusetts Institute of Technology took newborn ferrets and surgically rewired their brain.

Sur routed the ferrets' visual impulses to the regions of the brain normally associated with auditory processing. The ferrets soon began *seeing* the world with brain tissue normally used for hearing the world. The new wiring wasn't a

BREAKTHROUGH

before, until the monkeys easily fished the pellet out of the final cup. As the monkeys advanced from cup to cup, scans showed neural networks for their fingers had expanded significantly. But after the monkeys mastered the final cup, the networks shrank. Merzenich concluded that by then, the monkeys' cerebral cortex no longer had to activate so many neurons to process the same old sensory information from eyes and fingers. Instead, the information, once learned, moved to lower portions of the brain that control practiced motor skills. That freed space in the upper brain to process new things.

+ KELLER & PLASTICITY +

BEFORE SHE LEARNED to speak, Helen Keller (1880-1968) lost all vision and hearing to "brain fever." When she was six, teacher Anne Sullivan used her fingers to form signs in the girl's hands. One day at the pump, Sullivan spelled *w-a-t-e-r*, and Helen's brain made the connection. Keller learned to read Braille and to speak. Deprived of two senses, her brain compensated for the loss through touch. Plasticity let her learn language without visual or auditory input.

perfect substitution; the ferrets lost some of their visual acuity, seeing the world with perhaps 20/60 vision instead of 20/20. Nevertheless, the experiment raises the intriguing idea that blindness at birth, while the brain still retains the ability to respond to visual stimuli, could be corrected by surgery to reroute visual signals to healthy neural networks in regions other than those associated with sight. Sur told the *New York Times* his research team is exploring ways that neurochemistry might rewire both developing and mature human brains. Creating detours around damaged neurons would open new avenues for treating strokes and other injuries.

COMMON SENSE

The ancient Greeks observed that humans have many senses yet seem to possess an undivided experience of the world. They asked, how can the experience of unity arise out of separate entities?

Ancient Greek philosopher Aristotle argued for the existence of an unseen body function that integrates sensations. He called it *sensus communis,* or common sense, which perceives and integrates the common elements of sight, sound, and the other senses. Aristotle never got to probe for the physical site of this common sense, and anyway would have ignored the brain as a likely candidate. Yet his ideas had enough staying power to influence the thoughts of Thomas Aquinas, a 13th-century Italian theologian.

Philosopher and theologian Thomas Aquinas in the 13th century posited the existence of three chambers in the brain processing external stimuli. According to Aquinas, the first chamber contained the common sense, which integrated the information gathered by the five senses. The second chamber housed the faculties of reason, cognition, and judgment. It acted with the third chamber, which contained memories formed from prior perceptions, to evaluate information.

The notion of a common sense remained fixed in human understanding across time even though it could not be confirmed as a fixed region of the brain. Today, some neurologists suggest the limbic system as a candidate for integration of the senses because nerve fibers from exteroceptors converge there.

THEORIES OF PERCEPTION

Early in the 20th century, German psychologists proposed a gestalt theory of human perception. They argued that when you look at, say, the seashore through a window, you don't build up your perception based on thousands of bits of color, sound, and other sensations. Instead, you see it all at once and identify it as beach, sky, and ocean. Only when your attention zooms in on detail—what color is that bit of cloud?—do you see the fragments that form the brain's mosaic of external reality.

At mid-century, French philosopher Maurice Merleau-Ponty noted variance in gestalt perceptions. He

FAST FACT

A 1990s study mapped the brain regions blind people use in reading Braille and found that not only did feeling letters' raised dots activate tactile networks, it also activated the visual cortex. Those neurons, normally used for processing visual stimuli, got recruited to help decipher the shapes of raised letters on a page.

traced differences in gestalt perception to the individuality of the human body and brain.

Each body has its unique strengths and weaknesses in decoding information that is gathered by the senses, Merleau-Ponty said. The stream of impressions that flood the senses exist at a variety of levels, including some that are experienced below the level of consciousness. Only some sensations rise to the level of awareness, he said, and just as bodies differ, so do sensation-based perceptions that arise in the mind.

In the brain, neural networks act together to integrate sensations. Many networks actually do double duty, reacting with a primary response to one sense and a secondary response to another. Among cats, for example, cells in the superior colliculus respond to both sights and sounds. And in the human brain, the Aristotelian idea of some common threads among the senses is gaining new adherents. Yale University psychologist Lawrence Marks argues that all senses evolved from the tactile function of the skin, and still retain connections to it. He suggests, in a view Aristotle would have shared, that stimuli such as "brightness" exist across the senses.

CONJOINED SENSES

The human brain appears to naturally associate various senses. Sights and sounds, for example, often get paired. We think of sad music as being blue, while fast and furious high notes played by brass instruments in a major chord strike us as red-hot. The latter observation was noted by 17th-century British philosopher John Locke, who wrote

Proprioception forms the body's sense of its location in space. Lacking this sense, like wandering in a maze, proves unsettling.

FAST FACT When test subjects were shown curvy and jagged shapes and asked to name one *bouba* and one *kiki,* they almost always matched curves with the former and jagged with the latter. Psychologists see evidence of a common, mild synesthesia: Round sounds equal round shapes.

of a blind man who described scarlet as being like the sound of a trumpet. Today, in rare cases, neuroscientists recognize a condition known as synesthesia, in which the senses are conjoined. People who have this condition describe seeing colors when hearing music, and vice versa, as well as other sensual combinations.

In a sense, we're all born synesthetes. Sights, smells, sounds, and tastes are all ingredients in our nascent sensory soup, though we're unable to differentiate between them at first. Most are able to sort out the differences by three months of age. But for synesthetes, it seems, two or more normally independent areas of the sensory cortex remain cross-wired. Science has yet to fully explain why.

A conservative estimate puts the number of synesthetes at 1 in 20,000. However, since most synesthetes are unaware of their "condition," the number could be much higher. A recent test of 1,700 subjects suggests as many as 1 in 23 people has some synesthetic ability.

SIGNS OF SYNESTHESIA

Startling evidence of the inborn cross-wiring of senses emerged in 2008, when psychologists induced synesthesia among volunteers. Suddenly the test subjects could see numbers as colors. Neuroscientists believe newborns have multiple connections among the brain regions associated with the senses. One explanation for the rarity of these connections among adults is that most people grow out of them. However, synesthetes, in an unknown process, keep the connections active or grow new ones.

Another theory holds that synesthetic connections atrophy with age but can be prompted to reemerge. The experimental emergence of synesthesia through hypnosis supports the latter theory, although there remains a question about whether induced synesthesia is the same as the naturally occurring kind. "The fact that they induced it so quickly means that the brain's not sprouting new neurons or making new connections," said Marks, who studies synesthesia but was not involved in the 2008 research. "Maybe the connectivity always exists."

In the experiment, researchers Roi Kadosh of University College London and Luis Fuentes of Spain's University of Murcia hypnotized three women and one man, then instructed them to recognize the number one as red, two as yellow, et cetera.

Under hypnosis, the volunteers had trouble picking out numbers printed in black ink against a background that matched its suggested color. For example, they could not find the number one if printed against a red background because they saw the black numeral as red. Upon being released from hypnosis, however, the volunteers lost their synesthetic abilities. The researchers suggested that hypnosis broke through the barriers that segregate sensory regions.

If synesthesia has roots in brain development, who's to say it's a "wrong" perception? Perhaps the richness of integrated sensations compares with ordinary sensations as color vision compares with color blindness. "If you define it in a very basic sense as something beyond the ordinary that will light you up when you feel less than lit up, then that's what synesthesia is," said Marks. "And if I were to design the world, I'd give it to everybody."

+ COLOR OF MUSIC +

TO SOME SYNESTHETES, musical notes and letters have colors. Contemporary composer Michael Torke can't imagine music without color. The key of G major, for instance, appears bright yellow, while its G minor is a toned-down ochre. Russian writer Vladimir Nabokov saw the letter *q* as "browner than *k,* while *s* is not the light blue of *c,* but a curious mixture of azure and mother-of-pearl."

GLOSSARY

AMUSIA. Neurodegenerative condition affecting the ability to understand or express music. Often occurs with aphasia.

APHASIA. Neurodegenerative condition affecting the ability to understand or express language, generally as a result of stroke or similar brain trauma.

APRAXIA. Disorder of the nervous system in which an individual is unable to preform learned coordinated muscle movements although both muscles and senses work properly.

COCHLEAR HAIR CELLS. Hearing receptors located in the inner ear that allow the processing of sound. They bend in response to vibrations entering the ear and transmit signals to the brain stem via the audatory nerve.

CONES. Photoreceptive cells on the retina that provide color vision. They respond to the wavelengths of red, blue, and green.

EPITHELIAL CELLS. Cells that form the taste buds. Consist of supporting, basal, and taste receptor cells.

EXTEROCEPTORS. Sensory neurons that respond to external stimuli and are the basis for the five senses.

FOVEA. Center of the retina. This region contains the highest concentration of photoreceptors and is the area of highest visual acuity.

GLOMERULI. Cells located in the olfactory bulb that respond to particular odors.

HAPTIC TOUCH. The physical manipulation of objects in order to determine tactile qualities such as texture, hardness, and shape.

INTEROCEPTORS. Sensory neurons that register internal stimuli, such as chemical changes.

LATERAL GENICULATE. Located in the thalamus, this serves as a visual relay center for signals in transit to the visual cortex in the occipital lobes.

MITRAL CELLS. Neurons in the olfactory bulb that refine and amplify signals from glomeruli and relay the information to the olfactory tract.

OLFACTORY EPITHELIUM. A fluid-coated patch of sensory cells that is able to detect odor. It is located on the roof of the nasal cavity.

OTOLITHS. Crystals of calcium carbonate located in the otolithic membrane in the ear. Playing a vital role in balance, these structures both detect gravity and aid in the awareness of the head's spatial orientation.

PERCEPTION. The interpretation of the meaning of a stimulus.

PHEROMONES. Chemicals produced by insects and other animals that transmit messages to or affect the behavior of other individuals.

PROPRIOCEPTORS. Sensory neurons that are responsible for the sense of self, and the awareness of body position and movement.

RETINA. The sensory membrane that lines the interior of the eye. Composed of several layers, including one of rods and cones, it receives the image formed by the lens, converts it to signals, and transmits it to the optic nerve.

RODS. Photoreceptors that register dim light. Located on the retina, these allow vision at low-light levels.

ROOTING. A reflex that causes an infant to turn toward facial stimulation. Believed to facilitate feeding.

SENSATION. The brain's registration and awareness of a stimulus.

STIMULUS. A change in the environment that evokes a reaction from the brain.

SUPERIOR COLLICULUS. Located in the midbrain, this region adjusts the head and eyes to achieve maximal visual input.

SYNESTHESIA. A condition in which the stimulation of one sense is simultaneously percieved by another sense or senses.

TASTANTS. Chemicals that stimulate the sensory cells in taste buds.

TINNITUS. Hearing disorder characteried by chronic ringing or clicking in the ears in an otherwise silent environment.

TYMPANIC MEMBRANE. The eardrum, which conducts sound wave vibrations from the ear canal to the bones of the middle ear.

UMAMI. The fifth flavor, linked to foods containing glutamate and aspartate.

VESTIBULAR ORGAN. Nonauditory portion of the inner ear. Responsible for the body's detection of the spatial orientation and movement of the head and for maintaining balance and posture.

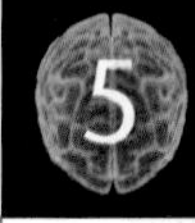

CHAPTER FIVE

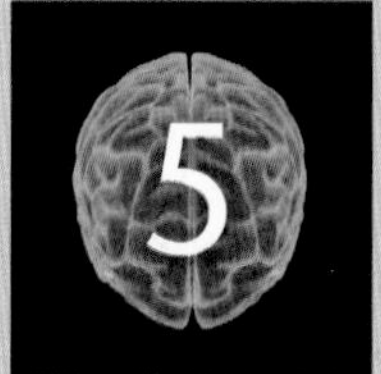

MOTION

NERVES CARRY out instructions to move the human body, making every sneeze or martial arts attack an extension of the brain. The brain rehearses many of the motions performed every day before signaling their execution. It learns through repetition to do many of them without your having to think about them at all. Researchers are learning that movement profoundly affects memory, learning, and emotion, underscoring the connections that bridge the gap between mind and body.

A jumping side kick executes a sequence of precise movements charted by the brain.

BRAIN IN ACTION

THE HUMAN BODY enjoys an astonishing variety of movements. The tapping of a finger, puckering of lips for a kiss, beating of a heart, and striding of a brisk walk are all carried out by muscles responding to instructions from the brain.

On the surface, the study of motion seems so simple: The brain sends out the appropriate stimulus through its network of nerves. And voilà, the muscle moves.

But motion has proved to be far more complicated than that. It may be voluntary or involuntary, or some mix of the two. It may involve skeletal muscles activated by the central nervous system, or, in the case of peristalsis (the rhythmic contractions that push food through the digestive tract), it may automatically call into play the smooth muscles of internal organs. It may occur consciously, take place at a level below awareness, or even, in the case of a reflex, not involve the brain at all. It may be executed through a hardwired set of instructions available at birth, or learned and refined after much practice.

It contributes vitally to learning, emotion, and memory. It helps bridge the gap between thoughts and actions. In short, as an extension of the brain, the body's movements contribute to the definition of who we are.

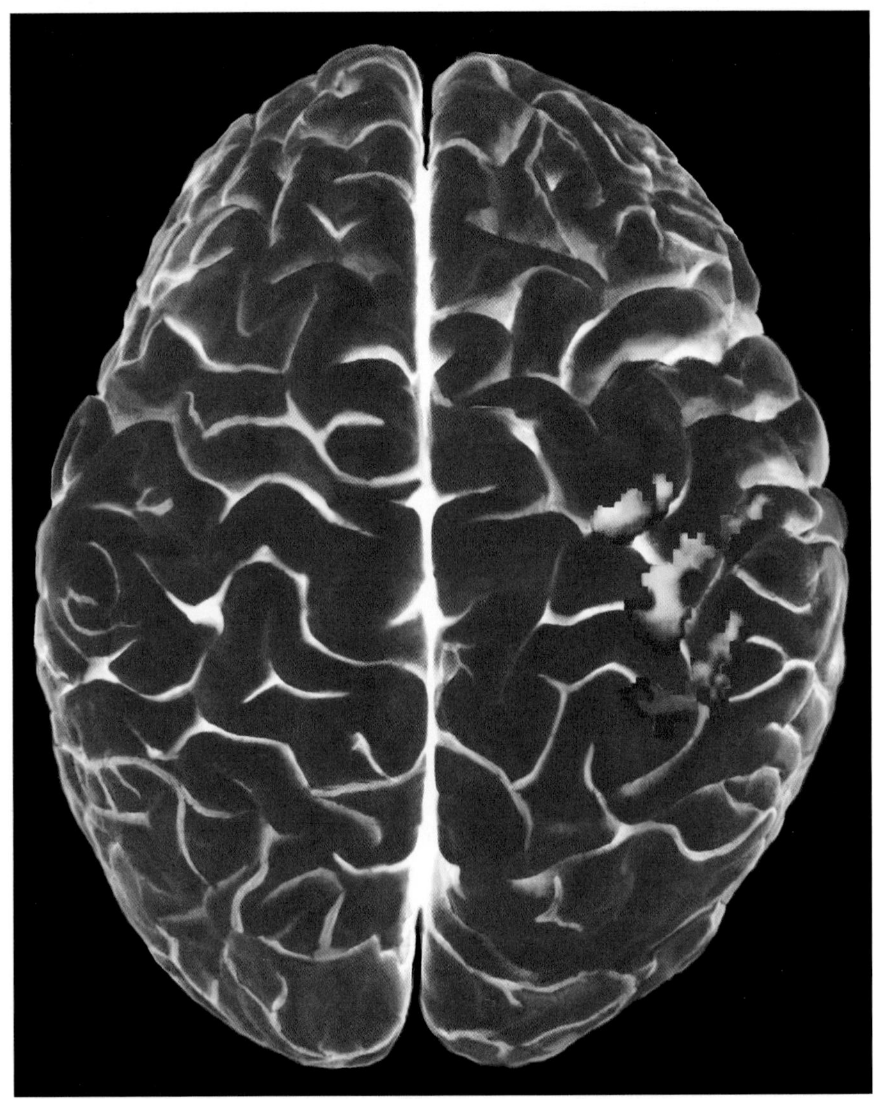

fMRI of the brain (frontal lobes, top) shows right hemisphere regions that move the left hand.

Nearly all motions of the human body occur because of the nerve-induced muscle contractions. Someone a long time ago observed that contractions beneath the skin look something like the scurrying of mice, and named the contracting tissues muscles, from the Latin for "little mouse." Muscles come in three types: skeletal, cardiac, and smooth. Skeletal muscles, which attach to the bones of the skeleton, are the only ones that respond to the dictates of the conscious mind, although they may also act involuntarily. When the brain tells the

CROSS REFERENCE: See "Messengers," PAGE 42

FAST FACT Neurologist Richard Restak says, "All proposed divisions within the brain are highly artificial and are based on our need to separate things into neat, easily understandable units. We must always remember that the brain functions as a whole."

fingers to type, it is sending electrochemical commands through networks of neurons to contract skeletal muscles in the shoulders, arms, wrists, and fingers. Cardiac muscles pump blood through the heart and into the network of blood vessels. And smooth muscle tissue forces body fluids and other substances through internal organs.

All such motion requires events in sequence. The brain orders one action after another. It gathers sensory feedback from motions in progress to refine them and prepare for others to follow. What may be surprising, however, is that the sequencing of thoughts also involves portions of the brain devoted to the sequencing of actions. As you call up memories, make plans, or get teary-eyed from a sad movie, your brain works in sequences. Neural circuits fire, one after another. The process moves electrical signals and neurotransmitters throughout the brain—and when the brain ponders action, those firing patterns appear similar to ones for execution of actual movements.

Even when the body seems at rest, it is always in motion. From the expansion of the lungs to the contractions of the intestinal tract to the electrochemical impulses that ping among trillions of synapses, the brain monitors and controls everything in its restless empire of the body.

MOVEMENT IN THE MIND

Think about the layout of your house. Create in your mind an image of the front door. As you step from your front porch and move through the door frame, what do you see? There may be a stairwell to one side, perhaps a living room with carpet and a fireplace surrounded by bookshelves. Off to one end is a hallway that leads to a kitchen or a bedroom. Step into that bedroom in your mind's eye and envision the details of the blanket on the bed, the books on the nightstand, the drawer where you put your glasses, and the peg in the closet where you hang up your pants. Now, figure how many shoes you have under the bed.

As you performed this mental task, you probably didn't move any muscles other than to scan your eyes across the page before you. But your brain "moved." Merely thinking about moving from room to room, as well as zooming in on details such as the placement of shoes or furniture in your bedroom, activated both cognitive and motor regions in your brain.

Your brain made mental images of your passing from room to room, and perhaps kneeling to pull back the covers and count the two sneakers and two loafers you have under the bed. Synapses fired in your occipital, parietal, and frontal lobes, as well as the cerebellum, just as they would have if you had actually gotten up to make the trip. To get the answer to the question in

+ BETA BLOCKERS +

YOU SIT AT THE PIANO for a recital in front of hundreds of people. Your hands shake and sweat, and you feel tightness in your chest—the classic symptoms of stage fright. Until recent decades, you would have had to cope as best you could. Now, you can take beta blockers. These drugs block the receptor sites for adrenaline, which the brain calls forth in response to stress. Adrenaline still circulates in the blood, but it can't find enough open receptors to alter heartbeat and blood pressure. The result: You are still ready to play, but you don't have to contend with the physical symptoms of stress.

STAYING SHARP

OVER EONS of competition for scarce food, natural selection favored individuals who had the body fat to survive famine better than their skinny neighbors. Today, the brain still recognizes body weight as important to survival. It devotes about a dozen neurotransmitters to increasing it and about the same number to decreasing it. Between the two, homeostasis favors a particular "set point" of body weight for each individual.

If you try to lose weight by eating less, your brain counters with a variety of tricks to return you to that set point. Your brain lowers your metabolic rate while you rest to burn less energy, and it also releases chemical messengers to tell you to take in more calories–in other words, you feel hungrier when you diet. Fat cells help communicate metabolic information to the brain by releasing a chemical called leptin, which circulates in the blood. As the nervous system senses changes in leptin levels, the brain reacts by creating feelings of hunger or fullness.

So, how can you overcome your brain's defenses and lose weight? These changes will reset your set point:

✓Exercise daily to raise your everyday metabolic rate.

✓Lower the number of calories your body stores.

✓Eat lower calorie foods.

✓Consume smaller portions more often instead of a few big meals, which tends to pack on the fat.

✓Make a lifetime commitment to diet and exercise. If you relapse into old patterns, your brain will return to a homeostasis that best fit our ancestors' lifestyles by building emergency reserves of fat.

the first paragraph, "Four shoes," you had to use portions of your brain that oversee not just counting but also movement.

THINKING IS FOR DOING

For decades, researchers believed the motor cortex functioned as a sort of in-out processor for movement—it executed orders for movements and then adjusted them in response to sensory feedback.

But more than a century ago, psychologist William James had a radical idea. "Thinking is for doing," he wrote. In that one simple sentence, he packed a lot of information. He meant, first of all, that thinking about an action expands the likelihood of doing it, as any dieter could attest when thinking about making a raid on the refrigerator and eating the last piece of apple pie. But James also argued a then-incredible notion that thinking about a particular motion activates neural networks in the brain that also fire up when carrying out the motion. In other words, the border between thought and action is at best permeable, and perhaps a mere illusion.

In the 1990s, PET scans of human brains proved James right. Neurologists pinpointed a brain region called the anterior cingulate that gets activated both by thinking about a word or action and by saying the word or performing the action. Other regions have similar electrical firings for both doing and acting.

Now, continuing research indicates that motion plays a crucial role in a vast array of cognitive functions. These include language, memory, and learning. Even emotion—whose name contains a hint to the *motion* on which it depends—is emerging as a mental state partially dependent on the brain's processing of movement.

Just as the cerebral cortex at the top and front of the brain has been shown to integrate thought

Exercising, eating right, and maintaining a healthy weight are excellent for mind and body.

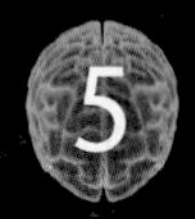

A teenager who follows a cake recipe is executing a series of motor functions, including analysis and prediction.

and action, so too does the cerebellum at its rear and base. It has long been recognized as a region that coordinates sequences of physical action, such as the motions necessary to maintain balance on a bicycle. However, it also appears to play a key role in the sequencing of thoughts. When you pictured moving from room to room in your home, your cerebellum activated to help form those memories in a certain order and allow you to take your virtual tour. Your behavior—your decision to act in a certain way, followed by the execution of those actions—is merely a sequence of motions both mental and physical determined by your brain.

FAST FACT Neurons need lots of energy. Fewer than one percent of them in the cortex can fire at any one time.

THINKING ABOUT COOKING

Take the decision to go cook a special dinner, for example. As you decide whether to cook it tonight, your brain processes facts ("There is a recipe here I've always wanted to try" and "I have all of the ingredients"), opinions ("I think I could do a good job"), thoughts ("This looks as if it will feed everyone for a couple of days"), memories ("My mother made something like this when I was a kid"), and predictions ("This will be a hit"). Your brain creates a plan for getting all of the ingredients on the counter and following the recipe, step by step. Every piece of that mental puzzle depended on motor functions—the weighing of options, the sequencing of events, the completion of the whole picture, and the predictions of a successful meal. The same sequence of neural networks begin to fire, in the same sequence, when you actually start to cook.

If this is your first time cooking the special meal, you might have had to actively think about the action with your frontal cortex. You would carefully read the recipe card, measure ingredients, and follow the directions as best you could. However, after you've made the dish many times, you would no longer have to think so much. You probably wouldn't even need to look at the recipe card, as your cooking skills would have become automatic. As you no longer sweat the details you've mastered, the skills involved in cooking get pushed from the frontal lobes, where decision-making takes place, to the lower regions of the brain where they are stored as automatic sequences.

Learned motor skills become unthinking habits. The first time you sit at a keyboard, you have to map out the location of each key. After months of typing, however, you no longer think, you just strike with your fingers. Your brain has relocated the motor sequences required for typing from the frontal cortex to the cerebellum, where they are recalled and executed upon demand. Furthermore, if you try to focus your active thoughts to locate the "R" key in your mind's eye, you may have trouble saying exactly where it is. But when you sit at the keyboard and start typing, your fingers fly straight to it—third row up, fourth from the left—thanks to your cerebellum's storage of that information for automatic retrieval.

PERFORMANCE

According to neuroscientist John Ratey, "Becoming a super athlete or piano player may require an efficient mechanism for the transfer and storage of these programs. A person who can push down

Kavya Shivashankar of Kansas moves her hands in the Scripps National Spelling Bee. Thought paired with motion enhances learning.

CROSS REFERENCE: See "A Memory Forms," PAGE 246

more and more intricate motor sequences [into the cerebellum, brainstem, and basal ganglia] can be engaged in complex motion and still have a quiet frontal cortex." Such a person could then devote more of the cortex to reacting to observations about the unfolding game or symphony.

If "muscle memory" handles all the basic needs, the star athlete and musician can concentrate on the extras—the anticipation of the next pass, or the interplay of instruments on a difficult stretch of notes—and elevate their performances far beyond the norm.

If you want to explore the intimate connections between motion and thought, take a walk. "I think better on my feet," says the person who paces a lot. That statement may literally be true. The motor activity of walking or jogging invigorates the body and brain, getting the blood flowing and ideas moving. Writers sometimes get their best ideas while hiking in woods, along streams, in mountains, or along the sidewalks of their neighborhood. The reason: parallel integration of brain functions. The cerebellum, the primary motor cortex in the cerebrum, and the midbrain work together not only to coordinate the movement of the body, but also the movement of one thought into another. Walking and running trigger patterns of firing in neural networks deep within the brain. It's likely that these firing patterns stimulate similar activity among patterns of complex thoughts. Creative ideas and answers to nagging questions sometimes walk into the conscious mind as the body takes itself out for exercise. If you have trouble with either mind or body, try working the other. Readers sometimes speak aloud when they encounter an unusual word for the first time, unaware that by doing so they activate multiple motor functions that aid cognition. In the 2006 movie *Akeelah and the Bee*, an 11-year-old girl from South Los Angeles becomes a spelling bee champion by learning to spell words to the rhythm of her jump rope. As Akeelah struggles for a moment in a major competition, she mimes jumping her rope and finds the letters emerging from the fog of her mind. The filmmakers based the ritual on repetitive motions such as kicking and stepping in half circles they observed among competitors in the National Spelling Bee.

+ DON'T CHOKE +

FOCUSING ATTENTION on potential bad outcomes during periods of stress—as in, "Don't miss this crucial field goal"—often brings about the very thing the thinker tried to avoid. Athletes "choke" because they fail to maintain concentration on positive performance and give in to the panicky fight or flight response, which causes physical and psychological impairment. Choking is all too natural, all too human, but little consolation to fans and teammates as Buffalo Bills kicker Scott Norwood discovered after missing a last-minute field goal that would have won Super Bowl XXV in 1991.

BRAIN REGIONS & MOVEMENT

Scientific observations into the brain's control over movement date to the 19th century. One of the first experiments raised many questions. During the early 1850s, England fell under the spell of spiritualism, including an occult fad called table tipping. Participants at a séance sat in a circle around a light table and rested their hands on top. Despite their promises not to move anything with their hands, the tables nevertheless rotated. Some observers

FAST FACT Young children are able to communicate in gestures and movements before they are able to speak words. For example, babies can shake their head "No" and wave "Bye-bye" before mastering the words and phrases themselves.

attributed the phenomenon to paranormal forces. Chemist and physicist Michael Faraday, inventor of the forerunner of the modern battery, set out to prove them wrong. He believed the séance participants unknowingly exerted force through their hands. He created devices to measure the application of lateral pressure and placed them between the table tippers' hands and the tabletop. The instruments didn't lie: The sitters' hands had pushed in one direction or another until the table began to spin even as they swore they pressed only straight down. Faraday labeled the phenomenon "a quasi-involuntary muscular action." What he had discovered was a divorce between willful intention to act ("don't shift the hands") and the unconscious execution of motion. The participants probably secretly wanted the table to turn and unconsciously acted on that desire. The experiment demonstrated what modern neuroscientists call the cognitive unconscious, the brain's processing of information at levels beyond awareness. Some actions—some movements—are voluntary, some are involuntary, and some occupy a gray area in between. Faraday's discovery didn't fit the age's understanding of free will, but it pointed toward future understanding of how patients with obsessive compulsive disorder or Huntington's disease can act and move despite their best intentions not to do so.

FAST FACT The brain works most efficiently when it focuses on one task at a time.

STUDIES IN MOTION

The first insights directly linking brain regions with movement arose in 1864. In that year, German doctor Gustav Theodor Fritsch treated the victims of the Prusso-Danish War. As he dressed a head wound, Fritsch touched one of the victim's cerebral hemispheres, which had been exposed by the injury. In response, the opposite side of the man's body twitched.

Fritsch shared his observations with a Berlin doctor named Eduard Hitzig, and together they decided to explore the phenomenon more thoroughly. They set up shop on the dressing table of Hitzig's wife and began electrically stimulating the cerebral hemispheres in dogs' brains. Dog or man, the result was the same: Stimulation to one hemisphere jerked the other side of the body. They hypothesized that movement of each side of the body is controlled by the opposite side of the brain.

Theirs was the first theory of cerebral localization for control of movement. They left refinement of the theory to their contemporary, English physician John Hughlings Jackson, who did groundbreaking work on the clinical description of epilepsy. He carefully observed the patterns of loss of muscle control as his beloved wife, Elizabeth, suffered a series of epileptic seizures that eventually led to her death. The seizures always progressed from one body part to another in a precise, repeated pattern. Jackson concluded that the pattern resulted from the electrical storm

WHAT CAN GO WRONG

MUSCLES take their marching orders from the brain. Nowhere is this more evident than in the range of neurological disorders known as cerebral palsy. The condition strikes newborns and young children and persists throughout life, getting neither worse nor better as they mature.

While cerebral palsy manifests itself as poor control or paralysis of voluntary muscle groups, its origin lies not in the muscles themselves, but in the portions of the brain that regulate movement. One of the most common causes of cerebral palsy is attributed to a temporary lack of oxygen to the brain during a

CROSS REFERENCE: See "Knowing Itself," PAGE 2

Séance participants circle a table in a 19th-century woodcut. Involuntary hand movements led to paranormal claims.

of the seizure moving from one brain region to another, stimulating a series of body parts along the way. That meant particular body parts were represented in the brain in regions now called the motor cortex. Today, the motor cortex is associated with Brodmann areas 4 and 6 (described in Chapter 2), with the primary motor region in area 4 of the precentral gyrus.

The conclusion rocked the neuroscientific world, which had considered the cortex the singular empire of cognition, and laid the groundwork for clinical neurophysiology. Jackson sniffed, "There seems to be an insuperable objection to the notion that the cerebral hemispheres are for movement. The reason, I suppose, is that the convolutions of the cortex are considered to be *not* for *movement* but for ideas."

Jackson argued that brain functions follow organizational lines. A proud man, he would have loved

WHAT CAN GO WRONG

difficult birth. Starved for oxygen, some brain cells die and cannot be replaced. Researchers are examining not only the mishaps of birth, but also investigating other possible causes, such as other genetic factors, epileptic seizures, circulation problems, and abnormal bleeding, as possible contributors to the disorder.

Affecting three in every thousand children, cerebral palsy ranks as the most common cause of childhood physical disability. Most have it from birth, although its presence may not be clear for months or even years. A few develop the disorder through later brain damage or infections.

Typical signs include lack of coordination, stiffness, exaggerated reflexes, a leg or foot that's dragged, or an unusual gait. Some also have difficulties with vision and speech. Half of cerebral palsy patients are mentally retarded, and half have seizures.

to have lived long enough to have seen solid scientific evidence of his pronouncements.

PENFIELD'S PROGRESS

That evidence came in the mid-20th century in the work and writings of an American neurosurgeon who immigrated to Canada, Wilder Penfield. He performed surgeries on the brain of patients who suffered epileptic seizures. As the brain contains no sensory neurons for pain, Penfield operated without anesthesia. Dulled only by local painkillers, his patients could listen to and answer his questions. As he electrically stimulated their brain to try to pinpoint troubled regions, he listened to their thoughts and memories and watched to see how their body moved.

He had his first aha! moment in 1934 when a female patient told him during her temporal lobe surgery that she felt as if she were reliving the moment she gave birth. Over the next two decades, Penfield acquired stacks of stories from patients connecting stimulation of particular brain regions with particular results, leading him to create the first maps of the motor cortex and associated areas. One patient heard orchestral music "when a point on the superior surface of the right temporal lobe was stimulated after removal of the anterior half of the lobe." The patient believed Penfield must

JOHN HUGHLINGS JACKSON

John Hughlings Jackson hit upon his theory of neural organization by watching his wife.

SELF-TAUGHT neurologist John Hughlings Jackson (1835–1911) owed his scientific success to his keen sense of observation and the woman he married. Jackson noted that his wife's disorder, now known as Jacksonian epilepsy, was characterized by seizing fits that seemed always to originate in the hand and crawled up the forearm, elbow, and shoulder to the face. The convulsions then marched down the torso before terminating at the leg on the same side.

Jackson hit upon the theory of motor cortex localization, radically suggesting that the brain's cerebral hemispheres are geographically arranged with regard to motion and not just to thought.

Each region of the brain, Jackson said, corresponds with an individual part of the body. Some regions, such as those that control hand and wrist, are as closely oriented as the body parts themselves.

The uncontrollable activation of neighboring cortical areas, as during a fit of apoplexy, explained how his wife's seizures appeared to travel throughout her body.

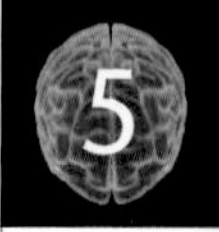

have had a record player in his operating room. When Penfield stimulated the same spot again, the patient heard the same musical piece, beginning on the same note. Others, he noted, twitched highly localized muscles on one side of the body when he stimulated the precentral gyrus on the other side of the brain.

FAST FACT Physical play lets young children develop social skills and master physical movements.

MAPPING MOTION

Inquiries into the mapping of motor functions led to the creation of a visual chart of the brain overlaid by the body parts they influenced. This so-called motor homunculus distorts the size and shape of the human body because body parts appear larger when they have more neural networks devoted to their control. Fingers and thumbs appear huge, in keeping with the fine motor control of eye-hand coordination. The torso and hips appear relatively tiny, as most people (other than, perhaps, ballet dancers) don't devote much energy to precise control of those body regions.

Maps that began with Penfield's are not the same as the pseudoscience of phrenology. The brain is much too complex and nonlinear to mark a spot with an *X* and say, "Here lies the site for 2 plus 2 equals 4." Updated charts of the brain divide it into areas of influence, with some functions more localized than others.

Overall, however, the mapmakers of this vast frontier realize that it operates as an integrated whole, and that movement is fundamental to its healthy functioning. An appreciation of the role of motion throughout the brain is an important key for ultimately understanding the most complex thing in the universe.

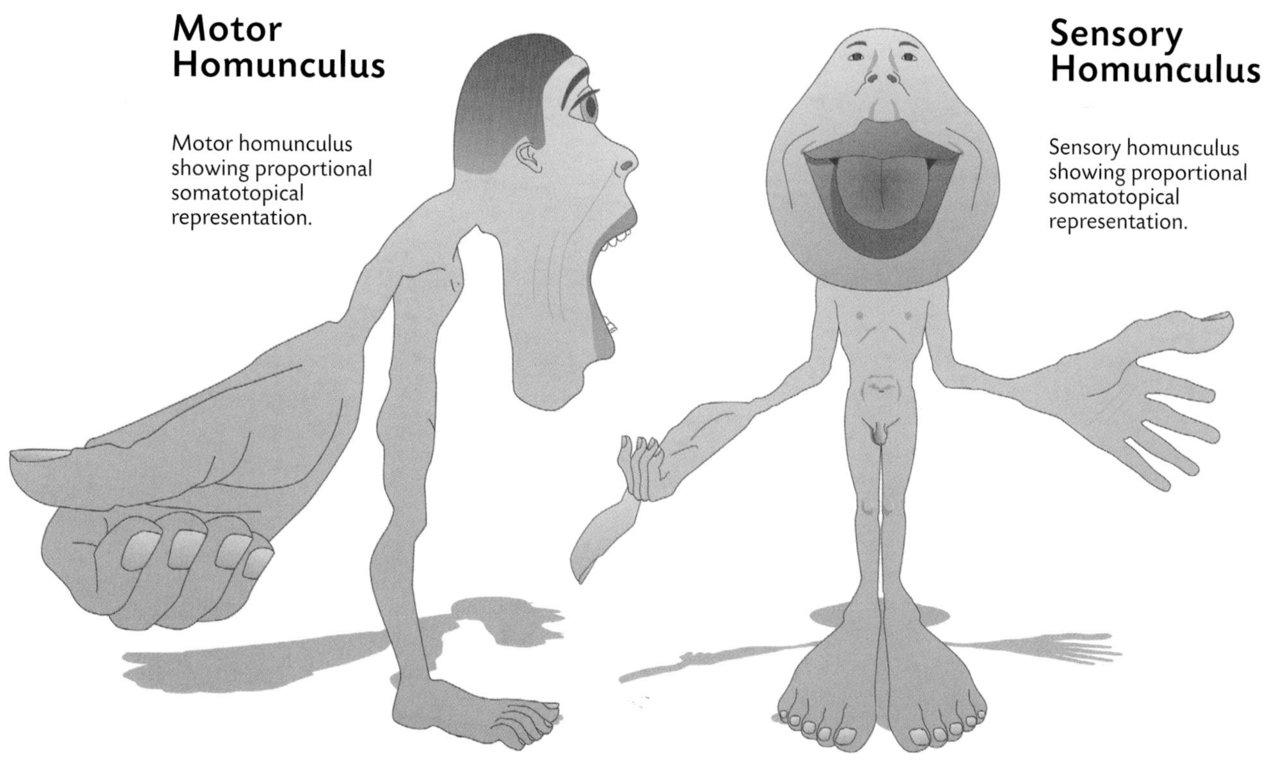

In motor and sensory homunculi, the larger the body part the more neural circuitry is dedicated to the area.

SILENT RUNNING

INVOLUNTARY MOVEMENTS

A swimmer underwater must not only will the arms and legs to move, but also suppress the natural movements of breathing.

THE HUMAN BODY carries out many motions without thinking—thank goodness. Imagine having to devote part of your constant consciousness to remember how to walk or drink a glass of water—not to mention how to breathe.

At the most basic level, things move inside your body through the action of the autonomic (involuntary) and somatic (voluntary) nervous systems. The autonomic nervous system regulates the body's internal state through webs of motor neurons that activate the heart muscle, organs, and glands. These internal tissues constantly

CROSS REFERENCE: See "Messengers," PAGE 42

WHAT CAN GO WRONG

MULTIPLE SCLEROSIS, a disease of the immune system, causes the body to eat away at the fatty coating of myelin that insulates nerves. As the protective sheath degenerates, the brain loses its ability to communicate efficiently with the body. The process is similar to the erosion of telephone wire insulation. When wires lose their protective sheaths, electrical messages sent along their length may get slowed, interrupted, or lost. In the human body, nerves that lose their coat start to deteriorate, and their myelin turns into hard, nonfunctioning lesions called scleroses.

Patients, usually diagnosed between the ages of 20 and 40, may lose their ability to walk or talk. They may suffer from visual impairments, weakness or clumsiness in their muscles, urinary incontinence, and, ultimately, paralysis.

The reasons for the onset of the disease are unclear, but genetics and

send information to the central nervous system, and as their name suggests, the autonomic nerves act on their own to adjust body activities to maintain homeostasis. Detection of a drop in temperature, for example, may cause the autonomic nervous system to speed the heartbeat and alter the diameter of blood vessels. Most of these actions take place beyond conscious awareness. Awareness occurs only when the autonomic action impinges on consciousness, as when you hear the pounding of your heartbeat after running or your painfully full bladder announces itself.

NETWORKS IN ACTION

Controlling these actions is spread throughout several regions of the brain, including the spinal cord, brain stem, hypothalamus, and cerebral cortex. The hypothalamus processes incoming stimuli and sends responses to the central nervous system. This tiny portion of the brain regulates the heart, blood pressure, the correct amount of water in the body's cells, endocrine activity, and body temperature, as well as playing a role in emotions and biological drives. Although the cerebral cortex usually is thought of as a source of consciousness, it also modifies the autonomic nervous system at a subconscious level, working through the limbic system. However, some conscious thoughts do alter the body through the autonomic nervous system. Consider, for example, how remembering a scary movie made your heart race or dwelling on the taste of your mother's homemade soup made your mouth water. Some studies have even demonstrated that test subjects can exert a measure of control over their heart rate and blood pressure through techniques such as biofeedback.

Involuntary and voluntary movements are controlled by different networks of the brain, even if they both connect to the same part of the body. Stroke victims who suffer paralysis to part of

FAST FACT An average adult at rest breathes between 12 and 20 times per minute.

+ MAJOR MUSCLE REFLEXES +

Reflexes produce rapid, automatic responses to stimuli. A particular stimulus always brings about the same reaction.

REFLEX	DESCRIPTION
Stretch reflex	Keeps a muscle at a set length by contracting it.
Golgi tendon reflex	Keeps a muscle at a set length through its relaxation and the contraction of other muscles.
Flexor (withdrawal) reflex	Causes quick withdrawal of a body part from a painful stimulus, such as a finger under a sharp knife.
Crossed-extensor reflex	Often works with flexor reflex in rapid withdrawal and redistribution of weight.

WHAT CAN GO WRONG

childhood infections have been targeted as likely contributors. Factors that heighten risk of developing multiple sclerosis include northern European ancestry and exposure to Epstein-Barr virus. In addition, women are twice as likely as men to get the disease. At least 300,000 Americans have been diagnosed. New drugs including interferon and glatiramer have shown promise in slowing the development of symptoms and decreasing complications.

Famous people with multiple sclerosis include actress Teri Garr, writer Joan Didion, actress and singer Lena Horne, and talk show host Montel Williams.

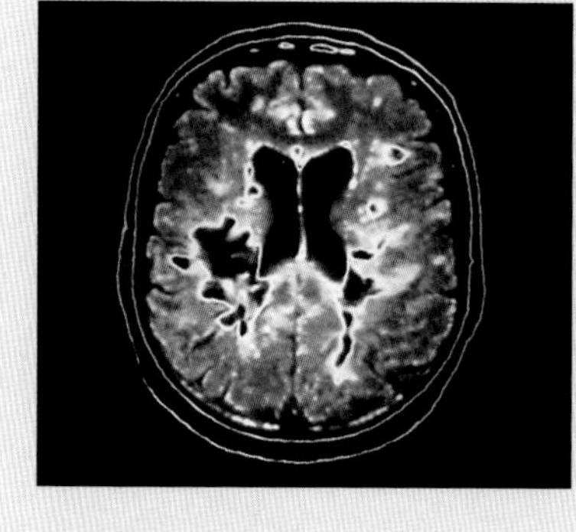

their mouth cannot smile on command with both sides of their face. However, when such patients hear a funny joke and laugh involuntarily, they smile equally with both sides. Scientists would explain this phenomenon by pointing out that even if the cerebral cortex has lost its control over voluntary action, the basal ganglia remain capable of autonomic response.

In contrast, the somatic nervous system works by activating motor neurons embedded in skeletal muscles. The cell bodies of these neurons reside in the central nervous system, and their axons stretch as far as the most distant toes and fingers. Their fibers are thick and well coated with myelin to conduct electrochemical impulses efficiently and quickly. All somatic motor neurons—the ones that move your body in response to external stimuli as well as mental instructions—work by releasing acetylcholine into their synapses. When acetylcholine concentrations reach a trigger point, surrounding muscles contract. When you stretch your legs after sitting at the computer, you may think you are elongating muscle fibers, but in fact the stretching occurs because fibers elsewhere have gotten shorter.

RESPONSES & REFLEXES

Some muscle contractions occur in response to extremely simple neural connections—so simple they never get close to the brain itself. One such reaction is the "knee jerk" reflex in response to the tap of a doctor's hammer. It works on a closed loop connecting the knee to the spinal cord, and it's as predictable as night following day.

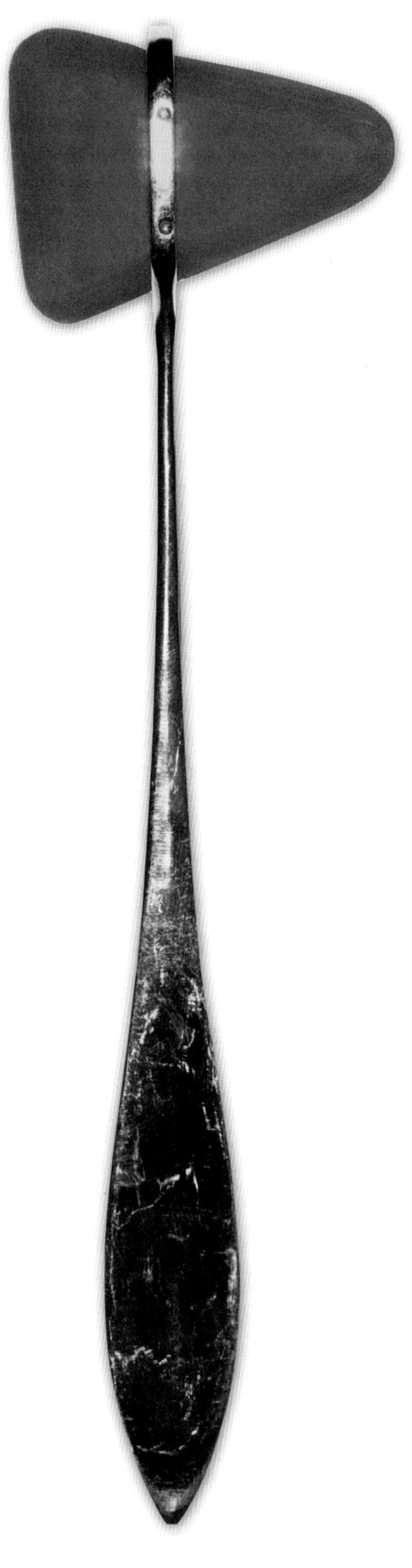

The tap of a reflex hammer initiates motion independent of the brain.

As more and more neurons get recruited to formulate responses to external stimuli, the reflexive response becomes less automatic. Primitive fight or flight responses get processed quickly for evolutionary reasons. More complicated stimuli such as the irritants of a sneeze take longer to process and get forwarded beyond the spinal cord to the brain.

FAST FACT Charles Darwin said the movements of expression "reveal the thoughts . . . of others more truly than do words."

Take the vestibular reflex, for example. It occurs in the brain stem and automatically adjusts muscle tone in the neck and body to maintain the posture of the head. Otoliths in the inner ears constantly send signals to the brain stem about the orientation of the head to gravity. Changes in that relationship, as well as sudden linear acceleration, send impulses through neural connections to the brain stem. Meanwhile, the ears' semicircular canals detect angles of the head's acceleration as it moves from side to side. If you are walking and you trip, the sudden movement of your head jerking forward and possibly to the

CROSS REFERENCE: See "Nerve Cells," PAGE 10

left or right causes the brain stem to automatically contract muscles in the neck and limbs to keep the head level and maintain balance.

Sensory information gathered by the peripheral nervous system gets incorporated into complex webs of potential responses. If you're lying on a beach and the sun's getting hotter and hotter, when and how do you decide to go inside? Chances are, given the incredible complexity of the brain, nobody could predict for certain the instant you would analyze changes in skin temperature, air temperature, wind, the nascent pain of sunburn, and other environmental factors, not to mention your state of mind—weren't you planning to stay out long enough to get some color?—and decide to get inside your lakeside cabin. Neuroscience has turned more and more toward examining the processes of cognition instead of simpler cause-effect reflexes.

INVOLUNTARY MOTION

But even our conscious thoughts, and the actions they influence, contain many unconscious movements. Awareness of the importance of acting without thinking has long been crucial for the development of athletic excellence. In an 1887 article in *Harvard Monthly*, A. T. Dudley wrote of the superior athlete: "Ask him how, in some complex trick, he performed a certain act, why he pushed or pulled

POLIO & THE PRESIDENT

President Roosevelt's memorial in Washington, D.C., depicts him appropriately seated.

IN HIS 30s, Franklin Roosevelt was "the handsomest, strongest, most glamorous, vigorous physical father in the world." That's how son James recalled a man who enjoyed tennis, golf, and horseback riding. Such vigor disappeared literally overnight. After swimming in August 1921, Roosevelt went to bed exhausted. The next day, after suffering fever, the future President was permanently paralyzed below the waist.

The culprit was polio, ingested in contaminated water. Polio attacks the neurons of the lower brain and spinal cord, and like a spy cutting a telegraph wire, it destroys the communication lines connecting the brain and extremities.

Vaccine programs begun in the 1950s have nearly wiped out the disease. Unfortunately, some survivors now suffer from a new disease, post-polio syndrome. Its trigger remains unclear, but speculation centers on ordinary aging. Humans lose neurons as they grow old. Survivors of polio, their store of neurons already depleted, struggle because they have fewer to lose.

Roosevelt tried to regain use of his legs through exercise and bathing in the spa at Warm Springs, Georgia. Nothing worked. However, the disease may have made him a better politician. Historian Doris Kearns Goodwin argues that paralysis stretched the wealthy man's empathy for the poor and underprivileged, "people to whom fate had dealt a difficult hand." Put another way, as Roosevelt's body withered, his soul expanded.

FAST FACT Irritants in the nasal passages—like pollen and dust—and illnesses—such as the common cold and hay fever—can trigger a sneeze. But sneezes can also be triggered by taking a quick look at a bright light, like the sun. Called the photic sneeze reflex or sun sneezing, the phenomenon and its causes are puzzles to researchers.

at a certain instant, and he will tell you he does not know; he did it by instinct; or rather his nerves and muscles did it of themselves. . . . "

Such involuntary movement is born out of much training and practice. Repeated motions, both self-initiated and in response to others' actions, eventually become automatic, even though they may contain complex sequences. The basketball center who fakes left, spins right, and lays the ball into the hoop with a finger roll at one time chose each of those actions. After much practice, the center carries out the sequence with little or no conscious thought. Thus, the lines between involuntary and voluntary actions become less clear.

FREE WILL?

Further blurring the boundaries between voluntary and involuntary action is the concept of free will, which recedes the more it is studied. Humans define themselves as creatures with the power to choose. Entire institutions, such as the church and the legal system, include the ability to make choices as mechanisms for reward or punishment. Yet brain scans raise intriguing questions.

In 1985, San Francisco neurological researcher Benjamin Libet fitted the scalps of volunteers with caps to record electrical activities in their brain, and then asked them to make a simple decision: move a finger at a time of their choosing. He discovered that his EEG recorded a brain readiness potential a half second to a full second before the subjects moved their fingers.

Libet decided to refine his experiment. He asked his volunteers to look at a clock face and note the exact moment they reached the decision to initiate the finger motion. They then moved their finger, and afterward gave Libet the exact timing of their decision. He compared those self-reported times with the readiness potentials recorded by his instruments and came to an odd conclusion: Electrical activity in the volunteers' brain manifested the decision 300 milliseconds before the subjects became aware of it themselves. The correlation became so precise and so predictable that Libet could watch the EEG scans for the telltale activity, then know with confidence that the subjects were about to move their fingers. Despite the results, the volunteers believed they acted when they chose to act. How could it be otherwise?

Libet's findings suggested that the brain knows what a person will decide *before* the person does. But if that's the case, the world must reassess not only the idea of movements divided between voluntary and involuntary, but also the very idea of free will itself. Libet showed that the conscious decision to act occurs only after the action, as measured by brain activity, is already under way. The brain produces motion, but part of the decision to move already had been made before the conscious mind becomes aware of it.

The paradox exists when observers consider the mind, the brain, and the body as separate entities. The problem vanishes, though, according to neurologist Richard Restak, once we realize, "We are our brain." The power to choose behavior lies in the influence of the brain on the brain.

+ SNEEZE ANATOMY +

IRRITANTS THAT REGISTER on the sensory neurons of the nose may trigger the explosive reflex of sneezing. The irritating sensation is passed along the trigeminal cranial nerve, which connects the face and the brain stem. In the brain stem, the sensation finally reaches the lateral medulla, which triggers the sneeze as an explosive burst to expel the irritants from the upper airways. Damage to the lateral medulla can cause animals to lose their ability to sneeze. Because the trigeminal nerve also carries signals from the eye, bright lights sometimes cause sensory impulses to spill over into other nerve fibers and set off sneezing in about one in four people.

CROSS REFERENCE: See "Awareness," PAGE 178

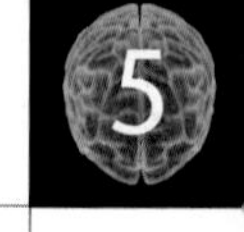

GLOSSARY

ACETYLCHOLINE. Neurotransmitter that causes muscles to contract.

ATHETOSIS. A movement disorder, linked to an overactive basal ganglia, characterized by slow continuous movement.

BASAL GANGLIA. A group of nuclei consisting of the caudate nucleus, putamen, and globus pallidus. Believed to play a role in movement regulation and coordination.

BETA BLOCKERS. A class of drugs that lower blood pressure by blocking the effects of epinephrine (adrenaline).

CEREBELLUM. Region of the brain that is most responsible for producing smooth, coordinated muscle movement.

CERVICAL NERVES. Eight pairs of spinal nerves that issue from the first seven vertebrae and supply movement and feeling to the arms, neck, and upper chest.

CHOREA. A movement disorder characterized by involuntary, irregular, jerking motions of the limbs and trunk.

CHRONIC TRAUMATIC ENCEPHALOPATHY (CTE). A degenerative brain disease found in individuals with a history of frequent concussions. Characterized by depression, memory loss, aggression, confusion, and the early onset of dementia.

COCCYGEAL NERVE. Single pair of nerve cells that issue from the coccyx, or tailbone. Supplies feeling to the skin between the coccyx and the anus.

COGNITIVE UNCONSCIOUSNESS. Takes in most of the information that allows individuals to consciously act within and know their environment.

CONCUSSION. A high-velocity impact injury to the brain that may interfere with movement, balance, speech, and memory and can have short- and long-term effects.

HEMIBALLISMUS. Uncontrolled movement of the limbs as a result of damage to the basal ganglia.

HUNTINGTON'S DISEASE. A hereditary condition causing degeneration of the neurons in the basal ganglia and cerebral cortex. It is ultimately fatal.

JACKSONIAN EPILEPSY. A type of epilepsy characterized by predicable fits confined to certain parts of the body.

LEPTIN. A hormone produced by fat cells that helps to regulate metabolism and food consumption.

LOCKED-IN SYNDROME. A neurological condition resulting in loss of voluntary muscle movement in all regions of the body except the eyes, though cognitive awareness and reasoning remain normal.

LUMBAR NERVES. Five pairs of spinal nerves that supply the lower back, fronts of the legs, and feet.

MIRROR NEURONS. Neurons that fire during a familiar action and when thinking of or observing others performing it.

MOTION BLINDNESS. Loss of the ability to detect changes in movement. Motion appears as a series of differing still images.

MOTOR HOMUNCULUS. A diagram linking body parts to the corresponding region of the motor cortex, with sizes of body parts shown in proportion to the number of neural connections.

OBSESSIVE-COMPULSIVE DISORDER (OCD). An anxiety disorder characterized by intrusive, unwanted thoughts (obsessions) and/or strong urges to perform countering actions (compulsions).

PATELLAR TENDON. Connects the kneecap to the shinbone and aids in leg extension. Site of the knee jerk reflex test.

SACRAL NERVES. Five pairs of spinal nerves that issue from the sacrum below the lower back. These nerves supply the backs of the legs and sexual organs.

SELECTIVE SEROTONIN REUPTAKE INHIBITORS (SSRIs). Drugs that inhibit serotonin reabsorption. Used to treat depression, anxiety disorders, obsessive-compulsive disorder, and eating disorders.

SUPERIOR TEMPORAL SULCUS. Brain region containing the neural networks responsible for motion detection/analysis.

THORACIC NERVES. Spinal nerves that issue from the 12 vertebrae of the upper back. They serve the trunk and abdomen.

TOURETTE'S SYNDROME. Neurological disorder characterized by repetitive and involuntary motor tics and vocalizations.

TRIGEMINAL CRANIAL NERVE. Connects the face to the brain stem. Responsible for facial sensation and motor control.

VESTIBULAR REFLEX. Automatic adjustment to the muscle tone in the body and neck to maintain the posture of the head.

VESTIBULOOCULAR REFLEX. Causes automatic adjustments to the eye muscles to maintain a stable gaze regardless of head movement.

SHARED ROLES

DIFFERENT REGIONS UNITED IN MOTION

BRAIN SCANS like Libet's and those of German neurophysiologist H. H. Kornhuber demonstrate that no single brain region is responsible for specific motions. Conscious movements that occur as you will your fingers to pick up a pencil are primarily governed by the cerebral cortex, which processes sensory information and chooses particular actions. However, the impulse to move apparently originates in networks of motor neurons in regions below the cortices. There, years of neural development have created pathways that carry out instructions for finely detailed movement beyond the scope of perception.

You may choose to pick up the pencil, but you cannot articulate the firing of neurons that make your fingers reach, bend, and grasp it. The mechanism that connects the thought to the movement lies beyond conscious thought. This does not mean that the connection between the brain and motion should not be studied. Far from it. Action, from facial gestures to bodily movement to vocalizing one's thoughts, is what defines an individual. And action is movement.

MULITPLE REGIONS

Brain scans indicate that the impulse of motion arises from a variety of regions working in concert. Neural networks discharge among the high and low regions of the brain in the four-story house metaphor described in Chapter 2. The brain has no single "decision center" for movement. Information constantly flows in and among all levels, from the basement rooms of the brainstem and spinal cord, through the first and second floors of the basal ganglia, cerebellum, and motor cortices, to the topmost floor of the executive function of the prefrontal cerebral cortex's frontal lobes.

Electrical discharges observed in brain scans occur in the cerebellum, basal ganglia, and cerebral cortex nearly simultaneously when a person decides to move a finger or a toe, and these discharges appear on screen well before the twitch of either digit. This finding shakes the foundation of theories by such pioneers as John Hughlings Jackson, who argued for a hierarchically organized brain with localized functions. Instead of a "higher brain" ordering movement, many brain regions—including some of the most primitive, in evolutionary terms—play significant if not equal roles in distributing the impulses for movement.

Cerebral Cortex Functions

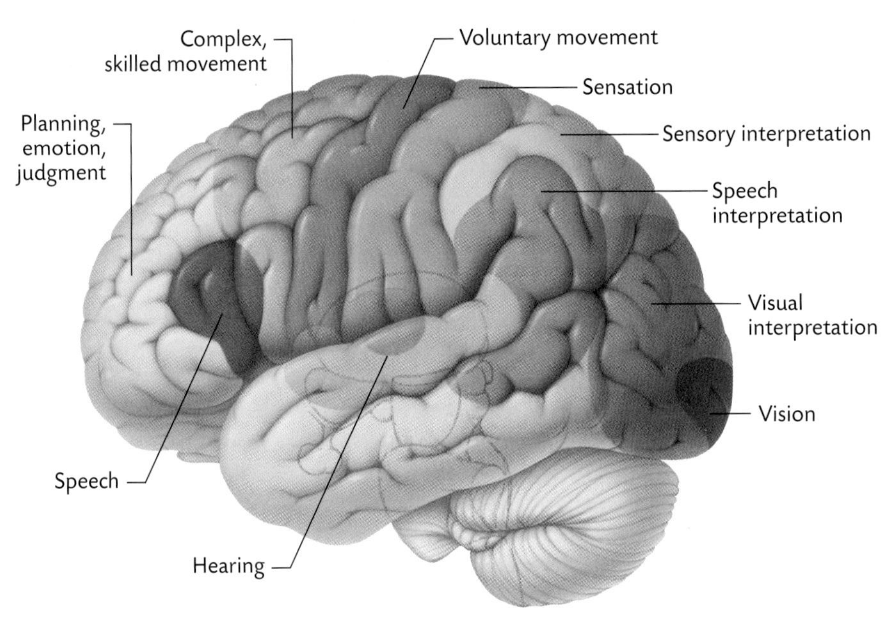

Areas of the cerebral cortex are designated, though not exclusively, to certain functions.

CROSS REFERENCE: See "In Harmony," PAGE 34

WORKING TOGETHER

The close integration of brain regions can be observed in the way neural networks work with each other through the marriage of movement and emotion: When you smile a genuine smile, you feel happy, and when you feel happy, you smile. Similarly, maintaining good body posture as you read or write helps you concentrate.

FAST FACT "Muscle memory" is the common name for how the neuromuscular system learns skills.

Restak uses the act of writing as an example to illustrate the advantages of the distribution of functions throughout the brain and the mix of conscious and unconscious movement. If you write by hand, you must activate neurons to control the muscles of your fingers, hand, arm, neck, and head—the latter two so you can turn your gaze between your notes and your paper. If you used a keyboard, you would call upon the same neurons that move your fingers, but the act of writing by tapping keys would differ from sliding the tip of a pen across a page.

You also have the options of dictating to a tape recorder, or to a secretary, as the dyslexic Agatha Christie did when writing her many mystery novels. You could even tap out your message using Morse code. No matter which method you choose, the communication would contain the same information because of the organizing purpose of your choosing to write.

However, each form activates completely different brain areas—neural networks for handwriting aren't the same as the ones for typing or dictating. The purpose of the act of writing calls forth whatever brain activities are required, from ritual skills such as typing to voluntary components—what should you write?—that cannot be localized in the brain. "The intermingling of voluntary and reflex activity lends majesty and power to the human brain," Restak said. "We are not simply reflex organisms . . . nor are we totally unrestrained in our behavior."

THE CEREBELLUM

How much each region of the brain operates during a particular movement depends on the specific form of the motion, its intensity, and the trigger that initiates voluntary or involuntary action. Although the brain works as an integrated whole, some regions seem to play a more important role than others in particular actions, such as the speech centers of the left temporal lobe. Neuroscientists focus much of their attention on the cerebellum for its role in coordinating movement at a level below awareness. The cerebellum lies amid the ancient, reptilian complex at the

STAYING SHARP

COMPETITIVE ATHLETES long have sought an edge over their opponents, but some modern competitors have substituted small doses of drugs: synthetic hormones known as anabolic ("tissue-building") steroids. These substances are dangerous, and when abused, can cause a host of mental and physical problems.

Synthetic anabolic steroids act like testosterone in the body by aping its action in binding to receptor sites in cells. Anabolic steroids bind to a high number of cell receptors, boosting protein production beyond normal limits and allowing the athlete to work harder, longer, and enjoy a shorter recovery time. Former professional athletes, such as football's Bill Romanowski and baseball's Jose Canseco, have admitted to steriod use and benefits from greater size and strength. But steriod use has a high mental and physical cost that can take a serious toll. Professional wrestler Chris Benoit, who killed his wife and son before committing suicide in 2007, had ten times the normal level of testosterone in his body and a cache of steroids in his home.

The National Institute on Drug Abuse calls anabolic steroids "dangerous drugs" and links their abuse to mental disorders including rage, aggression, mania, and delusions, as well as damage to the heart, liver, and kidneys. Studies conducted at Northeastern University in 2002 probed their possible impact on the brain. The researchers believe adolescent humans' developing brain would exhibit extreme sensitivity to steroids. Researchers found that hamsters given high doses of anabolic steroids had significantly lowered levels of the neurotransmitter serotonin. The decrease was especially evident in brain regions associated with aggression and violence.

back of the brain, near the brain stem, and connects with the cerebral cortex through neural fibers in the pons. The cauliflower-like cerebellum has two apple-size, heavily convoluted hemispheres connected by a wormlike vermis.

Like the cerebrum, the cerebellum has an outer cortex of gray matter, an inner body of white matter, and deep, paired masses of still more gray matter. The white matter looks like the branches of a tree, giving it the name arbor vitae, or "tree of life." The cerebellum contains representations of the motor and sensory networks of the entire body.

Each hemisphere is subdivided into three lobes: anterior, posterior, and flocculonodular. The anterior and posterior lobes coordinate motions of the body. Muscles of the trunk are influenced by the medial portions of those two sets of lobes, while intermediate regions influence hands and feet and skilled movments. The most lateral portions of the hemispheres work with the cerebral cortex to integrate information and appear to help plan movements. The flocculonodular lobes communicate with the inner ears to maintain balance during the process of standing, walking, and sitting.

Damage to the cerebellum, such as that caused by a wartime injury or a lifetime of alcohol abuse, may cause a person to stagger, lose balance, and flail an arm or leg when trying to do something as simple as scratch an itch.

FAST FACT Learned motor skills fall into two categories: fine and gross. Fine skills, like typing, primarily involve the hands. Gross skills, like rowing a boat, involve large groups of muscles all over the body.

CEREBELLUM FUNCTIONS

The cerebellum works to integrate information with sequences of events, which are essentially motions through time and space. A sense of sequence and of time is crucial to learning, thinking, and memory. Without a proper sense of timing, you would not know whether you can safely accelerate your car into the gap in oncoming traffic. Your correct assessment of whether to keep your foot on the brake pedal or stomp on the gas depends on your memory of previous episodes in the interstate on ramp as well as your own sense of timing developed over a lifetime of feedback from motor activities—of both the automotive and biological kind.

The cerebellum acts as a library for stored information about learned movements, the kind you don't have to think about to perform. Sports figures provide a good example of how the motor skills stored in the cerebellum work with the cerebral cortex to initiate action.

+ IGNORE THE PAIN +

DURING INTENSE competition, an athlete trained to focus on physical achievement can temporarily push through or ignore pain. This mental toughness temporarily dampens the pain pathways to the brain. Attitude plays a role, particularly in reassessment. "If you are distressed by anything external," wrote Roman emperor Marcus Aurelius, "the pain is not due to the thing itself but to your estimate of it; and this you have the power to revoke at any moment." After the sporting event's conclusion, when the need for concentration decreases, the athlete may feel the extent of the injury.

ANATOMY OF A GOLF SWING

When professional golfer Tiger Woods addresses a golf ball, he wouldn't even be able to spread his feet shoulder-width apart and maintain his balance without his cerebellum constantly receiving information from his eyes and ears and making involuntary adjustments to keep him stable. Woods selects his club for a particular shot and without thinking forms the proper grip. As he stands over the ball, waiting to begin his swing, he must make an intentional decision to start drawing the club head back in its characteristic arch.

Neuroscientists believe that this decision requires Woods's cerebral cortex to fire synapses in regions associated with movement, which then send impulses to the lateral cerebellum. The cerebellum creates a program of action and sends it via the thalamus to the motor cortex of the two hemispheres of the cerebrum. This program is a pattern of neuronal firing that Woods has played out many times before, so he doesn't have to think about the mechanics of a good golf swing.

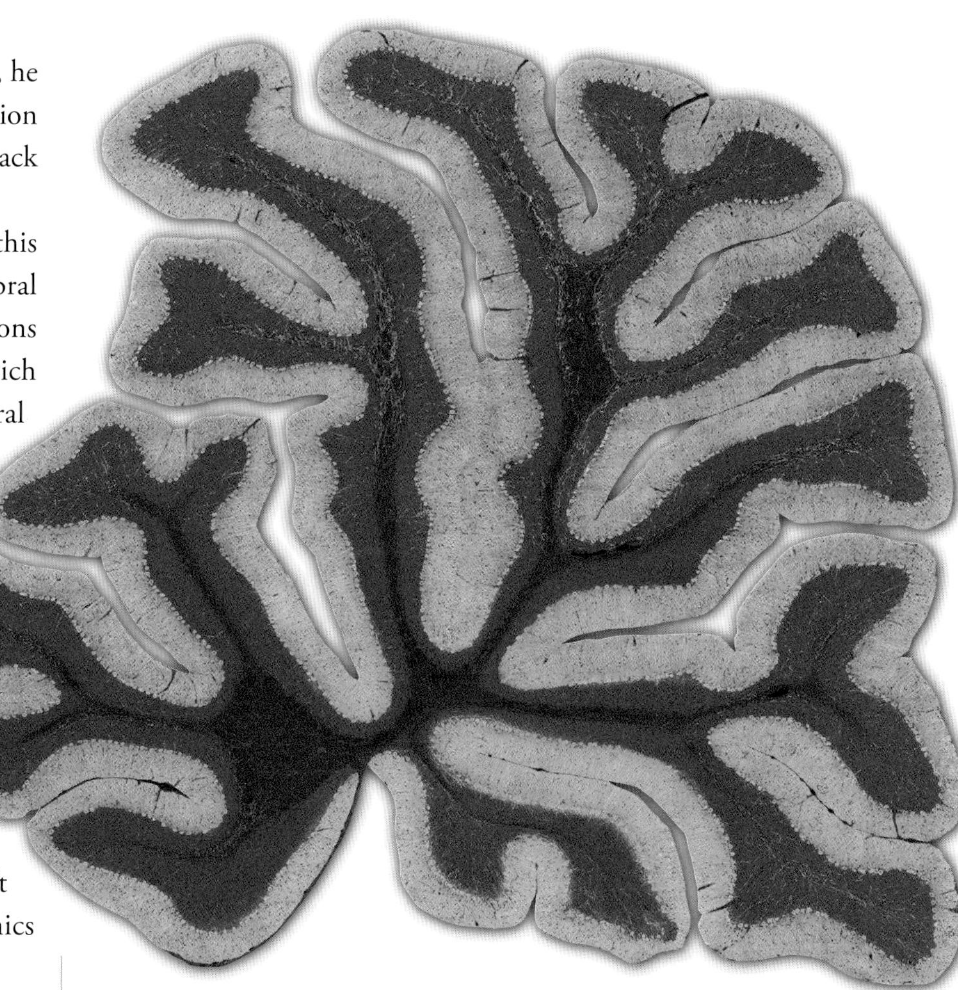

A section of the cerebellum exhibits its cauliflower-like structure.

However, each shot presents its own challenges. Perhaps there's a crosswind, or the ball is half buried in mud. Woods then may choose, using his cerebral cortex, to adjust the pattern, taking his usual swing but consciously attempting to fade or hook the ball onto the green. As he swings at the ball and follows through, his cortex adjusts the swing from instant to instant based on sensory feedback processed in the cerebellum.

Woods practices visualization techniques before he attempts any shot. He rehearses his swing in his mind and pictures the flight of the ball. It's always a detailed reconstruction of a perfect golf swing, with perfect results.

Woods being Woods, when he hits the ball, it's probably not much different from his imaginary version. There's now a scientific explanation for why such techniques improve performance.

VISUALIZATION

As a person visualizes an earlier experience, the mental imagery shares much of the same neural circuitry as the sensations of the original events. Imagining the breathtaking view from a mountaintop re-creates the sense of calm experienced on the summit. The same brain regions activated by being there get activated by reliving the experience, and the person may begin to relax and breathe slowly. Furthermore, if the person remembering the trip is asked to describe the valley below, her brain will tell her eyes to look downward. Scientists trace this phenomenon to "mirror neurons" in the cerebral cortex. They respond not only to actions but also to mental images of those actions, as well as action words.

SEEING SUCCESS

PROFESSIONAL GOLFER Jack Nicklaus always played a shot twice. The first shot took place only in his imagination, albeit in great detail. Once it felt right, he played the ball a second time, for real.

"I never hit a shot, not even in practice, without having a sharp, in-focus picture of it in my head," Nicklaus said in his book *Golf, My Way*. "It's like a color movie. First I 'see' the ball where I want it to finish, nice and white and sitting up high on bright green grass. Then the scene quickly changes, and I 'see' the ball going there—its path, trajectory and shape, even its behavior on landing. Then there is a sort of fade out and the next scene shows me making the kind of swing that will turn the previous images into reality."

Nicklaus went on to win 18 major tournaments. Close behind him on the all-time list is Tiger Woods, who borrowed Nicklaus's technique of seeing every shot in his mind's eye before hitting it.

MENTAL TOUGHNESS

The power of mental visualization has strengthened the games of many other athletes, who have applied similar imaging techniques. U.S. speed skater Dan Jansen, for example, had tremendous skill as a speed skater but had missed out on medaling at both the 1988 and 1992 Olympics.

At the 1994 Olympics at Lillehammer, Norway, Jansen tried again for a medal, but he slipped in his first event—the 500-meter sprint—and failed to win the race. When he lined up at the 1,000-meter event, in what was widely considered his last shot at a gold medal, he carried the mental weight of having failed repeatedly in previous events he could easily have won. But he won anyway, shattering a world record and then taking his eight-month-old daughter on a victory lap around the Olympic rink.

Speed skater Dan Jansen competes in the 1994 Winter Olympics at Lillehammer, Norway.

James Loehr, a sports psychologist who helped train Jansen, said he knew the skater had the mental toughness to accomplish what his body could do physically. He just needed to bring it to bear. "Mental toughness is the ability to bring to life whatever talents and skills you have—on demand," Loehr said. "That may come down to an ability to fight sleepiness, or to stay relaxed and calm or to not surrender your spirit when the odds are against you."

THINK POSITIVE

The power of positive thinking has helped many winners. If you apply visioning techniques to your own actions, you won't avoid failure forever because human beings aren't perfect. Still, you can turn failures into successes. Remind yourself of a previous performance where you achieved what you intended, and then create a new image of yourself once again performing at that same, high level. Stay focused on the positive and avoid the stress of negativity through relaxation techniques. It's a sure way to increase your chances

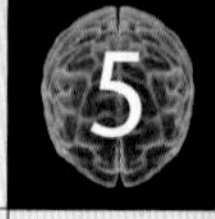

of winning, either on the athletic field or in the corporate arena.

In short, think like a champion. Your body will believe your brain and respond to it. If you run long distance and come to a steep rise, tell yourself how much you enjoy running uphill. You'll do much better than if you groan and tell yourself you'll never make it. If you rehearse hitting a curve ball in your mind before you swing your bat, you'll improve your chances of smacking a line drive when the baseball drops through the strike zone.

Loehr likens the effect of such thinking to software running a computer. Your body, your hardware, can have powerful tools to manipulate data, but it won't work right until you load and run the proper software, your attitude, in your mind.

SUCCESS STRATEGIES

Here are some mental exercises for improving performance from Richard Gordin, a professor of health, recreation, and physical education at Utah State University, and Michael Sachs, physical education professor at Temple University:

Be motivated. If you have motivation to perform well, you will do better than if you act for no good reason. Find a good reason to keep trying and working.

Feel good about your performance before you begin. Physical and mental fitness will improve your attitude, but something as small as wearing an attractive uniform or matching workout sweat suit, or having a new club or racket, can also make a difference. Looking good can help you to build your confidence before you ever take the field.

Feel as if you belong. Being in harmony with your surroundings, such as the comfort of feeling welcomed in a workout room by the others using it, can affect performance. Surround yourself with positive people who support your efforts and their "good vibes" will push you forward.

Confidently visualizing movements, such as steps required for climbing a wall, prepares the brain for making them happen.

Explore the memory of a previous success down to the smallest detail. For example, try to remember the feeling of sweat on your forehead and the tang of the drop that trickled into your mouth when you accomplished an athetic feat. Recall, if you can, the smell and temperature of the air, the feel of the clothes you wore, and anything else that could place you in the memory of a past triumph.

Star in your own mental movie. Envision the action that you want to create, making sure that the movie plays out from your point of view—as if the camera were in your head. Play the movie in real time, not slow motion, so that you see the action unfold at the right speed. Don't pause at critical spots; keep going until the end, and then play it again.

Set specific goals. Having something to shoot for will motivate you to reach it. Challenge yourself: You'll perform better if you tell yourself to do a high number of sit-ups rather than do the best you can.

Finally, don't obsess. The trick of applying all of the previous suggestions to achieve optimum performance is to know them and apply them without making them into tyrannical lists. When a specific physical activity becomes natural and fun, you'll do it well because it becomes its own reward.

Mirror neurons fire when physical activities that have already been learned are rehearsed in the mind. Rehearsing a physical exercise such as a martial arts kata causes the brain to prepare the muscles involved in that kata for increased activity. The more strongly the kata is imagined, the more the brain exerts control over heartbeat and breathing, just as if the chops, punches, and kicks were real.

Mirroring grows more powerful when the brain assigns special importance to the movements. For example, the brain of a black belt holder in karate would respond more strongly to observation of another's kata or to visualization of one's own than would the brain of a novice martial artist.

Similarly, the neurons in a piano player's brain fire more strongly when he is watching another person play Chopin than do the neurons of another observer with no musical training.

+ PRACTICE +

LEGENDARY track star Jim Thorpe played professional baseball and football. Acclaimed cellist Yo-Yo Ma showed early promise on the violin and piano. Folk musician Gordon Lightfoot, who studied piano as a young child, took up guitar at age 15 and went on to become highly skilled at acoustic fingerpicking. Why is it that a great athlete in one sport tends to be far above average in others? Or a musician tends to achieve on a second instrument beyond the skills of a person who takes up music as an adult? In a word, *work*. Great athletes and musicians take years to develop their talents.

PRACTICE MAKES PERFECT

If the human brain mirrors actions that it observes or imagines, it follows that visualizing an action before doing it will improve the performance. But practice is an essential part of the equation, as visualization alone won't get you there. Mastery of the skill will come in time, of course—nobody makes a perfect high dive the very first time off the board, and nobody will shoot 90 percent from the free-throw line one week after taking up the game of basketball.

Still, the expansion of physical skills requires mental discipline. "We learn to skate in summer and play tennis in winter," said William James, referring to the way the brain integrates experiences while the body is at rest.

When the mind is relaxed, it can rehearse actions that later can be called into play. When rehearsals let a competitor walk through actions without the stress of competition, they may encode patterns in the brain that later play out in real life.

SEEING IS DOING

Particularly important to the execution of any motor action is the maintenance of a stable field of vision. Without it, coordination between the body and the eye becomes difficult, if not impossible. If you wave your fingers in front of your eyes, they appear as a blur, but if you hold your hands steady and move your head back and forth, your vestibulo ocular reflex, which relies on the vestibular center of your inner ear, minimizes blurring. It's what allows athletes to orient themselves during even the most fast-paced games.

FAST FACT Social mammals such as wolves, dogs, and bears learn through play, which also reinforces social bonds.

The significance the brain places on recognizing and reproducing movement can be explained through evolution. Animal brains are hardwired to analyze motion and detect potential threats, such as a jungle cat creeping through the underbrush. True, the brain would recognize the shape of a panther as well as its characteristically slinky walk, but the brain recognizes the motion of the creature before the form.

Imaging studies suggest the brain's superior temporal sulcus

CROSS REFERENCE: See "Learning," PAGE 236

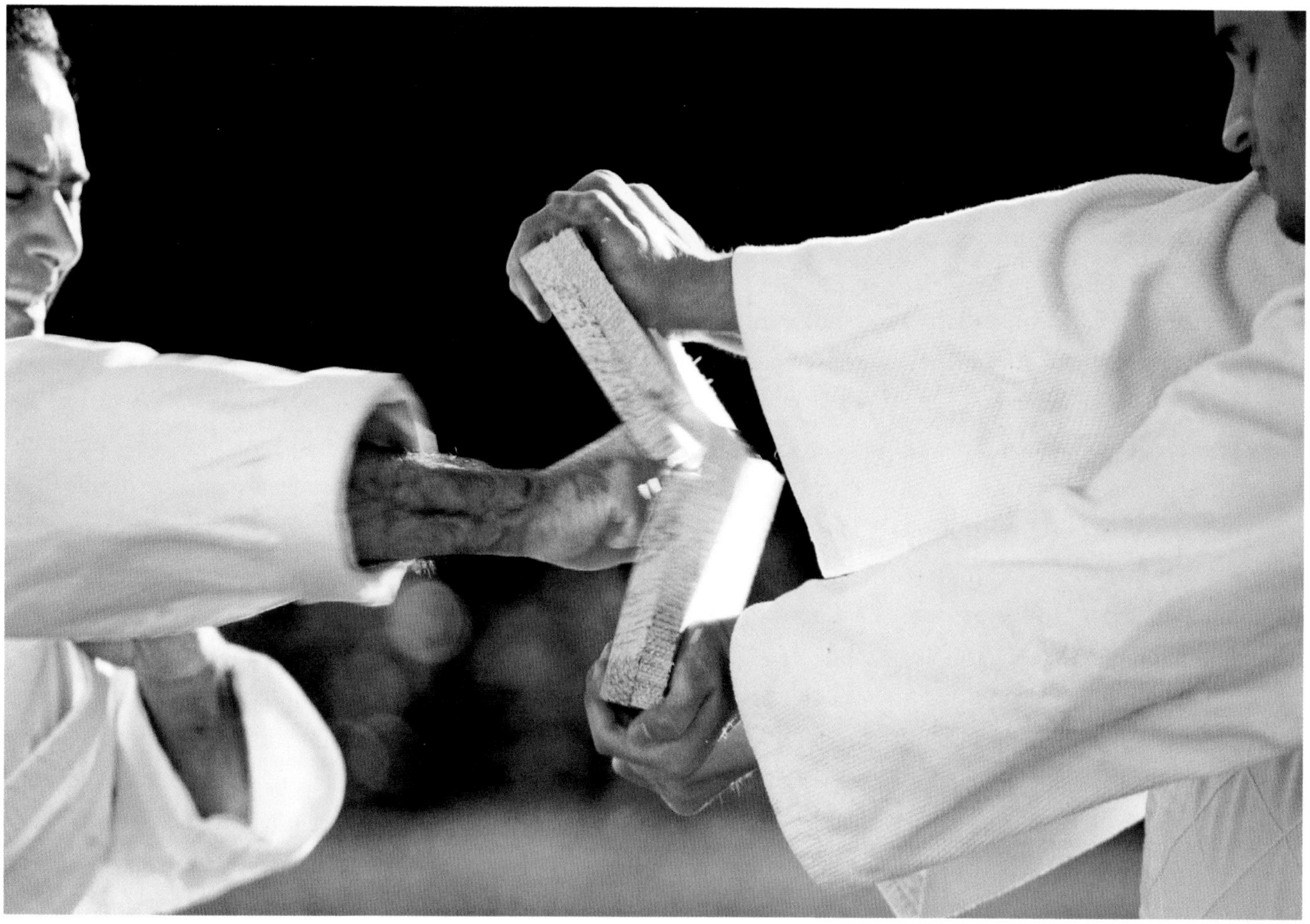

Mentally rehearsing a practiced motion, such as a martial arts punch, fires mirror neurons in the brain.

houses the neural networks most important for recognizing motion. It synthesizes two separate streams of information flowing from the visual centers in the occipital lobes. The top stream deals with motion and is informally called the "where" pathway. The bottom stream deals with recognition of objects by their forms; it is informally the "what" pathway and deals with recognition of things from moment to moment. Information from the "where" pathway reaches the superior temporal sulcus about 20 milliseconds before data from the "what" pathway, demonstrating the evolutionary preference of detecting movement over shape.

For further proof of the brain's hardwired attention to movement, try going alone into a bar, sitting down near a television, and ignoring the flickering images on the screen. As the commercials roll by, it will be hard to shut them out entirely. Your attention will be diverted, especially to those spots featuring motion. Television advertisers and the makers of promotional clips for blockbuster motion pictures know the importance of movement in capturing an audience. They fill their spots with action after action, lest a quiet moment allow the viewer to look away.

Damage to brain regions that sense motion can cause a condition known as motion blindness. The brain no longer processes physical changes from instant to instant. A person walking into the room appears to click from one position to another, like a collection of snapshots, instead moving fluidly, as in a motion picture.

MOTION SICKNESS

MANY DISORDERS and injuries can interrupt the normal functioning of the brain's control over movement. Some of these problems can be the result of physical injuries such as a fall or a blow to the head. Other problems have their sources within the brain itself, where neurochemical issues, like Huntington's disease and Parkinson's disease, can lead to different kinds of problems with movement.

BRAIN INJURIES

A physical blow to the head can bring on a concussion, an impact injury to the brain. It is an ailment caused by motion that also often affects motion. The brain is soft and sits amid a cushion of cerebrospinal fluid separating it from the protective shield of the skull. A high-velocity impact, such as a hard tackle in a football game or the bump of a head against the dashboard of a car that suddenly stops, jolts the brain against the inside of the skull, tearing nerve fibers and blood vessels. On occasion these tears can lead to tragic and even fatal consequences, as happened with movie actress Natasha Richardson after a skiing accident in March 2009. Athletes are at higher risk than most people because of the nature of contact sports. However, having one concussion raises the likelihood of having another, no matter how the first was inflicted. Furthermore, having a concussion doubles a person's chances of developing epilepsy within five years.

Damage to the brain can be severe or mild, and sometimes the recipient doesn't even realize he or she has suffered injury. The blow typically causes an immediate sense of confusion and short-term amnesia. Concussions also have longer-term effects, depending on the severity of the blow and the region of the brain that suffered damage. Concussions can interfere with movement, balance, speech, memory, reflexes, and judgment.

FAST FACT Parkinson's patients' symptoms may improve with regular practice of tai chi.

Most concussions are mild, and the brain usually recovers, but even routine sports concussions are nothing to trifle with. Researchers from the University of Montreal reported in 2009 that former athletes' mental and physical functions performed at subpar levels more than 30 years after they suffered concussions. The researchers found that those who had suffered concussions only once or twice had slower movements and decreased abilities in memory and focusing attention, compared with peers who had suffered no concussions. Symptoms include dizziness, headaches, slurred speech, ringing in the ears, and nausea. Anyone suspected of having suffered a concussion

WHAT CAN GO WRONG

DURING TEN SEASONS with the New England Patriots, former NFL linebacker Ted Johnson incurred more than 100 concussions before injuries forced him to retire in 2005. After retirement, Johnson fell into a deep depression, which he believes may have been caused by his chronic concussions. Johnson has since become a strong public advocate for more education on the long-term effects of head injuries on athletes like him.

Surrounded by spinal fluid, the brain normally floats inside its protective skull casing. During a sudden whiplash or collision, however, this natural

A boy jerks uncontrollably with the symptoms of Sydenham's chorea, also called St. Vitus' dance, sometimes associated with rheumatic fever.

should stop physical activity and seek medical attention.

LOSS OF CONTROL

When involuntary movement breaks free from conscious control, the results can range from embarrassing to dangerous. Several diseases of motion spring from damage to the basal ganglia or its overproduction or underproduction of dopamine. These include Parkinson's disease (described in Chapter 2), hemiballismus, chorea, and athetosis.

In hemiballismus, damage to a portion of the basal ganglia called the subthalamic nucleus, often

shield turns into a weapon. The cerebral cortex, which is the consistency of gelatin, bounces against bone, causing a concussion. The force of impact may twist or tear brain tissue. Concussions may be severe enough to cause loss of consciousness or so mild as to seem but a dream. Other symptoms include dizziness, ringing ears, and impaired balance, to name a few.

Recent studies by the Boston University School of Medicine have shed light on the long-term damage caused by concussions. Neuroscientists at that school analyzed the brains of several dead athletes, including Andre Waters, a onetime National Football League player whose deep depression and ultimate suicide at age 44 have been linked to chronic concussions. Waters's brain, the study found, bore microscopic protein tangles much like those of Alzheimer's patients nearly twice his age.

caused by stroke, can bring on uncontrolled movements of the arms and legs. No matter how the patient tries to will the motions to stop, they continue. Typically, the movements, which sometimes resemble the pitching of a baseball, grow weaker when the patient is at rest.

DANSE MACABRE

Chorea, the Greek word for "dance," is any of several disorders marked by uncontrolled jerking and twitching movements in the limbs and trunk. The most well-known condition is Huntington's chorea, a genetically transmitted illness in which the caudate nucleus atrophies. The uncontrolled movements of Huntington's disease prove even nastier than those of Parkinson's. Worse, while Parkinson's often responds to treatment, there is no relief or cure for Huntington's chorea—now called Huntington's disease—as the patient slowly loses control over movement and dies.

Athetosis is a slow, unbroken stream of writhing movements usually involving the hands and feet, but also affecting the face. An overactive basal ganglia has been identified as a contributor. Dopamine-blocking antipsychotic drugs have shown treatment results, as they have in some choreas.

The basal ganglia's crucial role in movement was underscored

WOODY GUTHRIE

Iconic songwriter Woody Guthrie sings and plays his guitar in an undated photo.

FOLK SINGER and songwriter Woody Guthrie (1912–1967) began acting strange when he reached his late 30s. The author of such American standards as "This Land Is Your Land" and "Roll On, Columbia" had always had a wry sense of humor and a scrappy personality. But as he aged, he began displaying depression, mood swings, and weird, uncontrollable jerky movements that got worse no matter how hard he tried to stop them. Dementia followed. Doctors diagnosed alcoholism and schizophrenia before settling on Huntington's chorea, which had killed his mother in 1930. Today the condition is called Huntington's disease.

The condition causes degeneration of neurons in the basal ganglia and cerebral cortex. It's genetically transferred; each child of a parent with Huntington's has a 50 percent chance of inheritance. The gene's location was pinpointed in 1993. As it expresses itself and damages the brain through a means not fully understood, it frees involuntary muscle movement from conscious control. Medication can treat the symptoms, but nothing can stop the neurological decline. Woody Guthrie had eight children. Of the five who lived to adulthood, two died of Huntington's, and three, including singer Arlo Guthrie, escaped the terrible legacy.

CROSS REFERENCE: See "Delicate Balance," PAGE 52

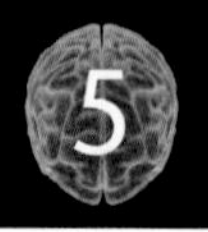

by an unusual disorder described in 1982. A 25-year-old Parisian nearly died of carbon monoxide poisoning. Breathing the noxious gas damaged a region in his basal ganglia called the globus pallidus. When the man woke up, he discovered that the damage had taken away his ability to will himself to move. Unless prompted to act by the touch or words of another person, the man would spend all day in bed, neither moving nor talking. He told interviewers that although he looked sluggishly inactive, his mind teemed with ideas he could not act upon.

FAST FACT The word choreography is derived from the same Greek root word as chorea.

Diseases affecting movement, such as Parkinson's and Huntington's, often interfere with a patient's memory and sense of timing. Among Parkinson's patients, for example, one study found a link between how much they had lost control of their motor functions and how difficult they found it to recall certain tasks. Such physical and cognitive links occur because the basal ganglia and cerebellum do more than coordinate motion. They also affect thoughts and memories through their roles in shaping sensory inputs on cognition and control of motor functions. When the basal ganglia malfunction, they may fail to dampen the activation of undesired movements and thoughts. Neural impulses normally overridden by the brain then lead to unwanted tics and jerky motions despite conscious efforts to control them. Even the behaviors of obsessive-compulsive disorder (OCD), in which patients fixate on unwanted thoughts and carry out ritualistic actions despite their best efforts to avoid them,

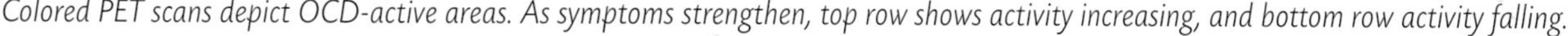

Colored PET scans depict OCD-active areas. As symptoms strengthen, top row shows activity increasing, and bottom row activity falling.

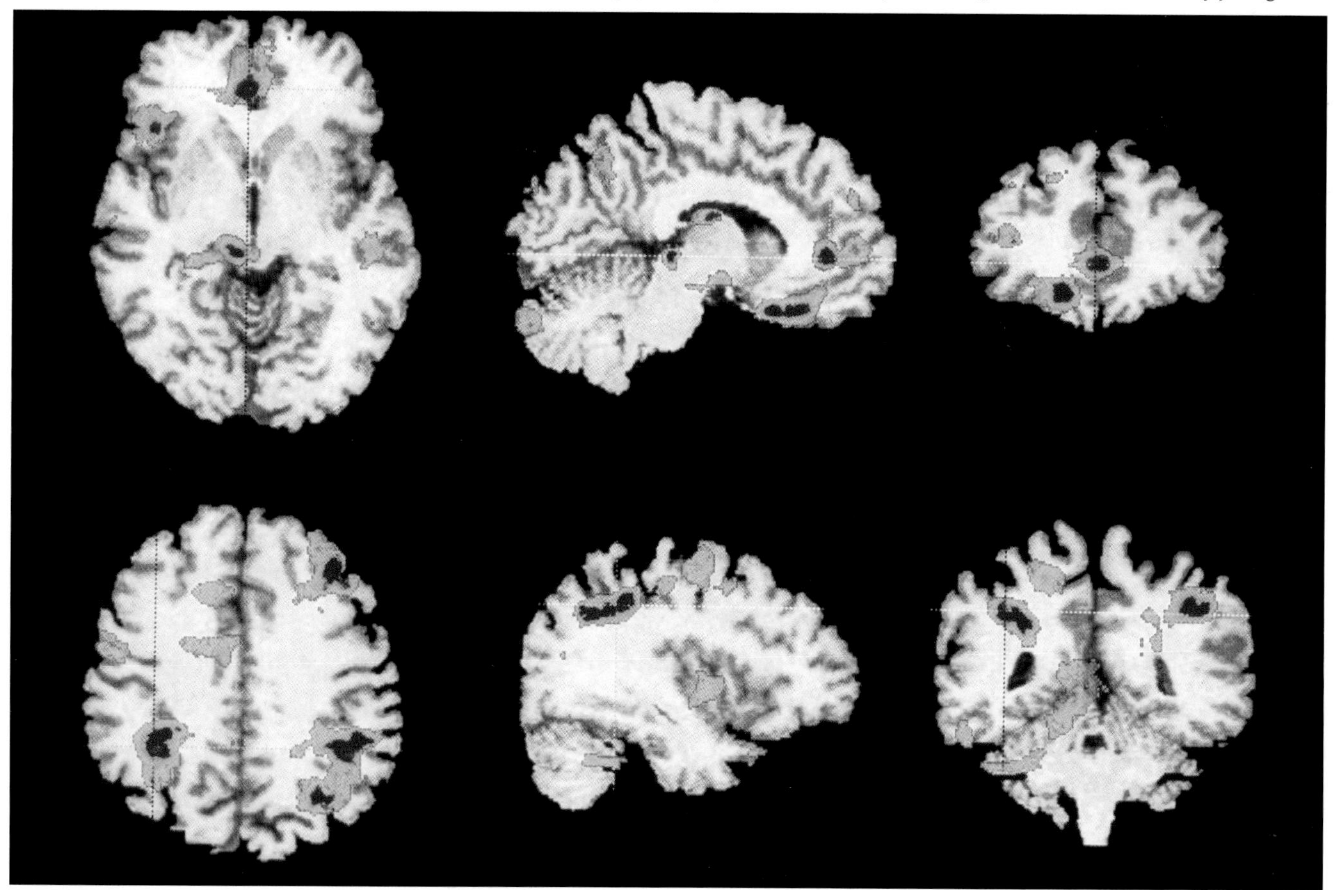

can be traced in part to abnormalities of the lower areas of the brain, although the exact cause has yet to be determined.

OBSESSIVE-COMPULSIVE DISORDER

In his book *The Mind and the Brain,* UCLA psychiatry professor Jeffrey M. Schwartz tells of a middle-aged woman who sought treatment from him at the Obsessive Compulsive Disorder Research Group at the university's medical center in Westwood. "Dottie" spilled out her sad story: Ever since she was a girl, the sight of the number 5 or 6 transfixed her with fear. After she grew old enough to drive, her seeing a 5 or 6 on another car's license plate prompted her to pull over to the side of the road to await the arrival of a car with a luckier number on its plate. If she didn't spot such a car, she could not move on without feeling her actions spelled some unspeakable misfortune to her mother. Later, after she gave birth to a son, she fell under the spell of a new irrational fear: that some action or inaction of hers would cause her son to go blind.

Dottie suffered from obsessive-compulsive disorder. Symptoms include series of intrusive, unwanted thoughts (the obsessions) that bring on intense urges to perform countering rituals (the compulsions). The obsessive thoughts can dominate the mind, yet patients with the disorder report that they seem to come from outside the self, as if an outsider had hijacked a portion of the brain. They may feel compelled to check over and over again to see if a door is locked, even though they know they just locked it.

FAST FACT British essayist and poet Samuel Johnson (1709–1784) is believed to have had OCD.

Like the fictional television detective Adrian Monk, they may count steps or obsess over germs or repeatedly make lists or insist on life's everyday rituals being performed in specified orders. The disorder isn't as rare as *Monk*'s creators might have you believe: Roughly one in forty Americans has some degree of obsessive-compulsive disorder. Typically, symptoms first appear in the years between adolescence and early adulthood. Unlike the repeated actions of addicts, OCD patients get no joy from their repeated, ritualistic behaviors.

PET scans revealed extreme levels of activity in the orbital frontal cortices of OCD patients' brain, right behind their eyes, as well as a tendency toward heightened activity in the caudate nucleus and cingulate gyrus. Studies about the role of the orbital frontal cortex suggest it acts as an error detector. It alerts the brain, for example, to receiving one stimulus when a different one was expected, as when you drink a glass of water that turns out to be gin. When expectations match reality, the orbital frontal cortex calms down.

Schwartz hypothesized that problems with the orbital frontal cortex could create difficulties with error detection, creating a feeling of something being out of order even when it's not, and bringing on unwanted corrective behaviors.

BREAKTHROUGH

FROM THE LATE 1960s through the early 1990s, the most common psychological treatment used to treat obsessive-compulsive disorder forced patients to graphically confront the source of their problems.

A behavioral therapy called exposure and response prevention, or ERP, was pioneered by psychologist Victor Meyer at Middlesex Hospital in London in the mid-1960s. Meyer exposed his patients to the trigger that brought on their obsessive thoughts and compulsive behavior. He began with triggers that the patient ranked low on the degree of distress they would cause. For example,

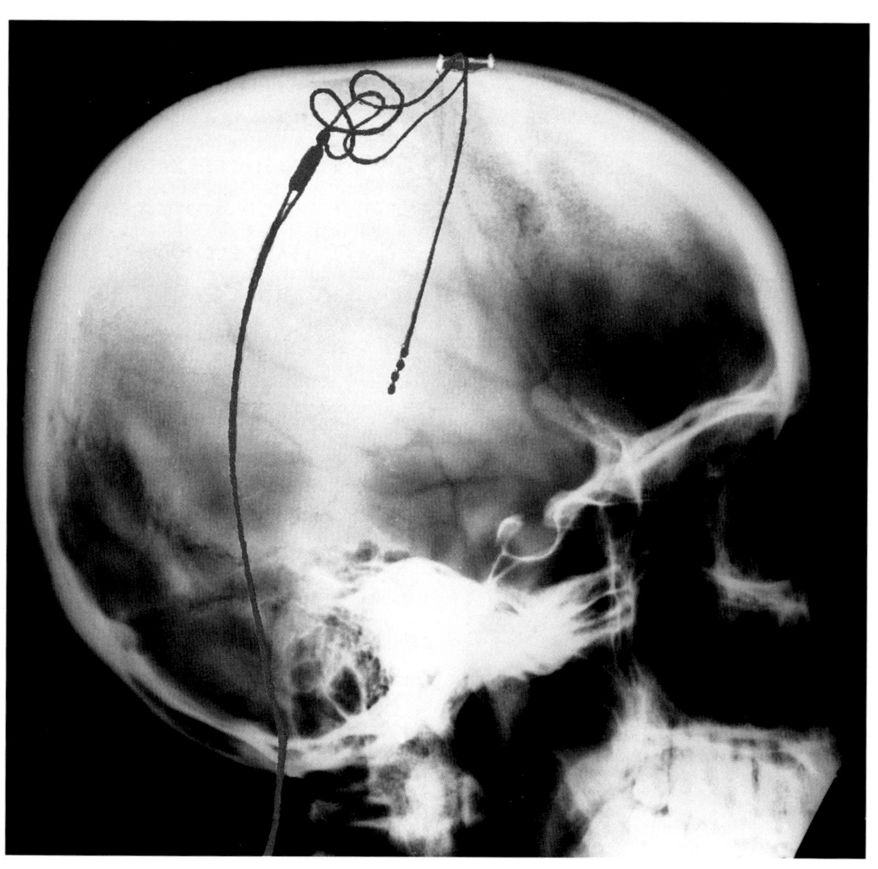

A Parkinson's patient receives treatment through implanted electrodes, revealed by x-ray.

When the so-called worry circuit locks on to the feelings associated with errors, OCD can result.

Freudian analysis considered obsessive-compulsive disorder to have roots in—surprise!—repressed sexual conflicts and childhood memories. Today, neuroscientists believe the disease springs in part from biochemistry, with genetics playing a role. Quite by accident, a drug called clomipramine hydrochloride, used in the treatment of depression, was found to relieve OCD symptoms in patients in the 1960s and 1970s. One of the drug's many reactions is to make serotonin linger in synapses. Researchers followed up by creating new kinds of drugs called selective serotonin reuptake inhibitors (SSRIs), which raise serotonin concentrations without some of the unpleasant side effects of clomipramine hydrochloride. SSRIs include Prozac, Zoloft, and Celexa. All seem to reduce symptoms in about 60 percent of OCD patients. Schwartz also has had good results using cognitive therapy, in which patients come to grips with the causes of their conceptual errors, relabel them as manifestations of brain functions, and reassess their behaviors in ways that bring out changes in their neural processing. He tells his patients, "The brain's gonna do what the brain's gonna do, but you don't have to let it push you around."

TOURETTE'S SYNDROME

An aristocratic Frenchwoman, the Marquise de Dampierre, began acting strange when she was seven. Her arms flailed in short bursts until she

BREAKTHROUGH

if the patient obsessed about germs but experienced a relatively low level of stress in response to germ exposure, Meyer might have the patient touch all of the doorknobs in a public building but prevent hand washing afterward. The treatment then progressed to triggers that caused greater amounts of distress—touching a half-eaten apple or shaking hands—as patients gained some acceptance of the triggers' existence and control over their reactions.

Prevention of compulsive action by the patient ranged from gentle coercion to actual physical restraints. Not surprisingly, the patients often became extremely agitated during the initial phases of treatment. Many refused to undergo the first treatment after learning what it involved. Practitioners claim high success rates, but that includes only those who agree to complete the treatment in the first place.

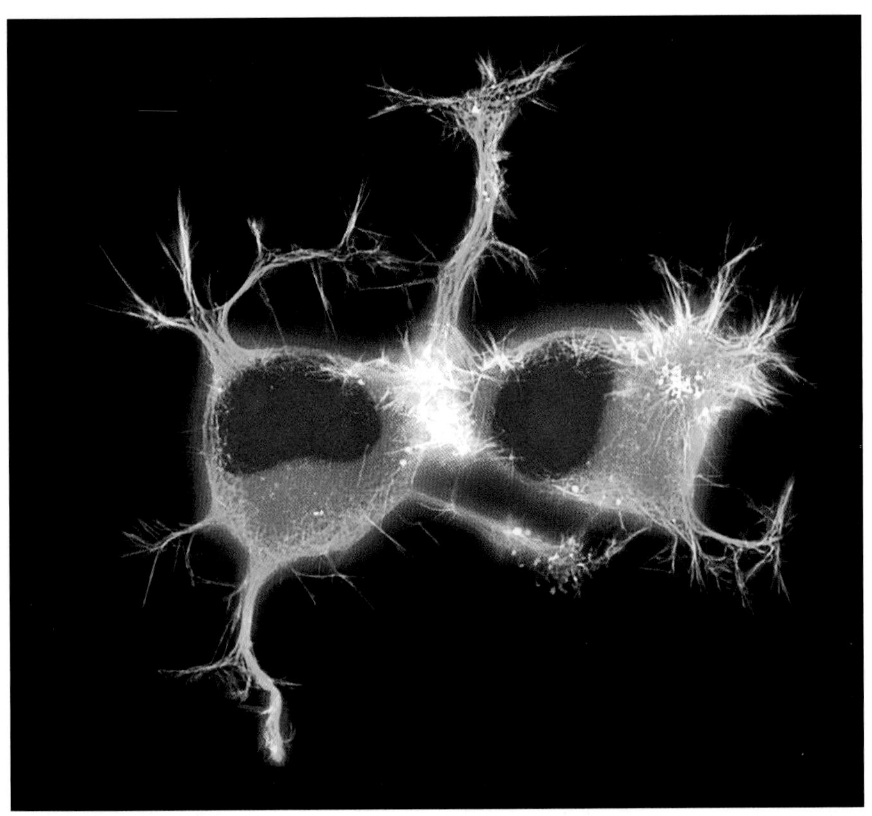

Neurons stimulated by nerve growth factor send out neurites that will be axons and dendrites.

could regain control of them, only to have them start moving again on their own. Her tics and jerks spread to her neck and face. She began screaming and speaking in gibberish too, yet retained awareness of her seemingly inexplicable actions. The marquise's affliction continued until her death at 80.

In 1884, physician Gilles de la Tourette described nine patients, including the marquise, afflicted with such compulsive motions and vocal outbursts. Today, the disease he described bears his name: Tourette's syndrome. In the last few decades it has been diagnosed as its own neurological disorder, although not without controversy over the details of separating it from other diseases. Symptoms include an excess of nervous energy, strange motions and mannerisms, grimaces, curses, and an outlandish sense of humor.

Neuroscientists in recent years have confirmed Gilles de la Tourette's hunch that the disease had an organic origin in the central nervous system. Tourette's excites both motion and emotion and seems to have its cause in the thalamus, hypothalamus, limbic system, and amygdala. The disease appears on the other end of the spectrum from Parkinson's. The latter stems from a lack of dopamine in the synapses of the brain, and the former appears to spring from an overabundance.

Most patients with relatively mild symptoms get through life without treatment. Drugs like haloperidol (Haldol) have proved effective in suppressing some symptoms.

THE FUTURE OF TREATMENT

The science of treatment for motor disorders stands at the edge of a

BREAKTHROUGH

RESEARCH at the University of Colorado at Denver has Stephen Davies believing that effective therapies for paralysis induced by spinal trauma lie only a few years in the future. "I can't promise complete recovery, but perhaps the major recovery of function," said Davies, whose work was financed in part by a donation from the foundation begun by actor Christopher Reeve. Reeve's foundation gave $150,000 to Davies, who used the money to help him get a $1.2 million grant from the National Institutes of Health. That money now is at work financing groundbreaking research on treating spinal injuries that, unfortunately, comes too late for the actor who played Superman.

Davies's research demonstrates that axons can be induced to grow great distances along the injured spinal cords of rats if not blocked by scar tissue. He has worked with a chemical called decorin

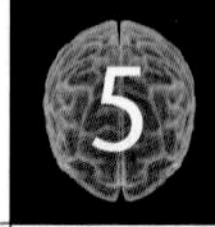

vast ocean, ready to set sail. Conventional wisdom once held that localized damage to the central nervous system ruined the physical functions associated with that region. Now, researchers have new models to give them hope. Patients with significant brain and spine injuries have made recoveries ranging from the startling to the miraculous. Neural plasticity and an astonishing array of new treatments for movement disorders is raising hopes.

Transplants of neural cells from one brain to another may be the most promising, and controversial, treatment. Surgeons can implant healthy neural tissues from aborted fetuses into the brain and spine of ailing patients to take over some of the functions of damaged nerve cells. Parkinson's and Huntington's patients are most likely to benefit from implants that help restore balance to the mix of neurotransmitters in their brain.

The ethical issues of using fetal stem cells in medical research have sharply divided American citizens and political parties. In March 2009, President Barack Obama made federal funds available to researchers working with embryonic stem cells. Meanwhile, researchers continue working on other lines of stem cells, including some created from adult human cells. Although adult cells show some promise, fetal stem cells still appear to be the most efficient at producing dopamine in the brain of Parkinson's patients.

Another potential source for stem cells exists within patients themselves. Stem cells have been discovered in adult brains and likely could be induced, given the right stimulus, to start producing new neurons. Under this scenario, healthy cells could be harvested from a patient with a motor disorder, grown in a lab, and then reinserted into the patient's brain with no chance of the body rejecting the cells as foreign.

FAST FACT

Children known as PANDAS (pediatric autoimmune neuropsychiatric disorders associated with streptococcal infections) became obsessive-compulsive as a result of strep throat.

IMPLANTS

Electronic implants soon are likely to be available for implantation into patients suffering from nerve damage. Radio-powered stiumulators may bring back voluntary movement to paralyzed limbs, and not long after that, nanotechnology may create microchips capable of firing like neurons and restoring channels of communication in damaged spinal cords. Already, a device called the VNS pulse generator can be implanted in a patient's neck to ward off depression and epileptic seizures through stimulation of the vagus nerve.

Electrical stimulation also shows promise in treating motion-related disorders. Surgical procedures for OCD patients have sought relief through destroying a collection of axons called the internal capsule, but electrical stimulation without the damaging invasion of surgery may provide similar results.

BREAKTHROUGH

that, when administered immediately after an injury, prevents scar tissue from forming and thus opening pathways for neural growth. And he has discovered how to make a specific type of astrocyte support cell from specialized, stem cell-like glial precursors that, when injected into rats suffering from spinal injuries, create new neural connections in the central nervous system. Treated rats return to nearly normal levels of spinal function.

Davies believes that breakthroughs in restoring movement to people paralyzed by spinal injuries will come about through the right combinations of effective and affordable treatment.

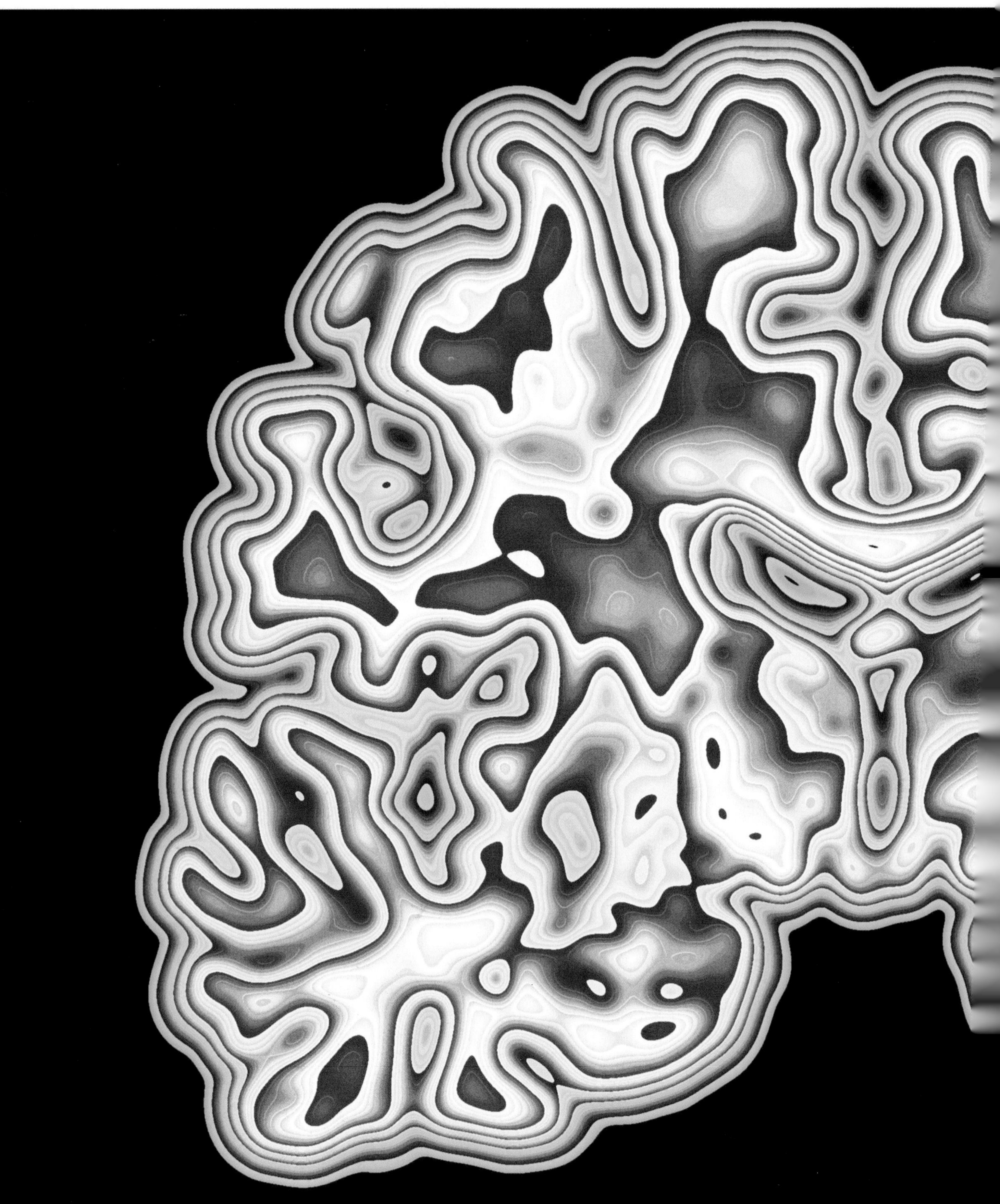

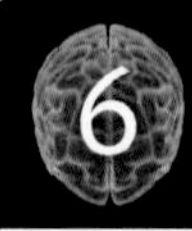

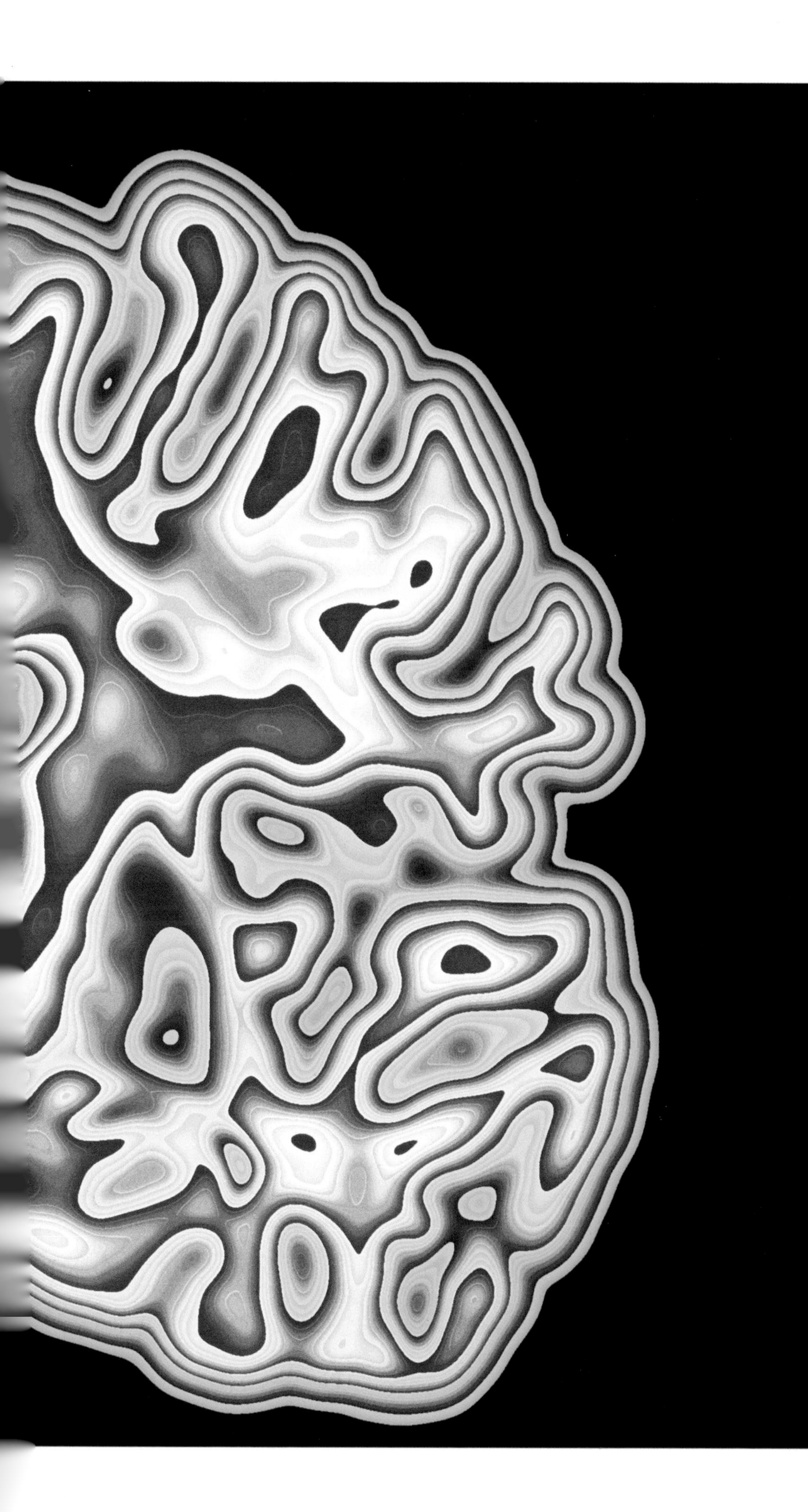

CHAPTER SIX

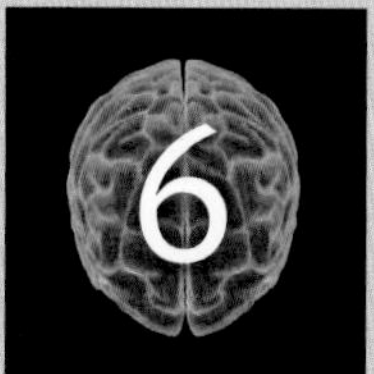

STATES OF MIND

ASLEEP OR AWAKE, the brain is always active. It works in times of razor-sharp alertness, in the chaos of dreams, and at levels below self-awareness. Even when the body rests deeply in a coma, beyond the reach of sensation and thought, the brain works to pump blood, move air into and out of the lungs, and digest food. At the other extreme of mental activity, drugs may push the brain into altered states of hyperactivity or distorted perception for good or ill.

A CT scan vertical cross-section of the brain depicts the cerebrum, home of consciousness.

DAILY ROUTINES

THE CONSCIOUS & UNCONSCIOUS MIND

Pedestrians on the Via Scarlatti in Naples shift attention as they move through the tasks of daily life.

When an iceberg calves off a Greenland glacier and drifts into the North Atlantic, passengers on passing ships see it as a rock-solid mountain of ice. But that's just the top 10 percent of the iceberg. Nine-tenths of it is lurking below the ocean's surface.

The brain is like that iceberg. Humans are the only animals capable of thinking about thinking, and when they do, they focus much of their attention on how the brain perceives the world and processes information to reach the state of awareness we call cognition.

However, research conducted over the past century has demonstrated that most of the brain's work resembles the larger, submerged portion of an iceberg. Throughout much of the day, the brain labors in states other than consciousness. And during periods of wakefulness, the brain still reacts to and incorporates the unconscious processing of information.

States of mind range from comas and unconsciousness, to awareness and the processing of sensory perception, to periods of hyperawareness and the altered states induced by drugs. All stem from the expressions of a functioning brain, a leap of understanding that first arose a little over a century ago. "There is not a single one of our states of mind, high or low, healthy or morbid, that has not some organic process as its condition," said psychologist William James.

A GRAY AREA

Although the brain acts mechanically, it is nothing like a motor or a lightbulb. Except in the most extreme circumstances, it doesn't simply turn on and off.

Consider anesthesia. Until 1964, doctors who gave patients anesthesia before surgery assumed they would be unable to perceive anything while unconscious. However, in that year a University of California at San Francisco physician, D. B. Cheek, studied patients who had given their doctors some trouble before surgery. While those patients were anesthetized, the doctors who operated on them freely shared some less than flattering comments with one another. Cheek discovered that after the patients woke up, under the influence of hypnosis they could sometimes recall word for word what their doctors had said while they were supposedly insensate. In a separate study a year later, doctors staged mock crises during surgery and voiced aloud their concerns that their patients might die. The experiment seems a bit unethical in hindsight, but it yielded fascinating results. When they awoke, some of the patients became quite agitated when asked to recall anything that happened during surgery.

FAST FACT Many U.S. schools start later in the morning to coincide with teenagers' circadian rhythms.

Such studies underscore the difficulty of making black-and-white distinctions about states of the brain. Many ordinary observers would draw a clear line between wakefulness and sleep, between the conscious and the unconscious mind. Yet if the supposedly unconscious mind can register sensations and store memories, it is difficult to say with certainty that it stems from a phase shift from the alert and conscious brain. Attention, consciousness, and memory work together to create states of mind.

In a typical day, every person goes through two obvious states, waking and sleeping, each of which has sublevels of awareness based upon controlled and automatic mental processes. The brain goes through many transitions every day. Some occur naturally. Others, such as the fatigue and irritability of jet lag, come in response to changes in the external environment.

THE RANGE OF MENTAL STATES

The alarm clock rings at 7 a.m. and you slam your hand down on the "off" button. As you get out of bed and wobble about to find your slippers and make breakfast, for several minutes your mind mirrors your stumbling body and performs its own clumsy dance as it slowly emerges from sleep. Quite possibly, in the preceding predawn hours as you slept, your brain played out a strange narrative or two combining images and sounds from your memories that seemed so real you felt real terror—or joy. In your

+ HEARTS & MINDS +

MANY ANCIENT SOCIETIES believed the soul and mind dwelled in the human heart. Science has long since proved otherwise, but the ancient idea lives on in the English language. To have learned something fully is to know it "by heart." To suffer emotional pangs at the loss of romantic love is to have a "broken heart." And to understand something at an unconscious level is to know it "in one's heart."

Those statements may actually hint at an unconscious understanding of layers of awareness in the brain—and that the mind works at levels beyond consciousness to understand the world.

half-awakened state, recalling how you romanced that Hollywood star in your dreams, your brain slowly resolves itself into an alert state. Finally, you can focus your attention on the sports section of the newspaper before working your way toward the news on the front page. Then, a quick shower, with a burst of cold water at the end, makes you yelp and feel as if you're finally, fully awake.

Off to work you go, where you concentrate on the information on your LCD monitor, alternating periods of intense concentration and analysis with moments of daydreaming and perhaps a mindless game of computer solitaire. A big lunch makes your thoughts slow and sticky, like honey, but after you take a walk around the block, you feel rejuvenated.

At day's end, back home you go, driving along the same route you normally take, allowing your brain to seemingly cruise virtually on autopilot. As you pull into your driveway, you find you can't really remember much about the trip home, but somehow your relaxed state of mind has come up with the perfect solution to the problem you were unable to solve that afternoon when you tried to wrestle it into submission with your most determined critical analysis.

Satisfied and relaxed, you eat dinner, do the dishes, watch some TV sitcom on television that doesn't tax your mind too much, and get ready for bed. As you drift off, you linger in oblivion for a while before your dreams return to haunt or charm you.

Now the question is, which of your mental states was the "real" you? Are you, in essence, the dreamer? The half-awake automaton? The deep thinker? Or the unconscious one? Indeed, you are all of the above.

Circadian rhythms, matched to passage of time, strongly influence states of mind.

+ BRAIN WAVES +

Electroencephalographs give real-time readings of brain activity through impulses detected by electrodes placed on the scalp. EEGs record these impulses as brain waves. Like radio stations across the AM or FM dial, brain waves exhibit various frequencies. From lowest to highest, they are:

NAME	EEG PATTERN	FREQUENCY	DESCRIPTION
Delta waves		0.1 to 4 Hertz	Occur most commonly during deep sleep, unconsciousness, and in newborns. As they increase, attention to the world dissolves.
Theta waves		4 to 8 Hertz	Occur during prayer, daydreaming, and some sleep stages and between waking and sleeping. May promote learning and memory.
Alpha waves		8 to 12 Hertz	When dominant, people feel calm, in control. Common during alertness when the brain is not focused on problem solving.
Beta waves		12 to 13 to about 30 Hertz	Associated with active mental states, including solving problems, exercising judgment, making decisions, and thinking analytically.
Gamma waves		36 to 44 Hertz	Continuous in nearly all brain states. May help synthesize various brain functions. Deficiencies linked to some learning disabilities.

CROSS REFERENCE: See "Looking Inside," PAGE 24

FAST FACT While their heart rate and breathing may slow, people placed under hypnosis generally retain normal brain wave patterns during their trance states However, hypnosis can alter brain activities, such as ability to remember—or forget.

DAILY BRAIN WAVES

Throughout every state, neurons in your brain constantly communicate with one another, even if they are not performing specific tasks. You must sleep, but a healthy brain never totally does.

Waking or at times while sleeping, all regions of the cerebral cortex hum with a background electrical energy of 40 cycles per second, or Hertz. This background pattern is one of many bands of "brain waves," which are rhythmic electrical pulses created by actions in various brain regions. Patterns tend to be relatively stable for individuals and reveal underlying brain states. As measured by an electroencephalogram, they range from slow to rapid cycles per second, with each band assigned a Greek letter for identification: delta, theta, alpha, beta, and gamma. Each set of brain waves has been associated with different states of mind and with different functions.

Electroencephalographs (EEGs) record these impulses as brain waves, each one exhibiting various frequencies measured in Hertz, or cycles per second. Delta waves (0.1 to 4 Hertz), the lowest, occur most commonly during deep, dreamless sleep; when you're unconscious, and in the brains of newborn babies. As delta waves increase, attention to the world dissolves.

Theta waves (4 to 8 Hertz) are commonly measured during times of deep reflection, daydreaming, spurts of creativity, and intuition. Theta waves are also exhibited during the groggy states between waking and sleeping. Next are the alpha waves (8 to 12 Hertz), which are associated with feeling calm and in control. Alpha waves are most commonly found during periods of alertness when the brain is not focused on solving problems, such as when adults are relaxed but are not drowsy.

Beta waves (12 to 13 to about 30 Hertz) are present when the brain is actively engaged in problem solving, decision-making, and analytical thinking. Beta waves at the high end of the scale have been found to accompany feelings of agitation. Last are gamma waves (36 to 44 Hertz), which are continuous in nearly all mental states and are believed to promote the synthesis of various simultaneous brain functions.

Stretching helps the body wake up. During sleep, the brain works hard to consolidate memories.

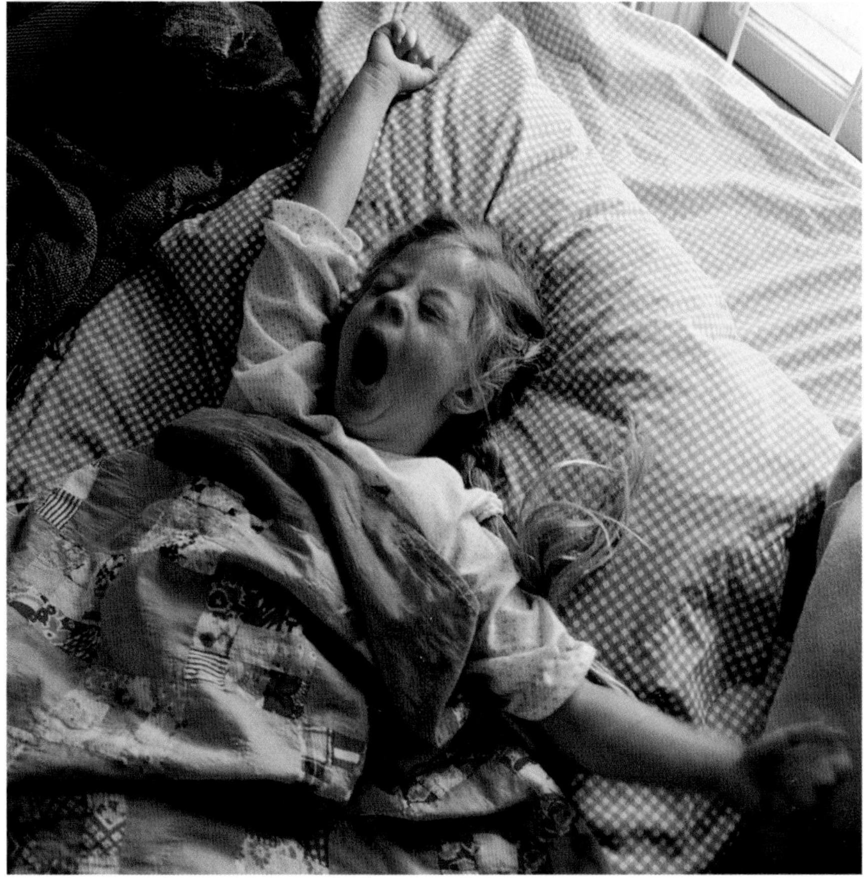

MANY MOODS

WHAT YOU GET out of life depends in great measure on what you bring to it. Moods, which are relatively long-lasting emotional states, strongly influence how much you enjoy life—or don't.

When you are in a good mood, small setbacks don't get you down, you feel as if you can do anything, and daily activities bring you pleasure. But when you are in a bad mood, you snap over minutiae that you normally ignore, or you get stuck in a pit of depression where nothing seems like fun.

Language is full of expressions tying mood to external events. "He got up on the wrong side of bed this morning," you say about the boss who grouches over the smallest mistake. Or, observing an upbeat friend, you remark, "Everything's going her way."

Such sayings underscore what seems to be common sense: Moods spring from external events. When things go right—you get a big tax return or find a ten-dollar bill—you are in a good mood. When they go wrong—your car breaks down or you lose your lucky T-shirt—you are in a bad mood.

HAPPY ON THE INSIDE?

But that's not exactly true. According to psychologist Robert Thayer, author of *Calm Energy,* moods arise more from internal than external circumstances. Think about it, he said: Events are so random that if you charted everything good and bad that happened to you last week, you'd find little correlation between events and feelings. Instead, moods link up closely with sleep, exercise, diet, and time of day. These variables cause changes in hormone and neurotransmitter levels, muscle tension, blood pressure, and other components of health. "Events and circumstances do influence mood," said Thayer, "but they happen on top of a biological edifice that gives them greater or lesser importance."

More than the world of rushing crowds or quiet moments, the brain's internal world shapes one's changing moods.

Thayer links moods to energy and tension. He argues that everyone feels four basic moods every day: calm energy, calm tiredness, tense energy, and tense tiredness.

Calm energy is the supreme state of mind for getting things done. In this mood, you feel energetic without any sense of being tired. You are in your most creative and productive zone. You get a lot accomplished and feel good about it.

Calm tiredness occurs after the prolonged expenditure of energy or attention. After an evening spent studying for a big exam, or hours of physical labor, or even while digesting a terrific meal, you feel satisfied, relaxed, peaceful, and tired.

Tense energy also can be a productive mood, but it comes without any feeling of calm contentment. This mood creates the sensation of being "wired" or on edge, even though you perform tasks with a high degree of skill. Mountain bikers careering down a rocky slope, on the brink of losing control and crashing, experience tense energy as exhilaration. They have a sensation of risk that makes fun seem sharper and brighter.

Tense tiredness is the pits. It's the black mood that envelops you when you're overworked, overstressed, tired, irritable, depressed, or otherwise incapable of facing the world. Think of a sleepy two-year-old fighting off a very much needed nap, and you get the picture: cranky, antisocial, and unable to focus.

When you're feeling tense tiredness, take a tip from the mother of that crabby two-year-old: sleep. If it's the middle of the day, find time for a brief nap. A short snooze after lunch can divert your body's energy stream from tense tiredness to calm tiredness. Rest is the only way to climb out of tense tiredness.

If you're in a period of calm tiredness, the easiest way to return to the peak efficiency of calm energy may be to take a brisk walk. According to Thayer, walking for about ten minutes boosts energy levels and keeps them at a heightened state for at least an hour.

MOOD MANAGEMENT

To figure out which moods you are likely to experience at particular times, first determine whether you're a lark or an owl. Larks are "morning people," who start the day fresh and alert. Owls are "night people," who like to stay up late and get their best work done after dark. Take advantage of your body rhythms. Time your most important work to coincide with your hours of calm energy and tense energy. If you're a lark who's at your mental peak from sunup to midmorning, that's when you should write your book, compose your sonata, or figure your income tax. Save routine tasks—washing dishes, bathing the dog, cleaning out the garage—for the hours when you enter the state of calm tiredness, as your body comes down off its mountaintop of energy.

After mental or physical exercise, a mood of calm tiredness, suggested by a Hawaiian beach, promotes feelings of satisfaction.

YOU'VE GOT RHYTHMS

All too often, circadian rhythms get out of sync. Nearly everyone who has flown several hundred miles east or west has experienced jet lag. It occurs when the brain's natural rhythms governing when to eat, sleep, and work get thrown off by rapid travel, changes in work shifts, and daylight saving time, all of which cause circadian rhythms to be out of phase with local cycles of day and night. Exercising before and after flights, as well as while on board a plane (where possible), has been shown to reduce jet lag.

Fight the temptation to chemically boost your energy levels with food and drink. The sugar of a candy bar snaps your metabolism like a rubber band. It creates a brief, artificial burst of energy, followed by a downward rebound that depletes your energy for a much longer period. Use caffeine with care. Moderate amounts, found in coffee, tea, and cola drinks, boost attention and concentration but can exacerbate the feeling of tense tiredness by bringing on anxiety and jitters. A heavy lunch also fights off the return of a calm energy mood, so don't plan to eat steak and potatoes and then tackle your toughest assignments.

If you can begin to assess your energy levels and know your moods at any time of day, you're on track to have your brain work in tandem with your body and get the most out of every moment.

OSCILLATION

In some cortical regions, the oscillation occurs in phase-lock, a state in which the waves match up in perfect unison. Researchers suggest the phase-locked oscillation occurs because a variety of brain regions respond to a neural feedback loop controlled by the intralaminar nuclei of the thalamus. This loop, composed of axons that connect many parts of the brain, synchronizes electrical activity at 40 Hertz, creating so-called gamma waves that can be detected by an EEG. The oscillation occurs during waking and during the dream stage of sleep but not in sleep's oblivious, nondreaming stages. During waking states, flashes of electrical activity tuned to slower cycles per second, like bass notes to accompany the steady 40 Hertz beat, light up brain regions in response to sensations, thoughts, motor activities, and other brain functions. The intralaminar nuclei go dormant during nondreaming sleep, temporarily shutting off the oscillation, but spring back into action with the onset of dreams.

Regional electrochemical activity also returns, even though the brain is not responding to sensory input from the eyes, ears, and other organs. During sleep, the cortex works to create its own meaning out of the electrical activity, making dreams out of patterns of electrical discharges.

Observation of the brain's electrical patterns during its various states suggests that there is no one place where separate streams of sensation come together to form consciousness. Cognition arises in all regions, working together, apparently to the synchronized 40 Hertz beat led by the intralaminar nuclei. Clearly, these neurons play a key role in cognition. Scientists have discovered that if they suffer severe damage, the patient enters an irreversible coma.

CYCLES & RHYTHMS

Humans like to think of themselves as creatures defined by mental abilities unique in the animal kingdom. Cognition, consciousness, awareness—these are what separate people from the animals. However, the human brain experiences varied states of mind, including some shared with animals, such as the nearly universal need for sleep. The sleep-wake cycle, or circadian rhythm, governs our daily routine, from the most subdued moments of brain activity to the most hyperactive, mountaintop experiences where thoughts flow lucidly and creatively, to the deepest, darkest sleep where dreams cease to play in our mind. Even during a coma, when cerebral activity plummets, the human brain still flickers with life and activity.

Circadian Rhythms

24 Hours

Day	Night
6:45 a.m. Sharpest blood pressure rise	**7:00 p.m.** Highest body temperature
7:30 a.m. Melatonin secretion stops	**9:00 p.m.** Melatonin secretion starts
10:00 a.m. High alertness	**10:30 p.m.** Bowel movements suppressed
2:30 p.m. Best coordination	**2:00 a.m.** Deepest sleep
3:30 p.m. Fastest reaction time	**4:30 a.m.** Lowest body temperature
5:00 p.m. Greatest cardiovascular efficiency	
6:30 p.m. Highest blood pressure	

Typical circadian rhythms match times to high and low performance levels.

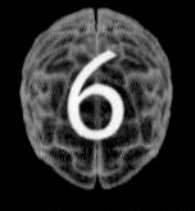

GLOSSARY

ALIEN HAND. A rare neurological condition in which one hand acts independently of an individual's conscious control.

ALPHA WAVES. Brain waves common at times of relaxed alertness. Important state for learning and using new information.

ASCENDING RETICULAR FORMATION. Part of a branch of interconnected nuclei located in the brainstem. Responsible for the waking state.

ATTENTIONAL BLINK. The brain's inability to detect a new target object flashed milliseconds after the first. Generally observed during testing involving rapid presentation of visual stimuli.

BETA WAVES. Brain waves associated with periods of active mental states, such as problem solving and critical thinking.

BRAIN DEATH. The brain's lack of electrochemical activity and loss of function. A brain-dead individual cannot recover.

CIRCADIAN RHYTHM. Any pattern an organism's body follows in an approximately 24-hour period, such as the sleep-wake cycle in humans.

COMA. A deep level of unconsciousness in which an individual cannot be awakened and does not respond to stimuli.

COMPLEX REGIONAL PAIN SYNDROME. A chronic condition causing swelling and pain in a limb and differences in skin coloration and temperature. Believed to be caused by a dysfunction of the central or peripheral nervous system.

CORPUS CALLOSUM. A thick band on nerve fibers that connects and allows for communication between the left and right hemispheres of the brain.

CORTISOL. A hormone released by the adrenal cortex in periods of extended stress. Has anti-inflammatory properties.

DELTA WAVES. Brain waves that occur most often during sleep or in periods of unconsciousness. Common in newborns.

ENVIRONMENTAL DEPENDENCY SYNDROME (EDS). A neurological condition in which an individual feels compelled to mimic others' actions or to use tools within their environment.

GAMMA WAVES. Brain waves that occur continuously in all states except for non-dreaming sleep. Believed to promote various brain functions, especially memory.

HIPPOCAMPUS. Region of the brain aiding in converting new information to long-term memory.

HYPOCRETIN. A neurotransmitter that promotes wakefulness.

INTRALAMINAR NUCLEI. Located in the thalamus, nuclei aresponsible for creating the brain's gamma waves.

JET LAG. A circadian rhythm disruption caused by rapid long-distance flights. Irritability, fatigue, digestive problems result.

KETAMINE. A drug that blocks NMDA receptors in the central nervous system. It is used as an anesthetic and in some cases to treat chronic severe pain.

LATERAL PONTINE TEGMENTUM. A region of the pons responsible for inducing REM sleep.

MONOAMINES. A class of neurotransmitters that includes dopamine, serotonin and adrenaline.

NARCOLEPSY. The inability to regulate sleep-wake cycles. Linked to the absence of or lowered amounts of hypocretin in the brain.

NUCLEUS ACCUMBENS. Region of the brain associated with feelings of pleasure and reward.

PERSISTENT VEGETATIVE STATE. A condition in which the brain maintains functions necessary to keep the body alive, but not cognitive function.

RAPID EYE MOVEMENT SLEEP (REM). The fifth stage of sleep, characterized by high levels of activity in the cerebral cortex. Stage in which dreaming occurs.

SEROTONIN. Inhibitory neurotransmitter that plays a role in sleep, mood regulation, memory and learning.

SLEEP APNEA. A sleep disorder in which an individual frequently stops breathing for short periods of time.

SLEEP REGULATORY SUBSTANCES. Proteins that accumulate in cerebrospinal fluid during wakefulness that induce sleep upon reaching threshold levels.

THETA WAVES. Experienced between waking and sleeping, during prayer, daydreaming, creativity, and intuition. Thought to promote learning, memory.

TRYPTOPHAN. An essential amino acid that aids in the body's production of serotonin and vitamin B_3. Prevalent in turkey and dairy products.

AWARENESS KNOWING THE WORLD AROUND US

CONSCIOUSNESS and unconsciousness often work in tandem with each other. But telling the difference between the two states is crucial to understanding the important roles they both play. States of mind are defined by electrochemical processing of information along neural pathways. To understand the nuances of such processing, it is useful to explore two sets of distinctions.

PROCESSING DATA

The first is between automatic and controlled processing. Much human behavior arises out of a necessary mix of these two. Automatic processes arise primarily in the back, top, and side lobes of the brain. Controlled processes rest primarily in the front half of the brain, with the executive, decision-making function centered in the region right behind the forehead.

For instance, when you drove to work, you probably didn't concentrate much on the route you took, the turns you made, or the speed you drove; these actions were monitored by automatic processing. Repetition made such actions nearly automatic, freeing your controlled mental functions to keep an eye out for emergency vehicles on the road and ice on the pavement. But when you sat at your computer and typed an email, your controlled processes kicked into high gear, analyzing and selecting concepts, words, and sentence structure. Even so, your operation of the email software and clicking of the computer keys contained many automatic elements. Likewise, most every action activates a multitude of brain regions and mixes automatic and controlled processes. The mix almost always favors the automatic.

Auguste Rodin's sculpture "The Thinker" (1902) depicts a man immersed in thought.

FAST FACT Masters of meditation, such as Zen practitioners, can lower their brain waves from the alpha to theta range. Studies of Tibetan monks locate intense responses to meditation in their left prefrontal cortices.

REGISTERING INFORMATION

The second distinction separates cognition from emotion. A definition in a popular textbook defines cognition as "the ability of the central nervous system to attend, identify, and act on complex stimuli." Neurologist Richard Restak suggests an alternate, shorthand version: Cognition encompasses all the ways we know the world around us. It ranges from daydreaming to figuring partial differential equations.

Emotions affect cognition but are believed to be generated automatically at a level below consciousness. For instance, emotions such as anger and fear arise from evolutionary programs that cause physical reactions to internal and

CROSS REFERENCE: See "Perception," PAGE 100

external stimuli. They're important for the survival of the species, but they tend to interfere with cognition until they subside.

During consciousness, three regions of the human brain constantly communicate among themselves. The prefrontal cortex, basal ganglia, and cerebellum work together to analyze sensations and time-stamp them. As these brain regions process the passage of time, they create images of the world in which causes produce events.

To become conscious of external stimuli, the brain must first attend to them, and then continue to attend to them. "Working memory" registers current events while communicating with long-term memory to shape awareness of the world. According to a theory by British neuroscientist and DNA researcher Francis Crick, "the mind" emerges when working memory, long-term memory, and expectations of the future link up to form thoughts. The so-called memory of the future allows the brain to compare current actions with future events, allowing for choices, judgments, and anticipation of consequences.

Other brain theorists, such as Gerald Edelman of the Scripps Research Institute, say consciousness arises from the brain's forming relationships between perceptions and prior experiences. This constant comparison, carried out over and over again, creates an awareness of the moment. Consciousness, although hard to define, apparently includes a component of understanding time.

Recent neurological research suggests that several mental disorders may be the result of the brain having a faulty timekeeper. Problems with the brain's internal

A Buddist monk teaches a pupil meditation to sharpen the mind in Shaolin, China.

STAYING SHARP

BRAIN SCANNING TECHNIQUES have let neuroscientists examine what happens to the brain of trained meditators. In a walnut shell, their findings indicate that meditation may sharpen the brain's attentional powers to more efficiently use its limited processing capacity.

For example, training in meditation techniques improved the ability of volunteers at the University of Wisconsin in 2008 to perceive numbers flashed for only a split second on a screen. They were told to look for two numbers flashed amid a series of letters. In a phenomenon known as attentional blink, the brain has trouble identifying the second number if it's flashed a split second after the first. When trained in meditation for three months, the volunteers experienced a significant increase in their ability to name the second number. Brain scans revealed the trained subjects devoted less brain energy on identifying the first number, freeing more attention to focus on the second.

Other research indicates that meditation can reduce the sensation of pain one feels when exposed to hot water and thicken the sections of the brain associated with focusing attention and processing information from the senses. It can even lower levels of stress hormones after a math quiz. Chinese scientists reported in 2007 that students given 20 minutes of meditation instruction each day for five days substantially reduced concentrations of cortisol in their saliva shortly after taking a stressful test of mental arithmetic. The group also reported having more energy and less anxiety. A control group taught a different method of relaxation also posted reductions, but they were not as great as those of the novice meditators.

COMA THERAPY

Rabies survivor Jeanna Giese greets Dr. Willoughby after her coma therapy.

LARGE DOSES of the drug ketamine, which blocks electrical signals between the brain and spinal column, bring on controlled comas that have potentially therapeutic value. In the past few years, the extremely painful and mysterious nerve condition complex regional pain syndrome has responded to ketamine comas in Germany and Mexico, where some doctors in the U.S. have sent their patients while the treatment awaits approval. The disease, first documented by Civil War surgeon S. Weir Mitchell, sends excruciating pain signals to the brain in response to the slightest touch. Studies suggest the disease arises after a dynamic change in the structure of central pain neurons causes a malfunction in their so-called NMDA receptors. Ketamine comas lasting five days are believed to "reboot" the nervous system and eliminate or lessen the malfunctions.

Rodney Willoughby, Jr., of the Medical College of Wisconsin in Milwaukee successfully used ketamine coma to treat Jeanna Giese, a Wisconsin teenager who unknowingly had been infected with rabies after being bitten by a bat in 2004. Too much time had elapsed before the deadly disease was discovered for usual treatment, a series of vaccine injections, and ketamine had demonstrated some anti-viral properties in animal studies. Willoughby gave Giese a cocktail of drugs to treat the infection during the week she remained in a coma. She became the first person on record to survive rabies without vaccine.

clock may play significant roles in everything from Parkinson's disease, autism, and schizophrenia to attention deficit disorder and the behavior of war veterans who suffered significant head wounds.

INTERNAL CLOCKS

A finely tuned internal clock affects how humans spend their energy. If they perceive enough time for a task, they're more likely to do it. Dividing the future into blocks of time, each of which gets allocated particular tasks, is an effective way to set goals and earn rewards.

The difficulty in studying the brain's time circuit is that there is none—at least, no independent region that acts as a biological clock. Rather, the complex circuitry of the brain makes the entire organ act as a timekeeper. Researchers suggest that the brain works most effectively when it focuses attention in bites of two and a half seconds. Call it the brain's construction of "now," based on perception, memory, and the unconscious mind.

When a person's interval describing "now" is significantly less than two and a half seconds, the result may be the person falling prey to easy distractions and other problems focusing attention on particular tasks. With longer spans of "now," a person may lock in too long on particular actions and not shift easily to new stimuli as the environment changes.

Neuroscientist David Eagleman of Baylor University says schizophrenics experience a form of "fragmented cognition"; they typically underestimate and overestimate the passage of time. Among Parkinson's patients, who often misestimate the timing of physical actions, the administration of medication to boost dopamine levels swiftly acts on the temporal anomalies and restores the patients' normal sense of timing. The advantage of a healthy temporal processing network is that it sets up the brain to react to events and predict results, staying one step ahead of whatever happens next.

FAST FACT

A Polish railway worker, Jan Grzebski, went into a coma after a train accident in 1988 and awoke in 2007.

UNDERSTANDING UNCONSCIOUSNESS

To better understand cognition, it may be useful define what happens when it is *absent*. In the 2003 movie *Kill Bill*, a female assassin awakens from a months-long coma and is able to kill man who is about to attack her. Fat chance.

The mass media often use the word *coma* indiscriminately, blurring the lines separating brain death, a persistent vegetative state, coma, and the so-called locked-in syndrome. Brain death encompasses a complete lack of electrochemical activity in the higher functioning regions of the cortex; patients in such a state will never wake up. A persistent vegetative state severely depresses brain functions, often as a result of trauma to the cerebral cortex, but while its patients lose higher cognitive functions their brain still works with some degree of effectiveness to pump blood, inhale and exhale air, and digest food. With good medical care, patients in a persistent vegetative state can live for months or years, albeit without the mental functions that many observers define as separating humans from lower animals. If a patient doesn't awaken in the first three months of a persistent vegetative state, he or she is unlikely ever to do so. Locked-in syndrome paralyzes all voluntary muscle control except for some facial movements. Patients remain aware of their environments but cannot interact with the outside

Jean-Dominque Bauby wrote a bestseller about his personal experience with locked-in syndrome.

+ BIG-SCREEN COMAS +

HOLLYWOOD FILMS depicting comas rarely do so realistically, concluded a neurologist who with his son viewed interpretations of comas in thirty Hollywood films and found accuracy in only two.

Eelco F. M. Wijdicks of the Mayo Clinic College of Medicine and his son, Coen, examined "realistic" movies released from 1970 to 2004 that included a comatose character. They found a common error, in which the filmmakers depicted the coma victim as tanned, fit, and seemingly asleep. The patients constructed by Hollywood often awake swiftly and get on with their life as if nothing happened. The researchers said a more accurate portrayal would show muscle atrophy, pallor, feeding tubes, and incontinence.

The films that got it relatively right? *Reversal of Fortune* (1990) and *The Dreamlife of Angels* (1998).

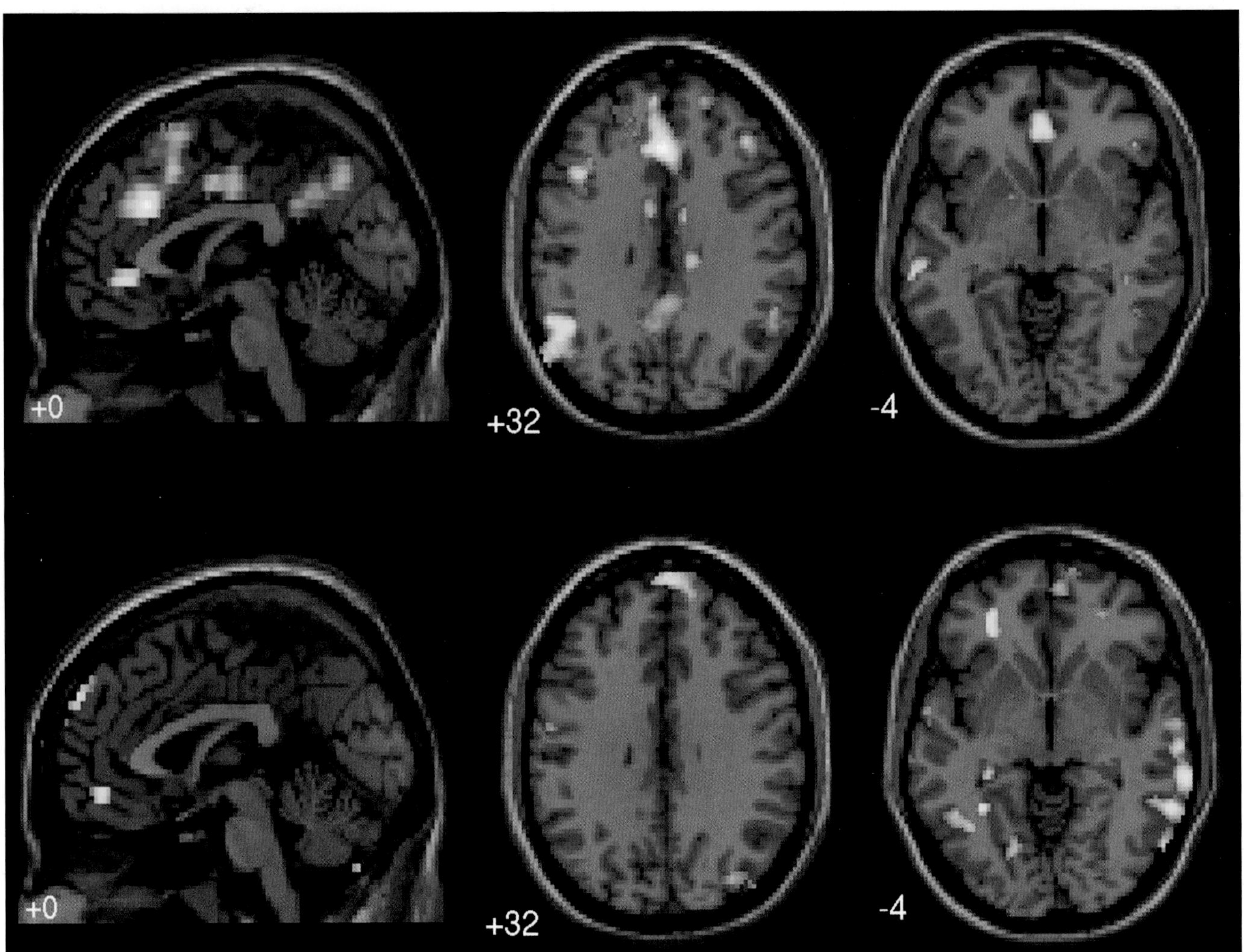

Brain activity in problem-solving regions increases when we daydream (top row) much more than when focused on a routine task (bottom row).

world except by blinking their eyes to communicate.

TRUE COMAS

A true coma is a deep level of unconsciousness in which a person cannot be awakened, even when subjected to extreme forms of stimulation. Causes of coma range from overdoses of alcohol or drugs, to epilepsy, infections, strokes, and insulin reactions. Coma patients retain noncognitive functions, and their brain goes through normal sleep patterns. While in a coma, patients may spontaneously move their limbs or eyes, or even grimace or cry. They may move their eyes in response to external stimuli. However, they have lost awareness of their surroundings; their reactions spring from automatic instead of controlled brain functions.

When a coma patient is out of danger from brain injury, medical care focuses on preventing infections, avoiding bedsores, and providing proper nutrition. Coma patients also often have physical therapy to maintain a minimal level of tone in their undertaxed skeletal muscle fibers. Coma patients often revive in a few weeks, but awakenings usually occur gradually, with

FAST FACT Journalist Jean-Dominque Bauby blinked one eye to communicate the text of *The Diving Bell and the Butterfly*, his 1997 book about his experience with locked-in syndrome. It took several months to "write" the memoir.

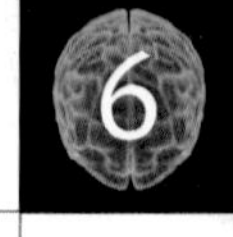

greater awareness and response building until reaching a threshold of cognition accompanied by clear communication.

Comas sometimes are induced medically as treatment for illnesses such as complex regional pain syndrome, a neurological disorder that causes extreme pain and in some cases has been shown to subside when the patient's central nervous system is "reset" by coma.

THE UNCONSCIOUS MIND

In a conscious state, the brain turns its attention to one thing after another, like a spotlight swinging through a dark night. In this metaphor, the unconscious mind takes in information at the edge of the light, and sometimes even in the darkness. Some neuroscientists believe the communication of the conscious and unconscious mind occurs across the corpus callosum, which connects the left and right hemispheres. Under this theory, normal consciousness requires both to function adequately.

The unconscious mind is constantly at work even though, by definition, we are not aware of it. Unconsciously received sensations influence thoughts and actions and can be quite powerful. According to psychologist Phil Merikle of the University of Waterloo, Ontario, "Unconsciously perceived information leads to automatic reactions that cannot be controlled

FREUDIAN ANALYSIS

Sigmund Freud, shown in 1938, sought specific reasons for troubling behavior.

MEDICINE once categorized a woman suffering from paralysis, pains, and other symptoms with no apparent origin as having a wandering uterus. The Greek for uterus was *hystera,* so these women were, clinically, "hysterical." But in the 1880s, doctors in Vienna observed men suffering hysterical symptoms.

An up-and-coming doctor, Sigmund Freud (1856–1939), and his mentor, Josef Breuer, noticed that when hysterical patients were urged to talk about early memories of symptoms, those symptoms generally abated. Freud developed theories linking many mental disorders to repressed memories and the unconscious. He created psychoanalysis to help patients find reasons for their mental troubles.

Freud tied basic human drives to generally repressed feelings about sex and death. He postulated the supreme importance of the unconscious mind on behavior. Detractors scoffed at his psychoanalytic method and his analysis of dreams. The worst wounds came in the 1960s from critic Karl Popper, who said a genuinely scientific theory must be subject to refutation, and later decades saw a rise in prescriptions as treatment.

But to dismiss Freud is to overlook his contributions. Freud insisted that neurotic behavior be treated as the product of a specific cause. Personality became something driven by mental energy, and the more science went looking for the source of free will in the human brain, the more difficult it became to pin it down.

by a perceiver. In contrast, when information is consciously perceived, awareness of the perceived information allows individuals to use this information to guide their actions so that they are able to follow instructions."

Merikle demonstrated the power of unconscious processing with an experiment. He had volunteers sit before a screen, upon which he flashed words, one after another, at a rapid rate. Some words came too swiftly to be perceived consciously, whereas others remained long enough to register. Merikle then presented the first three letters of a word that had been flashed and asked the volunteers to complete the word stem with letters that made a new word. For example, if the flashed word had been *dough,* they could correctly answer *doubt* or *double.* When the volunteers saw the word on the screen long enough to process it consciously, they had no difficulty completing alternative words. When the word *dough* had been flashed too briefly for conscious attention, however, they had great difficulty suppressing that word as their choice.

A classic test of unconscious processing dates to 1911. Swiss psychologist Edouard Claparede described an experiment he conducted on a patient who could recall nothing that happened to her only a few seconds before. Claparede hid a pin in his hands and stuck the tip into the woman's flesh when he shook her hand. After a minute or two, the woman completely forgot the incident. Yet when Claparede reached out again to shake hands, the woman held back. When he pressed her to shake, the woman protested that she thought people sometimes hid pins in their fingers but remained incapable of making a concrete accusation based upon her memory of being stuck. Her unconscious mind, unencumbered by the effect of memory loss on the conscious mind, had saved her.

FAST FACT

Ads often use unidentified celebrities' voices, which may unconsciously influence consumer behavior.

POWER OF SUGGESTION

Unconscious processing occurs all the time in advertising. Consumers respond defensively to overt advertising because they know advertisers stand to profit from their pitches and may provide less than objective information. That's why a new wave of advertising aims to tap straight into the unconscious mind, primarily through the use of product placements in movies and television. This is not the same as subliminal advertising, which briefly captured the nation's attention a half century ago before turning out to be a hoax. In 1958, television networks banned subliminal ads after reports claimed that movie audiences could be induced to buy soft drinks and popcorn by the flashing of written commands onto the projection screen for three-thousandths of a second. It turned out that such ads rarely had any effect on consumer behavior. Today, however, marketers who have plugged into the latest neurological research know that repeated exposure to

WHAT CAN GO WRONG

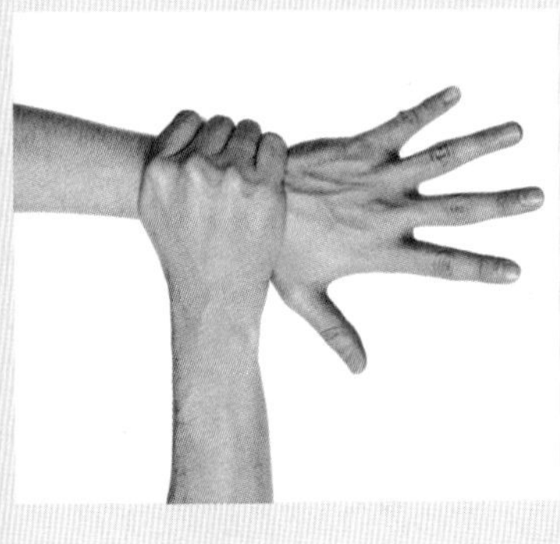

SOME BEHAVIOR happens outside conscious control. Two examples are "alien hand" (left) and environmental dependency syndrome, or EDS. Alien, or anarchic, hand causes one of a patient's hands to act independently of his or her willful control. Patients describe the hand's actions as foreign, or not part of the body. The alien hand may slap or pinch, much to the embarrassment of the patient. Or it may undo the actions of the other hand, as when one hand buttons a shirt and the other unbuttons it.

The syndrome usually arises after a person has undergone brain surgery,

CROSS REFERENCE: See "Silent Running," PAGE 146

Marketing agents know that sales decisions, such as whether to buy a dress modeled on a runway, often involve the subconscious mind.

consumer products affects a customer's desire for such products. A threshold level of exposure, such as a series of images of actors in a film wearing logo-branded clothing (or James Bond wearing a certain brand of wristwatch), raises observers' preference for the marketed object, even though they may not be aware of the branding, according to a 2008 study by Mark Changizi of Rensselaer Polytechnic Institute. Limited exposures suggest to the unconscious brain that the advertised product is scarce and thus has value. Changizi suggests this leads to the phenomenon of widespread adoption—the fad of new bands, new clothes, new cars. Once the exposure reaches a level where the unconscious brain perceives the object as common and thus easily obtained, it loses its value. Changizi's research points the way toward a potential wave of neuromarketing, where advertising campaigns are designed to sell products not on the basis of rational choice but rather through nonconscious appeal.

WHAT CAN GO WRONG

stroke, infection, or brain trauma. Grasping movements may occur after frontal lobe injuries, whereas complex actions such as unbuttoning are related to strokes, aneurysms, and tumors. No treatment currently exists; patients can try to keep the alien hand busy by having it grasp an object.

EDS occurs when a patient mimics others or feels impelled to engage in behavior triggered by environmental stimuli, even though he or she has no intention or idea of doing so. Patients feel compulsion to mirror the movements of those around them, combing their hair when someone else does, or standing or sitting in response to others. Or while undergoing a medical exam, they may pick up a tongue depressor and begin to examine the doctor. The syndrome has been linked to frontal lobe lesions; when EDS is brought on by acute stroke, symptoms may decline in a few weeks.

BRAIN AT REST

SLEEPING & DREAMING STATES

Japanese capsule hotels invoke streamlined mass production for business travelers seeking a single night's rest.

WHO CAN PUT a price on a good night's sleep? Without it, humans have trouble remembering what they've learned. They often experience emotional upheavals and make decisions that are less than optimal. They also may suffer a variety of health-related issues including increased risk of diabetes and obesity.

The brain treats lack of sleep as stress, resulting in the production of the hormone cortisol, which eventually damages the immune system. According to the Guinness Book of World Records, the longest certified bout of human wakefulness is 11 days. The reference book no longer recognizes attempts at the record because of the health risks involved. Menachem Begin, who served as prime minister of Israel from 1977 to 1983 and was tortured with sleeplessness by Soviet agents as a young man, said, "Anyone who has experienced this desire [to sleep] knows that not even hunger and thirst are comparable with it."

Even an extra day or so of good or bad sleep can have profound health effects. Researchers at Sweden's National Board of Health and Welfare examined their nation's comprehensive medical records from 1987 through 2006 and found heart attacks increased by 5 percent during the week after clocks sprang forward an hour in spring for daylight saving time, and dipped the same amount for the week in autumn when they "fell back." The lead researcher, Imre Janszky of the Karolinska Institute, said that he came up with the idea for the study after the spring adjustment left him groggy on a bus ride.

Despite evidence about the importance of sleep, science is only beginning to scratch the surface of why animals need sleep and how the brain regulates the mechanisms of sleeping and wakefulness.

FAST FACT Insomnia, once considered a disease, is now considered a symptom.

For centuries, observers considered sleep a somewhat passive process. The ancient Greek philosopher Aristotle traced the onset of sleep to the evaporation and condensation of food and drink in the stomach and bowels, carried upward to the head by rising heat.

Aristotle, who didn't think much about the brain, got his facts wrong: It's the brain, not the belly, that brings on sleep and wakefulness. He also erred in claiming that all animals sleep. Some fish, such as skipjack tuna, never fully sleep because they must constantly swim to keep oxygen flowing over their gills. And some amphibians, such as the bullfrog and salamander, never really sleep. Instead, these animals, which were among the first to evolve as land creatures, alternate between long periods of rest and short periods of motion.

Among reptiles and other lower vertebrates, sleep induces a slow, rhythmic pattern of brain waves, indicating the firing of many neurons in harmony. The patterns are somewhat like the slow-wave EEGs recorded when humans enjoy deep stages of sleep. As evolution created a more complex brain in

Hypnos, Greek god of sleep, rendered in a first- or second-century bronze.

STAYING SHARP

LACK OF SLEEP is bad for you. If you don't get enough, you're likely to have more stress, a lowered ability to make good decisions, and a propensity to gain weight. And that's just the beginning. Inadequate sleep has been linked to a higher risk of heart attacks. To get a good night's rest, try the following:

✓ Schedule regular times for going to bed and getting up. The body grows accustomed to daily rhythms.

✓ Don't consume caffeine within eight hours of trying to go to sleep.

✓ Get some aerobic exercise, such as walking, swimming, or jogging, during the afternoon. Don't do it within two hours of bedtime.

✓ Sleep in a room that has good ventilation, no swings between high and low temperatures, and windows that can be adjusted to block as much light as possible. If you can't darken the room sufficiently, try a sleep mask.

✓ Don't use an alarm clock that has a light or big, glowing numbers.

✓ Use a source of white noise, such as a fan, to block external noises that could disturb you. As an alternative, consider earplugs.

✓ Nibble on foods that induce sleep. Turkey is famous for being a source of tryptophan, a chemical that helps produce the neurotransmitter serotonin, which brings on drowsiness. Milk, cheese, and peanuts also induce sleep.

✓ If you can't fall asleep within 30 minutes of going to bed, get up and do some relaxing activity. Lying in bed stressing about the inability to sleep only makes matters worse.

birds and mammals, new kinds of sleep arose. Sleep stages expanded beyond the simple brain waves of snakes and fish to include patterns that appear on brain scans in ways that resemble wakeful stages.

SLEEP STAGES

It took a graduate student working on a degree in physiology in 1953 at the University of Chicago to detect and describe complex sleep patterns that finally destroyed the myth of sleep as a passive process. Eugene Aserinsky, working in a ward late at night, noticed that patients experienced jerky, symmetrical eye movements, accompanied by accelerated breathing and heartbeat, periodically as they slept. Aserinksy and his Ph.D. adviser, Nathaniel Kleitman, found that if they awakened the patients during these times of "rapid eye movement," or REM, they remembered vivid dreams but rarely recalled any if awakened when their eyes did not move. REM sleep accompanied a characteristic electrical signature that indicated activity in the cerebral cortex. The electrical firing patterns in the neural networks spiked in ways that suggested the cognition of someone who was awake, earning REM sleep its nickname, paradoxical sleep—so given because it doesn't seem very sleepy. Clearly, some mental stimulation occurred while patients snoozed.

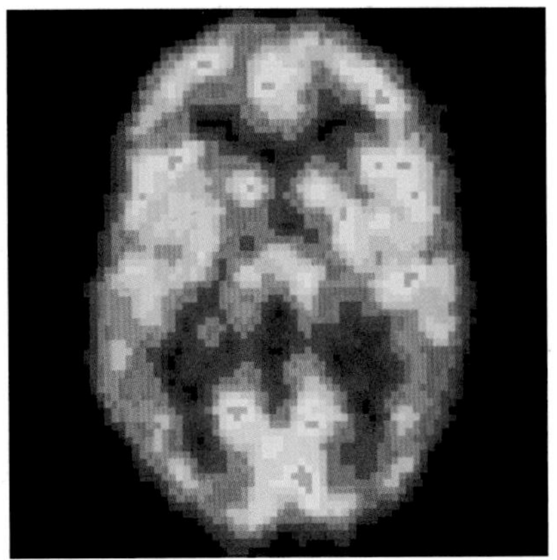

DREAMLESS SLEEP

A PET scan of a brain in a dreamless state shows relatively little activity.

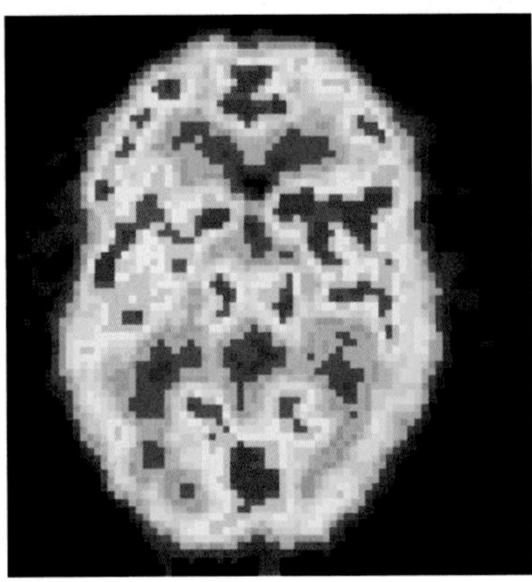

REM SLEEP

During rapid eye movement (REM), the sleeper's brain resembles wakefulness.

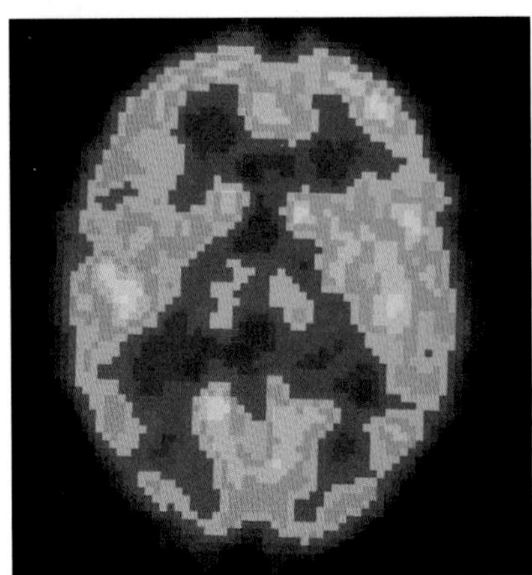

NORMAL SLEEP

The brain quiets when not in REM.

Human sleep has since been found to occur in five stages, with REM the fifth stage. After transitioning through all five, the cycle repeats itself. Stage 1 is light sleep, a level in which the sleeper can be awakened easily and may experience sudden muscle contractions. In Stage 2, which accounts for about half of the adult sleep cycle, brain wave activity decreases and eye movements stop. Extremely slow brain waves called delta waves appear in Stage 3 and predominate in Stage 4. These two stages form a very deep layer of sleep, in which it is difficult to awaken sleepers. Their muscles and eyes do not move, and if awakened they feel groggy for a few minutes. Stage 5, or REM sleep, accounts for about 30 percent of an adult's sleep time but 50 percent of an infant's.

Fish and reptiles don't experience REM sleep. As they rely on a comparatively primitive brain stem, they demonstrate the importance of the mammalian cerebral cortex in dreaming and make the case that dreams may not be necessary for life. Scientists believe that all mammals experience REM sleep, and evidence for this sleep stage also has been found in some birds. In humans, its onset usually occurs about 70 to 90 minutes after the beginning of sleep. Breathing becomes irregular, fast, and shallow, and the eyes flit about in various directions.

CROSS REFERENCE: See "Anatomy," PAGE 18

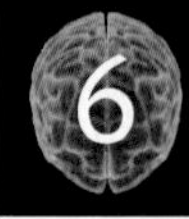

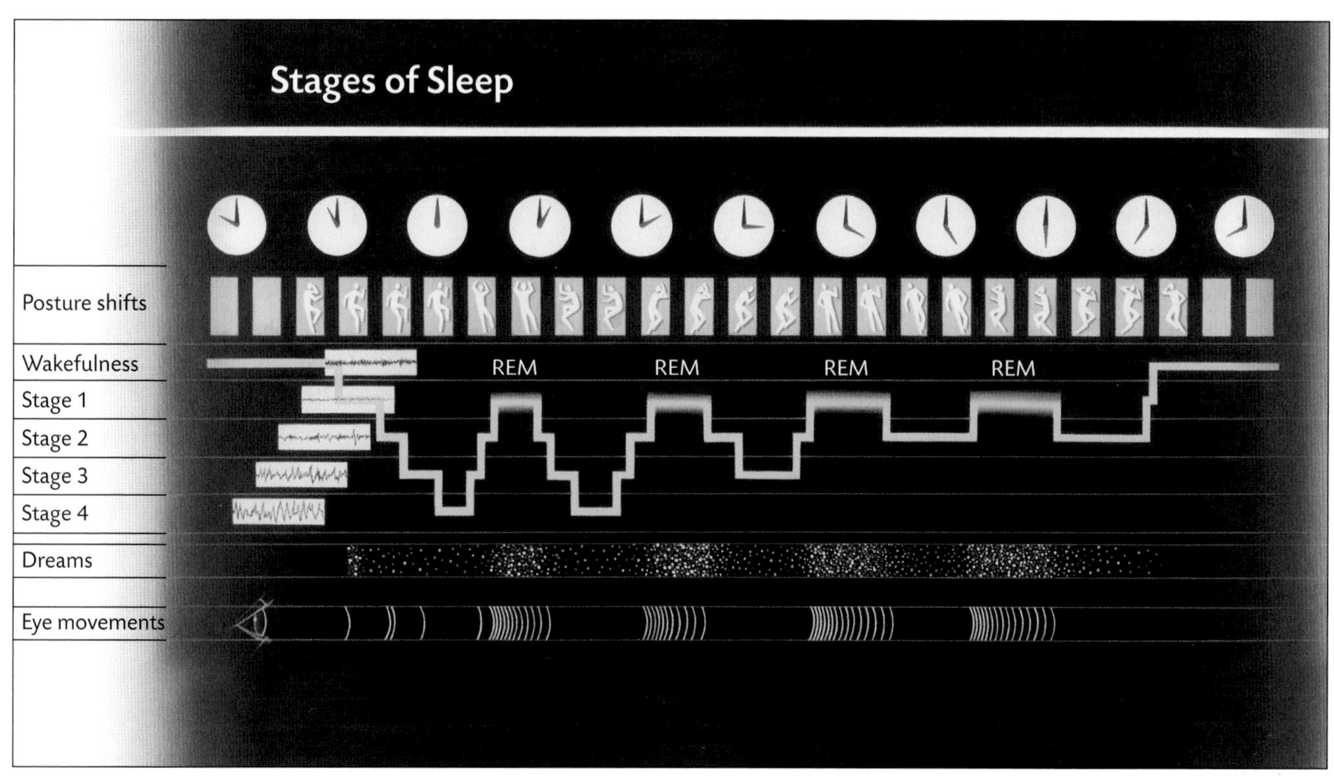

Changing brain waves correlate to shifts in sleeping patterns as recorded on an observational chart.

Limbs become paralyzed as the cerebral cortex inhibits the body's movements to keep it from acting out its dreams. A weak or nonfunctioning neural command to disable movement allows some people to sleepwalk. When the movement-inhibiting center is removed surgically from the brain of cats, they spit and claw in mock combat while they are experiencing REM. Apparently, cats dream a lot about fighting.

Sleepers often recall vivid dreams during REM. Afterward, the sleeper returns to Stage 1 to begin the cycle again. The longer the sleep continues, the more the REM stage lengthens and the deep sleep of Stage 3 and 4 decreases. By morning, sleep consists almost entirely of Stages 1, 2, and 5, with REM lasting about an hour at a time.

How much sleep each person needs on an average night depends on age, health, and prior sleep patterns. Infants need about 16 hours a day. Teenagers require about nine hours, and adult men and women seem to function best when getting seven to eight hours, although individual requirements vary for optimum sleep each night. Researchers at the Johns Hopkins University, in a paper published in 2007, suggest that gender may affect sleep patterns. Adult women have better sleep quality compared with men, the researchers said. Furthermore, men who wake up before getting seven hours of sleep do better at tasks than similarly sleep-deprived women. According to the study, 58 percent of men could function at their best on such short sleep schedules, compared with 43 percent of women. (Part of the difference may reside merely in how the two sexes described their sleep to the university observers. The researchers found that women

FAST FACT Thirty-five percent of Americans surveyed in 2002 reported one of the following occurred regularly: difficulty falling asleep, waking a lot, waking and not being able to get back to sleep, or waking unrefreshed.

were more likely to report sleep problems, whereas men were less likely to complain.)

For both sexes, the need for sleep increases when they have been deprived of adequate shut-eye. Think of going without sleep as taking out a loan from the sleep bank. The loan accrues interest, getting bigger and bigger the longer the debtor goes without sleep. Sooner or later, the brain demands that the debt be repaid. Until accounts are settled, the brain functions at a subpar level, with slower reaction times, impaired judgment, and emotional instability.

THE ONSET OF SWEET SLEEP

How and why the brain induces sleep is hotly debated by experts. Conventional wisdom about the brain, which emerged in the last half of the 20th century, suggests that certain regions of the brain stem regulate sleep and wakefulness in response to fatigue and diurnal rhythms. Scientists have noted, for example, that neurons in the ascending reticular formation send electrochemical signals via the thalamus, hypothalamus, and basal forebrain to the cerebral cortex during wakefulness. While the brain is in a wakeful state, the cerebrospinal fluid that surrounds it accumulates proteins called sleep regulatory substances, or SRSs. A threshold level of SRS concentration in the fluid induces sleep.

Neurons in the pons appear to be sufficient to generate REM. These neurons send signals to the thalamus, which in turn communicates with the cerebral cortex. A small region of the pons called the lateral pontine tegmentum, occupying just a few cubic millimeters of brain tissue, apparently initiates a wave of changes in brain activity that bring on REM sleep. In lab experiments, REM sleep can be induced by injecting the neurotransmitter acetylcholine directly into the pons.

A NEW THEORY OF SLEEP

In a 2008 challenge to this so-called top-down model, sleep scientists at Washington State University published a hypothesis that the brain has no control center for sleep. Instead, they proposed that

A polarized light micrograph reveals the inner structure of crystals of acetylcholine, a neurotransmitter linked to REM sleep.

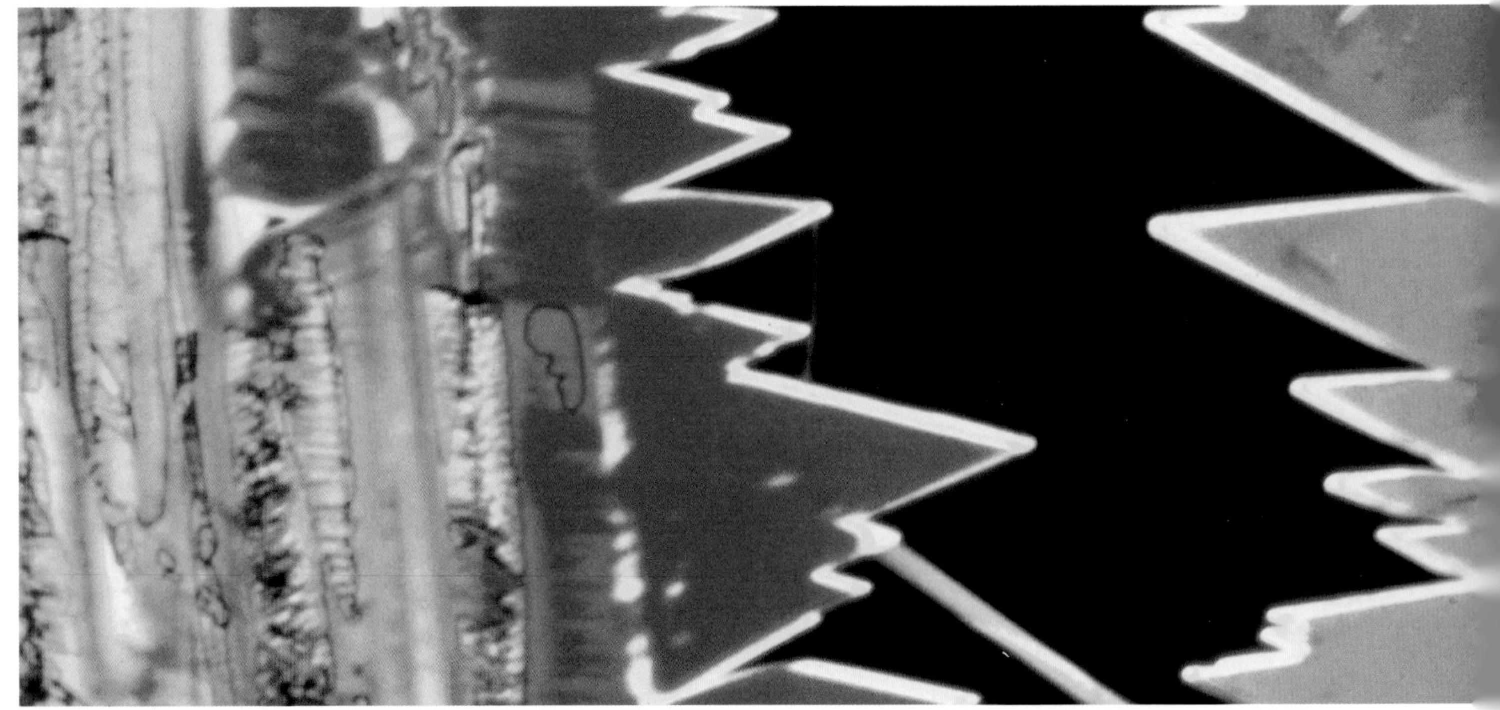

sleep occurs as an "emergent property" through the spread of fatigue among neural networks. In this view, sleep creeps up on the brain bit by bit, as first one group of neurons becomes tired and switches to a low level of activity, and then others follow like dominoes. When a threshold number of neural networks go dim with fatigue, sleep quietly arrives.

The idea of sleep arriving in bits and pieces explains such phenomena as sleepwalking, in which the sleeper's neural networks governing balance and movement are awake but those for consciousness are not, and the sluggishness of sleep inertia, which dulls cognition for a few minutes after awakening.

FAST FACT Sleepy people feel more energy at 4 a.m. or so, no matter how much they've slept.

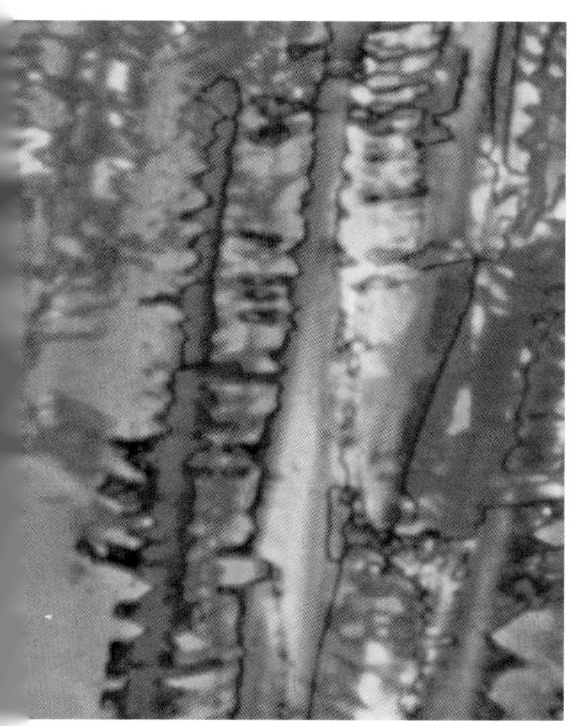

Researcher James Krueger noted that observers would expect the entire brain to react at once if specific regions dictated the conditions of sleep in a top-down model. What Krueger and his research team found more closely resembles an orchestra that does not have a conductor.

While most sections play in excellent time with each other, a few lag behind or zip ahead. In this analogy, if the "woodwind" neurons have been performing a fast and furious piece for some time, they are more likely to grow fatigued and enter the sleep state than the "brass" neurons, which had to work only during the overture and the finale. They noted that when a drop of SRS protein falls onto a few hundred neurons, they go to sleep without affecting other brain regions, demonstrating how sleep can be localized.

Neurons release adenosine triphosphate (ATP), the energy source for cells, when they communicate electrochemical signals to other neurons. ATP prompts glial cells to produce SRS proteins. Over time, these proteins enter nearby neurons and reach a concentration point in which they bring on a wave of chemical changes, altering how neurons react to neurotransmitters and bringing on sleep. Krueger and his co-researchers developed a mathematical model to show how the shift to a sleep state in one group of neurons could spread to others until the entire animal goes to sleep. Still, there's room for the pons, thalamus, and other "sleep centers" in this proposed model, Krueger said. He believes they coordinate the sleep-wake states of neuronal groups to achieve peak performance and react appropriately to external stimuli, such as nightfall and daybreak.

Scientists wrestle with the algebra of sleep because just how much a person really needs does

+ BEHIND THE WHEEL +

WHEN SLEEP-DEPRIVED drivers get behind the wheel of a simulator, their reaction times and judgment become as impaired as those with elevated blood-alcohol levels. Drinking alcohol *and* being sleepy severely affect the sensory, motor, and analytical skills of drivers. Every year, about 100,000 accidents occur in the U.S. because drivers were overly tired, according to the National Highway Traffic Safety Administration; about 1,500 drivers died. To avoid being a statistic, if you can't stop yawning, can't remember having driven the last few miles, or can't focus your vision, leave the road and get some sleep.

FAST FACT Alcohol may make you feel sleepy at first, but it has been shown to disrupt a good night's sleep. Alcohol interferes with sleep's rhythms, lowers oxygen consumption, and inhibits memory formation.

not seem to be a constant—it doesn't accumulate arithmetically over time. For example, if you stayed up past midnight reading a good book, you might have found that around 3 or 4 a.m. you felt more awake than you did a couple of hours before. The same goes for long-distance driving: A burst of wakefulness arrives just before the dawn, as you pull into a truck stop for breakfast after driving a couple of hundred miles.

"Since you start to get less sleepy as you approach the early morning hours just before dawn, it indicates that people don't get sleepy simply on the basis of missed sleep," said Thomas Wehr, a researcher of sleep and biological rhythms at the National Institute of Mental Health. "If that were true, you would get sleepier and sleepier the longer you stayed up. Instead, the pattern is typical of a cyclic internal program."

THE PURPOSE OF SLEEP

Some scientists believe that sleep allows the brain to "consolidate" memories. When you learn facts, faces, events, and other information about the world, those data initially are collected, sorted, and dispersed by the hippocampus. Gradually, in a process not fully understood, the data in those memories get redirected to long-term storage in various brain regions. How and where those memories get sent, as well as how they reconnect in neural networks, remains a mystery, but scientists are confident that sleep plays a role. Researchers at Israel's Weizmann Institute noted in the 1990s that when they awakened test subjects 60 times a night during REM sleep, the human guinea pigs completely lost the ability to learn new information. However, similar interruptions during non-REM sleep had no such effect, suggesting REM sleep plays a key role in organizing information and forming the links to make lasting memories.

Observers have noted that people almost never dream about things that happened to them in the hours before they went to sleep. Only after a delay of a few days do the memories of events work their way into dreams. Sam Wang, a neuroscientist and molecular biologist at Princeton University, suggests that this may be because sleep helps people process those events.

Humans repeatedly awakened during REM sleep, the period of vivid dreams, need to have more and more REM sleep to make up for their deficit. The greater the REM deprivation, the greater the need for substantial REM "rebound." Such experimental findings support the theories of psychoanalyst Sigmund Freud, who argued a century ago that people need dreams to make sense of the world and fulfill basic desires. If we forgo the dreams of REM sleep, we find ourselves rushing back to them as soon as possible. Freud would have interpreted

WHAT CAN GO WRONG

NARCOLEPSY targets about 135,000 Americans, even those who typically get a good night's rest. They have daytime attacks from a few seconds to a half hour, during which they may experience sudden muscle weakness, drowsiness, hallucinations, and periods of sudden sleep. The disorder appears to cause symptoms of REM to break through during periods of wakefulness. Humans typically enter sleep cycles in shallow stages that lead within about 90 minutes to REM sleep. Patients with narcolepsy enter the deep REM stage very soon. The disorder usually appears in people between 15 and

CROSS REFERENCE: See "Learning," PAGE 236

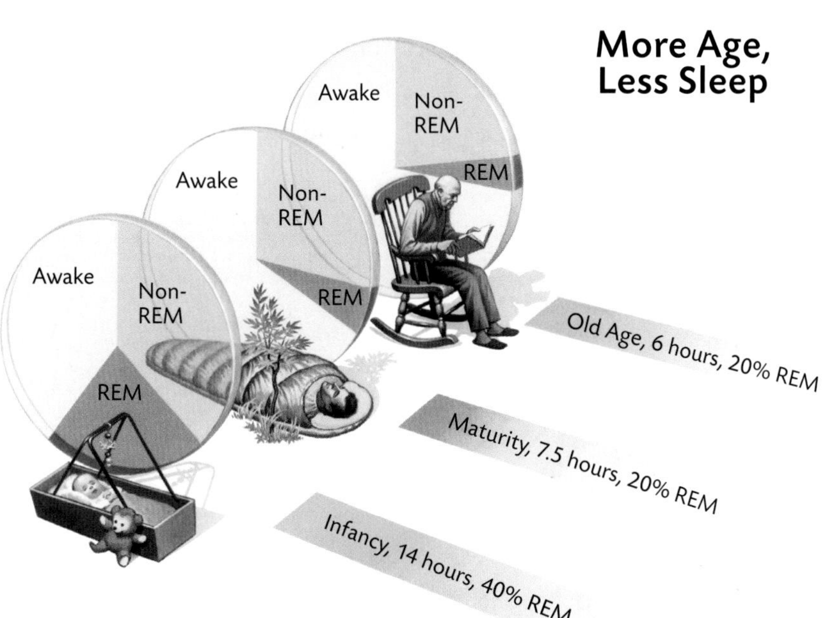

An infant spends over 40% of sleep in REM while adults spend 20% or less in REM.

that as a strong need for humans to work out unresolved issues through the unconscious medium of the dream.

DREAMS

Like the mechanics of sleep, the science of dreams is a vast jungle that science has only begun to explore. Dreams and their interpretation have carved out a huge space in human history, literature, science, and religion. In the Old Testament, Joseph and Daniel gained a measure of power by interpreting the dreams of kings. Mary Shelley had the germ of *Frankenstein* come to her in a dream; Robert Louis Stevenson got the idea for *The Strange Case of Dr. Jekyll and Mr. Hyde* in the same way. In 1845, Elias Howe dreamed of a mechanical improvement that would make his sewing machine practical. And a few days before he was shot in 1865, President Abraham Lincoln famously told his wife that he dreamed of his own assassination.

In modern science, Otto Loewi's grasp of how neurotransmitters carry information across the synaptic cleft came to him in a dream, and Belgian scientist Friedrich August Kekulé von Stradonitz deciphered the ringlike structure of the benzene molecule upon awakening from a dream in 1865. The molecule's structure had confounded chemists, who couldn't diagram it with a linear alignment of atoms. Kekulé fell asleep in front of the fire, as he had many times, and dreamed of atoms "gamboling" about and forming connections. Snakelike chains began moving about in his dream. "One of the snakes had seized hold of its own tail, and the form whirled mockingly before my eyes," Kekulé recalled. "As if by a flash of lightning I awoke; and this time also I spent the rest of the night in working out the consequences of the hypothesis." The snake grabbing its own tail in its mouth provided the clue: Benzene molecules form

WHAT CAN GO WRONG

25, although it can strike at any time and often goes undiagnosed. Stimulants and antidepressants often help lessen the effects. And if they take daytime naps of 10 to 12 minutes, avoid heavy meals, and keep away from nicotine, alcohol, and caffeine, they may reduce the disorder's impact on daytime activities.

Working with dogs in 1999, Stanford University researchers, including William C. Dement (opposite, holding a dog before and after a narcoleptic attack) and lead researcher Emmanuel Mignot, discovered a gene that causes narcolepsy. It leads to severe lack or absence of hypocretin, a neurotransmitter that promotes wakefulness. Mignot has since turned his attention to the same defective genes in humans. He asks: What makes neurons that create hypocretin die in narcoleptics and in late-stage Parkinson's patients? He believes that the answer will expand what science knows about how, and why, people sleep.

rings, not lines. Kekulé told his colleagues, "Let us learn to dream!"

MESSAGES TO THE SELF?

Such stories would appear to suggest that dreams open a doorway of communication between the conscious and unconscious mind. Freud would have agreed, as he famously called dreams the royal road to the unconscious. He chose to interpret dreams as manifestations of unconscious desires, many of which dealt with repressed sexuality. When the conscious mind is unwilling or unable to deal with important ideas or emotions, Freud said, the unconscious mind performs like a psychic safety valve, resurrecting the conflicts and playing them out to a resolution—albeit in symbolic forms that require interpretation.

Neurologist Richard Restak observed that Freud's description of the function of dreams relies on the metaphors of 19th-century engineering. According to this view, human psychology runs like a steam engine, building up pressure through powerful combustion, and unless the steam finds an outlet, it could explode. Freud had no room in his theories for the idea that dreams might arise from the normal, physical functioning of the brain. Most modern researchers have discarded Freud's theories, finding natural processes in dreams arising from basic brain functions. Harvard University researchers J. Allan Hobson and Robert McCarley developed the "activation synthesis" theory of dreams in 1977. It says dreams aren't stories with symbolic meaning but rather the brain attempting to impose order on the static caused by the random firing of neural networks connecting the pons and the cerebral cortex. The two researchers found a group of neurons in the pons that fire more frequently during sleep than during wakeful hours. When REM sleep begins, these cell groups send signals to the cerebral cortex. According to Hobson and McCarley's theory, the cortex then tries to create some measure of coherence out of the random information it receives. The reason so many dreams contain weird and fantastic elements is that the brain faces the task of constructing narratives out of chaos.

As elegant as their theory may appear, it doesn't explain everything. Who hasn't had a dream that came true, like Lincoln? And who hasn't found inspiration for solving a problem, like Kekulé? And, finally, who hasn't had a dream in which even an amateur, armchair psychoanalyst couldn't have found

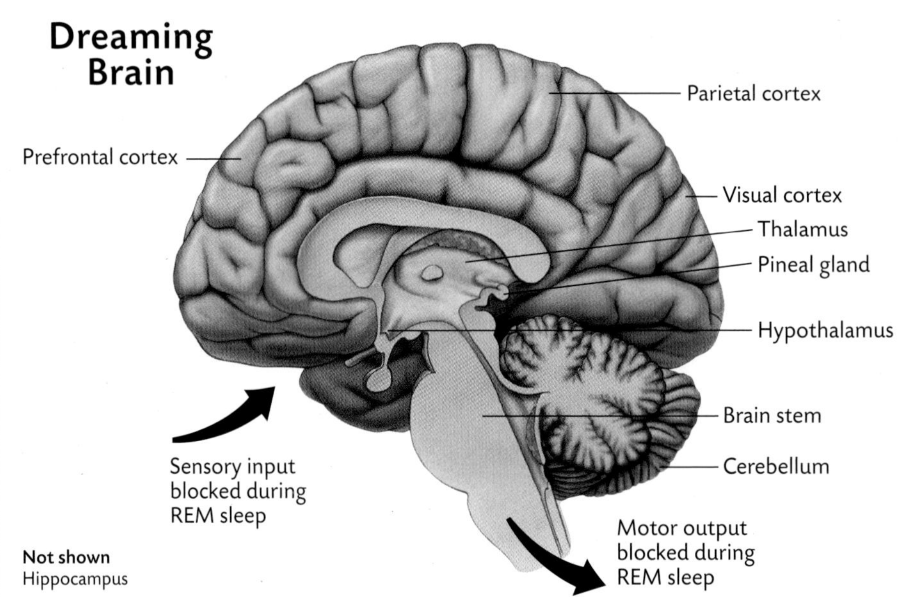

The purpose of dreaming is not fully understood, but the dreaming brain is very active.

+ INTERPRETATION +

THERE'S NO REASON why dream interpretation's biases can't creep into the work of dream interpreters. Freud found what he suspected he would in many dreams—sex. "Freud himself suggested that dreams of flying revealed thoughts of sexual desire," dream researcher Carey Morewedge said. "Interestingly, in the same text [*Interpretation of Dreams*], Freud also suggested that dreams about falling also indicate succumbing to sexual desire. One might interpret this as evidence that scientists are just as self-serving as laypeople when interpreting their dreams."

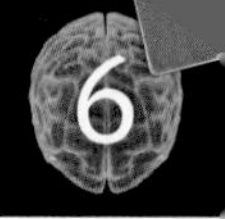

some symbolic, sexual meaning? "There are more things in heaven and earth, Horatio, than are dreamt of in your philosophy," said the melancholy Dane Hamlet.

FAST FACT The word *dream* appears 74 times in the King James version of the Bible.

WISDOM OF DREAMS?

Stanford University researcher William C. Dement, who began investigating sleep after being inspired by fellow student Eugene Aserinsky at the University of Chicago, has found that creative minds can use dreams like keys to unlock puzzles. In his college course on sleep and dreams, he gives his undergraduates brain teasers and encourages them to "sleep on it." In one test, he gave 500 undergraduate students the letters *o, t, t, f,* and *f,* and told them that they formed the beginning of an infinite series. He asked them to come up with the next two letters. Two students figured it out while awake, and seven got the answer while asleep. One student who dreamed the answer told of his dream-state body walking through a gallery and counting pictures. The student counted the first five paintings, then found numbers six and seven had been ripped from the wall. That was the solution—the letters represented the numbers one, two, three, four, and five. The next letters were *s* and *s,* for six and seven.

Obtaining solutions to problems while dreaming raises questions. Did the unconscious mind

Goggles alert observers that sleep researcher Stephen LeBerge of Stanford University has begun dreaming.

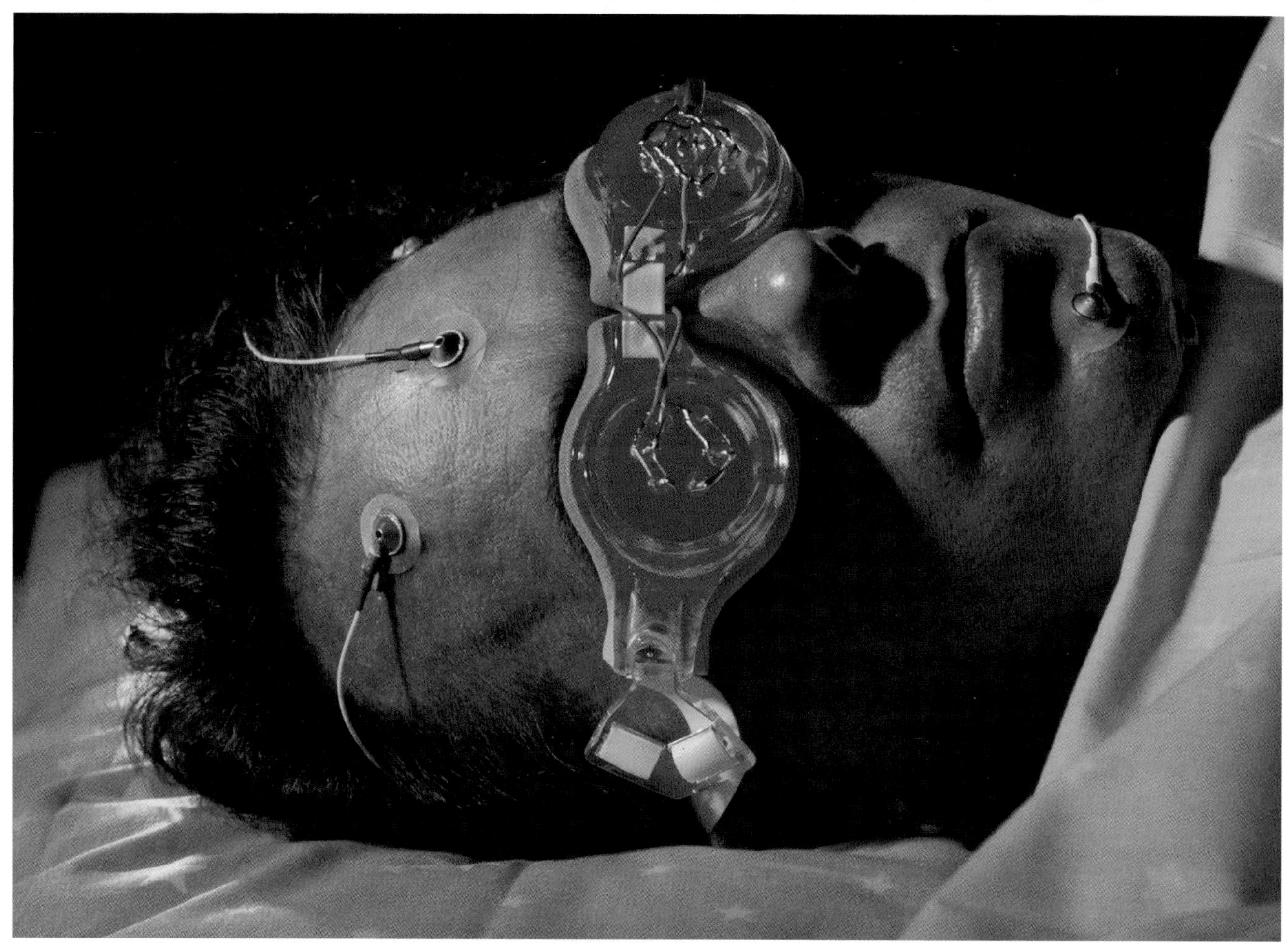

Surreal combinations of objects suggest dreams or alternate realities in 1953's "The Enchanted Domain III" by Belgian artist René Magritte.

have the answer while the conscious mind did not? And if the alert mind could not solve the problem, how could the sleeping mind do so?

DAILY REVIEW?

Some researchers believe dreams rehash fragments of daily life in a process known as incorporation. Incorporated events usually take a few days to appear in a dream. In a 1978 experiment, test subjects fitted with red goggles for several days began to see more and more red-tinted objects in their dreams.

Tore Nielsen, a dream researcher at the University of Montreal, hypothesizes that the delays of incorporation might be an indication of the unconscious mind "working through" problems. Dreaming, Nielsen said, "facilitates adaptation to the stresses and emotional difficulties of interpersonal relationships." Others, such as Mark Mahowald of the Minnesota Regional Sleep Disorders Center, scoff at the idea that dreams help the brain deal with emotional or psychological issues. He noted that scary dreams do not become more common after seeing a horror movie and that starving or thirsty people rarely report dreams of eating and drinking.

WHAT CAN GO WRONG

SLEEP APNEA, a blocking of the windpipe that temporarily cuts off oxygen to the sleeping brain, afflicts about 18 million Americans, most of them men. It occurs most often when loss of muscle tone or a buildup of fat partially obstructs the windpipe during sleep, when the body's muscle groups relax. Misfiring of neural networks associated with breathing also can bring on apnea. When the movement of air through the windpipe decreases sufficiently, the air entering the lungs gets cut off for a few seconds. The brain recognizes a drop in oxygen levels in the blood and awakens the sleeper enough to adjust the muscles around the windpipe and get air flowing again. The patient usually gasps or snorts and goes back to snoring. The cycle of closing, gasping, and awakening repeats itself throughout the night. Apnea patients typically are tired the next day, but the disorder can also lead

However, if dreams incorporate bits and pieces of everyday life, they pick up on stress and anxiety in disproportionate amounts. In the 1940s, researcher Calvin S. Hall of Case Western Reserve University cataloged the content of more than 1,000 test subjects' dreams. Anxiety ranked as the most common emotion of those dreams, and negative content outnumbered positive content.

Patricia Garfield, author of *Creative Dreaming,* said that the sensation of being chased—surely, a negative phenomenon—occurs in the dreams of about 80 percent of humans. Dreams of falling afflict about 60 percent. Why this is so may stem from the array of brain regions electrically firing during REM sleep.

According to Michael Stephenson, a sleep researcher at the University of Wisconsin, "During REM sleep, the parts of the brain that control emotional response and fear are more active. That facilitates having strange or bizarre elements in dreams."

On those occasions when strange and bizarre dreams seem to presage a creative breakthrough or act as premonitions of doom, they get remembered and written down. Far from being harbingers of psychic phenomena, such dreams may be mere relics of chance. Out of all the dreams we dream, many, if not most, get forgotten quickly. Only the ones that, by random chance, seem to correctly predict the future become the stuff of legend, such as Lincoln's death dream. Bizarre dreams that lead nowhere don't get counted on a statistical scoreboard of hits and misses.

A DREAMER'S MIND

The meaning of dreams, though, lies not in the dream itself, but in the dreamer.

FAST FACT Humans spend an average of more than two hours each night dreaming.

A series of studies conducted by psychologists Carey Morewedge of Carnegie Mellon University and Michael Norton of Harvard and reported in 2009 suggest that bias about dreams is in the eye of the beholder. They asked more than 1,000 people in India, South Korea, and the United States about the significance they placed on their dreams. Perhaps unaware of modern research into the "static" of random impulses, most responded with Freudian and Jungian notions of dreams opening pathways to unconscious emotions. Many saw omens. They said they would be more likely to cancel a plane trip if they dreamed of a plane crash than if they saw a news report of a plane accident along their intended flight path. Personal biases appeared during interpretations about people. Significance increased for dreams in two kinds of categories: negative dreams about people the dreamers disliked and positive dreams about people they liked. The researchers termed this the motivated approach to dream interpretation. In other words, dreamers tend to interpret dreams along the lines of their personal biases.

WHAT CAN GO WRONG

to stroke, hypertension, irregular heartbeat, and heart failure. It can contribute to poor performance in school or at work, as well as motor vehicle crashes from drowsy drivers.

Besides gender, risk factors include being older than 40 (although apnea targets people of all ages), being overweight, having a thick neck or large tonsils, and family history of apnea. Treatments range from weight loss, to adjusting sleeping positions (not on the back), to special masklike devices that alter the pressure relationships at the back of the throat, to surgery aimed at preventing obstructions from forming.

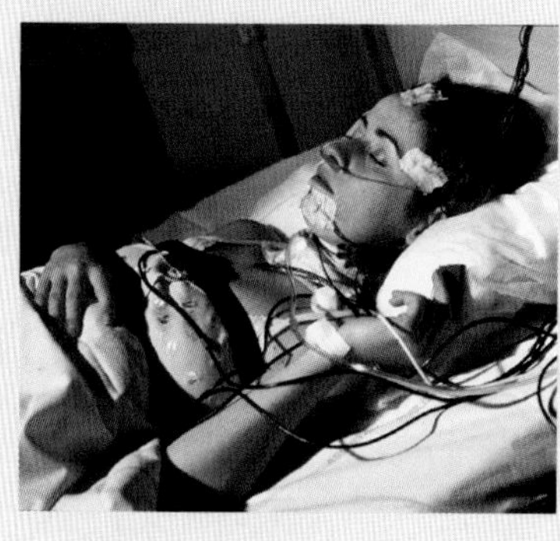

ALTERED STATES

EFFECTS & PROBLEMS

ALTERED STATES of consciousness arise from conditions that push the brain into something besides the normal waking state. Problems in the brain's ability to process information may arise from internal causes, such as schizophrenia and Parkinson's disease, or from external stimuli. People often induce altered states through substances; some are medications, prescribed to treat problems, while others, like drugs and alcohol, can also change brain functions to create an altered state of mind.

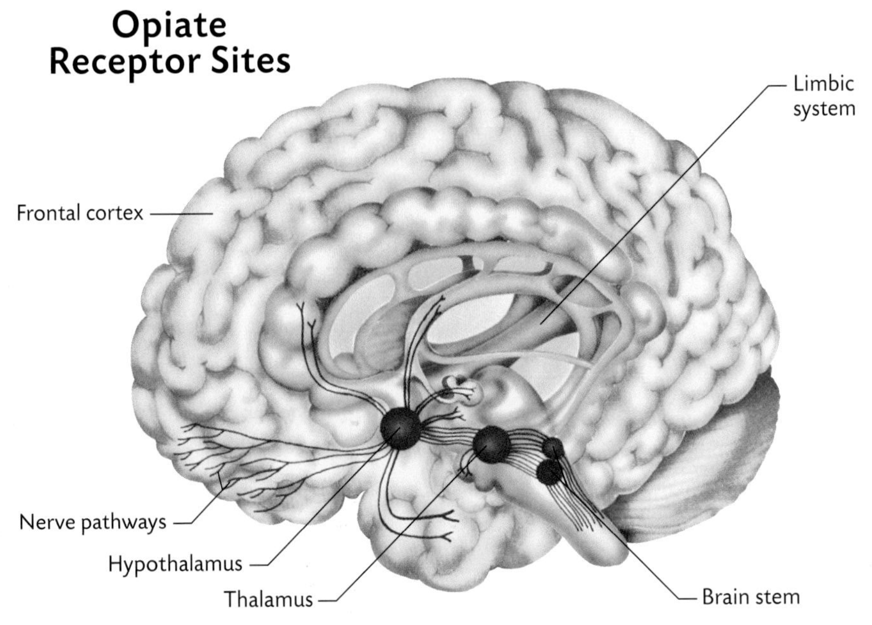

Opiate, or morphine-type, drugs involve many organs in the brain.

Most mind-altering substances change the functioning of neurotransmitters. Some mimic the work of the brain's stockpile of neurotransmitters while others prevent those neurotransmitters from doing their jobs. Common targets of mind-altering drugs are the brain's metabotropic receptors, a special type of neurotransmitter receptor—the "lock" to the neurotransmitter's "key"—that unleashes chemical signals to regulate the action of cells and the tissues they make up. Metabotropic receptors and their neurotransmitters interact within neural networks to regulate personality, movement, mood, attention, and sleep.

MEDICATIONS

Imbalances in a class of neurotransmitters called monoamines, including dopamine, serotonin, and adrenaline, play key roles in disorders such depression, Parkinson's disease, schizophrenia, and irregular sleep patterns. Of particular interest is serotonin. As serotonin interacts with more than a dozen receptors, it's difficult to predict how altering its function may affect someone. However, drugs that mimic or alter levels of serotonin typically affect mood and sleep.

Prozac, among the most commonly prescribed medications for depression, is one such drug. It needs to build up in a patient's brain before it takes full effect, and that can take several weeks of daily ingestion. When it does go to work, it prevents serotonin in the synaptic cleft from being reabsorbed by neighboring neurons, making Prozac a "serotonin reuptake inhibitor." As a result, serotonin molecules remain available in the spaces between axons and dendrites to forward electrochemical communications for longer periods of time. Exactly how

FAST FACT Foods rich in carbohydrates, like sugary snacks or potato chips, cause serotonin-producing cells to release the feel-good neurotransmitter. Serotonin, in turn, regulates appetite and helps prevent overeating.

CROSS REFERENCE: See "Nerve Cells," PAGE 10

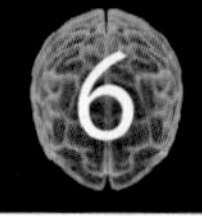

that lingering effect alters mood remains unknown.

A DANGEROUS HIGH

The street drug Ecstasy (technically known as methylenedioxymethamphetamine, or MDMA) brings about a euphoric "high" by working the same way as Prozac, except a single dose works almost immediately and lasts for three to four hours. It prevents a particular protein from transporting serotonin through the cell walls of nearby neurons, causing it to linger in the synaptic clefts. During the 1960s, some psychotherapists used it to bring about feelings of well-being in their patients. Some patients also experienced hallucinations.

Because of Ecstasy's ability to create mild euphoria and feelings of empathy, it became popular recreationally in the early 1980s. The U.S. Drug Enforcement Administration declared it illegal in 1985, and the National Institute on Drug Abuse has found that it damages serontonin-releasing neurons. Ecstasy's use has been linked to damage to the kidneys and the learning centers of the brain, but its likelihood of being abused appears to drop with long-term use because its emotional impact lessens with time. Why Ecstasy works so swiftly compared with Prozac is unknown. Possibilities include the rates at which Ecstasy and Prozac enter the brain, as well as Ecstasy's

DISCOVERING A DRUG

Albert Hofmann called LSD "My Problem Child" and spoke of it until his death at 102.

AS CHEMIST Albert Hofmann worked on the alkaloids of ergot fungus at his lab in Basel, Switzerland, in April 1943, he began to feel lightheaded. He went home and fell into a "kaleidoscope" of hallucinations. His curiosity aroused, Hofmann decided to deliberately try to re-create the experience three days later by exposing himself to lysergic acid diethylamide (LSD), a chemical he had been working on when he first felt dizzy. He ingested 250 micrograms in ten cubic centimeters of water. His limbs stiffened; he felt disoriented and experienced a "marked desire to laugh." In perhaps the most famous bicycle ride in history, Hofmann pedaled for home. He did not feel he was moving, but objects around him did—they spun and changed shape before his eyes. Faces morphed into colorful and hideous masks. After the horrors passed, he entered a state in which images opened and closed and sounds registered as colors. The drug had no lasting physical effect, but it powerfully opened the mind to new sensations. Perhaps LSD's greatest danger lies in lowering inhibitions and impairing judgment. LSD users have accidentally killed themselves by falling from buildings and walking into traffic, and the drug's effects have exacerbated the symptoms of users with preexisting mental disorders. The U.S. outlawed the drug in 1966 after declaring it unsafe.

side effect of partially blocking dopamine uptake. Such a chemical effect would make Ecstasy have a short-term impact on the brain like the dopamine-altering drugs cocaine and amphetamine.

CONTROLLED SUBSTANCES

Among drugs that affect serotonin, none acts as powerfully as lysergic acid diethylamide, or LSD. It binds so tightly to serotonin receptors that minuscule amounts—less than 50 micrograms, or one ten-thousandth the weight of an aspirin tablet—bring about profound alterations in the brain. Users report hallucinations and other altered states of consciousness.

Albert Hofmann, the Swiss chemist who synthesized LSD in 1938, accidentally exposed himself to about 25 micrograms in 1943 and discovered the drug's tremendous power to induce hallucinations. Usually, the images cause no lasting harm, but in some cases, LSD has been linked to psychosis, particularly when taken by someone with an already documented mental disorder. Unlike many other drugs, however, LSD has not been associated with death as a direct result of overdose; it binds so strongly and so specifically to serotonin receptors that an increase would have no effect. Many other drugs, on the other hand, affect a variety of neurotransmitter receptors and bring about a toxic stew of chemical reactions. However, because of LSD's powerful impact on the brain and its potential for abuse, the U.S. government classifies it as a Schedule I substance, the same as heroin and Ecstasy, under the Controlled Substances Act.

FAST FACT *In Health* magazine asked experts to rank drugs by their potential for addiction. Topping the list was nicotine.

Marijuana and caffeine act on metabotropic receptors through a different set of neurotransmitters, albeit in opposite ways. Marijuana's active ingredient, delta-9-tetrahydrocannabinol (THC), inhibits the release of the neurotransmitters glutamate and GABA. THC's inhibitory properties reduce communication among certain neural networks. Caffeine increases the likelihood that neurons will release GABA and glutamate. That's why taking a drink of coffee stimulates the brain and slightly raises cognitive function.

ADDICTION

Some drugs become habit forming, exerting powerful influence on the brain, particularly on neural networks associated with rewards. Users take the drug repeatedly to receive the neurochemical reward, an alteration in brain chemistry that, at least at first, creates feelings of well-being, euphoria, and calm, or shuts out unpleasant physical or emotional effects.

Brain scans allow researchers to observe the influence of addictive drugs. Morphine acts on the cerebral cortex but lets the lower, older portions of the brain continue unencumbered. Cocaine activates the entire brain but gives an extra stimulus to its emotional centers, a clue to users' reports that the drug

WHAT CAN GO WRONG

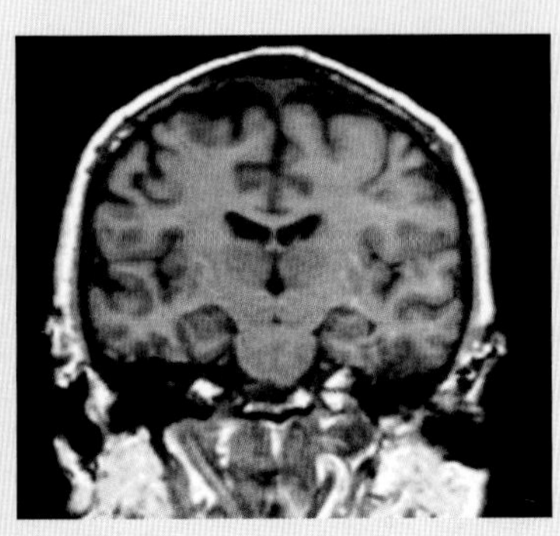

ALCOHOL ABUSE can lead to a plethora of problems: the inability to meet work, school, or family responsibilities, drunk-driving arrests and car accidents, and drinking-related medical conditions. But heavy drinkers also experience unusual shrinkage of their brain. Research on Japanese subjects showed that half of heavy drinkers in their 50s experienced brain shrinkage, while only 30 percent of nondrinkers exhibited shrinkage. Both white and gray matter were affected in drinkers' brain.

The reduction in gray matter volume appears to underscore the popular notion that alcohol kills neurons,

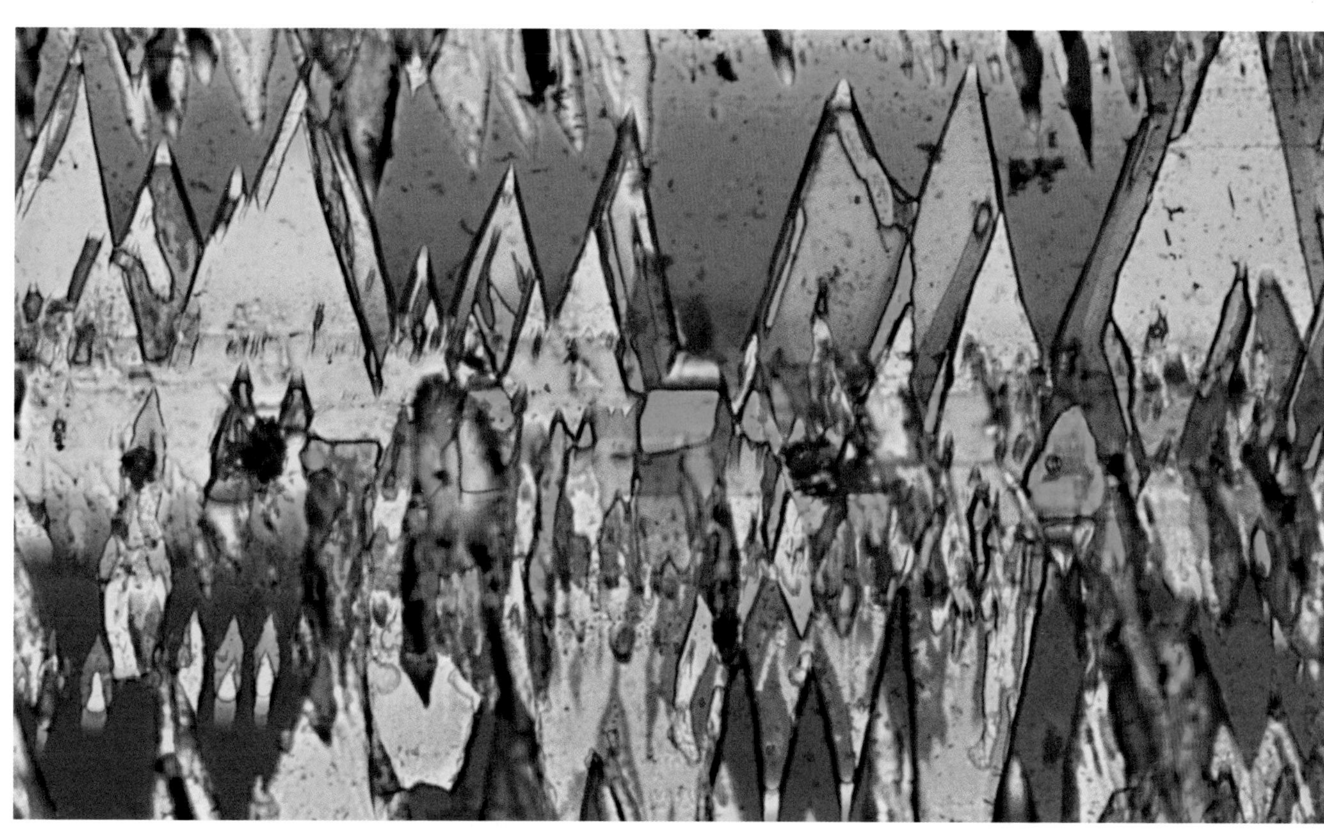

THC, in a polarized light micrograph, is the main psychoactive drug of marijuana.

enhances the experience of sex and food. In particular, cocaine been shown to release dopamine in the brain's nucleus accumbens, a region linked since the 1950s with pleasurable sensations. Heroin, an opiate like morphine, works on the brain's ventral tegmental area, which links to the nucleus accumbens through a bridge of neural fibers that produce dopamine.

One of the most addictive drugs is nicotine, present in tobacco. Laboratory research in the 1980s indicated that nicotine activates the brain's dopamine systems, much like cocaine and heroin. A 1986 study by the Harvard School of Public Health made the parallels explicit: Tobacco, like addictive drugs, creates physical dependence and withdrawal reactions, and like other abused substances usually is associated with overuse of other chemical stimulants and depressants, such as caffeine and alcohol. "Nicotine has a profile of

WHAT CAN GO WRONG

resulting in a loss of gray cells' volume. In fact, the loss of gray matter in the brain of a heavy drinker has nothing to do with the number of neurons being depleted; neuron counts remain constant. The loss of brain volume reflects a decrease in the size of dendrites, which compose a significant portion of gray matter. In general, when a heavy drinker gives up alcohol, dendrites begin to form more complex networks of branches, and brain function begins to be restored. However, chronic, heavy alcohol abuse may lead to high blood pressure, irreversible dementia, and other serious cognitive effects.

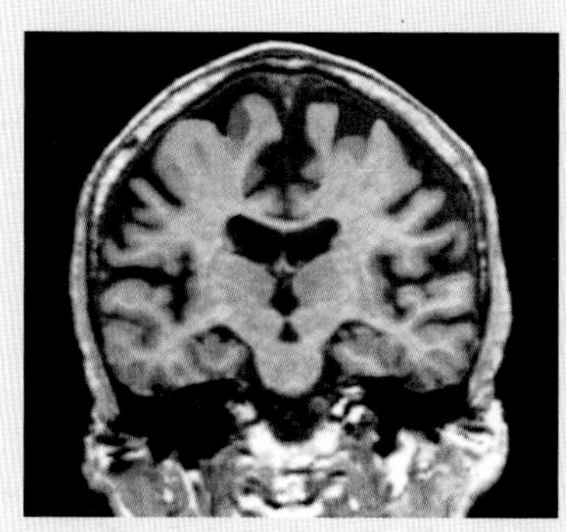

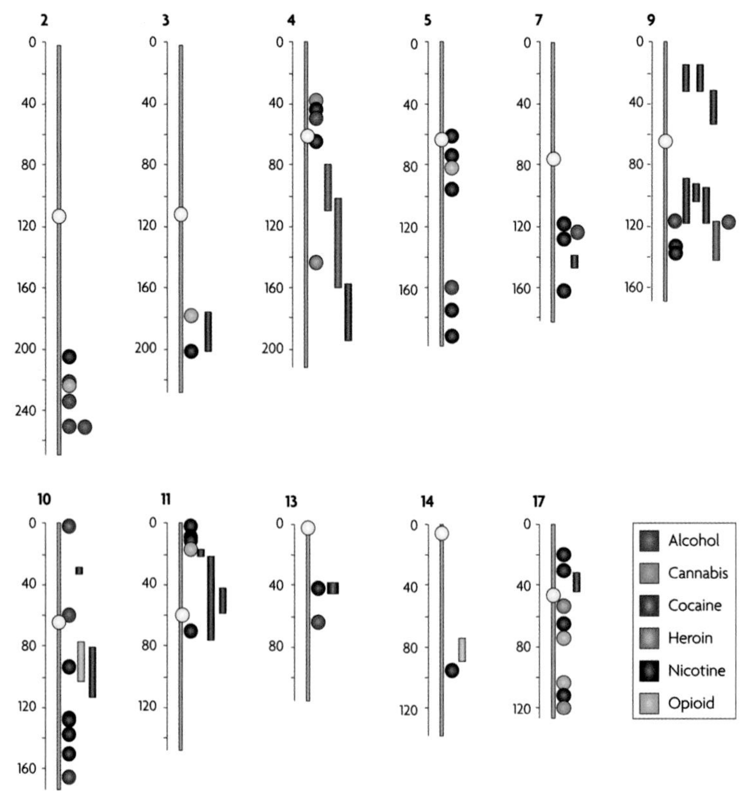

Markers on human chromosomes have been linked with specific drug addictions.

behavioral and physiologic effects typical of other drugs of abuse and, like heroin derived from opium, meets rigorous experimental criteria as a drug with considerable potential to cause dependence," the Harvard study said.

But drug use does not necessarily lead to drug abuse. Not every user of nicotine, alcohol, or an illegal drug becomes an addict. The extreme complexity of the brain, and the effect of small differences between brains, causes some people to have a higher or lower threshold for addiction. Genetics plays a role—perhaps 40 to 60 percent of the determination of whether a person becomes an alcoholic. Environmental and psychological factors also are important. Drug users who suffer from depression and anxiety, for example, are more likely to become re-addicted after giving up their habit.

Dopamine-enhancing drugs lose their punch over the course of time. They cause the brain to reduce the number of receptors for dopamine, which in the absence of dopamine-enhancing drugs lessens the reward response to normal reward-related behaviors, such as sex and food. Greater quantities of the drug are required to attain the level of the initial response.

WITHDRAWAL

When an addict tries to give up a dopamine-enhancing drug, the body compensates for the loss. Like a stretched rubber band snapping back, neurons not only return to their initial neurotransmitter

BREAKTHROUGH

A NEW ZEALAND study published in the *Archives of General Psychiatry* in 2009 suggests alcohol may trigger depression. Earlier research had suggested a link but did not specify the direction. Depression could lead to alcohol abuse when those suffering from it drink as a form of self-medication. Or, as happens in about 40 percent of cases of heavy drinking, alcohol use brings on symptoms of a depressive illness. Some third condition could cause both.

The New Zealand study looked at 1,055 people born in 1977 who were assessed for signs of depression and alcohol abuse at ages 17 to 18, 20 to 21, and 24 to 25. It found that problems with alcohol indicated a risk of depression nearly twice as high as for a control group of non-alcohol abusers. "At all ages, there were clear and statistically significant trends for alcohol abuse or dependency to be associated with increased

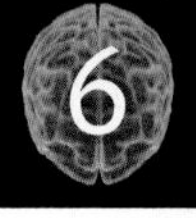

levels, they exceed them. Hyperactivity often results, contributing to the symptoms of withdrawal.

Some physical changes occur in specific regions of the brain. Scans of addicts' brains reveal lower levels of activation in the prefrontal cortex. In studies of lab animals' brains, long-term use of addictive drugs has been shown to lower the activity of cortical neural networks that link with the nucleus accumbens. These networks influence response inhibition and the forming of plans. Such findings suggest why drug addicts demonstrate impaired judgment.

The similarity of the brain's reward system responding to drugs and to natural, environmental stimuli such as food and sex makes it difficult to treat addictions. If the reward systems overlap, any medications that interfere with the feelings associated with drugs are also likely to depress other feelings that are a part of everyday life. They may kill the craving for heroin, for example, but at a cost of killing other appetites. However, such drugs, including rimonabant, may have a future as a treatment for overeating.

FAST FACT When a person is in love, brain regions associated with addiction light up in scans.

Meanwhile, ongoing research has found several of the genetic markers that indicate a greater likelihood of becoming an addict. Armed with such knowledge, future medical research could go forward on two fronts. First would be to identify prime candidates for addiction before they become addicted and steer them clear of their first encounters with drugs. Second would be to find ways to act upon the specific genes to modify or block chemical reactions that contribute to addictive behavior.

GENETICS & ADDICTION

Genetic sites linked to alcohol, nicotine, cocaine, and opiate addictions have been identified in the past few years, with a series of significant breakthroughs occurring in 2008.

At the University of California, San Francisco, researchers announced the discovery of a sequence of DNA on the 15th human chromosome that has a significant correlation to how strongly a drinker feels the effects of alcohol, a factor in his or her likelihood of becoming an alcoholic.

A 2008 study, funded by the National Institute on Drug Abuse (NIDA), found a genetic marker that increases susceptibility to nicotine addiction and raises the risk of developing lung cancer and peripheral arterial disease. More NIDA-funded research in 2008 found genetic markers for alcohol and cocaine abuse on the fourth chromosome, as well as DNA associated with nicotine and opiates. NIDA also reported sites linked to opiate dependence on the 17th human chromosome.

Once scientists understand the mechanisms of addiction, they may be able to find ways to repair the brains of addicts at the molecular level, or to immunize against addictive drugs.

BREAKTHROUGH

risk of major depression," wrote the research team, led by David Fergusson of the Christchurch School of Medicine and Health Sciences at the University of Otago. Analysis of the data indicated that the best model for which condition came first pointed to alcohol abuse or dependence leading to depression.

The researchers stressed that their findings "should be viewed as suggestive rather than definitive." They theorized that alcohol may trigger genes that encode for depression. They also noted that environmental issues, such as social and financial troubles, might bring on contributive stress reactions.

9
8

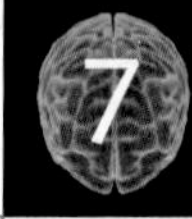

CHAPTER SEVEN

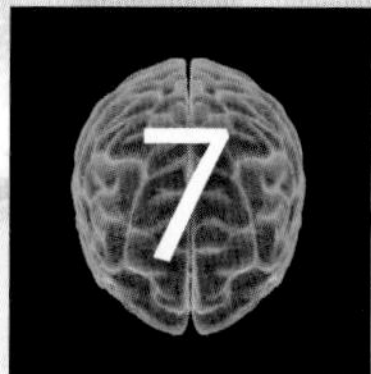

THE FEELING BRAIN

THERE'S MOTION inside the word *emotion,* and for a good reason. Emotions not only bring on highs and lows but also communicate via gesture and expression, such as facial changes that signal joy and anger. Powerful emotions can deeply carve events into memory, alter behavior and physical health, contribute to good (or bad) decision-making, and even cause a person to be literally scared to death. Science is only beginning to understand their importance.

The sensations of riding a roller coaster bring on internal and external physical reactions.

EMOTIONS

THE COMPLEXITY OF FEELINGS

NEUROSCIENCE has come late to the study of emotions. Scientific prejudices minimized the study of them during most of the 20th century, treating them as something too elusive and base for serious examination. Reason made humans stand alone as rational animals, making it the key to understanding the workings of the human brain. Emotions were too subjective, too far down the hierarchy of brain functions, to merit much attention. After all, dogs, cats, and other animals exhibit emotions, as any pet owner can attest. But emotions are moving to center stage as neuroscientists explore how and why the brain feels as it does.

Actress Halle Berry's emotions burst forth as she accepts her best actress Oscar in 2002.

REACTION TIMES

Until the late 19th century, reason held sway in the explanation of emotions. The logic went like this: The brain assesses a situation and assigns an emotion, which then moves the body to act. At the end of the century, psychologist William James flipped that idea on its head. He said the body reacts first, and then the brain responds with emotion. In other words, humans cry and then feel sorry, or prepare to run and then feel fear. Although James's theories later were discarded for a variety of reasons, he was the first to underscore the importance of physical movement to the appreciation of emotion.

In the early 20th century, Harvard researchers Walter Cannon and Philip Bard argued that humans become generally aroused by stimuli but must wait for cognitive assessment of the external world before the brain assigns an emotion. Other researchers added to theories of the brain playing an active role in deciding to create emotions. However, in the past few

FAST FACT

The Greek philosopher Aristotle split emotions into opposites: anger and calm, amity and enmity, fear and confidence, shame and pride, and kindness and unkindness, along with the unpaired pity, indignation, and envy.

decades science has done an about-face. Emotions now are believed to be processed in the brain at a level far below consciousness. They also are accepted as crucial to maintaining homeostasis, preparing for biologically appropriate responses to external stimuli, and even making reasonable and logical decisions. Researchers now view reason and emotion as intertwined, with too much or too little emotion detrimental to a healthy, rational mind.

Furthermore, some neuroscientists, such as Antonio Damasio, believe the advantage of being able to recognize emotions and feelings, and being aware of that awareness so as to make choices to maximize pleasure and minimize pain, gave humans the first glimmers of consciousness. If true, it is the human ability to know that it suffers joy or sadness, and to *know* that it knows, that allowed the species to step into the sunshine of consciousness.

DEFINING EMOTION

The struggle to apply science to emotions has ancient roots. It begins with the subjectivity of their description and classification. What are emotions? And how many are there?

To answer the second question first, observers have ranged from Aristotle, who classified more than a dozen emotions, including envy and pity, to medieval theologian Thomas Aquinas, who listed 11.

PHINEAS GAGE

WHILE WORKING with a railroad crew to clear rock from a Vermont gorge in 1848, Phineas Gage suffered a gruesome accident. He was pouring gunpowder into a hole in a boulder when a sudden noise distracted him and a nearby assistant. Gage thought the assistant had poured sand into the hole to snuff any sparks before Gage began to tamp the gunpowder with an iron bar. The distracted assistant had forgotten the sand, though. As Gage rammed the iron bar, it struck the sides of the hole and showered sparks on the powder, which ignited and exploded.

The explosion propelled the three-foot, seven-inch bar like a shell from a howitzer. It entered Gage's face below the left eye, passed through the cheekbone and frontal lobe, and zipped out of the top of the skull. It was a serious injury, one that could have taken Gage's life.

"The most singular circumstance connected with this melancholy affair," reported a nearby newspaper, "is . . . he was alive at two o'clock this afternoon, and in full possession of his reason, and free from pain."

Somehow, Gage survived. But he was no longer the same man. Whereas he had been a well-respected, energetic, and likable foreman of 25, Gage spent his remaining 12 years of life with an unpleasant, childlike personality. A doctor who observed and documented his personality changes described the new Gage as "fitful, irreverent, indulging at times in the grossest profanity (which was not previously his custom), [and] manifesting but little deference for his fellows."

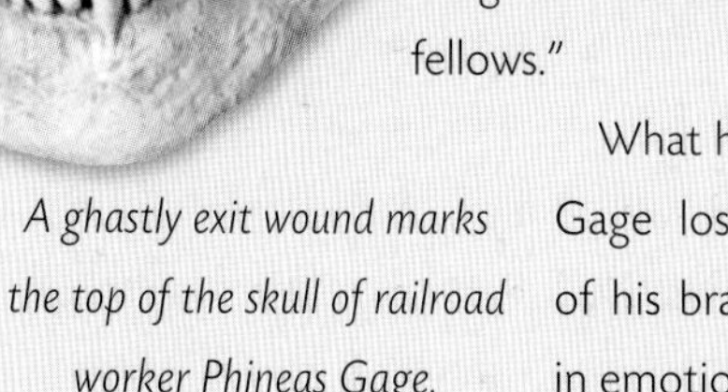

A ghastly exit wound marks the top of the skull of railroad worker Phineas Gage.

What had happened? Gage lost the portion of his brain that reined in emotion. Frontal lobe damage often interferes with the ability to synthesize information gathered through the senses and make sound decisions based on analysis. Without a judge to weigh and sift the raw, primitive impulses of the limbic system, a brain such as Gage's loses its balance between reason and emotion. His unfortunate accident gave some of the first glimpses into the specific tasks of brain regions.

The difficulty of taxonomy focused on shades of difference, as between embarrassment and shame. Aristotle included both shame and shamelessness as emotions, raising the issue of whether the absence of an emotion could be an emotion in itself.

The problem of classification endures today. Most scientists recognize either four or six basic emotions. The four most elemental are fear, anger, sadness, and joy. Damasio adds surprise and disgust, referring to the complete list of six as "primary" or "universal" emotions. Many casual observers would add love to the list, but researchers are divided over classifying it as emotion or drive. Emotions are easily recognized across cultures and around the world—a Mongolian farmer could look at a Dinka fisherman and recognize by face and body movements whether he was happy or angry. In fact, the movement provides the crucial distinction separating emotion from other behaviors. Emotions manifest themselves outwardly in visible changes to the body, such as muscle contractions, blood vessel dilations, and facial expressions including smiles and frowns. They occur at a subconscious level. The brain perceives stimuli that bring on a physical reaction, helping keep the body in homeostasis and providing an edge in the battle for survival.

+ HAPPY DOGS +

DOMESTICATION of the dog has resulted in friendly pets. But are there physical traits associated with domestication? Russian geneticist Dmitri Belyaev began an experiment breeding Russian silver foxes in 1959, choosing tolerance for human contact as his sole criterion for reproduction. His 40-year project yielded a group of kits as playful as golden retrievers. In 1985, the fox colony not only exhibited docility, but also floppy ears, rolled tails, and white patches of fur, which suggests that the genes that encode for human-pleasing behavior are associated with a variety of physical traits.

MAPPING EMOTION

Emotions appear to be processed in complex ways. There is no single emotion circuit; instead, a number of brain regions induce emotions, which then are processed by a variety of neural networks. Given their ancient evolutionary history, it's not surprising that most of the significant emotional centers lie below the cerebral cortex, which separates humans from all other animals. These "subcortical" emotional regions include the brain stem, the hypothalamus, and the basal forebrain. PET scans reveal the brain processes sadness, for example, mainly in the brain stem and hypothalamus, as well as the cortical region known as the ventromedial prefrontal cortex.

FAST FACT Emotions help memories form and stick, but they don't help with recall of facts.

Although emotions are not encoded in particular neurons, brain scans have led researchers to generally assign negative emotions such as sadness to the right hemisphere and positive emotions such as joy to the left hemisphere. For at least a century, neuroscientists have noted a link between damage to the brain's left hemisphere and negative moods, including depression and uncontrollable crying. Damage to the right, however, has been associated with a broad array of positive emotions. In the past two decades, University of Wisconsin researcher Richard Davidson has seen similarities in healthy, undamaged brains. Patients with more general activity in the left hemisphere tend to be happier than people with a more active right hemisphere.

EVOLUTION OF EMOTION

Charles Darwin believed that many of our emotions are inborn. The "chief expressive actions" of humans and animals are inherited, he said. "So little has learning or imitation to do with several of them that they are from the earliest days

CROSS REFERENCE: See "Evolution," PAGE 66

Newborns and their parents strengthen emotional bonds via facial expressions.

and throughout life quite beyond our control."

Infants are hardwired to express certain emotions. They laugh and cry at birth or soon afterward. Bodily expressions, particularly in the face, form the first means of communication between a child and its parents.

Other emotions are learned. Secondary emotions such as guilt require social conditioning through negative feedback; a child learns to feel ashamed for stealing, for example. Humans also can learn to react with primal emotions in modern situations. There were no guns in the ancient world, but the sight of one in the hand of a stranger today can bring on a fear response as solid as one created by predators in the African jungles.

Darwin reasoned that emotions must have emerged through evolutionary development by providing some edge in the battle for survival. For example, when the brain detects an unpleasant odor in a potential source of food, it causes the physical reaction of revulsion. The reaction prevents ingesting the possibly poisonous food or causes food already ingested to be vomited. Darwin noted the physical similarities in the face between disgust and the look of someone reacting violently to the smell or taste of rotten food.

PHYSICAL REACTIONS

Besides facial features such as smiles and frowns, emotions manifest themselves through body movements that evolution chose for particular purposes. We tremble when scared because of the onrush of adrenaline, which shakes up the muscles and organs for fight or flight. We cry when sad both to remove chemicals, such as manganese, that lower stress when expelled from the body and to signal our sadness to others. We blush with anger or embarrassment when adrenaline dilates blood vessels near the skin and flushes the face and neck with extra red blood cells. Blushing is perhaps

> **FAST FACT** Author of *How to Win Friends and Influence People* and self-improvement guru Dale Carnegie said, "When dealing with people, remember you are not dealing with creatures of logic, but creatures of emotion."

an evolutionary holdover from the time before human speech. No words are needed to communicate anger when an enemy flushes red, stands erect, and scowls.

When the brain reacts to stimuli by inducing emotion, it sets off a chain of body modifications through two channels to prepare for a physical response. One is to the central nervous system to prepare for instant action. In animals, this reaction may be to fight another animal, run away, or initiate sex. In humans, the reaction is essentially the same, although (one hopes) filtered through the executive function of the prefrontal cortex. The second reaction regulates the body's internal state, such as supplying more oxygen and glucose to the muscles. These emotional reactions lie beyond the control of the conscious mind. "We are about as effective at stopping an emotion as we are at preventing a sneeze," Damasio said. Who hasn't tried to speak at a particularly sad time, such as a funeral, and found it impossible not to choke up or tremble?

FEELINGS

Feelings differ from emotions in that they are inward and private. Feelings register on the mind at a conscious or unconscious level. They aid in survival by alerting animals to deal with the problems signaled by emotions. Feelings provide incentives to adapt and to act. A further advantage is conveyed by a higher brain function, which is awareness of the feelings and how responsive actions may alter them.

In most but not all cases, the brain becomes aware of its feelings as they register on consciousness. It is possible to go for some time feeling a vague sense of dread or anxiety before realizing the feeling exists. Usually, however, humans not only express the physical manifestations of emotions but also process their feelings and, ultimately, recognize they are happy, sad, or experiencing some other mood. Therein lies one of the key differences between humans and animals. While a dog may wag its tail with happiness, it lacks the consciousness to recognize the emotion.

PERCEIVING EMOTIONS

Humans communicate emotions through facial gestures. Nervous control of laughing and crying lies in the brain stem and amygdala, beyond consciousness. Evidence for this conclusion comes from patients who have pseudobulbar palsy, a disease that impairs voluntary control in the motor cortex. Such patients cannot control the muscles of their face. However, they still laugh and cry and show features of true emotions when moved by involuntary responses.

Physical manifestations of true emotions are hard to define precisely, but people know them when they see them. Sensory organs pick up on the finest of details when

BREAKTHROUGH

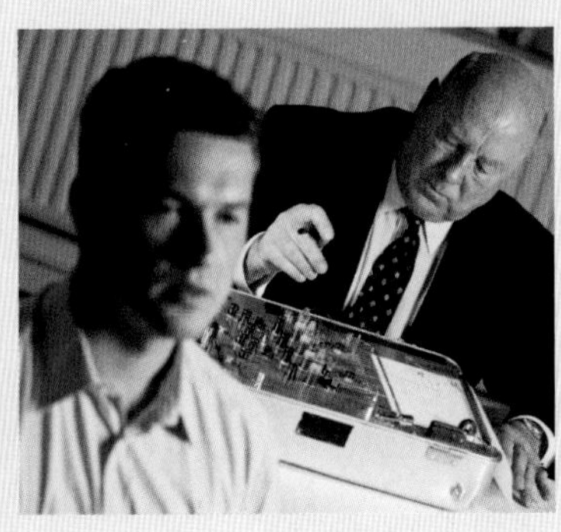

EFFORTS TO IDENTIFY lies through technology began in the 1920s with the invention of the polygraph machine. It measures common physical reactions associated with lying, including changes in breathing, heart rate, and the electrical conductivity of the skin. Polygraphs purported to follow the principles of emotion, measuring supposedly involuntary physical changes. They have been found to have two problems. First, some people have managed to gain a measure of control over their autonomic responses and can fool a machine. Second, polygraphs take their measurements not at the source of the lie—the examining the sights and sounds of emotions.

Take the simple smile: It originates in the cingulate region of the brain stem, which initiates a series of commands to contract the muscles in the face. When a smile reflects genuine joy, facial muscles shift in a way that observers recognize as joy. However, when someone tries to falsify a smile, tiny miscues in voluntary muscle contraction patterns give the faker away.

We begin to recognize emotions in others at an early age and copy what we see. Imitation helps create the parent-child bond during the first months of the child's life.

The urge to mimic the emotional behavior of others, called emotional contagion, continues throughout life. When you look at another's facial expression, you often take on aspects of that expression yourself. That's because the brain perceives an emotion in the face of another and automatically signals its own emotional circuits. A genuine smile or laugh in another person can quickly spread to nearby people.

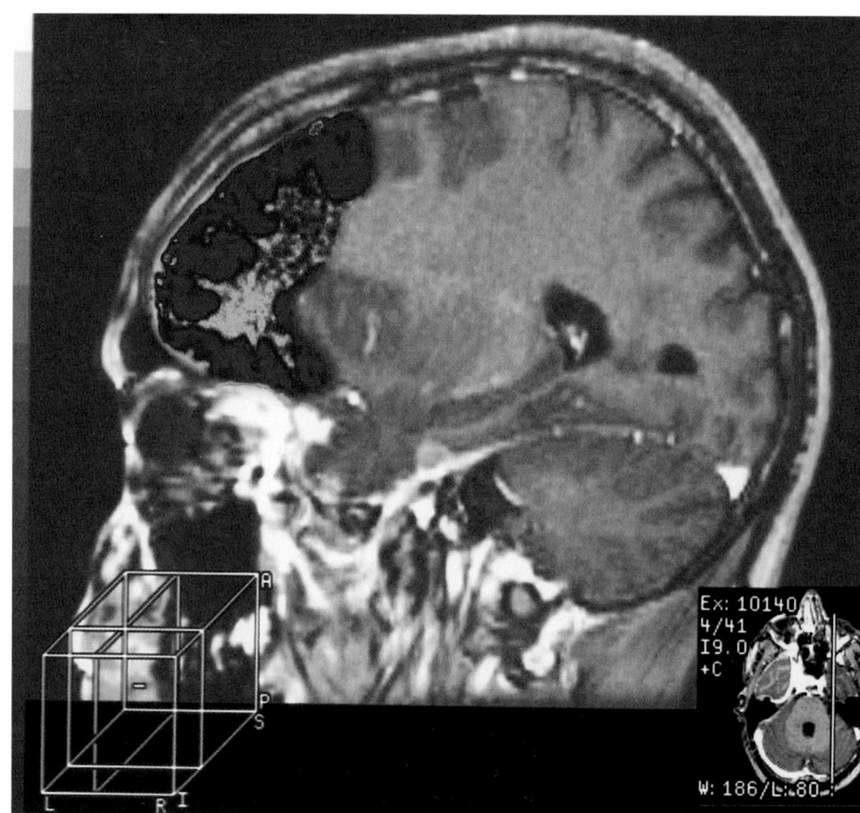

Emotional activity lights up the frontal lobe in this magnetic resonance image.

The ability to manifest emotion and recognize it in others lies primarily in the right hemisphere. This conclusion arose from the observation of patients who suffered disorders that compromised right hemispheric functions. They could not read faces. An angry face, a bored face, a disgusted face —they registered as mere collections of eyes and mouths, devoid of emotional meaning.

The task of decoding human emotions takes on an added dimension when sound is added. Prosody is the meaning imparted to language through means other than words. Tones of voice can convey

BREAKTHROUGH

brain—but far out along the peripheral nervous system.

Going straight to the source, Daniel Langleben of the University of Pennsylvania uses fMRI scans to examine the brain of people who choose to communicate the lie. When told to lie, the subjects show increased blood flow to their anterior cingulate cortex. It's a brain region associated with error recognition and inhibition. Suppressing the truth requires energy, and thus the increased oxygen associated with blood flowing to the anterior cingulate cortex.

Defense attorneys in San Diego introduced fMRI imaging from a private company, No Lie MRI, in a juvenile sex-abuse case in spring 2009, then withdrew the report. The company says its lie-detection techniques are accurate more than 90 percent of the time, and that its evidence eventually will find its way permanently into the courtroom.

A visitor at the Vietnam Veterans Memorial pauses for a reflective moment.

disbelief, sarcasm, and other shades of meaning. Imagine the difference between saying "I love you" as if you really mean it, and saying it in a clipped tone. As with the decoding of human faces, the brain's right hemisphere acts to evaluate the nuances of speech.

PROBLEMS WITH PROCESSING

Many autistic people struggle with decoding the emotional content of faces, bodies, and sounds. Doctors think of autism as a "spectrum" disorder, a group of disorders with similar features but a wide range of possible symptoms, including problems with language, repetitive motions, impaired social skills, and self-destructive behavior. However, autistic people also may have enhanced abilities such as powerful memories and artistic skills. Whatever their symptoms, autistic people typically have difficulty socializing with others. Their inability to see things from another's perspective gives them trouble in recognizing sarcasm or deceit.

Autopsies have found correlations between autism and cellular anomalies in the cerebellum, hippocampus, and amygdala, as well as shrinkage of the cerebral vermis. Other findings indicate that autism may result from disorders in the portions of the brain that specialize in imitation. These so-called mirror neurons are less active in autistic children than in nonautistic control groups when they are asked to imitate or merely observe facial expressions.

EMOTION & MEMORY

Emotion improves memory. Apparently, memories get encoded in different ways in the brain depending on whether they have emotional content. Simple memories, with no emotional content, get encoded by the hippocampus. However, when emotional content accompanies the memory, the amygdala takes on a significant role in memory processing. Sensations that promote strong emotional responses stimulate the amygdala, which communicates with the hypothalamus and sets off the release of hormones and other chemicals. So-called flashbulb memories linger in our mind because they were encoded with the most powerful of emotions. The shock affects brain regions far beyond the usual encoding circuits, and the brain reacts by recruiting many more neurons to encode the memory. Details such as the weather, the clothes you wore, and what you said or did become linked with the memory of the disaster, whereas on a normal day such mundane minutiae would quickly be forgotten.

FAST FACT

Pain is not an emotion. It may, however, help bring on reactions such as despair, depression, or anger.

CROSS REFERENCE: See "A Memory Forms," PAGE 246

GLOSSARY

ANTERIOR CINGULATE CORTEX. Region of the brain associated with error recognition and inhibition.

AUTISM. One of a group of developmental disorders characterized by significant communication and social interaction impairments and unusual interests and behaviors.

BIPOLAR DISORDER. A neurological condition characterized by extreme mood swings from mania to depression.

DYSTHYMIA. A chronic depression less severe than major depression, characterized by occurrence of depressive symptoms nearly daily for over two years.

EMOTION. A spontaneously occurring mental state characterized by strong feeling and often accompanied by physiological and behavioral changes.

EMOTIONAL CONTAGION. The process of emotion transfer from one individual to another by the unconscious mimicking of the emotional behavior of others in the vicinity.

ESTROGEN. Female sex hormone produced in the adrenal glands and ovaries. These feminizing hormones are found in both sexes, but in a much higher quantity in women.

GONADOTROPHIN-RELEASING HORMONES. A hormone produced in the hypothalamus that induces the production of luteinizing and follicle-stimulating hormones in the pituitary gland.

INSULA. Region of the cerebral cortex responsible for the recognition and perception of disgust in others.

LEVATOR ANGULI ORIS. Facial muscle that raises the angle of the mouth.

LEVATOR LABII SUPERIORIS. Facial muscle in the upper lip and cheek that raises the upper lip when one is smiling.

LONG-TERM POTENTIATION (LTP). The strengthening of neural connections, considered to be the cellular basis of memory.

MAJOR DEPRESSION. The most severe form, characterized by profound interference with normal daily activities. May occur as a single episode after significant trauma or repeatedly through life.

MELATONIN. A hormone produced in the pineal gland that helps regulate the sleep-wake cycle.

MONOAMIDE OXIDASE INHIBITORS (MAOIs). Antidepressant drugs that boost mood by preventing monoamide oxidase from metabolizing serotonin, norepinephrine, and dopamine.

NEGATIVITY BIAS. A psychological phenomenon by which individuals react more strongly to unpleasant situations and stimuli than to their positive counterparts.

ORBICULARIS OCULI. Facial muscles responsible for crinkling the eyes when an individual smiles.

OXYTOCIN. A hormone produced in the pituitary gland that is released during pregnancy and intercourse, promoting trust and pair bonding.

PHENYLETHYLAMINE (PEA). A neurotransmitter found in small amounts in the brain and in foods such as chocolate. This releases dopamine into the limbic system, creating feelings of pleasure.

PHOBIA. An unreasonable fear that may cause avoidance and panic.

PROSODY. Additional meaning imparted to language through the rhythm and intonation of speech.

PSEUDOBULBAR PALSY. A condition in which an individual is unable to voluntarily control the muscles of his or her face.

RISORIUS MUSCLES. Facial muscles responsible for the lateral movement of the corners of the mouth.

SEASONAL AFFECTIVE DISORDER (SAD). A type of depression that occurs most often with the onset of winter.

SEPTUM. Region of the brain associated with orgasm.

TESTOSTERONE. Male sex hormone produced primarily in the testes. This masculinizing hormone is found in both sexes, but in much higher quantity in men.

VAGUS NERVE. One of the primary communications pathways between the brain and the body's major organs.

VENTRAL PALLIDUM. One of the brain's main reward circuits; associated with attachment and stress relief.

VOMERONASAL ORGAN (VNO). A small olfactory organ found in most vertebrates that detects pheromones and sends corresponding signals to the brain.

ZYGOMATICUS MUSCLES. Facial muscles involved in smiling. These lift the corners of the mouth.

DARK EMOTIONS

UNDERSTANDING FEAR, ANXIETY, & ANGER

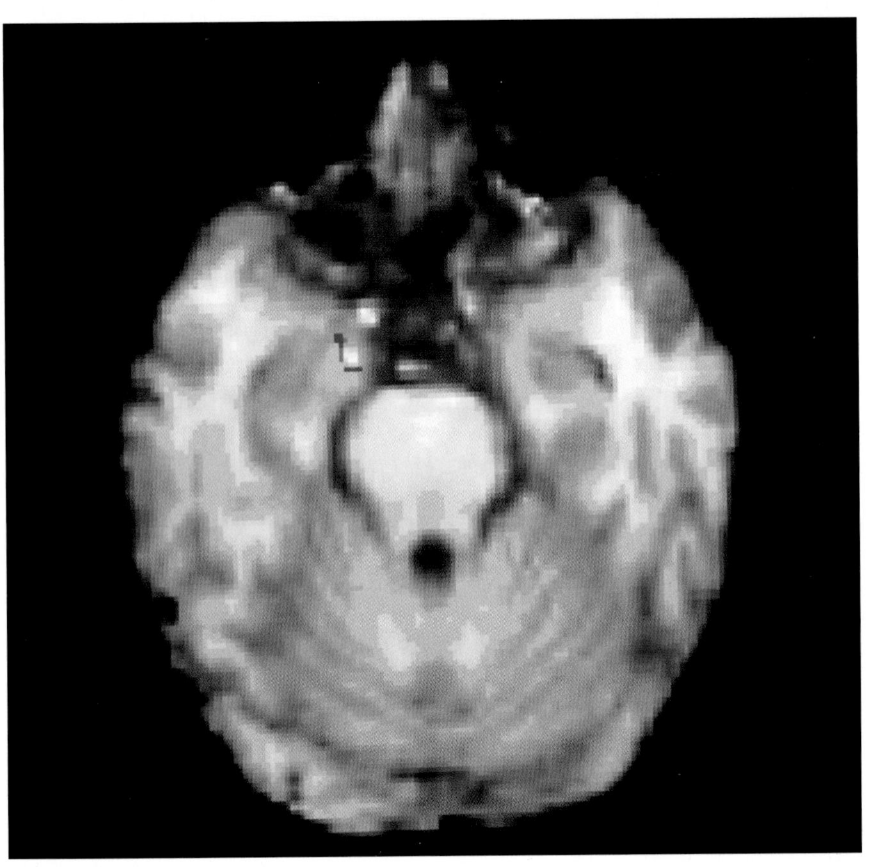

The left amygdala glows yellow and red in a colored PET scan to indicate recognition of fear.

THE HUMAN BRAIN exhibits what psychologists call a negativity bias. Bad news sticks longer in memory than good news, and unpleasant encounters affect the brain more powerfully than pleasant ones. The brain's supersensitivity to negative emotions emerged at the dawn of the human species. Fear, anxiety, and anger prepare the body with what it needs to survive, creating the fight or flight response.

Unfortunately, the evolutionary wiring to respond to negative emotions does not always work appropriately in the modern age. Overreacting to life's ordinary troubles can bring on panic attacks, phobias, migraines, ulcers, high blood pressure, and even heart attacks.

DIFFERENT RESPONSES

Not everyone has the same bias toward negativity. Everyone has friends who seem plagued by anxieties and dark thoughts, but they also probably know people who are chronically cheerful. A theory developed within the past few years by Edward and Carol Diener of the University of Illinois explains different temperaments by positing the existence of emotional "set points." Each person has an equilibrium point for mood that marks a basic level of happiness or sadness. Events may push one higher or lower, but eventually the person returns to his or her baseline.

People who are prone to negative emotional states are generally pessimistic, anxious, and likely to avoid other people. Those prone to positive states generally are active, outgoing, and confident. Brain scans reveal the two groups' brains react differently when confronted with other people's emotions. Photographs of fearful faces provoke stronger reactions in the amygdala of the former group, whereas smiling faces produce the opposite reaction, with stronger responses in the brain of upbeat people.

Research indicates that negative emotions play an important role in making rational decisions. Antonio Damasio described a patient of his with abnormal calcium deposits in her amygdala as suffering from lack of an appropriate fear response. Neurons in her amygdala could not function normally. Patient S, as he called her, had no problem learning new information. However, without the fear- and anger-generating

CROSS REFERENCE: See "Messengers," PAGE 42

mechanisms of her amygdala intact, she approached new people and new situations with an invariably rosy attitude and could not feel or recognize fear: She could not identify telltale signs of potential threats or unpleasantness in social encounters, and she could not learn coping mechanisms for scary or intense situations. Without fear, anger, and other negative emotions supplying information about danger and risk, the brain makes poor decisions.

RECOGNIZING FEAR

The physical manifestations of fear are well known. Triggering fear activates the autonomic system and releases stress hormones, including adrenaline. The amygdala and thalamus mobilize the body by increasing heart rate and blood pressure, as well as sharpening the focus of the senses. The body is immediately ready for fight or flight. Meanwhile, slower sensory signals move from the thalamus to the frontal cortex to identify and reassess the source of fear's signal. The fast response of the amygdala has the evolutionary advantage of preparing the body for the worst-case scenario, while the slow response of the cortex keeps the body from running away from every shadow. The slow response of the cortex, if it finds no actual threat, overrules the amygdala and the rest of the limbic system and inhibits the fight or flight response. Blood pressure,

A fear of heights can be chemically blocked or treated with relaxation techniques.

STAYING SHARP

A HEALTHY RESPECT for potentially dangerous creatures, like spiders and snakes, can save your life. Having butterflies in your stomach before performing in front of an audience isn't anything special. But sometimes everyday anxieties can turn into intense, paralyzing fears called phobias. If left untreated, these fears can prevent a person from enjoying everyday things in life.

To an observer the cause of a person's fear may not appear to be much of a real threat, like thunder or public spaces; nevertheless, the fear becomes every bit as real as if caused by genuine danger. These phobias, such as fear of heights, open spaces, spiders, or public speaking, usually start in childhood or adolescence, although most of the people who suffer from them cannot recall specific triggers. They appear to have at least some genetic component.

Modification of phobias is one of the most successful psychiatric treatments. Therapists use behavioral therapy, sometimes with fear-blocking drugs, to get their patients to slowly face and reassess their fears in order to eventually overcome them. Repeated, controlled exposure without negative consequences lessens the fear. Small steps are crucial.

To treat the fear of heights, for example, a therapist might begin by showing a patient an image taken from the roof of a low building. Gradually, as the patient becomes more comfortable with the initial steps, the therapist raises the intensity of the exposure in controlled environments. The patient might be asked to imagine a balcony and then be taken to safe, elevated places on which to stand. Eventually, the patient's anxiety fades to an acceptable level, and the extreme, paralyzing fear dissipates.

MODELS OF THE MIND

MODELS OF THE brain and human activity rely on the popular science of the day, from Newtonian ideas to the advent of quantum physics. As scientific knowledge about the way the world works has increased, knowledge about the brain's inner workings has followed

PHYSICS & THE BRAIN

First came mechanical models. The classically deterministic science of Sir Isaac Newton held out hope that neuroscientists could understand the physical rules that governed the working of the mind. Human deeds became the result of actions that preceded them, like a baseball soaring through the sky after being struck by a bat.

If researchers could find each link of the causal chain, they could trace every thought and action back to hardwired brain circuitry. To take the matter back even further, they might trace brain circuitry to a person's DNA, which itself is a mechanical combination of two strands of parental DNA.

Such a model poses certain difficult questions. Where is free will? Where is responsibility? If humans are nothing but machines carrying out encoded instructions, society's entire system of rewards and punishments, to say nothing of religion and politics, becomes moot. Nobody is responsible for negative emotions and bad behavior. Nevertheless, behavioral scientists such as psychologist John Watson, working in the early 20th century, embraced mechanical explanations for human actions. According to Watson, instead of expressing love, humans exhibit a "conditioned love response," and instead of fear, a "conditioned fear response."

Chaos theory holds that minute stimuli, such as the flap of butterfly wings, can profoundly alter physical systems.

UNPREDICTABILITY

Seeds of a new paradigm to explain human behavior began sprouting early in the 20th century, even though its advocates initially applied it only to the realm of physics. The classical view held that any action could be explained if observers could only see, and measure with sufficient precision, its causes and effects. German physicists Max Planck and Werner Heisenberg, among others, shot that theory full of holes. Planck presented a paper in 1900 proposing that electromagnetic radiation, such as light, travels not in continuous waves but in tiny, discrete bursts of energy. He called them quanta. He developed quantum theory to explain the behavior of heated, glowing objects, and it soon proved better than Newtonian physics in a variety of applications. Heisenberg expanded on this theory in the mid-1920s with a theory that has come to be known as the uncertainty principle. Precise measurement—the basis for a physically deterministic world—becomes impossible at the smallest scale. One can never determine the exact location and motion of an electron simultaneously because observation changes reality. There is *always* uncertainty, and thus we can never fully know the world—or the brain.

CHAOS

A new science called chaos theory, discovered in 1961 by MIT meteorologist Edward Lorenz, demonstrated that physical systems such as the weather react deterministically but cannot be predicted accurately. The movement of air masses and water vapor over the Earth's surface is constantly affected by forces too small to measure accurately. Under the right conditions, a butterfly flapping its wings in China could set off tornadoes in Kansas. Scientists call this phenomenon sensitive dependence on initial conditions, and it is the cornerstone of chaos. Chaotic systems are found everywhere, from forest fire patterns to human enterprise such as stock market activity. Like quanta, chaotic systems have room for possibilities, such as that one-in-a-trillion-trillion butterfly, to effect change.

Physicists in the 20th century came to some strange conclusions as they wrestled with the implications of the new science. Experiments have demonstrated that subatomic particles behave both like particles and like waves, depending on how they are observed. Quantum theorists talk of electrons as clouds of uncertainties that snap into definite existence only through the act of observation. According to this theory—actually, not just some pie-in-the-sky idea, as it has been bolstered by countless experiments done with extreme precision—there is no "real" world independent of the defining act of observation.

These new, unsettling theories are taking neuroscience away from determinism. Chaos theory has applications both to the study of the brain and to human action. In 2009, scientists at the University of Cambridge in England discovered chaos theory-like patterns in the brain's synchronization of electrochemical activity among different functional regions. Chaos's telltale clusters—called strange attractors—also have been found in EEG patterns. Computer models suggest that the brain's self-organization along chaotic lines maximizes its ability to store information and recognize sensory inputs.

Jeffrey M. Schwartz describes how patients overcame OCD using fundamentals of quantum physics and chaos theory.

Chaos and quantum theories suggest that life is not predetermined. If you're in a bad or hopeless mood, there's nothing in the scientific world that says you must stay there. Small changes in your life, like the butterfly flapping its wings, can have a huge impact. Quantum theory suggests that electrochemical actions in your brain are possibilities, not certainties.

Using quantum theory, Jeffrey M. Schwartz, an expert on obsessive-compulsive disorder, posited in his controversial 2002 book *The Mind and The Brain* that the act of the brain observing itself—the force of attention to one's own thoughts and feelings—could alter brain circuitry at the molecular level. Concentrating on altering one's own feelings of obsession or depression could push them from one path to another, just as observing an electron alters its course. He reported results not only in his patients' behaviors, but also in their brain scans. Negative feelings (and their neural networks) grew quieter, and positive ones became stronger. If Schwartz is correct, there is new life for the concept of free will and a new burst of hope for patients who feel stuck with negative feelings.

STAYING SHARP

ANGER MANAGEMENT aims at controlling the physical reactions caused by the onset of the emotion of anger. It doesn't change the source of anger, or its ability to irritate. And it avoids the discredited advice to "let it all hang out"—it's been shown to be counterproductive to relationships and mental health to vent anger like a volcano spewing lava. Rather, anger management reduces the emotion's impact on the body and allows a person to control his or her experience of the emotion.

You can take an entire course or read a series of books on anger management. Typically, they elaborate on the following simple steps:

✓ Breathe deeply from the belly. Repeat a calming word or phrase, such as "relax," as you breathe.

✓ Visualize a calm scene from your memory or imagination, such as a seashore or mountain meadow.

✓ Exercise with slow, nonstrenuous movements.

✓ Use imagery; visualize a relaxing experience, such as lying in a hammock, from your memory or your imagination.

✓ Remind yourself that acting out your angry feelings and impulses won't fix your problem. Purposeful, calculated action will.

✓ Learn to communicate better with those around you. Listen. Talk calmly.

✓ Reassess your situation and use a positive vocabulary to describe your situation. Instead of cursing or describing your life as hopeless, look at your situation logically. Say things aren't so bad, and you'll get through them.

heartbeat, and other body changes return to normal.

The fear response can be overcome by blocking the receptor sites for adrenaline in the heart muscle. Drugs that plug the appropriate neurotransmitter "keyholes" prevent circulating adrenaline molecules from initiating the fight or flight reaction. So-called beta blockers stop the fear without interfering with rational thought.

In addition to chemical curbs, the fear response can be countered by cultivation of the relaxation response through breathing exercises, which can lower the heartbeat and enhance slow-wave brain activity. Some fears can be reduced or eliminated through conditioning.

ANGER

Anger is natural. Violent aggression evolved among animals to produce strong males through combat, and to protect offspring when mothers direct anger at potential threats. Genes that encode for a propensity toward anger are evident in the rat colonies bred by Dmitri Belyaev. Mating to produce one colony of docile rats and another of rage-filled, aggressive rats, Belyaev created creatures that throw themselves at the bars of their cages in fits of rage and "go crazy" if humans try to pick them up.

The modern world presents different challenges to humans today. Anger still plays an important role in human society, though. People get angry to defend a mate, territory, possessions, and themselves.

Most humans learn to exercise some control over their anger at an early age. Still, about one in five adults reports difficulty controlling rage. In small bursts, anger can alter behavior in advantageous ways. But too much can contribute to coronary heart disease and dysfunctional relationships.

FAST FACT Anger is one of Christianity's seven deadly sins. (The other six are lust, greed, gluttony, sloth, envy, and pride.)

Psychologist and anger expert Mike Obsatz has identified eight different kinds of anger. Chronic anger manifests itself as ongoing resentment that is directed outwardly. Volatile anger comes in waves, building to crests of rage that explode in aggression. Judgmental anger belittles or shames others. Passive anger comes out through sarcasm and avoidance. Overwhelmed anger occurs when people resort to aggressive outbursts to deal with feelings of stress. Retaliatory anger is, just as its name suggests, a way to get back at another person. Self-inflicted anger targets the sufferer as punishment for some negative behavior. And finally, constructive anger turns the negative emotion toward positive ends, perhaps as motivation for a change.

Anger explodes into rioting by supporters of an opposition political faction in the West African nation of Togo.

Anger, like fear, manifests itself through a variety of pathways starting with the limbic system. Researchers are still investigating its triggers and its controls. However, anger releases cortisol in the brain, which promotes feelings of stress and frustration. People who habitually get angry appear to have reduced neural activity in their frontal lobes, which communicate with the amygdala as the mind seeks a balance between reason and emotion. A weakened prefrontal cortex fails to exert a normal level of control over the aggressive behavior programmed in the deep, ancient regions of the brain. High and low levels of serotonin and high testosterone levels contribute to aggressive behavior, through mechanisms that are not entirely understood.

COPING WITH ANGER

Therapeutic treatment of anger includes talking about it. It sometimes serves to help violent people if they are able to verbalize the causes of their anger. Verbalizing about aggression becomes less successful, though, if the perpetrator has learned to receive a psychic reward for aggressive acts, in which the action brings a temporary relief from physical stress.

Three main strategies allow people to cope with anger: expressing, suppressing, and calming. Expressing anger in a calm, nonaggressive way communicates a person's needs in a healthy way.

Suppressing occurs when the emotion of anger is inhibited. Sometimes, suppression redirects anger toward constructive behavior. In other cases, suppressed anger may turn inward, causing depression and heart disease, or passive-aggressive behavior.

Calming involves the attempt to physically slow the body's internal responses, as with the relaxation response reducing fear.

GOOD FEELINGS

JOY, LOVE,
& HAPPINESS

Falling in Love

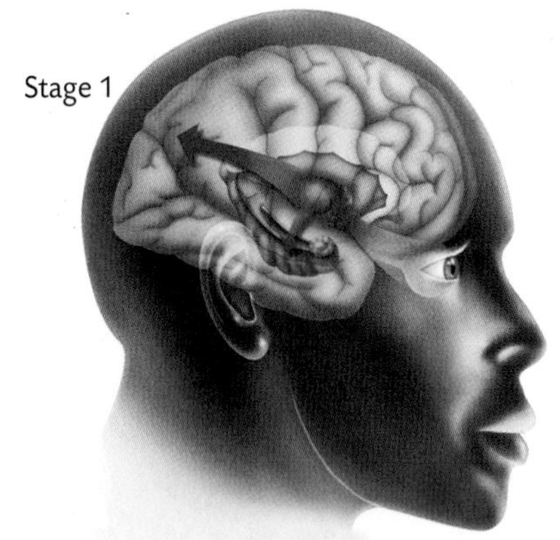

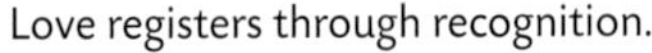
Love registers through recognition.

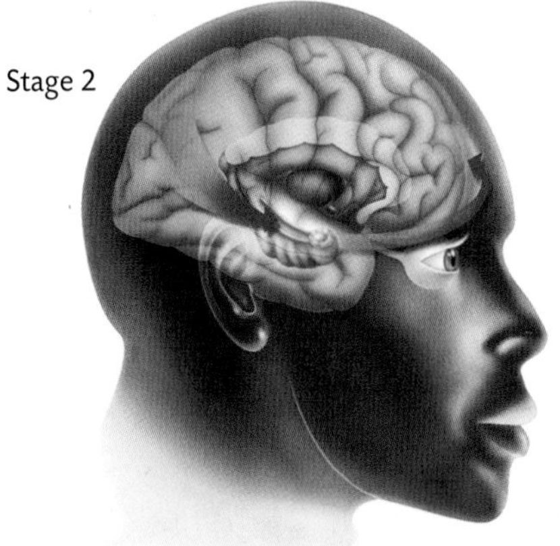

Emotions in the amygdala.

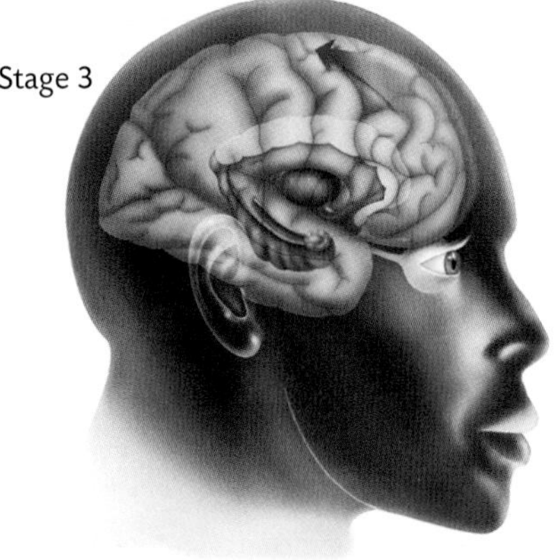

Decision and strategy for action.

WHAT IS YOUR motivation to experience love and joy? You might as well ask why you need to breathe. The physiological sensations of satisfaction and warmth, coupled with the awareness of how right the emotion feels, are their own reward. Joy, love, happiness—they are what make life worth living.

LOVE

Of all the positive, pleasurable states, none has commanded so much attention from scientists, artists, and poets as love. In the laboratory, love has gone under the microscope, to be dissected into categories and probed for possible clockwork mechanisms. Helen Fisher, a Rutgers University anthropologist, charts three kinds of physiological and emotional kinds of love: lust, attraction, and attachment. All arose through evolution to promote the continuation of the species through mating and parental bonding.

According to Fisher, each type of love has its own special purpose and chemistry. Lust sends people out into the world looking for a mate. It's associated with the hormones estrogen and androgen. Attraction focuses that physical energy on one person instead of spreading it too thin. Its link to the desire for emotional connections is believed to be associated with serotonin. Attachment keeps Mom and Dad together for the evolutionary advantage of having two parents. The neurotransmitters that keep two people together have been harder to find, but experiments with animals have turned up neurochemicals that promote pair bonding. The different effects of biochemistry during stages of love make sense when viewed through the lens of personal history. Who hasn't felt the butterflies and racing heartbeat of the early stages of love, only to see those feelings be replaced with a calm confidence as relationships develop?

FAST FACT In 2005, researchers found that some mice pass pheromones through the male's tears.

CHEMICAL ROMANCE

Brain scans of people in love show activation in the caudate nucleus, putamen, and insula, along with the anterior cingulate and the

CROSS REFERENCE: See "A New Brain," PAGE 72

cerebellum. Other areas of the brain became deactivated by love. Specifically, and surprisingly, the depressed brain regions have been associated with sadness, anxiety, and other negative emotions. Love apparently not only makes you giddy but also smothers feelings that might topple your elation.

Maternal love may share some of the same neural circuits as romantic love. The neuromodulator oxytocin, released in the brain during female orgasm, also is released during childbirth and promotes mother-child bonding. Men also have their reward areas of the brain activated during orgasm, including those that contain receptors for oxytocin and arginine vasopressin, or AVP.

Oxytocin seems to increase the general level of trust people have for each other, including strangers. That would explain why some people make social decisions that seem stupid when the chemicals wear off. Consider the impaired judgment found in a 2009 study conducted in Britain. Researchers gave men and women a whiff of oxytocin and asked them to rate the attractiveness of others. The experimental group rated strangers as more attractive than control groups did.

+ THE CHEMICALS OF LOVE +

Hormones and neurotransmitters play important parts in how the brain and body function when we fall in love.

NAME	DESCRIPTION
Estrogen, testosterone	Hormones responsible for the sexual drive. Testosterone increases in women when they're in love, but declines in men.
Dopamine	Neurotransmitter released in the brain's reward systems. Brings on a feeling of bliss.
Norepinephrine	Causes racing heart, flushed skin. With dopamine produces elation, focused attention, higher energy states, and craving.
Serotonin	Lowered levels of this neurotransmitter during stages of love may contribute to feelings of obsession with another person.
Oxytocin	A hormone that builds trust and helps form social bonds.
Arginine vasopressin (AVP)	Found in reward areas of the brain activated by romantic love in both sexes and orgasm in men.
Phenylethylamine (PEA)	Releases dopamine in the limbic system, causing pleasure. Occurs naturally in the brain, but also is found in chocolate.
Sex pheromones	Chemical messengers passed between males and females of the same species to induce mating.

+ RODENT ROMANCE +

PRAIRIE VOLES are small, brown rodents. They mate for life, and if one partner dies, the other usually refuses to mate again. Meadow voles are small, brown rodents too. They don't form pair bonds, and they mate promiscuously.

Scientists at Florida State University discovered that different levels of two neuromodulators, oxytocin and arginine vasopressin (AVP), led the two species toward their different mating habits.

When the researchers induced the expression of AVP in the ventral pallidum of the promiscuous meadow voles, they converted to monogamy.

GENETIC TENDENCIES

Today, most cultures limit people to one marital partner. Historically, polygamy has been much more common, from the harems of the Old Testament to the multiple wives taken by some men in the Mormon Church in the 19th century. It's that way in the animal kingdom too, with only 3 to 5 percent of mammals mating for life. Children raised in cultures where bonding occurs (in pairs or otherwise) become conditioned to their environment; they see relationships around them as normal, whatever they are. However, research published in 2008 discovered a genetic code in men that strongly correlated to whether they lived in monogamous relationships with women, had multiple relationships, or never married. Variations in a gene that codes for a receptor for the AVP receptor predicts whether they will shy away from long-term commitment or be devoted husbands, according to researchers at the Karolinska Institute who studied 552 Swedish men.

They focused on AVP after reading of previous research that found that variations in AVP receptors in two species of prairie voles make one mate for life and the other promiscuous. Human males can have zero, one, or two copies of a gene section called RS3 334. The higher the number, the lower the men performed on measurements of pair bonding. Men with two copies were more likely to be unmarried, or if married, to have had a marital crisis. The researchers had no immediate explanation for how the extra copies of RS3 334 may alter the probability of someone enjoying a happy, monogamous relationship.

According to evolutionary theories developed in the past decade, women tend toward monogamy for reasons of survival. A million years ago, on the plains of Africa, women worked to prevent male infidelity because of worries about the potential loss of their mate, which would leave them without resources to raise a family. On the flip side, males developed jealousy over their mate because they did not want to waste resources feeding children who were sired by other men.

SEXUALITY

Testosterone, a steroid hormone, plays a key role in sexual and aggressive behavior.

REWARD

Dopamine, here in crystal form, gets released by the brain's reward system.

REPRODUCTION

Progesterone is crucial for preparing the womb for pregnancy.

LUST & ATTRACTION

Sexual behavior is programmed deeply in the brain. The drive for sex is controlled by the hypothalamus and the pituitary gland, two ancient parts of the brain. The hypothalamus, which also plays a crucial role in eating, drinking, and the regulation of body temperature, stimulates the creation and release of testosterone in males, and sex hormones such as estrogen in females. In men, the tuberal region of the hypothalamus secretes gonadotrophic-releasing hormones, discovered in 1971, into the anterior pituitary gland. The pituitary then releases luteinizing and follicle-stimulating hormones. They act upon the testes to stimulate the release of testosterone and the creation of sperm. In women, gonadotrophic-releasing hormones prompt the ovaries to release sex hormones. Estrogen makes a loop through the bloodstream and back to the brain, stimulating the ventromedial hypothalamus to produce a sexual response. Sexual activity also gets a boost from the medial preoptic area of the

CROSS REFERENCE: See "Smell & Taste," PAGE 116

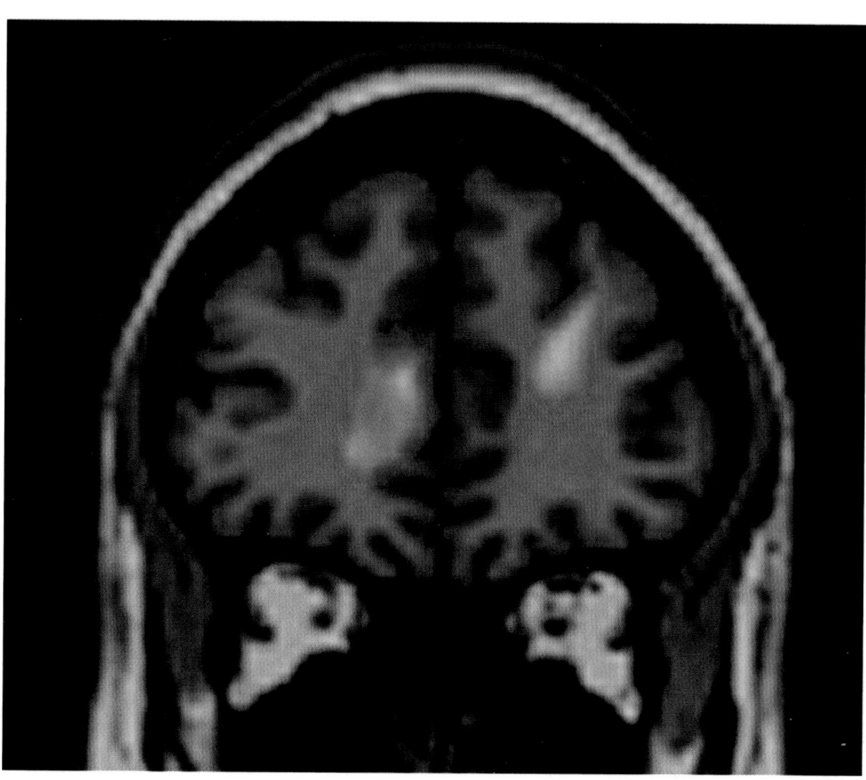

A male brain reacts with sexual excitation to erotic images of women.

hypothalamus when prompted by the introduction of testosterone, primarily responsible for the libido of both sexes. The medial preoptic area is more than twice as large in men as in women.

FAST FACT Stimulation of the medial preoptic area in animal brains induces copulative behavior.

Rhythmical applications of gonadotrophic-releasing hormones have restored the sex drive of patients who had suffered hypothalamus damage. When the hormones were cut off from communicating with the pituitary, the loss halted testicular functions necessary for sex drive and sperm production. The start of sexual behavior among adults seems to rely on the production of testosterone. But sex is more than just pure chemistry. Anthropological studies have revealed patterns of flirtation across cultures that contain surprisingly similar behaviors. That suggests flirtation, a sexually loaded form of communication, could have a significant biological component.

R. Douglas Fields, a neuroscientist at the National Institutes of Health, is convinced that hardwired biochemistry strongly influences sexual attraction. He believes that a little-known cranial nerve, called nerve zero, may be a missing link in research on human pheromones—the chemicals that carry behavioral messages between insects and other animals far below humans on the evolutionary scale and thought to play a possible role in human attraction as well. Fields said in 2007 that he considers nerve zero, which connects the nose with the regions of the brain involved in sexual reproduction, to be a primary candidate for lust-inducing chemical communication.

Smell is also known to have a role in sexual communication. Claus Wedekind, a researcher at the University of Bern in Switzerland, asked an experimental group of women in 1995 to smell a variety of T-shirts worn by men they did not know. Women preferred odors produced by men whose immune system differed widely from their own. Mating between such couples would be likely to produce healthy children.

Nerve zero, first described in 1913, often gets overlooked in medical schools because its thinness

+ PHEROMONES +

THE JURY IS OUT on whether humans respond to sex pheromones. Among reptiles and mammals, a tiny olfactory area known as the vomeronasal organ, or VNO, detects pheromones. Humans cannot smell pheromones, but they do have a VNO. One study done in 1998 on 38 men suggested that synthesized male pheromones brought about more physical contact with women, including sexual intercourse, but more studies are needed.

Hormonal changes during pregnancy, childbirth, and breast-feeding promote mother-child bonding.

and its position at the top of the brain cause it to be removed as brains are prepared for dissection. Some scientists consider nerve zero a likely branch of the olfactory system or perhaps an evolutionary dead end, like the appendix. Some scoff at Fields's claims, or suspend belief until they see more evidence. Fields, however, defends nerve zero with evidence from 1987 experiments on hamsters. When their nerve zero was severed, they stopped mating.

FORMING BONDS

Sex may keep the world filled with people, but attraction and attachment provide it with families.

Brain researchers describe attraction as a period in which a person fixates on another person and his

BREAKTHROUGH

THE JOURNAL *NATURE* carried an intriguing headline in 2005: "Trust in a bottle."

It was accurate, to be sure. The accompanying story reported on a Swiss research team that created a nasal spray containing oxytocin, a powerful hormone associated with the forming of social relationships, and induced experimental subjects to take a whiff. The subjects and a control group then played out a make-believe scenario in which they gave their money to investors. After sniffing the oxytocin, the volunteers handed over more money, knowing that the trustee could either invest it and share big returns with the "investor" or keep all of the money.

The researchers, at the University of Zurich, said that out of 29 subjects given oxytocin, 13 gave the trustee all of their cash. The control group, which sniffed a placebo, had only 6 of 29 hand over all of their investment credits.

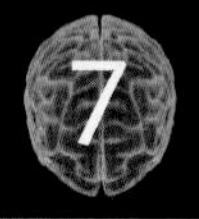

or her special attributes. Self-reports of emotions felt by the afflicted include hope, elation, fear, and uncertainty. During this early stage of love, the brain exhibits heightened levels of phenylethylamine (PEA), a chemical cousin of amphetamine. It accounts for the "high" feeling of early love, as well as the loss of appetite. The rush continues for up to three years. Couples bond by sharing the rush of love. At some point, however, the fireworks of love settle into steadier patterns. Scientists speculate that the change may reflect the fact of the brain becoming accustomed to heightened levels of PEA, much like a drug addict's failure to get high after habitual doses of neurostimulants.

Once the couple enters a period of calm confidence, they begin a phase known as attachment. During this phase, the brain boosts the production of endorphins as well as the hormones oxytocin and AVP. Strong bonds form as these hormones increase, particularly during the time of childbirth. New parents nurture their children, creating strong parental-child bonds, and also increase their nurturing of each other. They see themselves as mutually supporting and find comfort in each other's company.

FAST FACT Buddhists seek a state of bliss through meditation—an application of the brain upon itself.

The existence of a third stage beyond the first two has drawn the support of some neuroscientists, who call it the detachment phase. Separation and divorce occur with all too much regularity, as anyone can attest. Brain chemistry may play a role in this post-attachment phase. It is possible that endorphin receptors lose their heightened levels of sensitivity and that individuals with a temperament for novelty and risk may react to feelings of too much security. Divorce statistics suggest biological drives contribute to the breakup of relationships. Most divorces occur around the fourth year of marriage, after the attraction phase has worn off. Typically they occur among no-child or one-child families, with most divorced people remarrying before they exit their procreative years. This suggests the significance of the reproductive role of bonding.

Family bonding with newborns begins at birth. Oxytocin, which promotes trust and feelings of closeness, must be present for mother-child bonding to occur. Experiments with rodents reveal that when females who have never had pups are injected with oxytocin, they approach others' pups and attempt to mother them. When rodents' oxytocin receptors are blocked during the birth of offspring, the mothers fail to form parental bonds. Female mammals, including humans, release oxytocin during the vaginal stimulation of sex as well as childbirth. Initiating feelings of closeness at the very moment child and parent are introduced to each other makes sense from an evolutionary point of view—mothers feel compelled to care for their infants until they can care for themselves. Later in human relationships, oxytocin

BREAKTHROUGH

The effect disappeared when the subjects interacted with a random-number generator, demonstrating that oxytocin promotes social trust instead of the willingness to gamble.

Altering oxytocin levels promises breakthroughs in treatments of social disorders where patients trust too much or exhibit social phobias. Or the technique could be used by unscrupulous marketing agents to promote good feelings about inferior products. Imagine a political candidate seeking your vote in an auditorium spritzed with oxytocin, or a used-car salesman trying to unload a lemon after spiking the water fountain.

STAYING SHARP

HAPPINESS AND HARMONY can extend your life.

That's the message from author Dan Buettner, who visited world regions where people live a long, long time. He sought the formulas for longevity in the Nicoyan Peninsula of Costa Rica; Sardinia, Italy; Okinawa, Japan; and Loma Linda, California. In those pockets of long life and happiness, people live on average a decade longer than their peers and have a fraction of the rates of cancer and cardiovascular disease.

In a 2009 newspaper interview, Buettner advocated gardening as a way to add up to 14 years to one's life. "The world's longest-lived people tend to do regular, low-intensity physical activity like walking with friends and gardening," he said. In contrast, Americans tend to overdo strenuous exercise, he said.

In his book *The Blue Zones,* he proposed nine rules, gathered from his observations, for extending life:

✓Move naturally. Do enjoyable activities every day. Get your exercise in fun ways, such as walking instead of driving.

✓Cut caloric intake by 20 percent. Try using smaller plates, bowls, and glasses.

✓Increase the proportion of fruits, vegetables, and nuts in your diet.

✓Have a glass or two of red wine each day.

✓Find the purpose of your life.

✓Seek relief from stress. Relax.

✓Join a spiritual group.

✓Put your family above other things.

✓Surround yourself with people who share your values. Be likable.

contributes to every type of bonding imaginable: sexual, parental, fraternal, and even the love of one's self. The brain has adapted biochemistry to create social groups, which allow humans to care for one another and support the continuation of the species.

Once the child is born, the sensation of touch is crucial to long-term health and the creation of parent-child bonds. Infants respond instinctively to the sensation of touch, as when they root toward a breast to feed. As the infant suckles, oxytocin in the mother's blood makes the breast's milk ducts contract to expel milk into the infant's mouth. Breast-feeding thus underscores mother-child bonding.

JOY

Happiness, or joy, is one of the universally recognized emotions. As with other emotions, it has eluded simple categorization. However, it seems to be more a matter of nature than nurture. In the past few years, neuroscientists have determined that more than 60 percent of an individual's tendency to have a character dominated by positive emotions comes from his or her genetic makeup. The rest is what a person learns through experiences, emotions, and thoughts.

States of happiness are associated with top physiological functioning and the belief that one's life is running smoothly. Joyful states also are associated with calmness and clarity of thought, warmth and relaxation, and ease in deciding how to act. Such feelings promote not just survival, but richness in life that makes people want to continue living. The Dutch philosopher Baruch Spinoza associated joy with the achievement of states closer to perfection. Joyful people have a greater sense of freedom and of power.

FAST FACT Is happiness a warm puppy? A 2004 study found that playing with dogs increased levels of joy-inducing hormones.

The right to pursue happiness is enshrined in the Declaration of Independence. The definition of what brings happiness, however, probably varies from person to person. Neuroscience has only recently turned to the investigation of happiness by focusing on the related states of pleasure and desire. Both involve reward behaviors.

INVESTIGATING HAPPINESS

As with the darker emotions, the brain is hardwired for elation. Researchers James Olds and Peter Milner discovered this by accident in the 1950s when they implanted an electrode in the hypothalamus of a rat and connected the wiring to a bar so the rat could administer small electric shocks to itself. The rat hit the bar up to 4,000 times an hour, forgoing food and sex. Small wonder Olds and Milner considered that the rat most likely took great pleasure from the stimulus.

Further research, on human brains, has pinpointed some of the regions for happiness. They include the hypothalamus along with the nucleus accumbens and septum. Each pleasure region releases neurotransmitters and endorphins, as well as dopamine, which has gained the most attention as an inducer of positive emotions. Dopamine plays a key role in reward mechanisms in the brain.

Follow-up research on the pleasure centers of the brain in the 1960s branched out into some ethically questionable directions. Researchers at Tulane University in New Orleans tried to manipulate the brain's pleasure circuits to "cure" patients of homosexuality through activation of electrodes. Such research was halted.

Research by Kent Berridge of the University of Michigan pointed to the likelihood that the electrically stimulated regions in the rats' brains in Olds and Milner's study may have been associated with desire instead of pleasure. Based on his studies, Berridge proposed a difference between desire and pleasure both in the brain regions affected and in neurochemical triggers. He associated dopamine with desire and the opioid system and its morphine-like neurochemicals with pleasure. In such a system, happiness might exist as a state of contentment, bringing pleasure without desire.

The appeal of illegal drugs is their ability to boost the availability of dopamine in neural networks artificially to promote desire. Their chemically induced highs promote addiction and, all too often, physical rebounds that leave their abusers feeling worse than before.

Mirror neurons can spread emotions, such as the laugh-filled joy of these California children, from person to person.

IMPACTS OF LOSS

SADNESS, GRIEF, & DEPRESSION

No translation is necessary—a homeowner's face and posture say it all after she lost her home to an Oklahoma tornado.

THE EMOTIONS of sadness and grief serve practical purposes. They may have evolved to let the brain slow down and recognize the impact of loss. Sadness may also call attention to negative behaviors and contribute to the motivation to change them. In its mildest form, sadness can be a simple melancholy that spurs a change in action or scenery. In its fiercest manifestations, sadness can bring on everything from terrible grief to crippling depression and suicide.

WHAT CAN GO WRONG

DEPRESSION that sets in during midwinter, after the holidays, may be a holdover from our evolutionary past. Animals caught in the depths of winter, cut off from bountiful food, may have trimmed their metabolism or hibernated to survive. Despite central heating and refrigerators stocked with food, humans perhaps still face winter as a long, hard slog toward survival.

Human coping mechanisms with winter may include eating and sleeping more. Among some people, onset in winter of seasonal affective disorder (SAD) may play havoc with emotions. Patients may experience depression, fatigue, social withdrawal, irritability, and extreme lethargy, brought on by the shortening of days and long periods cooped up indoors. The brain chemistry responsible for such feelings arises from an overabundance of melatonin, which the brain produces during periods of

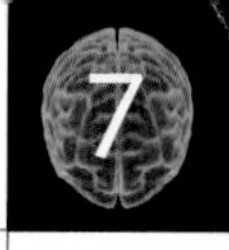

Everyone feels sad from time to time. That's only normal. Illness, death of loved ones, financial setbacks, and divorce—all forms of stress—rank among the most common triggers of this emotional state. Causes need not be tremendous life changes. Sadness can be caused by the smallest of life's disappointments as well as genuinely life-altering events. Like other emotions, sadness can be triggered not only by things as they happen, but also by the memory of events, such as a funeral or the recollection of a missed loved one.

Sadness that accompanies a serious illness, such as cancer, may interrupt normal sleeping and eating patterns; interfere with the ability to concentrate; inhibit social behavior; contribute to greater feelings of impatience and irritability; and generally disrupt established patterns of life. With time, however, sadness naturally fades in the rhythmic cycles of homeostasis unless it changes into the chronic, debilitating illness known as depression.

FAST FACT Depressed people suffer the most in the early morning, which also is a common time for suicide. Suicides are most common on Fridays and Mondays, when work-related issues and loneliness loom large.

SIGNS OF SADNESS

In keeping with its power to force the body to slow down and reassess the world around it, sadness induces flaccid muscles and a general passivity in the body. Other signs of sadness include crying, visceral feelings of gloom and sorrow, drooping eyelids, a protruding lower lip, and a rise in the inner edges of the eyebrows.

In the brain, sadness seems to manifest itself in a left-right split in neural activity. Brain scans associate the emotion with increased activity in the left hemisphere of the amygdala and the right hemisphere of the frontal cortex, and decreased activity in the right hemisphere of the amygdala and the left hemisphere of the frontal cortex. Long bouts of sadness may inhibit the bank of neurotransmitters in the frontal lobe and amygdala, leading to a numb and empty form of depression.

DEPRESSION DIFFERENCES

Sadness can turn into depression, but there's a big difference between the two. Depression lasts longer—two weeks or more—and it hurts more. It interferes with the day-to-day routines of life.

Depression overwhelms an individual, causing withdrawal from family and friends. It can lead to chronic thoughts of death, as well as self-injury and suicide. Symptoms of depression, as opposed to sadness, include persistent sad or empty moods, a chronic drop in energy, loss of pleasure in things that usually bring enjoyment, guilty and helpless feelings, and a general feeling of not being one's usual self. Physical problems that do not seem to have any physical causes, such as chronic headaches, also may be bodily signals of depression.

Efforts to find the causes of depression are thousands of years

WHAT CAN GO WRONG

darkness. The excessive amounts of the neurotransmitter reset the body's internal clock and cause havoc with diurnal rhythms. Treatment includes bright lights, administered in specific wattages and for specific periods of time. The light, mimicking the longer hours of sunlight in spring and summer, tricks the brain to alter the body clocks and ameliorate the symptoms of SAD. Possible side effects from light therapy include headaches and eye strain, with the possibility of insomnia if the exposure occurs too late in the day. Patients who are bipolar also have some risk of a manic episode being triggered by the light.

old. The Greek physician Hippocrates traced depression to "black bile," one of the body's four humors. When black bile attacked the body, it produced epilepsy, he said, and in attacking the brain it created depression.

A depression patient in the 17th century, Robert Burton, added social factors to the causes of depression, blaming lack of parental affection (he included himself among the unfortunates) for the inability to express love. Biochemical causes didn't emerge until 1931, when two researchers in India discovered that a medication called reserpine, which is derived from the tranquilizing rauwolfia plant, calmed psychotics at the cost of drug-induced depression.

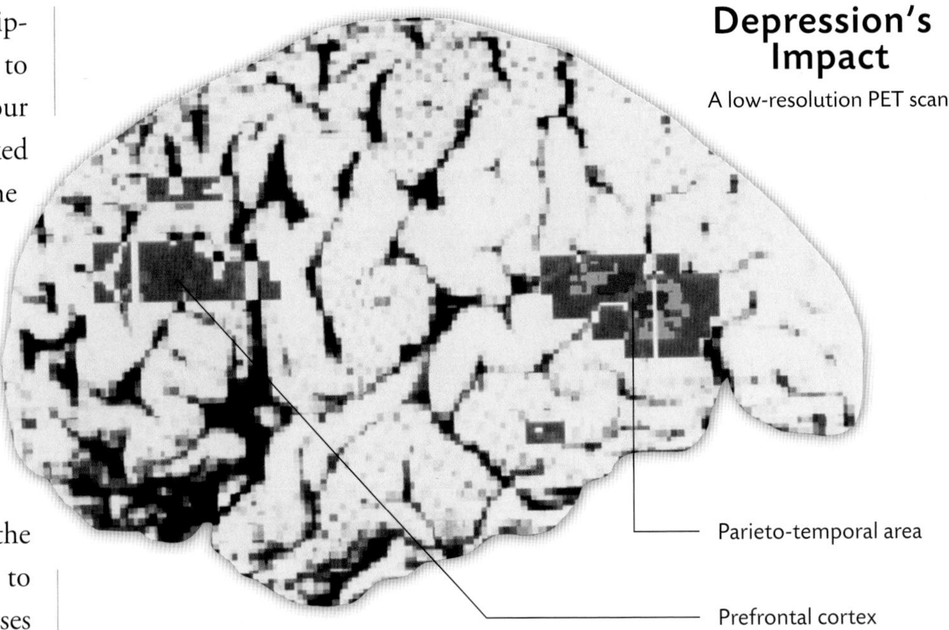

Depression's Impact

A low-resolution PET scan

+ FAMOUS PATIENTS +

DEPRESSION is all too common. A 2001-02 survey released by the National Institutes of Health revealed a little over 13 percent of Americans had experienced "major depressive disorder" at some time in their life. Famous patients include:

+ WINSTON CHURCHILL. Inspirational British prime minister called his depression his "black dog." Feeling unloved, he channeled his energy into ambition.

+ PATTY DUKE. Television and movie actress titled her memoir *A Brilliant Madness*. She attempted suicide and became hooked on drugs before getting help.

+ ABRAHAM LINCOLN. The 16th President fought against depression his entire life. "His melancholy dripped from him as he walked," said his law partner, William Herndon.

+ COLE PORTER. Witty composer of popular songs transformed from a bon vivant to a virtual recluse in his final years. He suffered horrible pain after a horseback riding accident and loneliness from the death of his wife.

DEPRESSION TODAY

Depression can strike virtually anyone. Among 43,000 American adults interviewed in 2001 and 2002, 5.28 percent had experienced major depressive disorder in the previous 12 months, and more than one in eight had experienced it at some point. Demographic analysis found some people were more likely to be depressed than others. Being of middle age and of Native American ancestry boosted the risk of depression, as did low income, divorce, separation, or the death of a life partner.

Children as young as five or six can experience symptoms that resemble depression in adults. The onset of true depression increases sharply during the early teenage years, then gradually rises to peak around age 40. The average age of onset was about age 30, with treatment usually beginning about three years later.

Heredity plays a role in the likelihood of developing depression. If one of a pair of identical twins is diagnosed with clinical depression, the other has a 70 percent chance of following with the same diagnosis. In some families, depression can span and virtually fill generations. The great 19th-century British poet Alfred, Lord Tennyson was one of 11 children who reached adulthood. Nine of the eleven suffered from bipolar illness; rage, unstable moods, or

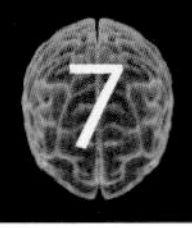

insanity; or, like Alfred, recurrent depressive illness.

Gender also plays a role in how likely a person is to be depressed. Women were more than twice as likely as men to suffer depression, and a bit more likely to seek treatment. However, male patients are more likely than females to report feelings of fatigue, irritability, and sleep disturbances. Men also are more likely to hide their depression from themselves and others by abusing alcohol or drugs or by working excessively long hours.

DIAGNOSING DEPRESSION

Doctors consider depression a medical condition when it lasts longer than two weeks and has a noticeable impact on a person's day-to-day ability to function. There's a long list of potential symptoms, and clinical depression—the medical term for the physical disorder—may include all, many, or just a couple. The most prevalent forms include: major depression, dysthymia, and bipolar disorder (formerly called manic depression).

With major depression, some combination of chronic physical and mental symptoms interfere with work, sleep, appetite, and the ability to find joy in activities that formerly were considered fun. A major depressive episode can occur only once in a lifetime or repeat over and over again. Dysthymia is a less intense form of long-term

VINCENT VAN GOGH

Vincent van Gogh levels a melancholy stare in one of many introspective self-portraits.

NINETEENTH-CENTURY Dutch painter Vincent van Gogh filled letters to his brother, Theo, not only with descriptions of his struggles as an artist but also with his depression. "[H]ow miserable the 'dregs' of work are, that depression after overexertion," he wrote Theo in 1883. "Life is then the colour of dishwater."

Van Gogh failed at many things—during his lifetime he sold only one painting. Ultimately he went into the hospital for his mental illness and committed suicide—failing at that, too, lingering for hours after a self-inflicted gunshot wound.

Johns Hopkins University psychiatry professor Kay Redfield Jamison has suggested causal linkages in the overlapping moods and creative temperaments of brilliant authors and musicians, including not only van Gogh, but also Lord Byron, Virginia Woolf, and Robert Schumann.

depression that nevertheless prevents normal functioning. Bipolar disorder occurs when phases of depression and lethargy alternate with times of elation and bursts of activity, called mania. Mood changes can occur rapidly but typically move at a gradual pace.

SYMPTOMS & SIGNS

Depression is believed to have many causes, and this intangible quality makes some neuroscientists refer to it not as a disease, on a par with the common cold, but rather as a syndrome—a mix of signs and symptoms. Depression strikes all over the brain, involving the cerebral cortex, amygdala, hippocampus, hypothalamus, and other regions.

One common symptom is that depressed people exhibit shrinkage of the hippocampus, a brain region that regulates stress. It's unclear which comes first—the depression or the shrinkage—but the result is the same: a reduced ability to regulate stress, which can stoke the fires of depression.

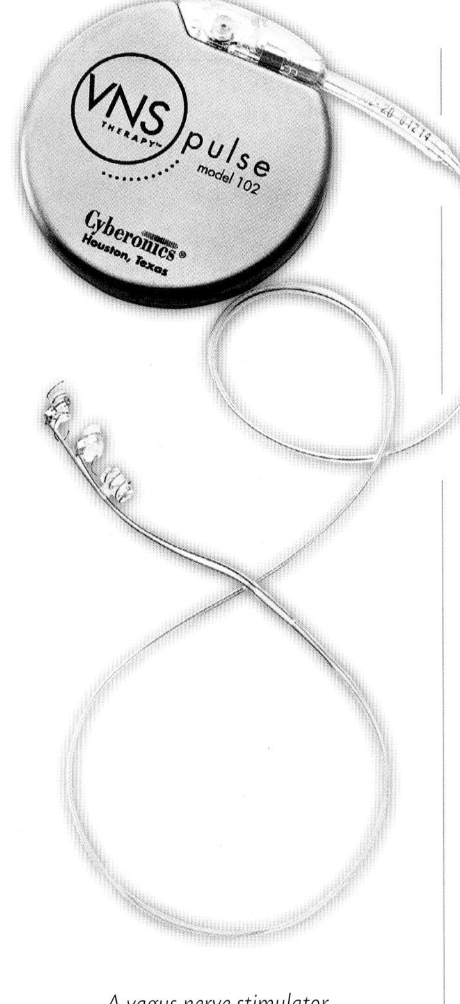

A vagus nerve stimulator

Major depression also affects the frontal lobes, lowering the ability to reason while ratcheting up the emotional limbic system. "In depressed people the turnoff switch by which thinking controls emotions isn't working properly," said University of Toronto neurologist Helen Mayberg. "As a result emotion overrides thinking." After a regimen of taking an antidepressant such as Prozac, PET scans reveal the brain of depressed people starting to reset itself to a normal balance of limbic versus cortical activity.

CHEMICAL TREATMENTS

Research suggests depression arises from imbalances of neurotransmitters. Many pharmaceutical treatments target serotonin in particular. The antidepressant Prozac, for example, increases the amount of serotonin in the synapses. Normally, after one neuron communicates with another through serotonin, the neurotransmitter molecule is destroyed, or reabsorbed by the neuron that initially released it. Prozac blocks the releasing neuron's ability to reabsorb the serotonin molecule, making the neurotransmitter stay longer in the space between neurons.

BREAKTHROUGH

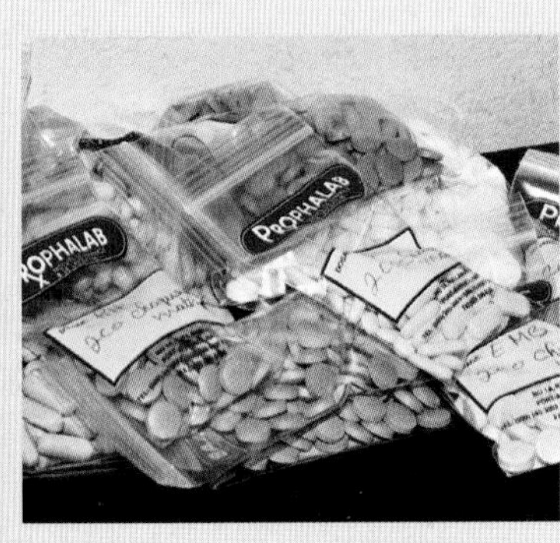

TWO STATEN ISLAND physicians putting new tuberculosis drugs to the test in 1951 got some unexpected results. Not only did the drugs, isoniazid and iproniazid, dramatically improve the patients' physical health, but they also created what the doctors, Irving Selikoff and Edward Robitzek, described as a general boost to their mood—so much so that some had to be put under psychiatric care.

The following year, a Cincinnati psychiatrist, Max Lurie, decided to try isoniazid as a stimulant for patients suffering from depression. Amazingly, two-thirds exhibited improvement. Lurie and his

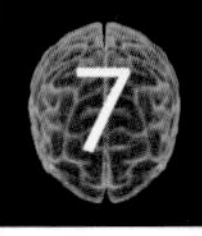

As it lingers there, the molecule is more readily available for another electrochemical communication. The release of other molecules of serotonin joins with the molecules already in the synaptic cleft, creating a higher serotonin concentration than normal.

Because Prozac affects serotonin principally, it's called a selective serotonin reuptake inhibitor, or SSRI. Other SSRIs are Zoloft, Celexa, and Paxil.

It takes a while for SSRIs to have a noticeable effect on a depressed person's symptoms because the drug needs time to build up in the patient's system.

After two or three weeks, the enhanced release of serotonin causes the receiving neuron to become more sensitive to its presence. Electrochemical communication through that particular neurotransmitter grows swifter and more efficient, and the patient usually experiences the desired relief. Several more weeks typically are necessary to feel the full effects of the drug.

FAST FACT The melancholy title character of Shakespeare's well-known play *Hamlet* may have had literature's most famous bout of depression, stemming from his suffering "the slings and arrows of outrageous fortune."

TALK ABOUT IT

Often usage of SSRIs is coupled with talk therapy to help treat the symptoms and causes of depression. The simple act of talking through one's problems is time honored and still showing much success. Cognitive therapy works by getting depressed patients to mentally reframe their problems in ways that put a more positive spin on things. This strategy makes it easier to get their brain to heal itself by reframing an issue. In short, talk therapy boosts the brain's cognitive functions while lowering emotional ones.

Researchers at UCLA reported in 2007 that a series of brain scans on volunteers found that when they verbalized their feelings the intensity of negative emotions such as sadness and anger decreased. When someone is sad or angry, getting that person to talk or write may provide emotional benefits no matter what the person says. Putting emotions into words activates the right ventrolateral region of the cortex and lessens the activity in the amygdala.

TACTICS

Coping with depression can be a catch-22. Taking decisive action can help bring about change and restore emotional balance, but one of the symptoms of depression is lethargy, which severely dampens the ability to take such action. Trying to effect change can seem overwhelming, as if the problems are too big to face.

If that's the case, you can work at setting and achieving small goals. Try calling someone you love or taking a walk. As you start to get more energy, try taking bigger steps. Reduce your stress. Talk to friends and family members. Take care of your health by eating, sleeping, and getting exercise.

BREAKTHROUGH

colleague, Harry Salzer, coined the term antidepressant for the drug's impact.

Tests on isoniazid and similar drugs found they lowered the action of monoamine oxidase (MAO), an enzyme in the brain that breaks down the neurotransmitters dopamine, serotonin, and norepinephrine. These drugs are called MAO inhibitors. After a period of common use, they are rarely prescribed today because of potentially dangerous side effects, including spikes in blood pressure. They worked by helping beneficial neurotransmitters avoid breakdown in the space between neurons, contributing to an elevated mood.

Another class of antidepressants acts not by preventing neurotransmitters' destruction but rather by preventing them from reabsorption—called reuptake in neuroscience—by the neurons that released them. Hence, these antidepressants are called reuptake inhibitors.

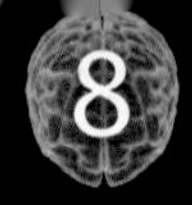

CHAPTER EIGHT

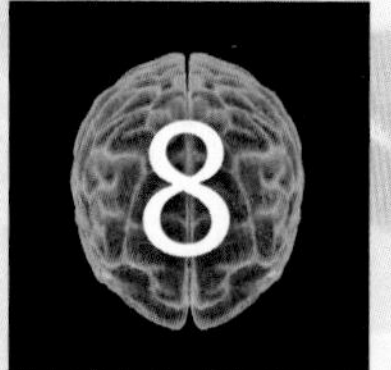

LEARNING & MEMORY

THE BRAIN constantly rewires itself to become an organ that is physically different from what it was in the preceding moment. Never resting, it churns through new experiences daily, incorporating some into its network of stored information and discarding others. The creation of memories, along with the application of electrochemically coded information toward the flowering of wisdom, creates personal and social identity. When disease and disorder rob the brain of memory and language, they take away what makes each person unique.

University graduates end a stage of formal education but will learn from each experience that follows.

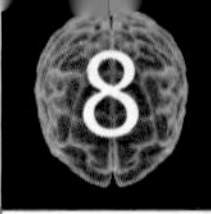

LEARNING

ACQUIRING & STORING INFORMATION

LEARNING AND memory make each human unique. Even before birth, the human brain takes in sensations, processes them, and begins to encode them into trillions of synapses. Those connections, and the electrochemical firing patterns that unite them, make the brain an organ different from what it was a moment before. As new connections form in response to stimuli, and then become strong through repeated use, the brain integrates new information and stores it until it is needed. Without learning and memory, the human brain would be little more than clockwork.

Learning and memory work together. Some learning is transformed into lasting memories; other experiences prove ephemeral.

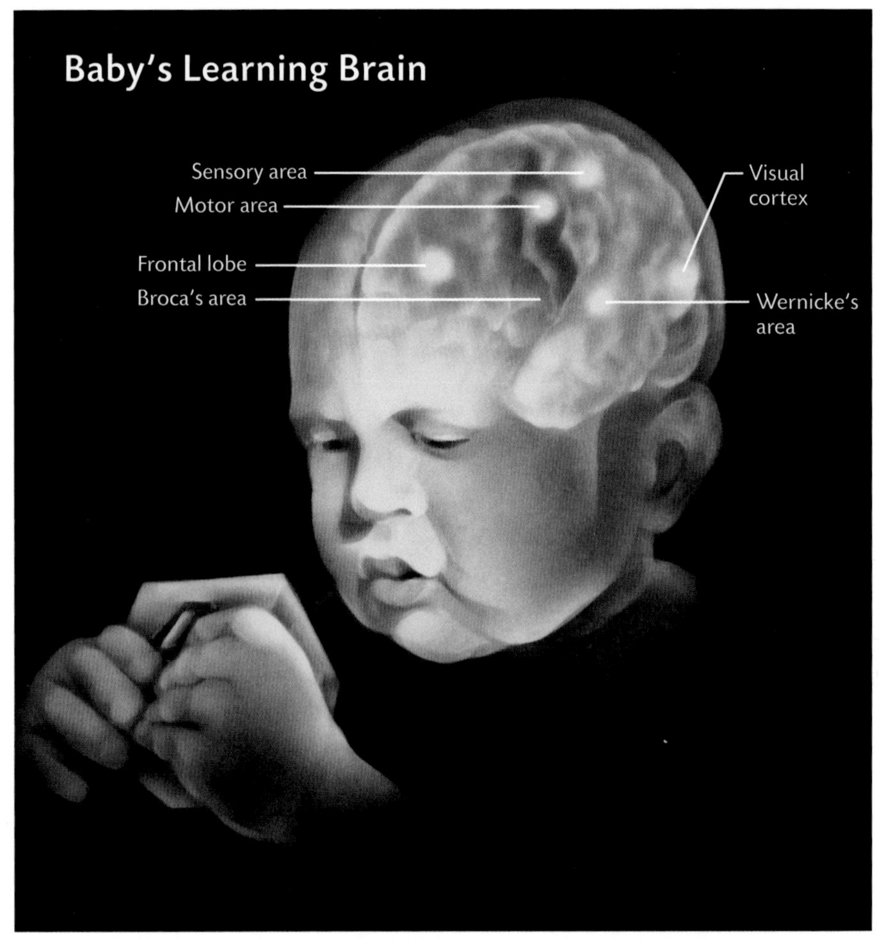

Learning activates a wide range of regions in a child's cerebral cortex.

Eric R. Kandel, who received a Nobel Prize for research on the molecular foundations of memory, draws this distinction: "Learning is how you acquire new information about the world, and memory is how you store that information over time. " Learning includes cognitive components, such as solving quadratic equations; motor components, such as tying a necktie in a perfect Windsor knot; and affective components, such as feeling shame at a social faux pas. Kandel traces learning to physical alterations in the brain's neurons, specifically to the interaction of neurotransmitters and their receptor sites. With repeated stimuli, Kandel and other neuroscientists believe neurons physically change the number of neurotransmitters released and the sensitivity of receptor sites across the synaptic clefts.

Cognitive, motor, and affective learning are sometimes called KSA domains, for knowledge, skills, and attitude.

DIFFERENT LEVELS OF LEARNING

The cognitive domain includes levels of mental development ranging

In the animal kingdom, imitation is not only the sincerest form of flattery. It's also a educational strategy. For example, young chimpanzees learn how to use tools by watching older chimps manipulate tools. This is a form of social learning.

CROSS REFERENCE: See "Shared Roles," PAGE 152

from the simple recall of data to the formation of judgments. Each level requires mastery of those that came before. These levels are: knowledge, the ability to recall information; comprehension, the understanding of meaning; application, the ability to apply a concept to a new situation; analysis, the logical examination of the concepts or parts of a problem to better understand the whole; and synthesis, the creation of a whole concept from individual parts and patterns.

The motor domain deals with movement, coordination, and the application of motion to specific tasks. Motor learning occurs with repetition, such as the eye-hand-foot coordination of driving. Mastery is evident with increased precision, speed, and other measures of excellence, such as a swimmer's execution of a kick-turn.

The affective domain encompasses the brain's ability to deal with emotions and feelings, as well as behavioral issues such as attitude and motivation. It ranges from the ability and willingness to focus attention to the recognition and internalization of social values.

EARLIEST LESSSONS

When a healthy baby is born, his or her brain already is wired for the basic functions of survival, such as regulation of heartbeat, breathing, and digestion. That's nature, and it continues to execute orders for brain development that include the growth and myelination of neurons. But as soon as the baby enters the world, nurture begins.

Reading aloud, a Sudanese man builds his children's vocabulary and prepares them to read.

STAYING SHARP

PSYCHOLOGIST Janellen Huttenlocher of the University of Chicago reported in the early 1990s that the amount parents speak to children during the sensitive second year of life significantly affects vocabulary later. Just as important, according to the nonprofit literacy organization Reading Is Fundamental (RIF), reading aloud is the most effective means of preparing children to learn to read on their own. RIF offers these tips:

✔ Have a special time set aside every day to read aloud. Before bedtime is an obvious choice, but other options may better fit your schedule.

✔ Read not only simple books, but also whatever is at hand. Road signs, cereal boxes, and other everyday items help children connect printed words with objects and concepts.

✔ Read Mother Goose and other rhymes. Young children pick up on the cadence of the language and love to join in on favorite verses.

✔ Read slowly and expressively. Don't be afraid to play the ham.

✔ Sit so the child can see the pages of the book, especially if it's a picture book. Point to new words and say what they mean.

✔ Read old favorites, especially if the child joins in, but also offer new readings.

✔ Be flexible. If the child seems bored, try another book. Ask the child what kinds of things he or she would like to read.

✔ Take time to answer questions.

✔ Children like to end what they begin. Finish what you start to read, or stop at a good ending point for the day, such as the end of a chapter.

The baby's environment enhances neural connections through learning. New connections may grow strong and become permanent, or grow weak and get pruned.

The first signs of memory emerge around the age of two to three months, when babies cry less often and begin to smile at familiar faces. The smile suggests recognition, and recognition requires comparison with experience. This probably corresponds to the rapid development of the brain's frontal lobes, which grow dramatically between ages three and eight months, at which time children can briefly recall objects that get hidden from view. "Separation anxiety" occurs when children feel vulnerable at the departure of a parent or fearful at the introduction to a stranger. Such feelings must accompany memory, as the brain must recognize a difference between the present and the past.

Between ages 18 and 24 months, children begin to develop a sense of self, including feelings, intentions, and interactions with other people. Vocabulary expands exponentially between two and six years as children interact with the world.

Children younger than about seven learn about the world in simple ways based on repetition. In associative learning, children learn that two things go together. A child who drops his spoon and sees it fall, then repeats the action again and again, gains a rudimentary understanding of gravity.

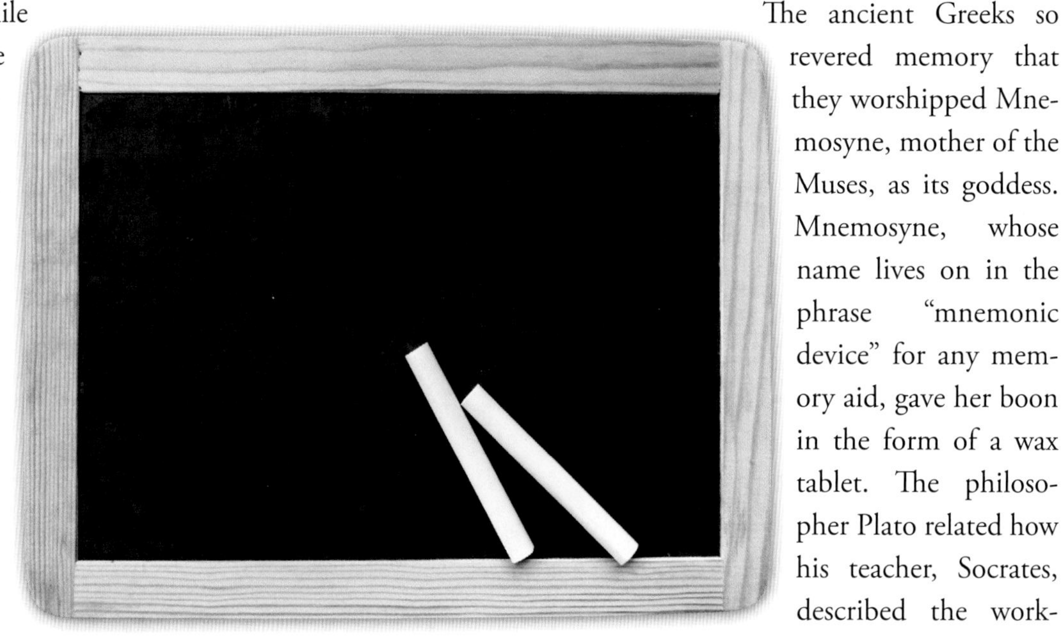

Far from a blank slate, a newborn brain is hardwired to learn, particularly language.

Associative learning includes classical conditioning, the kind Ivan Pavlov used, and operant conditioning, which requires voluntary behavior that brings rewards or punishments. Nonassociative learning demonstrates that two things may or may not be related.

Play is an important method of social learning. When children play, they experiment with how things work. In groups, they learn to interact. When play imitates the actions of adults, children begin to learn their culture, especially when using toys that represent materials in the adult world. Playing games, children learn the importance of rules—and, eventually, when it is appropriate to ignore them.

HISTORY OF MEMORY

The ancient Greeks so revered memory that they worshipped Mnemosyne, mother of the Muses, as its goddess. Mnemosyne, whose name lives on in the phrase "mnemonic device" for any memory aid, gave her boon in the form of a wax tablet. The philosopher Plato related how his teacher, Socrates, described the workings: "[W]henever we want to remember something we've seen or heard or conceived on our own, we subject the [wax] block to the perception of the idea and stamp the impression into it." Plato's student Aristotle argued that deficiencies of memory occur from imperfections in the wax.

From wax tablets in ancient times, through books and libraries in the Middle Ages, to photography and telegraphy and computers, explanations for memory have evolved to keep pace with the times and their science. English philosopher

CROSS REFERENCE: See "A New Brain," PAGE 72

David Hartley (1705–1757) drew on the works of Sir Isaac Newton, who believed all matter contained subtle vibrations, to postulate on the encoding of memories through hidden motions in the nervous system. More than a century later, German physiologist Ewald Hering (1834–1918) suggested that all organic matter contains memory. He defined heredity as memory that passed from parent to child through germ cells.

FAST FACT The neuropsychologist Donald Hebb considered forgetting to be a function of lost communication among cell assemblies.

Psychologist Hermann Ebbinghaus (1850–1909) attempted the first systematic studies of memory in 1885. In one experiment, he forced himself to memorize long strings of nonsense syllables—baf, dak, gel, kim, wauch, and so forth—and then say them in order as quickly as he could. He thought, correctly, that memory included the creation of associations in the brain, and he used nonsense syllables because they had no preexisting associations. He discovered a correlation between the number of times he repeated the syllable list on one day and the speed with which he repeated the sounds the next.

EXAMINING ASSOCIATIONS

If the creation of memories involved associations, where might these associations exist? German evolutionary biologist Richard Semon (1859–1918) proposed in 1904 that experience leaves a physical trace on specific webs of neurons. He called this ghostly trace an engram. American neuropsychologist Karl S. Lashley (1890–1958) tried to find Semon's engrams and gave up in frustration. Lashley proposed a counterargument that said memories become encoded through a form of "mass action" throughout the brain.

As a third alternative, Canadian neuropsychologist Donald O. Hebb (1904–1985) argued for some localization of memories through modification of groups of "cell assemblies" in the brain. To Hebb, learning and memory formed real, physical changes in the neuronal circuits of cell assemblies. Under this theory, cell assemblies work in chains, so that one element of memory—say, the name of an object—excites another assembly that encodes its image.

Hard evidence of physical alterations in neuronal connections arose in 2007 when University of California at Irvine scientists demonstrated concrete changes in the synapses of rats' brain critical to learning. A high-tech scanning technique called restorative deconvolution microscopy found expanded synapses in the rats' hippocampus regions associated with learning after the rats had mastered new tasks.

WHAT IS MEMORY?

Memory has three parts: encoding, storage, and retrieval. Without encoding, the brain has nothing to process for storage. Without storage, the brain would always live in the present. Without retrieval, memories stored in the brain would remain there to no practical purpose. Memory must be stable enough to allow the brain to build upon experiences, but flexible enough to adapt to changes in those experiences. Some memories (What is my name?) need to remain

+ MOZART MEMORY +

AT AGE 14, Wolfgang Amadeus Mozart stole a secret from the Vatican in 1770. He visited the Sistine Chapel and heard Gregorio Allegri's 12-minute *Miserere* performed twice. Mozart re-created the score from memory, defying the Vatican's attempts to prevent duplication. It seemed a miracle at the time, but not so much today. Anyone with a musical ear, good memory, and a technique of memorization called chunking could break the work into small parts and tie them to mental images for sequential recall.

constant, while others (My clothes are a size larger now) must be modified to navigate through life.

Encoding requires paying attention. The strength of the memory may depend upon the type or amount of attention paid to stimuli. Attention to the physical characteristics is encoded more shallowly than the sounds of words, which are not as deeply encoded as the meaning of information processed by the cortex. In addition, emotional content enhances encoding and can lead to so-called flashbulb memories, which may include minute details of extremely emotional moments.

A process called elaboration associates new information with other information and adds strength to encoding. That's why associating words with images, a common mnemonic device, increases the likelihood of their recall. The brain's executive function then would have two ways to grasp the memory: visual and verbal. Strategies to maximize encoding-include minimizing distractions, managing study time effectively, analyzing the material instead of merely trying to memorize it, and using associative memory techniques. Frequency of exposure helps too.

Storage retains the information gathered in the initial stages of encoding. The information passes into sensory memory as the brain processes sensations such as sights and sounds. Some gets processed into short-term memory, and a fraction makes it into long-term memory. Sensory memory lasts a fraction of a second, just long enough to register a perception. Short-term memory lasts for about 20 to 30 seconds but can be extended with practice. Psychologist Alan Baddeley proposed "working memory" as a variation of this concept. Working memory holds information during cognition, including sounds, images, and thoughts.

Memory Formation

Rehearsal

Stimuli/input

Short Term
20–30 seconds

Long Term
1 second–lifetime

Sensory
<1 second

Forgetting

Stimuli work their way through complex neural circuits to enter memory . . . or oblivion.

HOW TO REMEMBER

Long-term memory includes many types of stored information, from facts to autobiographical bits to motor skills developed through repetition. Memories called up by the conscious mind are called explicit, or declarative; those automatically recalled during physical actions are implicit, or nondeclarative. Storage capacity of long-term memory is believed to be virtually infinite. It's not clear whether forgetting occurs because some long-term memories disappear, or whether they exist but cannot be recalled.

Visuospatial memory prompts remembrance of the placement of objects in space.

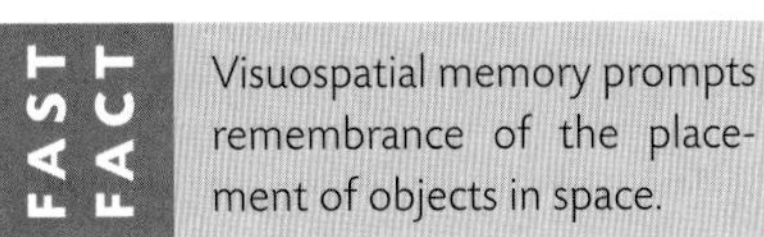

Retrieval may seem random, but it follows an orderly, cause-and-effect process. So-called retrieval cues, to dredge up memories from long-term storage, may come from specific thoughts or be suggested from outside, as during hypnosis. Associational organization of information in the brain may cause one thought to naturally lead to another, which was encoded at the same time, or in the same way, or with links to other information. Moods and physical environments also can affect retrieval.

Two types of retrieval are recall and recognition. Recall involves memories of previously digested information. On a high school biology test, recall would help you list the major phyla of the animal kingdom. Recognition is mere identification of learned items. On that same test, recognition would allow you to pick out animal phyla from a list that includes plants.

ORGANIZING YOUR MEMORIES

The brain's medial temporal lobe, which includes the hippocampus and parahippocampal regions, forms webs of rich connections to the cerebral cortex, helping create, organize, and store memories. The cortex itself is crucial to the long-term storage of information such as events, facts, and daily routines.

Remembrance of facts and events is called declarative memory, whose retrieval requires conscious effort. Brain scans indicate widespread regions of the cerebral cortex interact to support these memories. Declarative memory's major forms include working, semantic, and episodic. It's balanced by nondeclarative memory, which does its work without conscious direction.

Working memory is what the brain keeps handy to navigate through the world from second to second. One way to think about it is to compare it to a computer monitor, which displays a portion of the information held in the computer's memory for easy access and manipulation. New experiences register through the senses and cognition and reside in the current moment

A MEMORABLE AMNESIC

H. M. lost his ability to form new memories because of damage to his hippocampus.

MORE THAN any poet, mystic, or lover, H. M. lived in the moment. It was all he had.

Until his death at age 82 in 2008, Henry Gustav Molaison was written up in scientific literature only as H. M. to protect his privacy. As a nine-year-old in Connecticut, he banged his head in an accident that later led to convulsions and blackouts. By age 27, Molaison could no longer earn a living fixing motors. Neurosurgeon William Beecher Scoville tried large doses of anti-seizure drugs before settling on a radical option: removing two slivers of Molaison's brain.

Immediately after the operation, while H. M. was still in the hospital, it was clear that his memory had been adversely affected. Scoville had cut into the hippocampus unknowingly. Molaison could remember much of what had happened before the surgery, yet he could not form new memories. Brenda Milner, a psychologist in Montreal, began traveling to Hartford regularly to test Molaison. Each visit seemed the first for the patient. Nevertheless, Molaison welcomed her tests, as well as studies by others. Milner's first revelation was that damage to the hippocampus could have such profound results. The second revelation, published in 1962, was that Molaison had retained a form of memory based not on cognition but on motor skills—his "motor learning" evidently took place below the level of consciousness and involved other brain regions.

"The study of H.M. . . . opened the way for the study of the two memory systems of the brain, explicit and implicit, and provided the basis for . . . the study of human memory and its disorders," neuroscientist Eric Kandel told the *New York Times*.

Kayla Hutchinson (left) suffered complete amnesia after a collision with a teammate.

of consciousness, just as open documents fill the screen. Working memory depends on the prefrontal cortex's interacting with other regions of the cerebrum. If the prefrontal cortex's executive function calls upon long-term memories to enrich current experience, it interacts with posterior cortical areas, where words, sounds, and images are stored for retrieval.

STORAGE UNITS

Semantic memory involves knowledge about general facts and data. That includes much of the learning done in the schoolroom but also recognition and naming of people, animals, places, and things. Neural networks appear to be specialized for storing particular data and appear to be widely scattered throughout the cerebral cortex.

Episodic memories are the kind of biographical bits in which people recall what happened to them at particular times and places. Initial processing and storage of such episodic memories are believed to rely heavily on the medial temporal lobe and parahippocampal region. Portions of the parahippocampal region process different bits of episodic memory through different streams, bringing together neural networks that encode for "what," "when," "where," and other details. These links then are stored in various cortical regions. Dredging up a long-term memory brings together the individual bits—images, sounds, times, places, and so forth—to complete the memory.

Memory that is nondeclarative involves the rote memory of deeper and older parts of the brain. It's the memory of how to perform learned habits and skilled actions, such as typing, skiing, and dancing. Processing and storage occurs in the cerebellum and basal ganglia.

Memories that contain emotional content get an extra kick from the amygdala, encoding them more powerfully than emotionally neutral stimuli. When emotional memories are retrieved and expressed, the hypothalamus and the sympathetic nervous system become activated, prompting physical changes in the body that are linked to the expression of emotion.

MEMORY & INTELLIGENCE

To heck with a better mousetrap. Science can build a better *mouse.* All it takes is a gene that alters the brain's receptors and improves memory.

FAST FACT Got a song stuck in your head? Your brain is repeating "sequence recall," a function crucial to remembering everyday tasks. Try retrieving another song sequence to break the loop.

CROSS REFERENCE: See "Emotions," PAGE 206

In 1999, Joe Z. Tsien, a biologist at Princeton University, created a new strain of mice called Doogies—after the whiz-kid doctor in a fictional TV series—by inserting a gene into their hippocampus that extended receptor function. Normally, the receptors of nerve cells stay active for a fraction of a second. The insertion of the new gene lengthened the activation period by 150 percent, improving the mice's memory, because of the role of the hippocampus in memory encoding, and their intelligence. Doogie mice easily outperformed regular mice in intelligence tests. Doogies also had greater curiosity.

Mice brains and human brains have the same biological circuits, so it's no stretch to assume that increasing memory in humans also would increase intelligence. Neurologist Richard Restak believes that makes perfect sense. "An increased memory leads to easier, quicker accessing of information, as well as greater opportunities for linkages and associations," he said. "And, basically, you are what you can remember."

+ MOVIE AMNESIA +

IN 2004, clinical neuropsychologist Sallie Baxendale examined film depictions of amnesia. She found common misconceptions. These included the atypical ability of patients to learn and retain information without impairment; the loss of particular memories as temporary, and likely to return in time; and the fallacy that a second blow to the head may undo amnesia caused by an earlier blow.

Memory supports not only intelligence, but also language, motion, and everyday cognition. Using working memory, the brain interacts with the world in the here and now. More sophisticated communication requires an expansion of memory far beyond that of other animals. A speaker and a listener need extensive short- and long-term memory to communicate through symbols, using words that they have agreed to assign to particular objects. Memory must keep track of the words and their meanings, as well as the syntax that puts them together.

As communication expanded to include long speeches and written documents in human history, both short- and long-term memory adapted. Short-term memory allows the brain to track the progress of individual phrases and bring order to sentences and paragraphs. Long-term memory calls up the meaning of simple words as well as more abstract forms such as figures

+ MEMORY TYPES +

MEMORY TYPE	DESCRIPTION	LOCATION
Short-term / Working	The brief time of keeping something in mind before dismissing it or pushing it into long-term memory.	The hippocampus and subiculum store short-term memories.
Long-term / Procedural	An implicit memory, allowing action to be performed unconsciously; "how to" knowledge.	Stored first in the motor cortex, then sent to the cerebellum.
Long-term / Priming	An implicit memory, which biases the brain to nonconsciously recall recently experienced information quickly.	Stored in cerebral cortex regions that process original stimuli.
Long-term / Episodic	A declarative (explicit) memory, in which conscious thought recalls personal experiences.	The prefrontal cortex and the hippocampus.
Long-term / Semantic	A declarative (explicit) memory, in which conscious thought calls up learned knowledge, such as facts about the world.	Perhaps the same regions as episodic memory.

STAYING SHARP

THE SAME OLD ROUTINES bring comfort in old age, but they do little good for the brain. Novelty appears to be a key to keep the aging brain acting young.

Many adults in their 70s and 80s, and beyond, remain sharp. They play bridge (a game that requires an excellent working memory), do puzzles, attend plays, and take college classes. All in all, they seem to enjoy youthful brains. Yet autopsies performed on many of the agile-brained elderly reveal abnormalities such as those associated with Alzheimer's disease. As many as two-thirds of people whose autopsies revealed some evidence of Alzheimer's demonstrated cognitive acuity until death.

Searching for an explanation of the conflicting evidence, Columbia University Medical Center researchers Nikolaos Scarmeas and Yaakov Stern, working over the past two decades, have settled on "cognitive reserve." According to their theory, the brain's development of extra neurons and significantly more axon-to-dendrite connections over the course of a lifetime may provide a cushion of cognitive power against the effects of dementia and even delay its onset.

The loss of brain function is extremely complex. However, many studies suggest that it's good to expose the brain to new stimuli to create a cognitive reserve. Evidence points to the benefits of engaging in leisure activities; complex stimuli are good, especially if they introduce new problems for analysis. If you do crosswords, graduate to Sudoku and acrostics. If you enjoy opera, learn the librettos. Read books not found on your usual shelf at the bookstore. Try learning a foreign language and chatting with native speakers. Turn off vapid television programs and substitute classic plays and poems.

of speech and metaphors; it suffered a blow when narrative works could be put on paper instead of residing in oral traditions.

PERSONAL CONNECTIONS

One of the most important social functions of memory is the recognition of faces. Most people can remember as many as 10,000, and can identify as many as 90 percent of their school classmates by their yearbook photos after an absence of 35 years. Memory functions for facial recognition are not perfectly integrated, however, as evidenced by the all too common ability to recognize someone but not remember a name or how that person is familiar. Three regions of the brain activate when you recognize a face. First is the inferior occipital gyri, which lie at the back of the brain where visual processing occurs. This pair of structures analyzes the bits and pieces that make up the face, including the shape of the nose, lips, and eyes; wrinkles; skin tone; and other characteristics. Second is the right fusiform gyrus, which recognizes the face. And third is the anterior temporal cortex, where memories of specific faces are stored, allowing comparison between the recognized face and the storehouse that provides identification. If the inferior occipital gyri fail to do their job, the brain may miss important identifying information. Breakdown of the right fusiform gyrus may lead to believing a variety of faces are those of one person. And problems in the anterior temporal cortex may result in a profound inability to name people who are recognized. A well-tuned ability to recognize faces promotes social bonds and social order as people recognize each other and the roles they play.

A learning-enhanced Doogie mouse works its way through a memory test.

GLOSSARY

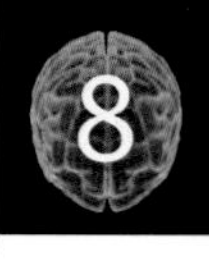

ANTERIOR TEMPORAL CORTEX. This region of the temporal cortex stores facial memory and plays a key role in facial recognition and identification.

ANTEROGRADE AMNESIA. Loss of the ability to create memories of events and experiences following a trauma that causes amnesia.

ARCUATE FASCICULUS. A bundle of nerve fibers connecting Broca's and Wernicke's areas.

ASSOCIATIVE LEARNING. A process in which learning occurs by associating an action with its consequence.

BROCA'S AREA. Region in the left frontal cortex of the brain responsible for motor movements in the production of speech.

CHUNKING. The technique of dividing a large amount of information into smaller groups to facilitate memorization.

CLASSICAL CONDITIONING. A process of behavioral training in which a previously neutral stimulus evokes a particular response through repeated pairing with a stimulus that naturally evokes it.

CONFABULATION. A disorder in which an individual unintentionally fabricates occurrences to fill gaps in his or her memories, believing them to be accurate.

DISSOCIATIVE FUGUE. A psychiatric disorder, often induced by stress, that is characterized by amnesia of self or personality.

DYSCALCULIA. A learning disability that is characterized by severe difficulty in understanding math.

DYSGRAPHIA. A learning disability affecting an individual's ability to write. This may affect both fine motor hand control and idea processing.

DYSLEXIA. A learning disability of neurological origin that impairs the ability to process language, leading to difficulties in spelling, reading, and writing.

ELABORATION. A memory technique in which new information is associated with previously learned material, aiding in long-term storage.

ENGRAM. A term for the physical trace that memory formation may leave on participating neurons.

EPISODIC MEMORY. A type of declarative memory that consists of stored autobiographical remembrances of personal experiences.

EXPLICIT MEMORY. Memories that are consciously recalled.

HYPERTHYMESTIC SYNDROME. A condition in which an individual has a superior autobiographical memory.

IMPLICIT MEMORY. Memory recalled unconsciously during physical activity.

KORSAKOFF'S PSYCHOSIS. A form of amnesia, often caused by severe alcoholism, in which an individual is unable to form or store new memories and is much given to confabulation.

MEDIAL TEMPORAL LOBE. Region of the brain that includes the hippocampus and amygdala. This area is crucial to the formation, storage, and organization of memory.

NONASSOCIATIVE LEARNING. Learning that occurs through repeated exposure to a stimulus without the result of either positive or negative consequences.

OPERANT CONDITIONING. A process of behavioral training in which a voluntary action is reinforced through reward or diminished through punishment.

PAPEZ CIRCUIT. A system of interconnected brain regions, including hippocampus, hypothalamus, and cingulate gyrus, that participates in short-term memory formation and emotional processing.

PROSOPAGNOSIA. A condition also known as face blindness in which an individual is unable to recognize a person by his or her facial features or to differentiate between faces.

RECALL. Memory process that involves retrieving previously stored information.

RECOGNITION. Retrieval process of memory that consists of the identification of learned items.

SEMANTIC MEMORY. Stored knowledge of general facts and data.

SEPARATION ANXIETY. The distress found in some young children at the departure of a parent or the introduction to a stranger.

VISUOSPATIAL MEMORY. A type of declarative memory, allowing the remembrance of the location of objects in space.

WERNICKE'S AREA. Brain area located in the posterior region of the temporal lobe; responsible for ability to understand and produce intelligible speech.

A MEMORY FORMS ENCODING, STORING, RETRIEVING

WHEN WE experience something—perform a task, read a book, pick out notes on a piano—many parts of the brain are activated. Sensory memory involves input from vision, hearing, and other senses, which are flashed into the sensory regions of the brain. Those sensations last only for a moment unless we pay attention to them. Then they register on the conscious mind and get transferred to short-term memory in the cerebral cortex. The thalamus is especially important to the ability to focus attention on the sensory stimuli.

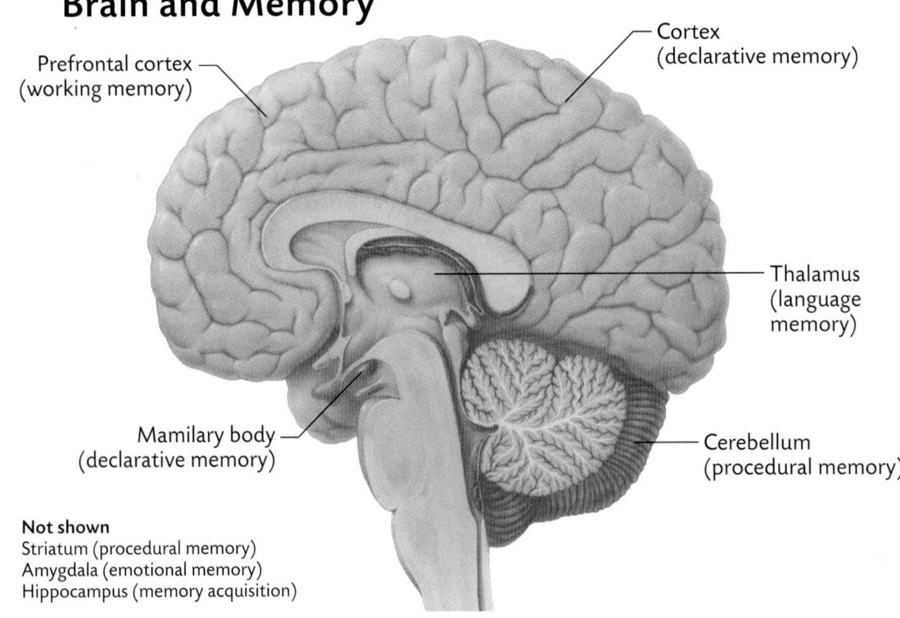

Memory gives us the ability to store and retrieve information.

ENCODING MEMORIES

By focusing on events, the brain engages short-term and working memories. Working memory, in the prefrontal cortex right behind the forehead, processes different kinds of incoming information and plays an important role in consciousness. It's working memory that allows us to remember decisions long enough to carry them out. For example, working memory lets a waiter at a restaurant remember a dinner order long enough to write it down and remember to ask about drinks and dessert. Short-term memory is a temporary storage area where information is held for up to 30 seconds. Working memory may manipulate that information though conscious effort.

FAST FACT "Brain fingerprinting" detects the electrical activity accompanying recognition.

Information in working and short-term memory exists because of temporary electrochemical connections among neurons. It disappears unless it can be transferred to stable neural connections in long-term memory, which lies along the medial (inner) side of the temporal lobe. The transfer process can take up to 24 hours, and sleep has been shown to improve the process of consolidation. Important regions for long-term memory include the hippocampus and amygdala, as well as sensory pathways associated with the particular sensations such as sight and sound.

CONSOLIDATION

The consolidation of ephemeral memories requires one of several actions. The information can be repeated, a strategy that lets children learn their ABCs. Or it can be analyzed so that it has meaning and is linked to information in long-term storage, as when communism is remembered by the ways in which it differs from democracy. Both kinds of encoding involve work in the frontal lobes and appear to engage a kind of protein synthesis in neural pathways that locks information into long-term memory.

CROSS REFERENCE: See "Anatomy," PAGE 18

+ MEMORY BOOST +

THE NATIONAL INSTITUTE of Mental Health found increased memory performance after volunteers inhaled vasopressin, a peptide hormone manufactured in the hypothalamus. Ampakines work on the cortex by concentrating neurotransmitters. In experiments in the 1990s in Germany and Sweden, subjects who took a version of the drug called Amaplex scored twice as well as control subjects on short-term memory tests.

A third way to encode memories for long-term storage lies in a strong emotional reaction. In "implicit" encoding, the amygdala reacts to an emotional jolt by more powerfully encoding the memory and linking it to the emotional response. And finally, physical skills get implicitly transferred from short-term to long-term memory with repetition. That's why the first time you ride a bike, you must concentrate to keep your balance, but with repeated rides atop the two-wheeler the skills of balance and coordination are pushed from the cerebrum to the cerebellum and motor cortex. Practiced bicycle riders no longer have to think about riding a bike. Similarly, people who practice their ABCs soon can recite them without having to pay serious attention to them.

CHEMICAL ENHANCEMENT

Drugs can enhance the performance of memory in the short term. Students cramming for exams sometimes take amphetamines or other dopamine-enhancing drugs to put huge amounts of information into memory. Safer, experimental drugs are already being tested to enhance short-term memory. Success in producing pills that enhance long-term memory are not an "if" but a "when," neuroscientist Tim Tully of Cold Spring Harbor Laboratory, predicted in 2002. His research focuses on two molecules, known as cyclic-AMP (c-AMP for short) and CREB (c-AMP response element binding protein). The former relays nerve signals to the nucleus of a neuron and activates the latter molecule, which sets off a chemical reaction that increases protein production in the synapses. Pills also could be developed to block the action of c-AMP and CREB, as well as other molecules involved in memory storage. It's not inconceivable to imagine swallowing a pill after a traumatic event to ensure it doesn't get stuck in long-term memory and cause long-term emotional problems. The lag time of up to a day for the memory to move from short-term to long-term storage means that a doctor could give the pill several hours after a traumatic event and still have it erase the painful memory.

CELLULAR STORAGE

The physical process of storing a memory is incredibly complicated. It has many steps that lead to the storage of information chemically through the manufacture of

Pausing at a dead end, a hamster gets its bearings in a maze in Germany.

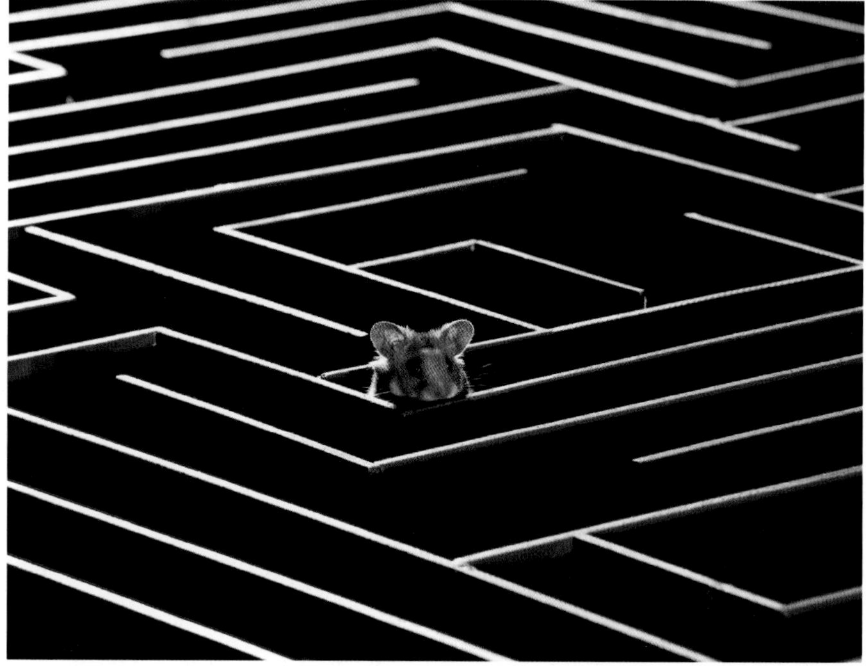

proteins that make certain synapses more apt to fire. That creates new patterns of neural networks, and thus a stored memory.

Frequent jolts along the membrane of the presynaptic neuron excite the membrane enough to elevate the voltage for a sustained period. Because of the heightened voltage, a glutamate receptor called N-methyl-d-aspartate (NMDA) receptor shifts position in the postsynaptic neuron and allows calcium ions to flood into the postsynaptic neuron. Those ions modify proteins in the walls of both neurons, the one that sent the signal and the one that received it. (The change in the sending neuron occurs through backward movement of chemical messengers such as nitric oxide, which increases the synaptic response to follow-up stimuli.)

+ EDITING MEMORY +

AT LEAST 117 kinds of molecules play a role when neurons construct a memory. One of them, PKMzeta, appears in neurons that get primed for quick connections with neighboring cells, a linking process called long-term potentiation. Studies at the SUNY Downstate Medical Center in Brooklyn have shown that the administration of a drug that blocks the action of PKMzeta prevents memories from forming in lab rats. After a single dose of a drug called zeta inhibitory peptide (ZIP) that interferes with PKMzeta, the rats almost immediately forgot their training.

The calcium ions activate c-AMP molecules, which switch on genes in the nucleus of the postsynaptic neuron, creating more proteins for deposit at the synapse where the ions entered the cell. The message to turn on the genes to boost protein production and synaptic growth is carried out by CREB.

The end result of this chemical complexity is an increase in receptor sites. The increase makes the neural connection more sensitive to the presence of neurotransmitters, and thus more apt to fire. More firings mean stronger connections, as neurons that fire together wire together. The change in the strength of the response of the receptor neuron after stimulation is called long-term potentiation, or LTP. It occurs in all brain regions involved in memory. Long-term potentiation in networks of associated neural nets creates memories.

TAKING IT ALL IN

To take an example of how memories get stored in the brain, try

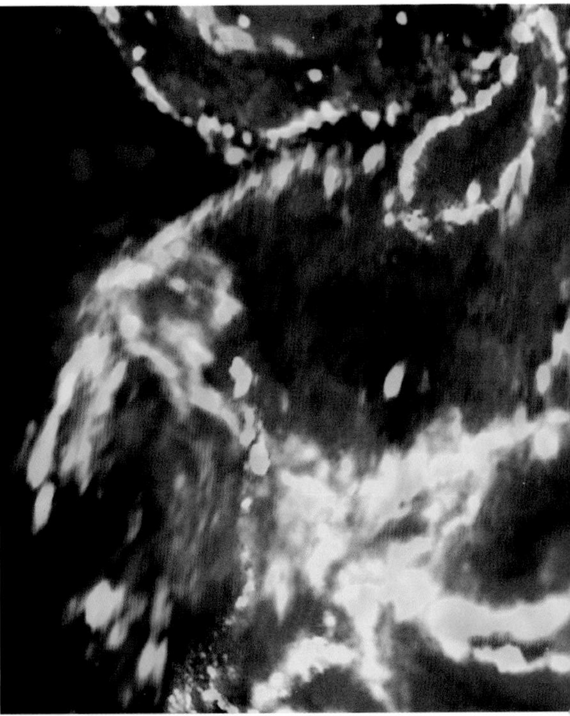

looking at a news story on a computer screen. The brain takes in all sorts of information from the monitor. The story may have a digital photo and colorful graphics, which is visual information. It surely includes words strung together in text, headlines, photograph captions, and links to other stories, which is language. It also may have a video or audio clip, providing auditory information.

The information gathered by the senses gets channeled along neural pathways initially leading to the hippocampus. Neurons in the hippocampus, like those everywhere else, communicate along networks by sending electric signals along the length of their axons until they terminate at synapses. There the electric signal may or may not be enough to release chemical

CROSS REFERENCE: See "Perception," PAGE 100

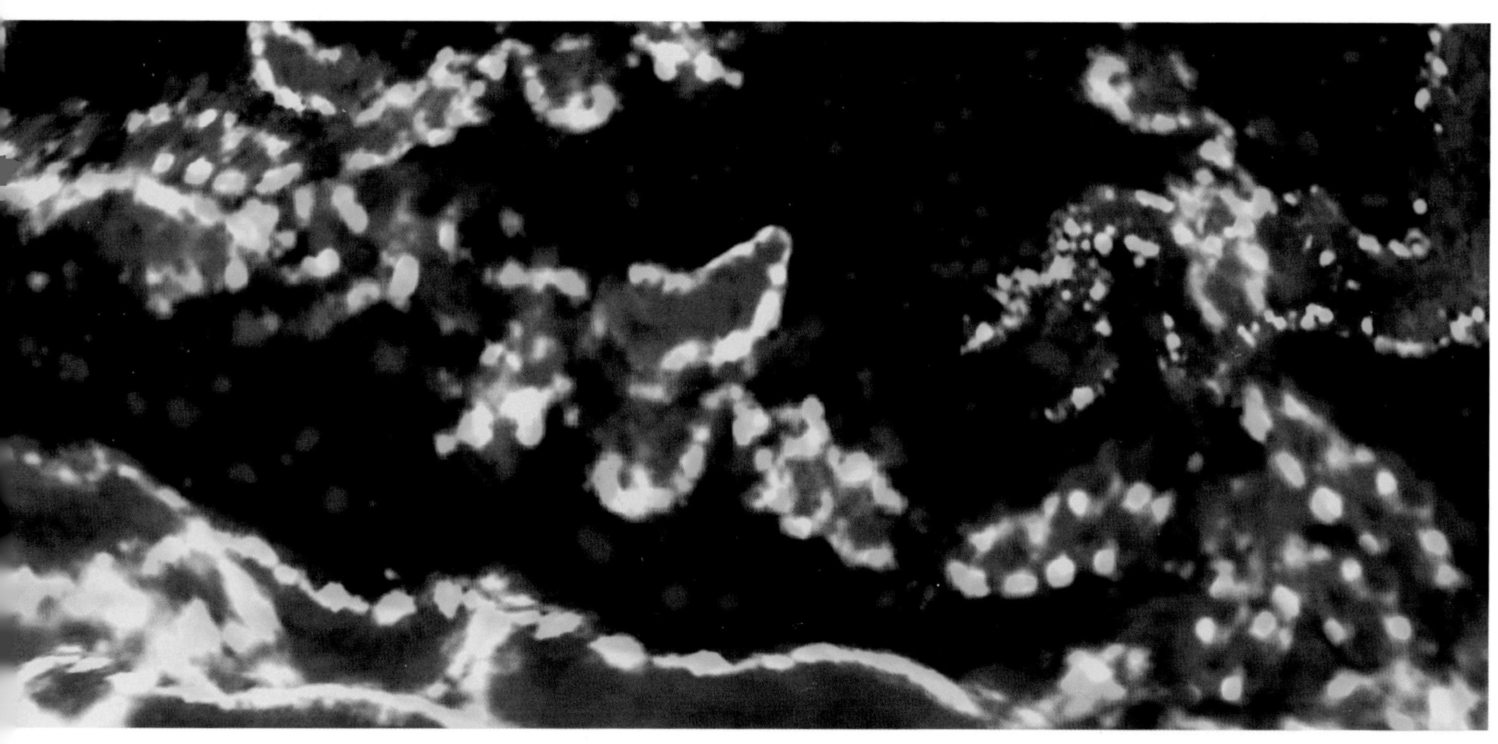

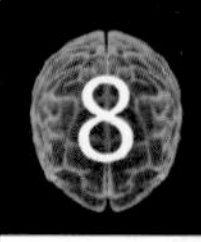

Synaptic transmission links acid-sensing ion channels in Purkinje neurons (in a composite immunofluorescence micrograph) to memory.

neurotransmitters that travel across the synaptic gap to dock at receptor sites on nearby neurons.

When memories get stored, the chemical reactions taking place across the synapses stabilize. When thousands or even millions of neurons strengthen their chemical connections in response to stimuli such as the words, pictures, and sounds on the computer page, they form a memory at the cellular level.

Studies have determined that the molecular cascade touched off by c-AMP molecules is not crucial to learning or short-term memory. It is essential to long-term memory, though, as demonstrated by experiments in 1999 in which genetically engineered mice received enhanced NMDA receptors and developed greater powers of learning and long-term memory.

Studies of mice, humans, and all sorts of animals in between indicate that memories are stored in specific places. Canadian neuropsychologist Hebb theorized that cell assemblies worked in harmony to store and retrieve memories. One cell assembly might store the words on the computer screen, and another the images, and a third the sounds. Cell assemblies, consisting of neurons that could be packed tightly or scattered throughout the brain, excite one another in chainlike fashion to bring forth the components of memory, which then get synthesized into a more coherent whole. This theory explains how dementia can rob some aspects of memory but leave others intact. For example, memory could recognize the human face in the digital photo but forget the name that goes with it or fail to decode the text beside it.

FAST FACT In 2004 a team at MIT discovered mitogen-activated protein kinase, which increases synthesis of memory proteins.

HOW RECALL WORKS

When we recall a memory, we call upon many of the very same neural pathways that sensed the event. Recall nearly re-creates the event, as evidenced by the warm, fuzzy feelings you get from a good kiss and the memory of a good kiss. Sensations link into memories

BETTER MEMORIES

ABOUT 2,500 years ago, a Greek poet named Simonides of Ceos attended a banquet thrown by a nobleman named Scopas, who asked for the recitation of a long lyric poem. As recorded by the Roman orator Cicero, Simonides left the banquet hall for a moment. While he was outside, the roof collapsed, killing all of the guests. The corpses were so mangled that family members could not identify them for burial. But Simonides knew where each had been sitting.

Cicero explained in his *De oratore* that Simonides inferred "that persons desiring to train this faculty [of memory] must select places and form mental images of the things they wish to remember and store those images in the places, so that the order of the places will preserve the order of the things, and the images of the things will denote the things themselves, and we shall employ the places and images respectively as a wax writing tablet and the letters written on it."

Simonides' technique has come to be called the Roman Room, or loci, method. It works on the principle of inventing images for things you want to remember, and situating those images in places (loci) that naturally form a progression.

MODERN APPLICATIONS

The technique still works astonishingly well, as demonstrated by National Basketball Association Hall of Fame member Jerry Lucas.

Memorization works well when data are "chunked" into groups of seven—conveniently, the number of digits in a phone number.

He has used the Roman Room to memorize the content of various magazines and books, including the entire New Testament, which he said took him a year to commit to memory. In television appearances, he astounds audiences by greeting them in large numbers and then recalling all of their names. He explains his techniques in *The Memory Book,* co-written with magician Harry Lorayne, and shares the rationale behind it in speaking engagements.

"All children have very active minds, get bored easily and want things to do, and I was no different," Lucas said. "I came up with all sorts of mental games to keep my mind occupied when I had nothing to do."

He described the key concept of his memory systems as "automatic learning," and linked it to childhood mastery of language. Parents teach children by pointing to and identifying objects as a way to light the lamp of understanding. Whereas teachers in grade school often impart lessons through brute repetition, Lucas puts his focus on making intangible objects, such as the words on a page or spoken aloud, register concretely on the mind.

REMEMBERING WORDS

To remember single words, Lucas recommends using a system of substitution. "When you hear or see a word or phrase that seems abstract or intangible to you, think of something—anything—that sounds like, or reminds you of, the abstract material and can be pictured in your mind." Minnesota and Mississippi, for example, might become a tiny bottle of fizzy water (mini-soda) and a married woman drinking from the bottle (Mrs. Sip). To remember the French word for father, *pere,* picture a yellow pear holding a baby.

PEOPLE, PLACES, & THINGS

Images, words, and spatial orientation can go together to help you remember lists. Try imagining your living room. Within the room, so familiar to you, are the objects you have collected and come to know. Perhaps as you look around in a clockwise direction, there's a bookcase, fireplace, piano, and sofa. Now associate images representing the information you want to remember with the objects scattered around the room. If you are trying to remember the names of American Presidents, in order, you associate their names with concrete objects. For example three Presidents before Lincoln—Fillmore, Pierce, and Buchanan—may be hard to remember because they've left relatively little mark on American history. Assign each a vivid image. Fillmore could be a pitcher pouring a half inch of beer into a glass ("fill more") sitting on the bookcase. Next to it, on the fireplace, is a purse with a knitting needle stabbed through its side ("pierce"—which is also a close verbal association with the word *purse*). Next to the fireplace, on the arm of the sofa, is a small statue of a ghost pulling the lanyard of an old-fashioned artillery piece ("boo-cannon"). This technique can be expanded by adding more details, or by moving into other rooms and assigning them other objects. If you run out of rooms, you can move to other houses or build new wings in your imagination, filling them with objects that would naturally be there, such as sink and bathtub in a bathroom or bed and dresser in a bedroom.

Chester Santos, 2008 USA Memory Championship winner, can memorize a shuffled card deck in five minutes.

NUMBERS

Even extremely long strings of numbers, objects, or words (in case you're wondering, there are 180,000 or so words in the New Testament, depending on the translation) can be remembered by being broken into "chunks," or small groups of objects, and then the groups retrieved in order. This principle was promoted in 1956 by Princeton University psychologist George Miller in a now famous paper titled "The Magical Number Seven, Plus or Minus Two," which tested the observations of William Hamilton, a 19th-century Scottish philosopher, that if you throw a bag of marbles on the floor, your brain most likely cannot grasp the location of more than six or seven at a time. Seven is about as many numbers as the average brain can remember at a time. Conveniently, that's the number of digits needed to make a local telephone call. However, Hamilton said larger groups of objects can be broken into small, meaningful patterns as a memory aid.

Medical school students use a version of chunking and rhyming to remember the names of the 12 cranial nerves. The first letter of each word in the rhyme represents the first letter of a cranial nerve, and the sections are easily committed to memory: On Old Olympus's Towering Tops, a Finn and German Vied at Hops.

FAST FACT

"Confabulation" of verbal memory can be caused by damage to the frontal lobes. This disorder, which was first described by neurophychiatrist Sergei Korsakoff in 1889, confuses truth and lies.

because we take information from sensations and use it in the construction of ideas. Science writer April Holladay of *USA Today* says understanding storage makes retrieval easier to grasp. The hippocampus, which consolidates new memories, is the starting and ending point of a loop that stores memories. When thinking about a delicious red apple, she said, the "red" part gets stored in the visual centers in the occipital cortex, and the sound of the crisp bite into the apple's flesh gets stored in the auditory center in the temporal cortex, where the sensations of the apple were first observed.

"When I remember the new fact, 'delicious apple,' the new memory data converges on the hippocampus, which sends them along a path several times to strengthen the links," she said. The information traced the so-called Papez circuit, starting at the hippocampus, then picking up any emotional information in the limbic system, such as a glorious October day in an orchard, and spatial associations, such as the arrangement of trees in the orchard. The circuit goes to the various cortical regions and then back to the hippocampus. Repeatedly turning to the memory completes the circuit again and again, making the neural links stable enough to exist without the intervention of the hippocampus. The individual attributes of the memory of the apple are stored separately but linked with an overall neural connection. The enhanced pathways become long-term memories.

When recalling the memory, the brain retrieves the information by firing the network connecting color, sound, emotion, spatial orientation, any other facts stored with the memory, and an overall network that integrates all the information.

FOUR CATEGORIES

Although a keen observer could no doubt create a very long list of attributes for a delicious red apple, the information would fall into four categories of memory: sensory, motor, visuospatial, and language.

Sensory memory involves the five senses, with smell the most powerful memory trigger. As the senses create our appreciation of the world, it's not surprising that many memories can be recalled via sensory cues. A certain song, for example, might bring up memories of a wedding or family reunion. People with powerful memories often create visual cues in their imagination to increase the strength of their long-term memories.

Motor memory provides fine motor control for practiced actions. These include everything from the subtle controls over the vocal cords to produce speech to the motor control that underlies the complicated act of walking without losing one's balance. Motor memory is tied to the learning of skills. Damage to brain areas that affect one

WHAT CAN GO WRONG

WHEN JILL PRICE, left, a school administrator in Los Angeles, pictures just another day—say August 19—she is invaded by the sights, smells, sounds, and tastes of August 19ths from 10, 20, or 30 years ago.

Brad Williams, a radio reporter from La Crosse, Wisconsin, has the same capacity for autobiographical memory. Asked by one journalist what he ate for breakfast on December 26, 1962, he answered unflinchingly: Frosted Flakes.

Price and Williams are two of only four people in the United States known to have hyperthymestic syndrome, or a "superior memory" for one's own life.

invariably affect the other. Recent research suggests that motor memory's application to learning new skills takes place in two stages. The first stage recruits neural networks that best represent the motions required for the skill, such as eyes, ears, and fingers for the playing of a piano. The second stage occurs after the basic motions are mastered and implicitly memorized, when the brain recruits additional neurons to refine the motions. That's the difference between the adequate performance of a weekend musician and one who practices for a seat in a symphony orchestra.

Visuospatial memory combines the neural pathways of the visual cortex and the spatial orientation of the temporal lobes. The left hemisphere is significant in perceiving details, while the right hemisphere works to integrate the details in a whole. Together, they let you see the trees *and* the forest.

Language memory leads to the ability to associate words with objects, the crucial foundation of communication. The grounding of communication in the brain's hardwiring to acquire language can lead to serious consequences from verbal memory disorders. Not only can they interfere with communication, but they also can distort the perception of reality of someone who may be unable to discriminate between statements that have their source in true memories or those that draw on fantasies. Unable to tell the true stories from the false in his head, the person may struggle with the concept of truth.

Attributes of an apple—such as color, taste, name—get stored in separate neural circuits.

ARE MEMORIES RELIABLE?

Nineteenth-century British wit and dramatist Oscar Wilde called memory "the diary that we all carry about with us." If that's so, memory never leaves a page unedited. The process of remembering by retrieving information from storage, and then restoring it, colors memories with additions, subtractions, and substitutions. Memory can change fluidly and dramatically over time, and the more that has elapsed since an event, the more likely the brain has rearranged the memory of it. Fragments of actual memories can be combined with bits of information available during retrieval, creating false memories that seem true.

Elizabeth F. Loftus, a psychology professor at the University of

Williams can pick and choose which memories to suppress, but Price's mind is stuck on autopilot. She says she views the world through a sensory split screen in which the past and present are constantly competing for attention.

Since 2000, neuroscientists at the University of California at Irvine have studied Price, who was the first documented case of hyperthymestia in the country. Scans of her brain, when compared with images of thousands of "normal" brains, have revealed several regions that are substantially larger than average.

Specifically, the caudate nuclei, which are responsible for the formation of habits, as well as a part of the temporal lobe that stores facts and figures, are abnormally large. Scientists believe these regions may act in tandem, which might explain why Price's memory for minutiae is as automatic as your remembering how to tie your shoes.

Emotions, such as shock and grief at the Challenger *explosion, create "flashbulb" memories.*

California at Irvine, even suggests, perhaps half in jest, that courtroom oaths given to witnesses should be amended to read, "Do you swear to tell the truth, the whole truth or whatever it is you think you can remember?"

What's likely to stick correctly in memory is the central fact of an event, especially when encoded with emotion. Americans who were old enough to remember things in 1986 have no problem recalling the fact that the space shuttle *Challenger* exploded. They may even remember it was a cold day for a rocket launch in Florida because the frigid weather contributed to the disaster, or recall seeing pictures of ice on the launch towers on television newscasts. Memory for details is much more likely to fade. Despite a widespread belief in the permanence of memory, people "may forget where they were and what they were doing when they heard" about *Challenger* or the twin towers collapsing in 2001, said Harvard University memory expert Richard McNally.

Even "flashbulb" memories, the kind that feel so immutable, are subject to reconstruction over time. Behavioral neurologist Heike Schmolck tested this idea by asking people to share their memories of the not-guilty verdict in the 1995 murder trial of actor and football star O. J. Simpson. Schmolck asked for details of subjects' memories three days, 15 months, and 32 months after the jury announced its decision. After 15 months, she found that half of the memories matched the three-day accounts very closely. Major discrepancies appeared in only 11 percent of the accounts. But when reinterviewed at the 32-month mark, most people exhibited memory drift. Only 29 percent matched their original accounts; 40 percent had major discrepancies in the various versions of their memories.

The power of suggestion can strongly affect memory. In a process called memory morphing, marketing agents manipulate information after an event to convince consumers that they have experienced things that never happened. In controlled experiments, psychologists have been able to get 25 percent of adults to accept the idea that they were lost in a mall at age five. In another manipulation, 16 percent of adults who read a rigged Disneyland advertisement became convinced that they had met Bugs Bunny at the theme park when they were children, even though Bugs is a Warner Bros. cartoon.

Marketers use "backward framing" to suggest positive reactions to a product or event, such as a new Hollywood movie, that a consumer has already experienced. The suggestions can be so powerful that

CROSS REFERENCE: See "Emotions," PAGE 206

the consumer may forget initial negative reactions. Advertisements and information on television and in books, magazines, and newspapers can become combined with retrieved memories and be stored with the original memory.

If you want to improve your memory of objects or events, use acronyms and the first letter of words in sentences to remember words in order, "chunk" large groups of items into bunches of seven or fewer, repeat the information to be stored, or incorporate rhymes and songs into memorization.

Rhymes and songs link words with music and other verbal associations, creating an obvious and fun way to remember difficult information. Kevin Roose, a 21-year-old student from Brown University in Rhode Island, learned this when he went undercover for a year at the conservative, Christian Liberty University run by evangelist Jerry Falwell and ended up writing a book in 2009 about his experiences. "On one exam, we had to name all the twenty-seven books of the New Testament, in order," Roose told a reporter. "I was up all night. Finally I went to one of my hallmates and said, 'Dude, this is killing me!' He said, 'It's so easy, just sing the song!' He taught me this song he'd learned in Sunday school. The next day, I heard this hum all around me of all the other students singing this song."

MEMORY DISORDERS

Doctors see plenty of patients suffering from memory disorders. They range from the temporary loss of memory resulting from concussion to devastating erasures of a lifetime's worth of experience through dementia, including Alzheimer's disease.

Head injuries, encephalitis (infection of the brain), and brain malformations affect memory. Concussions have provided significant clues about the process of transferring short-term to long-term memory. The process apparently takes several minutes, as a concussive blow wipes out not only the memory of events immediately afterward, but also for a few minutes beforehand, too.

+ FALSE MEMORIES +

IN EXPERIMENTS begun in the 1970s, researcher Elizabeth Loftus discovered that, given the right prompting, about one-quarter of adults would assert they had childhood experiences that never actually happened. A researcher pushed the subjects to recall made-up events. Imagining them made them seem familiar, and more likely to be called out of memory. Loftus had her own false memory of her mother's death suggested by an uncle. Some false memories are innocent, but the power of suggestion opens doors to false—yet heartfelt—court testimony.

FAST FACT Ads are more likely to stick in memory if they combine information and subtle emotion.

Physical causes inhibiting the formation of new memories include things as common as depression and lack of sleep. Depression leads to a lack of focus and attention, which lessen the power of sensory memory, the first step to forming and storing memories. Treatments to lift depression usually improve memory. Insomnia also inhibits the transfer of memories from short term to long term. Getting a good night's rest helps strengthen memories.

AMNESIA

A bizarre but not uncommon disorder of memory is Korsakoff's psychosis, an incurable form of amnesia that eliminates the brain's ability to create and store new memories. The most common source of the disorder is long-term alcohol abuse. Named for the doctor who first diagnosed the disorder, Korsakoff's seems to affect only memory, leaving intelligence and emotional responses untouched, but prompting a high level of suggestibility. If you meet a Korsakoff's patient, chat for a few minutes, and then leave for a few minutes, the patient won't recognize you when you

Speech pathologist Maj. Ava Craig plays a game with Sgt. Dan DaRosa to help his memory at the brain injury clinic on an Alaskan base.

return. However, if you suggest that you've met before, he or she will fabricate a story to describe the previous meeting. The brains of Korsakoff's patients reveal a substantial loss of neurons in the thalamus, near the midline.

A Korsakoff's patient named Jimmie G. lived forever in the final days of World War II, even after he was admitted to a New York City hospital in 1975. Asked what year it was, Jimmie responded, "Forty-five, man. What do you mean? . . . We've won the war, FDR's dead, Truman's at the helm. There are great times ahead." He swore that despite his head full of gray hair, he was only 19 years old—going on 20. Yet Jimmy was quick-witted, observant, and adept at doing puzzles.

When asked for advice on dealing with Jimmie, the eminent psychologist Aleksandr Luria responded with compassion: "There are no prescriptions in a case like this. . . . There is little or no hope of any recovery in his memory. But a man does not consist of memory alone.

WHAT CAN GO WRONG

FOR 20 DAYS in 2008, New York's Hannah Upp was afflicted with dissociative fugue, a form of amnesia so rare that its most famous case involves a fictional secret agent.

Jason Bourne has total recall of a dozen or so languages and an uncanny intuition for hand-to-hand combat, yet he can't remember his own name. That's because, for all intents and purposes, he's someone else.

Dissociative fugue, a psychiatric disorder characterized by amnesia of the personality induced by stress, may last hours or days, or, in the case of Ansel Bourne—the 19th-century preacher who loosely inspired the fictional Jason—months or years.

During the fugue, muscle memory is retained: Surveillance footage from an Apple store captured Upp logging on to her email account, fingers automatically keying in her user name and password.

He has feeling, will, sensibilities, moral being—matters of which neuropsychology cannot speak." Over many years in the New York hospital, Jimmie knew the sibling who shared his childhood and youth but could not understand why he looked so old; Korsakoff's took away Jimmie's memories but did not make an impression about the nature of his loss. Losing one's memory, one's deep identify of self, takes away the part of the brain that recognizes the self. Jimmie's memory loss removed his ability to realize he had a memory loss.

FAST FACT Biologist Richard Semon is an advocate for "retrieval cues" to aid in remembering.

Amnesia also can occur when oxygen is cut off to the brain at birth or during an accident or convulsions, damaging the regions that encode and store memories, and by physically or emotionally traumatic events. The former may interfere with episodic memory yet leave semantic memory virtually intact. According to research in 1997 by the Institute of Child Health in London, three British children who suffered hippocampal damage from oxygen deprivation lost their episodic memory. They could not tell you what television program they had just seen or how to get around their neighborhood, yet they went to school and learned how to read and write. Their "fact" memory had been spared, even while their biographical memory disappeared. The latter kind of amnesia, reported since the 1800s, occurs when the brain suppresses conscious memory of a traumatic event while maintaining the emotional and sensory memories. Cues that trigger negative emotions related to the trauma can create a condition known as post-traumatic stress syndrome.

Damage to the regions of the brain particular to language—on the left hemisphere in about 97 percent of the population—can cause difficulties in speaking and recognizing words.

+ TYPES OF MEMORY LOSS +

There are many different forms of amnesia, with different causes and symptoms.

TYPE	DESCRIPTION
Alcohol blackout	Causes partial or total memory loss for events occurring after rapid, heavy consumption of alcohol.
Dissociative fugue	Creates confusion about identity and life events. Commonly accompanied by wandering. May last from hours to months.
Korsakoff's psychosis	Strikes some chronic alcoholics. Causes inability to form short-term memories. Patient may invent memories.
Post-traumatic amnesia	Occurs after coma. Causes disorientation, agitation, inability to remember anything for a while before being injured.
Repressed memory (dissociative amnesia)	Arises as reaction to early trauma. Memories are later recovered. Validity of claims has divided researchers.

WHAT CAN GO WRONG

One explanation of the fugue state is that the individual is subconsciously seeking escape. Doctors theorized that the stress of a friend's terminal illness led to 40-year-old Jeffrey Ingram's month-long fugue state.

Ingram, right, disappeared en route to visit his friend. Four days after his departure, he turned up in Denver, Colorado, with no idea who he was and then spent the next month searching for his identity. A television appearance viewed by his family reunited Ingram with them and his past.

With the passing of the fugue state goes any memory of its occurrence.

LANGUAGE

UNIQUELY HUMAN COMMUNICATION

READING WORDS on a page—understanding their meaning, speaking them aloud—involves language, but the functions don't exist in just one part of the brain. Language involves the front and back, as well as right and left hemispheres. Its importance to human survival accounts for the massive amount of cranial space devoted to language production and comprehension.

TASKS & REGIONS

Reading the letters silently on the page initiates action in the primary visual cortex at the back of the brain to sense the shapes of the letters and words. It also places demands on working and short-term memory to hold those words under an umbrella of attention long enough for the reader to make sense of how they go together to form sentences, as well as analytical functions to pull those sentences apart for meaning.

A child hearing words spoken aloud activates the auditory association area of the brain. Understanding speech requires encoding and decoding of language in the receptive language areas, which take in signals from auditory association areas and, if the words are read on a page, from the visual association areas. These primary language areas lie in the left hemisphere for nine of ten people who are right-handed and about two-thirds of those who are left-handed.

+ HOMESIGNS +

UNTIL THE SANDINISTAS took power in 1979, Nicaragua's deaf children had no formal education. They developed "homesigns," individual gesture systems, to communicate. When the children first went to school, they thus had no common language. Yet by 1986, when American linguist Judy Kegl visited a school for the deaf, the students had transformed pidgin homesigns into a sophisticated language, Idioma de Señas de Nicaragua.

The parent reading aloud uses a different brain circuit from that of the child being read to. Although there can be nuances in the processing, reading generally begins with activation of the visual cortex, which sends signals to Wernicke's area, named for German neurologist Carl Wernicke, who described it in the late 19th century. Patients with lesions in Wernicke's area can speak, but what comes out is a jumble of sounds sometimes called "word salad," and they have difficulty understanding language.

Speaking aloud activates an adjacent region called Broca's area, after the French scientist's discovery that lesions in that section of left hemisphere interfered with speech. Speech requires the cooperation of Wernicke's and Broca's areas. Words come together through processing in the former and get relayed to the latter along a collection of nerve fibers called the arcuate fasciculus and a brain region called the

BREAKTHROUGH

FOR A LONG TIME, scientists debated whether language was a matter of nature or nurture. Most observers believed newborns began learning all attributes of language from their parents. However, in 1959 MIT linguist Noam Chomsky, left, provided evidence that children's brains are hard-wired to acquire language. He reasoned that a "universal grammar" in a child's brain begins to specialize in the local tongue upon repeated exposure.

Chomsky proposed the concept of universal grammar after examining and cataloging the structure of many languages. While vocabulary obviously

CROSS REFERENCE: See "Evolution," PAGE 66

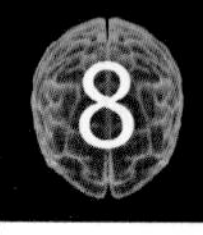

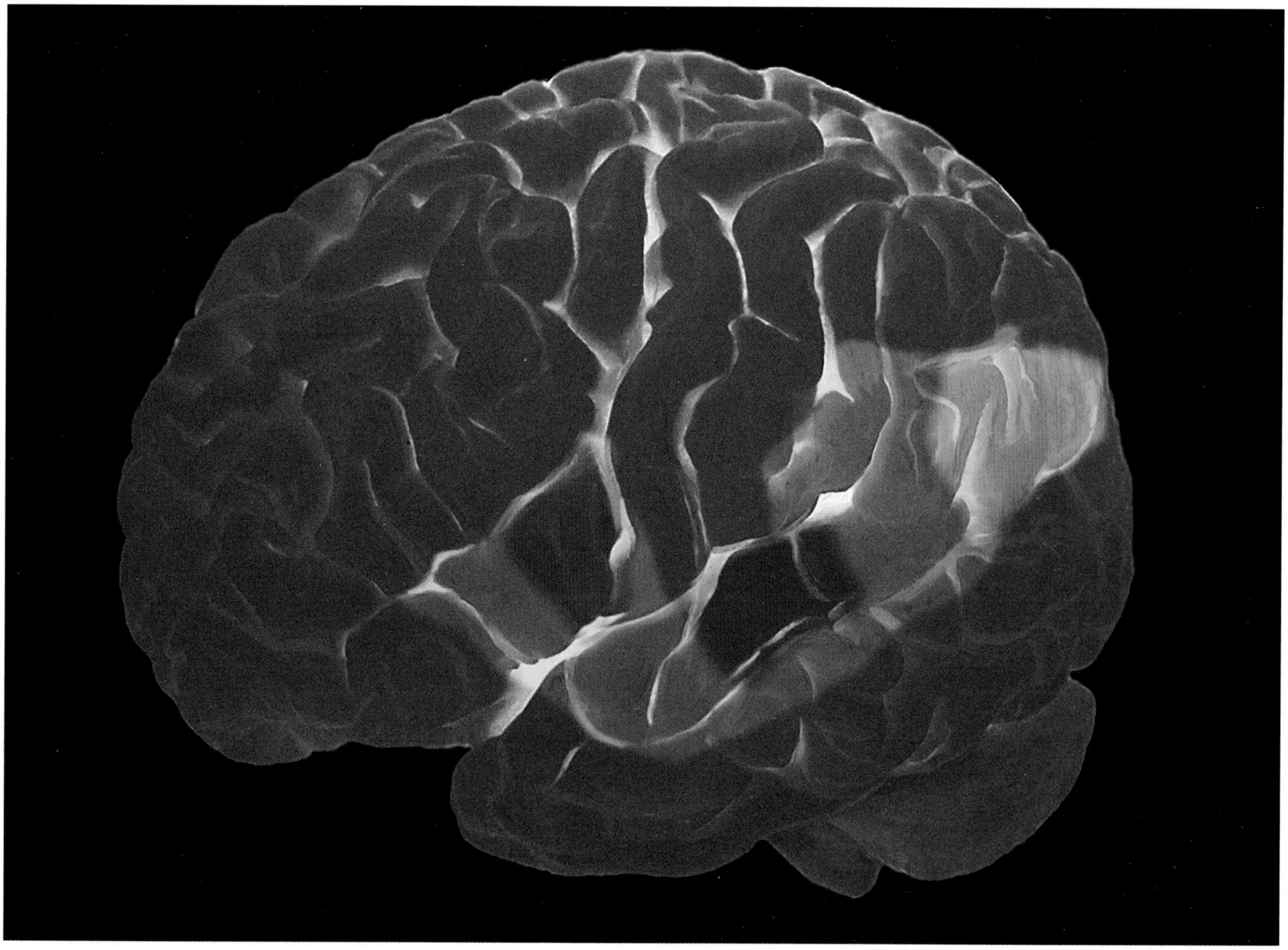

A three-dimensional MRI highlights language regions: Wernicke's area (pink), Broca's area (orange), and the primary visual cortex (green).

angular gyrus. From Broca's area, the nerve impulses for speech zip along to the motor cortex, which controls the muscle movement of the lips, tongue, and face.

Broca's area, Wernicke's area, and the basal ganglia work as a single unit to analyze incoming sounds and process outgoing language. Regions of the cerebral cortex that surround these areas act to connect the processes of language with cortical regions that hold ideas and concepts. That makes it possible to speak an intelligent, grammatically

BREAKTHROUGH

differs from language to language, there are only a few ways to put words together to make sentences. In English, for example, adjectives come before nouns, while in French they come afterward. Babies catch such differences in structure and adapt to their local grammatical complexity, Chomsky said.

The concept of universal grammar conquered all challenges until 2005, when Illinois State University linguist Daniel L. Everett claimed exceptions in a language called Pirahã, spoken in the Amazon rain forest. As is often the case in science, other linguists have since challenged Everett's findings.

Fresh evidence of universal grammar comes from attempts to create artificial languages. Educators of deaf children sometimes invent languages. Typically, the children don't learn them well—until they change the rules so that a language conforms to the universal standard.

A LIFE WITHOUT LANGUAGE

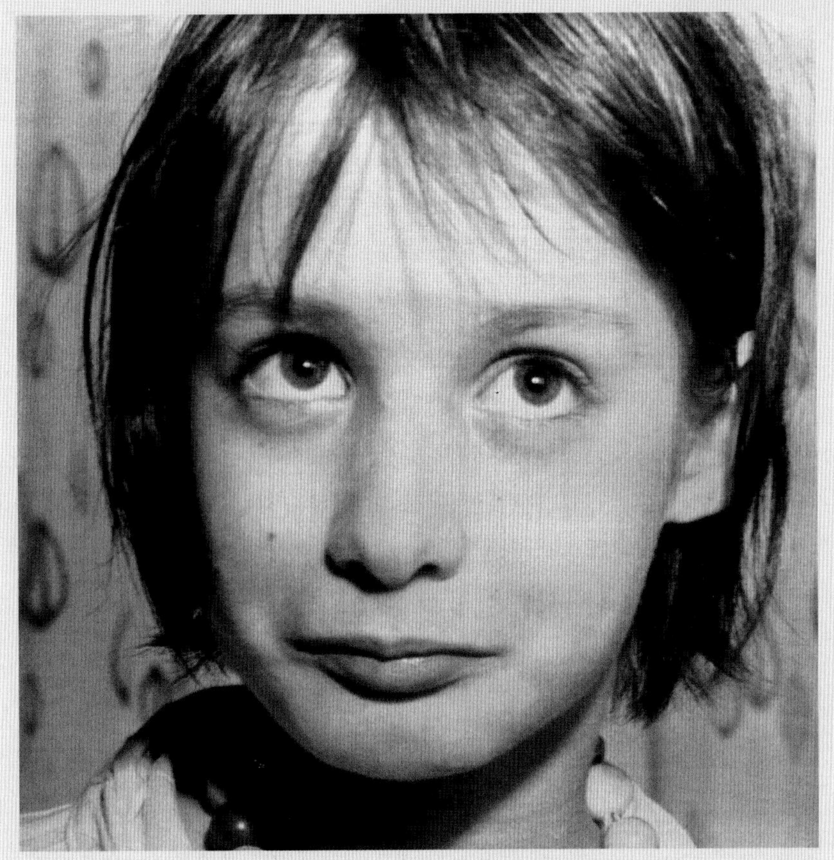

Rescued from an abusive home in 1970, "Genie" suffered severe language impairment.

LANGUAGE must be acquired at an early age. So wrote linguist and neurologist Eric H. Lenneberg in *Biological Foundation of Language.* And so linguists believed, but they were unable to test that hypothesis.

Then, in 1970 a 13-year-old girl was rescued from her captivity in a home in Los Angeles. From birth "Genie" was kept locked in a single room and beaten if she made noise. Muffled sounds from beyond her walls gave her an inkling of speech, but she never developed normal language.

Four years of training gave Genie the basics of communication, including sign language. She also developed a basic vocabulary, pairing words with objects. But Genie could not articulate sentences. The left hemisphere of her cerebral cortex had not received the sensations required for normal speech development. Starved for stimuli, Genie's speech centers had suffered irreparable damage. After years of rehabilitation, Genie wound up in a series of foster homes and her language skills regressed.

correct sentence while thinking about what to say next.

The child being read to processes the words in the primary auditory complex. Signals sent to Wernicke's area allow comprehension. If the child reads along too, the brain forwards the signals for the words to Broca's area and finally to the motor complex.

In the nondominant hemisphere—the right, for about 97 percent of healthy adults—the corresponding brain areas decode nonverbal communication elements. The right hemisphere also plays an important role in nuances of spoken language. A healthy brain normally can easily tell the difference between "dark room" and "darkroom." People with damage to the right hemisphere have difficulty making such distinctions.

EVOLUTION OF LANGUAGE

Although some scientists, such as behavioral neurologist Antonio Damasio, believe the mind can exist without language, others argue that language produces mind. "Without language, I wouldn't say that it is impossible to have mental experiences, but I'd say the mental experiences would not be very coherent," said Derek Bickerton, an expert on creole languages.

According to Bickerton, pidgin is the first developmental step in the creation of language. Pidgin languages form when people come

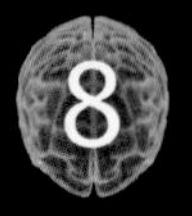

together who cannot communicate in a common tongue, such as slaves taken from various tribes of West Africa and placed in the New World. Pidgin assigns words to objects but lacks grammar and thus complexity. "Arrow . . . deer" might be a command to help kill a deer or an announcement that a hunter has just done so. Interestingly, children of pidgin speakers, whose brains are still plastic for speech, add the grammar to pidgin to create creole languages.

FAST FACT Koko the gorilla typically makes sign-language sentences that are three to six words long.

Examples of pidginlike speech occur in very young children, who develop a limited vocabulary before the subtleties of syntax. A child might say "Juice . . . me" instead of "Please pour me some apple juice." Bickerton believes that the increasing complexity of the child's brain and its accompanying ability to master speech may mirror the evolutionary history of language. Language remains one of the primary differences between humans and animals.

ANIMAL COMMUNICATION

However, animals have developed their own complex ways to communicate. Acoustic communication includes the chirping of birds and the songs of whales. Visual communication involves decoding of light waves. Chemical communication spreads information through substances that one animal leaves for another to find. And tactile communication sends signals through physical touch. Each animal's choice of communication method relies on its strongest and most sensitive neural receptors.

Irene Pepperberg of MIT works with one of the gray parrots she studies for insight into animal communication and learning.

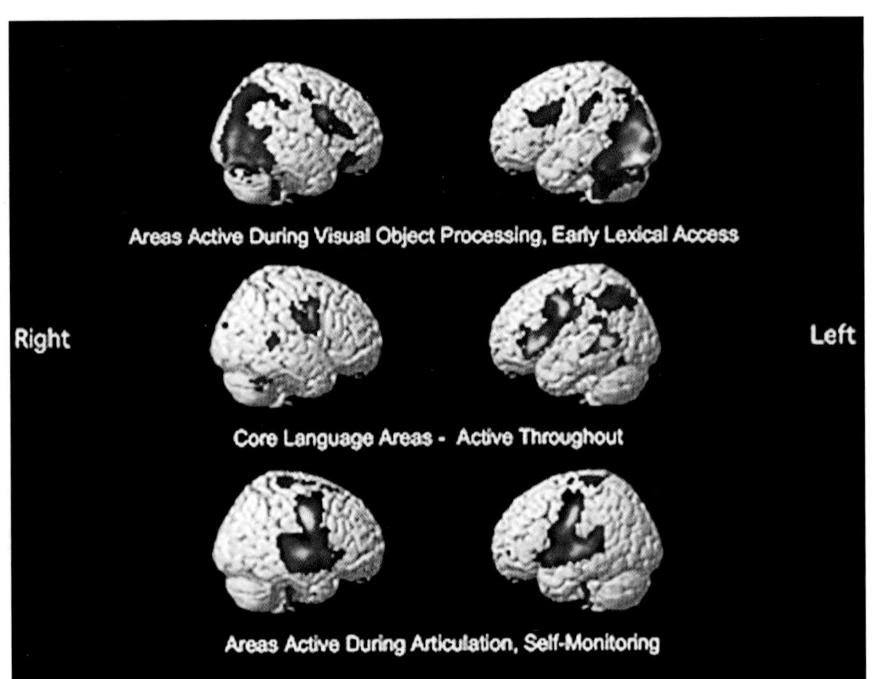

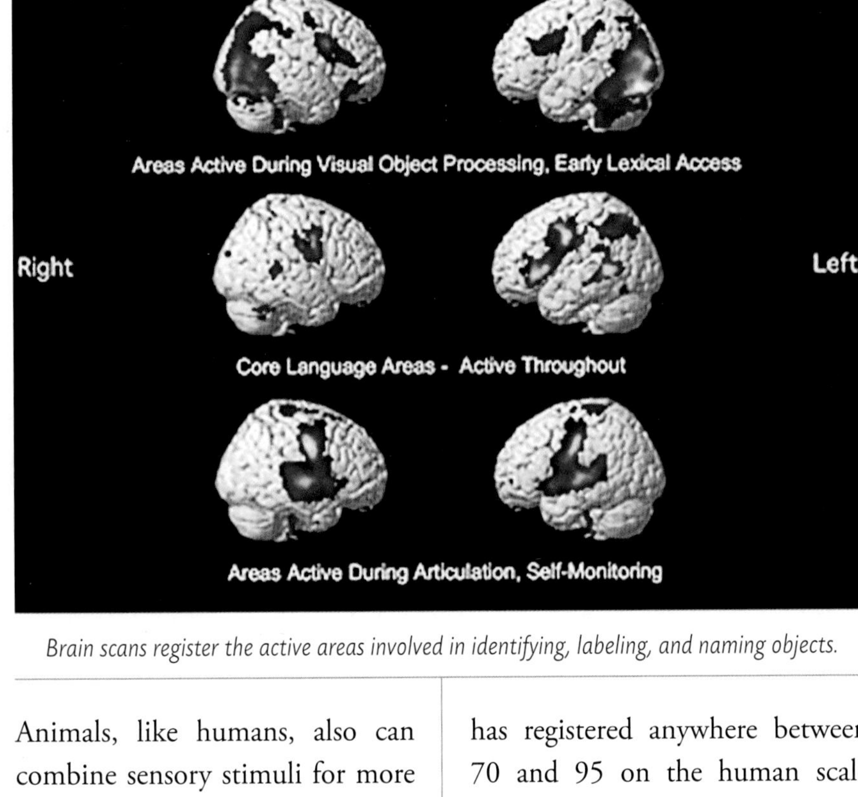

Brain scans register the active areas involved in identifying, labeling, and naming objects.

Animals, like humans, also can combine sensory stimuli for more sophisticated communication. The great apes, closest to humans in evolutionary development, possess a communication system that incorporates gestures, postures, expressions, and vocalized sounds. The Gorilla Foundation taught a lowland female named Koko, born in 1971, about 1,000 words of American Sign Language. Her IQ has registered anywhere between 70 and 95 on the human scale (100 is average).

LANGUAGE INSIDE THE BRAIN

Paul Broca relied on autopsies to examine the impact of brain damage on language and speech. Fortunately, advances in high-tech brain imaging techniques such as PET scans and functional MRIs have opened a window on the primary means of human communication.

In experiments using PET scans, volunteers first have a radioactive isotope of oxygen injected into their bloodstream. They rest quietly, looking at a blank screen until a single word appears, and then read the word aloud. In a final step, they produce another, related, word—such as the verb "eat" if the noun on the screen was "cake." Each stage produced distinctive images of where the brain took up needed oxygen from the bloodstream as registered by the radioactive isotope. Looking at the word activates the visual cortex at the back of the brain. Saying the word aloud activates motor neural networks, required for the muscular contractions of speech. Prompting the subjects to come up with their own related words lit their brain in many regions, including the language areas of the

FAST FACT Tonal languages are processed in the left hemisphere, not in regions associated with music.

CROSS REFERENCE: See "A New Brain," PAGE 72

WHAT CAN GO WRONG

IN 1894, BRITISH PHYSICIAN W. Pringle Morgan gave the first description of a case of dyslexia. Morgan wrote of a patient, "Percy F has always been a bright and intelligent boy.... His great difficulty has been—and is now—his inability to learn to read."

Early treatments for dyslexia, from the Greek for "impaired language," focused on visual processing. Doctors thought a visual impairment might explain why dyslexics sometimes transpose letters when writing. In fact, dyslexics typically see just fine—in fact, some see better than average. What they usually struggle with is decoding phonemes into words that have meaning. Some dyslexics also experience difficulty holding sounds in short-term memory to combine them into words. Others can decode phonemes but only at a snail's pace. Nevertheless, dyslexics score as well as other groups on IQ tests.

left hemisphere, motor complexes, and regions associated with stress.

Reading and writing are perhaps 5,000 years old. Silent reading is even more recent. The practice of reading all written material aloud was at one time so common that St. Augustine, writing in the fifth century about his contemporary, St. Ambrose, expressed astonishment that when he was reading, "his eyes glided over the pages and his heart sensed out the sense, but his voice and tongue were at rest."

Children today begin by sounding out words on paper, like St. Augustine, but progress to silent reading, like St. Ambrose. These recent developments appear not to be evolutionary brain developments so much as the co-opting of existing brain regions for new purposes; reading and writing are not encouraged by genetic programming in the way speech is.

Observational evidence suggests that reading and writing use different neural networks. Some people can identify letters and write but cannot read. Reading has its own quirks; some stroke victims can read normally except for particular kinds of words, such as adjectives.

The sites associated with reading vary, but people extremely adept at verbal communication tend to have reading functions located in the superior temporal gyrus and storage of names for objects in the middle temporal gyrus

+ DYSLEXICS +

MANY NOTABLE figures—artists, scientists, musicians—were dyslexic:

+ Leonardo da Vinci
+ Andrew Jackson
+ Hans Christian Andersen
+ Thomas Edison
+ Pablo Picasso
+ Agatha Christie
+ Ansel Adams
+ John Lennon

Given the complexity of understanding words on paper, it's no wonder children take different routes to learning to read. Some perform best with phonics, which sounds out words. It works well with many words, but not so well with English words that have silent letters or letter combinations that don't say their names. (Think of the words *cough, colonel,* and *cello.*) Phonics emphasizes the mechanics of putting letters together and the rules of how letters make sounds in combination.

Other children perform better with the "whole language" method, which teaches recognition of the shapes of words and relies on visual circuits. American schools tend to emphasize whole language in reading instruction and teach reading by getting the kids interested in understanding words in context, such as in stories.

It's not an either-or proposition, however. Most people learn to read by employing both systems, calling on a wider array of neural circuits. A substantial fraction struggles with mastering reading. Developmental dyslexia, a failure related to reading that cannot be explained by intelligence or problems with the child's learning environment, touches 5 percent to 15 percent of the population.

WHAT CAN GO WRONG

Researchers focus on a variety of possible causes. Physical impairments of the angular gyrus at the rear of the left hemisphere, active during reading, have received attention. Others have looked at parallel processing in a variety of neural networks, including the visual system. Functional MRI scans of dyslexics reveal decreased activity in both parietal and posterior temporal lobes. Some have abnormalities in the thalamus, where sensory information gets channeled.

Many dyslexics are artists and musicians. Research suggests they may have enhanced abilities to hear bass notes and see sharper colors in peripheral vision.

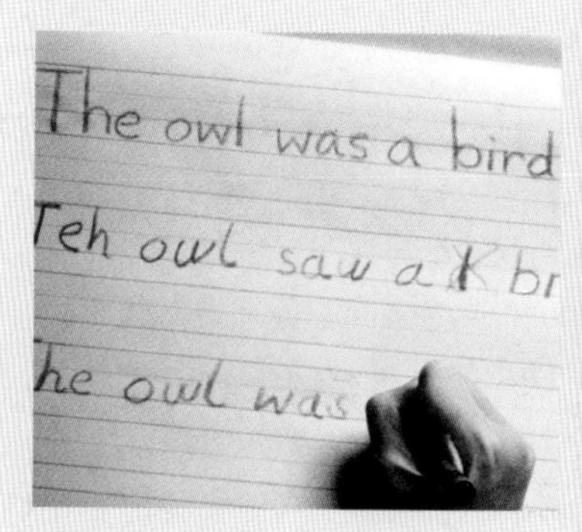

BROCA'S PATIENT

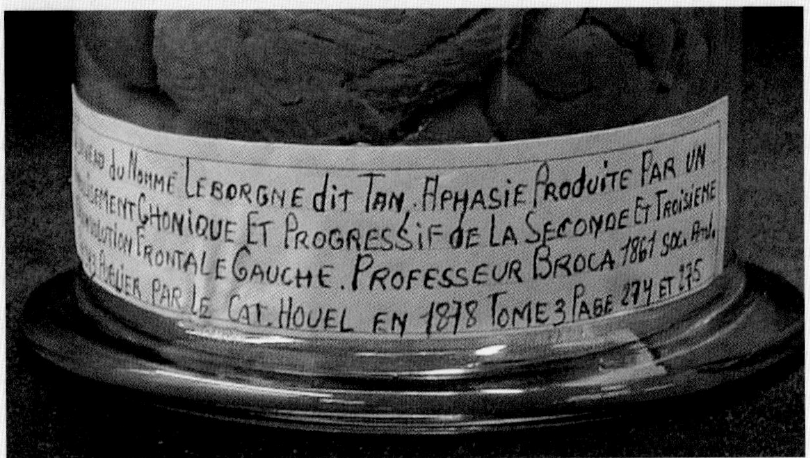

Neuroscientist Paul Broca preserved his patient's brain for future study.

A MAN named Leborgne entered a hospital in the suburbs of Paris. Leborgne had lost the power of speech, a condition known as aphasia. All he could say when asked a question was "Tan, tan," gesturing wildly. The staff took to calling him Tan.

Tan's health slowly declined. Paralysis spread throughout his right side until his lack of movement confined him to bed. Still, he said "Tan, tan" and apparently understood much, if not all, of what was said to him. After being hospitalized for more than 20 years, Tan died in 1861.

Surgeon and neuroanatomist Paul Broca examined Tan's brain during an autopsy. The dura mater had become thick and the left frontal lobe unusually soft. A section of the left hemisphere the size of a hen's egg had been destroyed, leaving a hole filled with fluid. Broca argued, convincingly, that Tan's language impairment stemmed from the anomaly. The damage also accounted for the paralysis of Tan's right side.

Today, neuroscientists take it for granted that damage to the left hemisphere carries the risk of language impairment. In 1861, such a conclusion was stunning. Two camps had been arguing over whether the brain's hemispheres shared functions such as speech or whether some might be localized. Tan had given Broca the key to settle the argument. Although brains differ, and localization doesn't fit inside strict borders, the site of the Parisian aphasiac's brain, in the left temporal lobe above the Sylvian fissure, today carries the name Broca's area, and Tan's unfortunate condition is known as Broca's aphasia. Tan's brain—and Broca's—have been preserved for science.

SPEECH DISORDERS

Until a few years ago, doctors considered stuttering a nervous or emotional condition. Now, the condition that affects about three million Americans falls squarely in the realm of neurology. A new drug, pagoclone, has begun clinical trials as potentially the first medical treatment for stuttering.

Stuttering usually begins between the ages of two and six, as children accelerate their learning of language. Three-quarters of those children spontaneously lose their stutters. Among adults, stuttering affects four times as many men as women. The exact cause of stuttering has eluded scientific research, but it's believed to have a genetic component. Brain scans of non-stutterers show speech processing most often performed in the left hemisphere of the cortex. Stutterers have an unusual amount of activity in the right hemisphere. This may indicate problems in coordination of the two hemispheres across the corpus collosum, and investigators suggest an overabundance of dopamine may contribute to the disorder. Before the drug trials, the most common treatments for stuttering included speech therapy and the use of an in-the-ear device that sends a person's voice into the ear canal at a slightly different pitch and after a brief delay. The changes are believed to activate the so-called choral effect, which suppresses stuttering when people speak or sing with others.

LEARNING DISABILITIES

Learning disabilities result from faulty reception, processing, and communication of information within the brain. Pediatricians can supply charts of developmental milestones, allowing parents to compare their children with norms in major stages of growth.

FAST FACT Lip-reading activates both the visual and auditory complexes of the brain.

Learning disabilities include: auditory processing disorder, which makes it difficult to distinguish between sounds, and visual processing disorder, which causes problems with reading, map interpretation, and other work requiring analysis of visual information. Dyscalculia interferes with math ability and can cause problems using money. Dysgraphia is a difficulty with writing that may include problems with handwriting, spelling, and organization of ideas. Dyslexia is a disorder of language processing. Dyspraxia/apraxia is a sensory integration disorder that interferes with motor coordination or speech. People with apraxia or dyspraxia of speech, two names for an absent or diminished speech ability, can understand what's said to them, but have trouble articulating what they want to say. The root cause isn't in the muscles of the lips or tongue but rather in the brain.

James Earl Jones overcame his stutter to achieve fame as an actor and a recognizable voice.

Adults can acquire either disorder through a brain injury, stroke, or tumor. A form of the disorder called developmental apraxia exists from birth. Scientists believe it arises either from a problem in overall language development, or through faulty neural communication between the brain's language centers and the muscles that produce speech. The fact that children with developmental apraxia have a greater likelihood of family members with communication or learning disabilities points toward a genetic component.

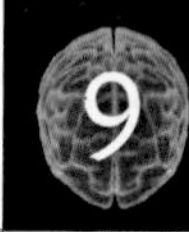

CHAPTER NINE

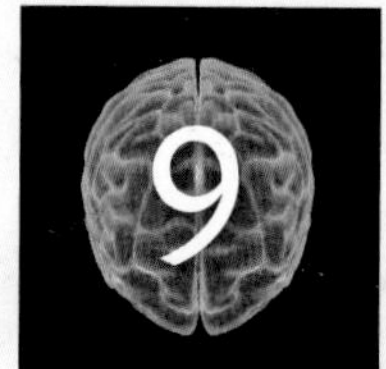

THE AGING BRAIN

THE BRAIN DOESN'T transform overnight. With the exception of a lightning-like stroke, the aging brain changes gradually in its abilities to perceive sensations, process information, create and store memories, and learn. It may move slowly, but it moves with greater purpose. Except for a decrease in processing speed, the healthy mature brain performs about as well as a youthful one in any task requiring planning, analysis, and organization of information. And with the wisdom of a lifetime, the elderly brain usually outperforms the youthful brain in judgment.

Sunrise finds Arcadia Tagawili, an elderly Filipino peasant, at work in her vegetable fields.

MATURE MINDS

BLESSINGS & CHALLENGES

THE ELDERLY BRAIN loses an edge in some functions but generally stays active and even gains a bit in others. Some neurons die through normal aging or through disease or injury. Thanks to plasticity, however, a healthy elderly brain can actively redesign itself to respond to a decrease in neurotransmitters and loss of some neurons. Connections among neurons get pruned, leaving those that remain the most useful. The only cost is the brain's ability to come up with information quickly.

Some areas of mental ability actually increase with age. In the absence of disease, the elderly brain enjoys a larger vocabulary and sharpened language skills. The aging brain requires greater concentration to take in and process sensory information, but the result may be a focus on the important stimuli and disregard for the peripheral.

An elder receives a traditional greeting as a gesture of respect in the Philippines.

A host of disorders and diseases can affect the aging brain, from hearing loss to dementia. Depression and anxiety also are common among the elderly, especially when they lose some of their memory and muscle strength, and friends and family members die. Suicide and alcoholism aren't uncommon. However, maintaining a healthy brain through mental and physical exercise, as well as medical treatment, can add life to one's final years as well as years to one's life. Fortunately, the brain gets stronger the more it's worked. Education, sensory challenges, cognitive puzzles, and exercises to improve blood flow, balance, and muscle mass all support the most important organ in the body. Nobody can guarantee that doing everything to maintain a healthy brain

FAST FACT The brain never loses its ability to absorb new information. Plasticity and experience mean a healthy brain always has the capacity to be creative, even though its ability to be nimble inevitably declines.

will shield it from the slings and arrows of aging. Letting the brain wallow and stagnate in old age, though, invites trouble.

ELDERLY WISDOM

Keeping a healthy brain into old age has long been seen as a blessing. The Old Testament Book of Leviticus commands respect and honor for the elderly, and the ancient Chinese philosopher Confucius said, "A youth who does not respect his elders will achieve nothing when he grows up." It was common in traditional, ancient cultures, where the average life span was much less than it is today, for the elderly to be revered because they had the knowledge and skills to survive. They also accumulated a lifetime of wisdom, living long enough to build up a much larger bank of personal experience than youth could possibly possess. In contrast, some modern, Western attitudes toward aging disregard it as outside mainstream culture. Even as wise a man as Sigmund Freud, the founder of psychoanalysis, dismissed the elderly as unworthy of his attention. "Psychotherapy is not possible near or above the age of 50," he said; "the elasticity of the mental processes, on which treatment depends, is as a rule lacking—old people are not educable—and, on the other hand, the mass of material to be dealt with would prolong the duration of treatment indefinitely." On the other hand, Roman orator Cicero urged fighting old age—not because it was to be shunned, but rather because it opened new doors to the mind and soul, "for they, too, like lamps, grow dim with time, unless we keep them supplied with oil. . . . intellectual activity gives buoyancy to the mind."

Scientific observation of aging and death may have begun with Aristotle. He viewed the decline and death of the mind and body as somehow built into the mechanisms of life. Violent death occurred through external means, he said, while natural death is "involved from the beginning in the constitution of the organ, and not an affection derived from a foreign source. In the case of plants the name given to this is withering, in animals senility." But just because aging and death are inevitable doesn't mean they should be passively embraced. The aging brain, if free of disease and disorder, can bring decades of experience to bear on the appreciation of moments past, present, and future.

PHYSICAL CHANGES TO CNS

The central nervous system experiences a series of natural changes as the body ages. Both brain and spine lose some of their nerve cells, resulting in a decrease in weight. This decline in mass begins in the brain of a young adult and continues steadily for the next six decades or so. Overall, the loss is relatively small and the plasticity of the remaining neurons so high that surviving neurons reconnect in response to new learning and offset many of the losses. Still, the remaining nerve cells in the aging

+ WISDOM CIRCLE +

THE 600 ADVICE GIVERS of the Elder Wisdom Circle answer questions from folks in their teens, 20s, and 30s via a nonprofit website created in 2001. His site made seniors feel more appreciated, founder Doug Meckelson said, but its popularity made him realize that there was "a whole slew of people who were interested in what the elders had to offer."

Contributors range in age from 60 to 105. Some work in groups at retirement centers, others from home. The elders' advice, gathered over decades of personal experience, tends to emphasize unselfish actions and self-acceptance.

STAYING SHARP

RESEARCHERS AT THE University of Illinois demonstrated in 2008 that when the elderly play video games, they improve cognitive skills and maintain those improvements for weeks. Even better, the skills transfer to tasks in the real world.

The study had 20 adults older than 60 play a game, while another 20 served as a control group. The game players outscored the nonplayers on measures of alertness, working memory, and ability to shift between tasks. The takeaway?

✓Strategy-based games could become a way for the elderly to maintain and perhaps even improve some of the cognitive abilities that decline with aging, according to Art Kramer, author of the study.

✓Video games can help seniors have fun and create or support social networks. Those that require body movement encourage physical exercise and eye-hand coordination.

brain begin to slow down in their transmission of electrochemical impulses. A fatty brown pigment called lipofuscin can build up in nerve tissues. Waste products in brain tissue can collect and, in the case of Alzheimer's disease, form plaques and tangles. As nerves lose mass and some begin to break down, reflexes may slow or disappear, and stimuli may not register as easily upon the five senses, leading to difficulties in moving and interacting with the environment. Age also causes some decline in reaction time, the speed of perception, decision-making, and other functions of the executive centers of the frontal lobes.

Some people experience a host of debilitating changes, while others seem to escape virtually unscathed.

Seniors are turning to handheld video games to challenge their brains.

The chances of suffering from dementia, a group of brain diseases including Alzheimer's, appears to increase until the 80s, when the risk actually decreases. Furthermore, the mere presence of the plaques and tangles of Alzheimer's disease doesn't necessarily cause dementia.

FAST FACT Mental decline occurs from disease, normal aging, and lack of mental exercise.

True dementia may exist in only about 5 percent of the population over age 65. The rest of the cases may involve "reversible dementia," a mental impairment caused by low blood pressure, a reaction to prescription drugs, poor diet, depression, hormone imbalances, and other problems that could be treated if properly diagnosed.

The loss of some of the nerve fibers in an elderly spinal column may affect not only the transfer of sensations and voluntary muscle commands but also involuntary signals, including those that affect muscle coordination. The spine contains a posterior column of nerve fibers that provide the brain with information about the location of the legs, feet, and toes. Age-related malfunctions in this spinal bundle impair the dynamic actions of maintaining equilibrium. Where the body actually is, and where the brain unconsciously *thinks* it is, may be two different things. The resulting impact on

CROSS REFERENCE: See "Maturity," PAGE 92

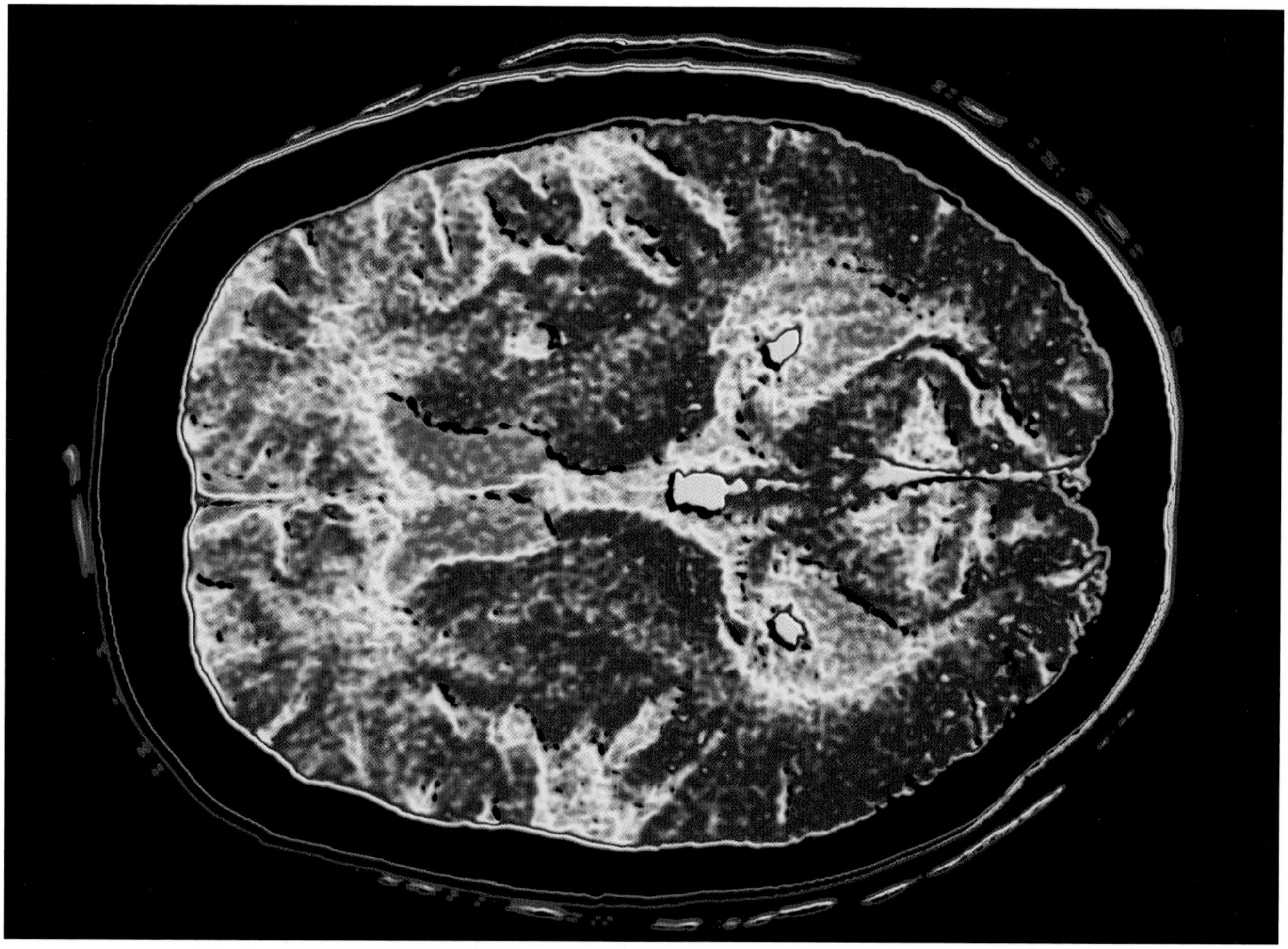

Color-enhanced CT scan of a 90-year-old woman reveals typical cerebral atrophy that occurs as we age.

muscle coordination contributes to jerky movements and falls.

LOSS RATES

In the brain itself, neurons disappear at different rates in different regions. A half century ago, studies suggested that the cerebral cortex lost about 30 percent of its neurons as it aged, leading to the widespread claim that the brain loses 50,000 to 60,000 neurons every day. Those studies gave highly inflated figures; the cortex loses only a fraction of its neurons. However, some subcortical regions do lose substantial percentages of neurons as the brain reaches old age. The acetylcholine-producing nucleus basalis loses about half of its neurons, which perhaps is the key reason for slowed reaction time in the elderly brain. The dopamine-producing substantia nigra loses 35 percent to 40 percent. The raphe nucleus—site of cells that produce the neurotransmitter serotonin, crucial for appetite and mood—loses about 35 percent to 40 percent. Finally, the hippocampus loses about 5 percent of its neurons in normal aging. Among healthy seniors, neurons compensate somewhat for the loss of their neighbors by lengthening their dendrites.

Subcortical regions contain fanlike collections of neurons that project upward into the cerebrum. These neurons, which neuroscientist Paul Coleman calls the brain's "juice machines," help create the mental energy necessary to maintain alertness. Age-related disorders in these neural fans can

DEMENTIA

TODAY'S scientists know dementia is not a normal part of the aging process, although many people who live to a ripe old age experience a decline in mental faculties. Dementia, from the Latin for "apart" and "mind," describes a variety of symptoms that stem from as many as 50 disorders of the brain. All involve neuron destruction.

DIAGNOSIS

Physicians diagnose dementia if two or more brain functions, typically including memory and language skills, are significantly impaired without the patient losing consciousness. Alzheimer's disease is the most common cause; others include Huntington's disease, Creutzfeldt-Jakob disease (whose variants include "mad cow" disease), and vascular dementia, which decreases blood flow to the brain. In the degenerative disease Lewy body dementia, neurons die in the cortex and substantia nigra through the buildup of Lewy body proteins. The same protein accumulation affects Parkinson's patients, but in Lewy body dementia it strikes at a wider range of neural systems, causing memory problems, poor judgment, and shuffling movements that mimic Parkinson's disease. Infections, nutritional deficiencies, reaction to medications, brain tumors, and sharp declines in oxygen to the brain can also bring on dementia symptoms.

Other forms of dementia may be accelerated by physical injuries to the brain. Former President Ronald Reagan suffered a head injury in 1989 when he was thrown from a horse in Mexico. Doctors believe his concussion and subdural hematoma may have hastened the onset of Alzheimer's disease, which was diagnosed five years later.

Seated in a wheelchair, a woman with Alzheimer's disease waits alone in a hospital corridor.

Dementias commonly are classified by the location of affected brain regions, symptoms, and causes. Cortical dementia affects the cerebral regions associated with language, memory, cognition, and social behavior. Subcortical dementia, affecting the lower regions of the brain, influences emotion and movement, as well as memory. Progressive dementia gets worse as time goes by. Primary dementia is caused directly by a disease, while secondary dementia gets triggered by another physical cause, such as injury. Alzheimer's is a cortical, progressive, and primary form of dementia, and its onset can be accelerated by other factors, such as physical injury.

SYMPTOMS & SIGNS

There are many common signs of dementia, but perhaps the most common is memory loss. Dementia patients may raise the same questions over and over, not realizing they've already heard and forgotten the answers. They may forget parts of daily routines, such as failing to serve a meal that's been cooked. Patients may also establish a pattern of misplacing objects. Everyone misplaces car keys, but the dementia patient may put them in the oven or the refrigerator.

Deterioration of certain cognitive skills can also indicate dementia.

Dementia patients may forget common words and have difficulty communicating because of language impairments. They may also experience disorientation and become lost in familiar places or believe it's a different time or year.

Personality changes can also be symptoms of a larger problem. If a person begins exhibiting extremely poor judgment—for instance, a lack of proper clothing in winter or summer—that behavior may provide a clue to the patient's confusion. A drastic change in outlook, going from happy-go-lucky to paranoid or fearful, is also a symptom. Dementia patients can also exhibit extreme passivity and may not feel like going out and doing things.

TREATMENT

Medication typically cannot cure dementia, but it can relieve symptoms temporarily. Perhaps 10 percent of dementias can be treated or even reversed because the cause is a temporary reaction. Substance abuse and negative reactions to prescription drugs, as well as biochemical imbalances, can bring on symptoms that disappear once the trigger is removed. Severe depression can also bring on dementia-like symptoms.

Drugs are now available to treat progressive dementias including Alzheimer's disease. For now, they offer only relief from symptoms and a slowing of deterioration—not a halt or a reversal of disease. They aim to extend the quality of life and possibly delay the need for the patient to enter a nursing home. Common types of anti-dementia drugs are called cholinesterase inhibitors, which work by retarding the breakdown of the neurotransmitter acetylcholine, which has been linked to the formation of memories, particularly in the hippocampus and the cerebral cortex. These drugs also appear to help patients retain the ability to do routine tasks and to stave off radical changes in behavior. Other drugs aim to relieve the seizures, depression, and other side effects of dementia.

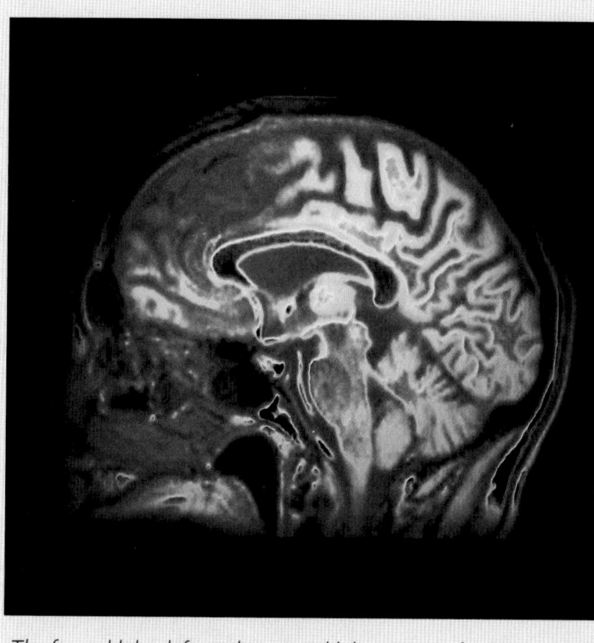

The frontal lobe, left, and temporal lobe, center, of a 50-year-old patient exhibit shrinkage from a rare form of dementia.

PREVENTION STRATEGIES

To ward off a decline in memory, patients in the early stages of dementia may benefit from using mnemonic devices or taking notes. Systems of rewards and punishments also may help modify dangerous or unacceptable behavior.

There's no magic bullet to prevent the onset of dementia, and genetic factors are believed to play a role. Some studies suggest that practicing a variety of intellectually stimulating tasks may lower the risk. In the so-called cognitive reserve theory, these activities enrich neuronal connections. Other studies have looked at physical traits associated with lowered risk of developing dementia, and Alzheimer's in particular. These include low levels of the amino acid homocysteine in the blood, which can be controlled with folic acid and B vitamins; low cholesterol levels; lowered blood pressure; and physical exercise, which optimizes blood flow to the brain. The long-term use of nonsteroidal anti-inflammatory drugs (NSAIDs) also has been linked in some studies to lowered risk and delayed onset of Alzheimer's. Researchers aren't sure why but are looking at the reduced incidence of inflammation and the drugs' possible interference with the formation of amyloid plaques.

alter moods, attention and anxiety levels, and states of arousal. For example, the frontal lobes of the cerebral cortex lose some of their ability to maintain working memory and manipulate information as they age. When young people are shown a small picture fragment and asked to hold it in their mind for four seconds, they activate not only the frontal cortex but also a focused region of their parietal lobe. Given the same task, elderly people exhibit less action in the frontal lobes and a more diffuse action in the temporal and parietal lobes. Visual recognition also is more evenly distributed in the two hemispheres of the elderly brain. As a result of these cortical changes, maintaining concentration tends to become harder with age. However, it can be helped with the artificial introduction of energy stimulants such as caffeine.

Darkened by a fatal explosion of blood, a cross section of a stroke victim's brain is preserved.

BLOOD & BONE

The brain's coordination of body movement declines with age as joints and muscles become weaker. Nearly all elderly adults suffer some degeneration of the joints, which act as cushions between bones and allow flexibility in the skeleton. Joints grow drier and thinner, allowing cartilage and bone to rub together and cause pain. In some cases, the loss of muscle and tendon tissue slows the reflexes even though nerve conduction remains intact. Movements begin to creep, and the length of stride becomes shorter and slower. Overall energy levels drop too.

FAST FACT The third leading cause of death in the U.S., strokes hit more than a half million Americans every year.

Like the spine, the brain also shrinks with age. It loses about 5 percent to 10 percent of its weight and 15 percent to 20 percent of its volume between ages 20 and 90. This shrinkage occurs through reduction of gray matter. As it decreases in volume, the amount of fluid cushioning the brain from the inside of the skull increases, by as much as 30 percent in men but only 1 percent in women.

SUPPORT STRUCTURES

Physical changes in the body also directly affect the brain. Without oxygen from the blood, the brain dies. Two processes affecting the flow of blood that begin in youth often take a toll on the aging brain. First is atherosclerosis, the thickening of artery walls through the buildup of fatty deposits. This degenerative process is hastened by smoking, inactivity, and poor diet. It rarely has serious consequences until middle or old age, when the buildup cuts off circulation and causes a heart attack or stroke.

Blood pressure also changes with age. A newborn typically registers an arterial blood pressure of about 90/55, with the first number measuring the millimeters of mercury displaced when the heart contracts and the second taking the same measurement when the heart relaxes. Blood pressure rises through childhood to typically reach an adult value around 120/80. Among the elderly, blood pressure commonly reaches levels around 150/90. High blood pressure is defined as having the first number register at 140 or higher, and the second number at 90 or higher. According to the *Journal of the American Medical Association*, more than three-quarters of

HER STROKE OF INSIGHT

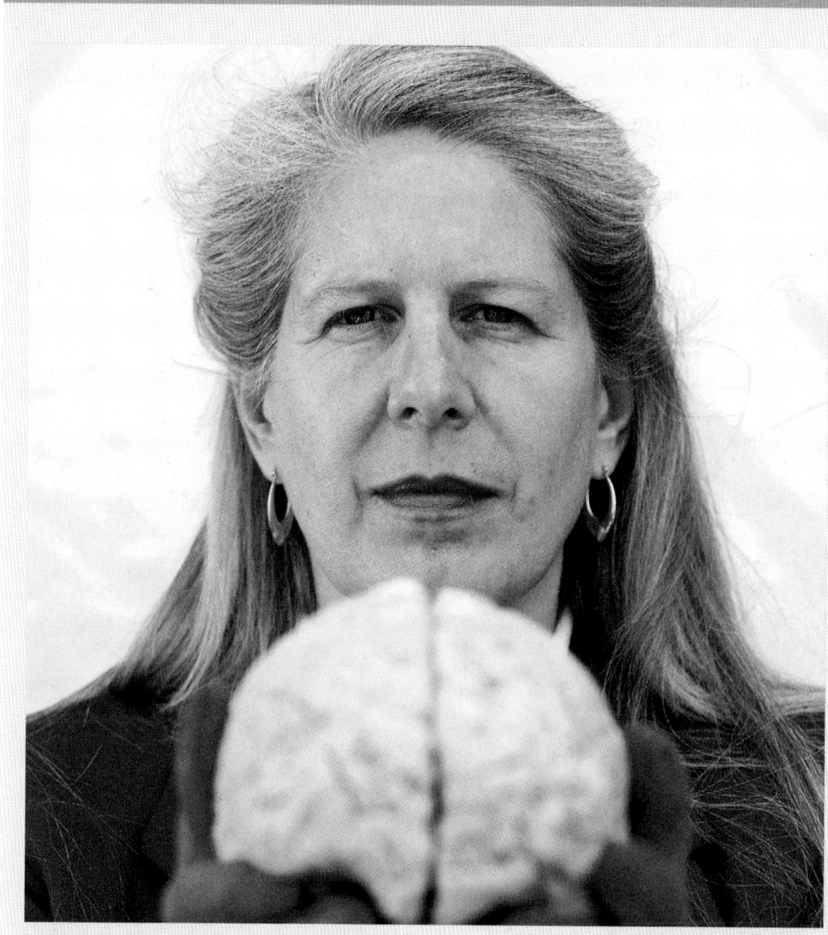

Stroke survivor Jill Bolte Taylor holds a human brain, which she uses in her presentations.

WHEN SHE AWOKE December 10, 1996, to a sharp pain behind her left eye, neuroanatomist Jill Bolte Taylor became the subject of her own research. She felt as if connections between her brain and body had become compromised yet found herself enveloped in peace.

"As the language centers in my left hemisphere grew increasingly silent, my consciousness soared into an all-knowingness, a 'being at *one*' with the universe, if you will," she said.

It took Taylor eight years to recover her speech and her professional life. In a sense, she's never recovered from the revelations of having a dominant right hemisphere. She wrote a best-selling book, *My Stroke of Insight*, went on TV talk shows, and gave lectures.

She said her right hemisphere put her in the "magnificent present moment," integrating life's big-picture landscape. Left-hemisphere skills are important, but a healthy brain needs a balance, she said.

GLOSSARY

ALZHEIMER'S DISEASE. The most common cause of dementia, primarily affecting memory, thinking, and reasoning. Nearly all brain functions are affected.

ATHEROSCLEROSIS. A blood vessel disease characterized by hardening of artery walls due to the buildup of fatty deposits.

BETA-AMYLOID. A protein that forms the characteristic plaques found in the brains of Alzheimer's patients.

CATARACTS. A clouding of the eyes' natural lens due to protein buildup.

CHOLINESTERASE INHIBITOR. A common type of anti-dementia drug that slows the breakdown of the neurotransmitter acetylcholine.

COGNITIVE RESERVE THEORY. Suggests that building and strengthening neural connections through mental stimulation may offset dementia symptoms.

CORTICAL DEMENTIA. Dementia induced by damage to the cerebral cortex. Results in impaired social and behavioral skills, thinking, memory, and language.

CREUTZFELDT-JAKOB DISEASE. A rare and fatal degenerative brain disorder characterized by rapid progressive dementia.

DEMENTIA. Term for a group of symptoms caused by disease, infection, or trauma resulting in a loss of mental functions, interfering with normal daily life.

EMBOLIC STROKE. Stroke occurring when fatty deposits, detached from an artery wall in another region of the body, lodge in an artery of the brain, cutting off blood supply.

FREE RADICALS. Molecules in the body with an unstable electric field. They take an electron from a neighboring molecule, making it unstable and creating a chain reaction that may result in cell damage.

HAYFLICK LIMIT. The term given to the discovery that cell division occurs a finite amount, the number of which may determine an organism's life span.

HEMORRHAGIC STROKE. Stroke that occurs when a brain artery ruptures, generally as a result of high blood pressure.

GLAUCOMA. An eye condition that develops as fluid pressure builds inside the eye. This may damage the optic nerve, leading to vision impairment or blindness.

LEWY BODY DEMENTIA. The second most common form of degenerative dementia; characterized by abnormal structures in certain areas of the brain.

LIPOFUSCIN. A naturally occurring fatty brown pigment that accumulates in cells as the body ages.

MACULAR DEGENERATION. A chronic eye disease characterized by the degeneration of the eye tissue responsible for clear center vision.

PRIMARY DEMENTIA. A dementia, such as Alzheimer's, that does not occur as a result of another disease.

PROGRESSIVE DEMENTIA. Dementia that becomes worse through time.

RAPHE NUCLEI. A cluster of neurons in the pons, medulla, and midbrain primarily responsible for production of serotonin.

REVERSIBLE DEMENTIA. Dementia brought on by an unrelated, treatable condition.

SECONDARY DEMENTIA. Dementia resulting from injury or another disease.

STEM CELL. An unspecialized cell with the ability to grow and develop into other types of cells and tissues.

SUBCORTICAL DEMENTIA. A dementia affecting the lower regions of the brain, resulting in changes in movement and emotion as well as memory problems.

SUBSTANTIA NIGRA. A cluster of cells at the base of the midbrain responsible for the production of dopamine.

TAU. A protein that aids in the formation of microtubules, which help transport nutrients within neurons. Becomes irregular in Alzheimer's disease.

THROMBOTIC STROKE. Stroke occurring when an artery supplying blood to the brain gets clogged by fatty deposits on its inner walls, cutting off blood supply.

TISSUE PLASMINOGEN ACTIVATOR (TPA). A clot-destroying drug that when administered soon after a stroke improves chances of recovery or minimal damage.

TRANSIENT ISCHEMIC ATTACK (TIA). A stroke occurring when an artery serving the brain becomes temporarily blocked. Often precedes more severe, acute stroke.

WILLIAM'S SYNDROME. A rare genetic condition characterized by mental retardation or learning difficulties, an overly friendly personality, and a distinctive facial appearance.

Americans age 80 or older had high blood pressure that was not being adequately controlled. Heart attack, heart failure, and stroke occur all too often when blood pressure rages out of control. High blood pressure contributes to about 70 percent of strokes by narrowing arteries to the brain, breaking off fatty deposits that block circulation, or rupturing the walls of blood vessels.

Billy Perry plays a Nintendo Wii game as a form of stroke rehabilitation therapy.

ALL ABOUT STROKES

Unlike many brain diseases, a stroke occurs in an instant. The ancient Greek physician Hippocrates described stroke as *plesso,* meaning "to be thunderstruck." A stroke occurs when a blood clot or broken artery cuts off the flow of blood to the brain. Without oxygen-rich blood, brain cells die, taking with them the cognitive and motor functions they make possible. Stroke has an immediate impact on the brain because the brain uses so much oxygen, fully one-fifth of the body's supply.

Thrombotic strokes occur when an artery serving the brain closes through the buildup of fatty deposits on its inner walls. Embolic strokes occur when a fatty clot forms elsewhere in the body, such as in the walls of the heart, and flows through open arteries until it lodges like a dam in a blood vessel of the brain. Together, these two account for 80 percent of strokes. Hemorrhagic strokes occur when an artery ruptures in the brain, usually as a result of high blood pressure.

+ STROKE SYMPTOMS +

When a stroke cuts off blood to a portion of the brain, neurologists can begin to pinpoint the location of the damage by the patient's symptoms.

DAMAGED BRAIN REGION	SYMPTOMS
Right hemisphere	Weakness or paralysis on left side; reduced vision to left; confusion; disorientation; denial of paralysis; impaired judgment or ability to reason; emotional instability
Left hemisphere	Weakness or paralysis on right side; reduced vision for objects to right; impaired thinking; difficulty speaking or understanding others; depression
Cerebellum	Impaired balance; nausea, vomiting; dizziness; extreme weakness of arm and leg on same side of body as cerebellum injury
Brain stem	Unstable blood pressure and pulse, leading to coma; difficulty swallowing, pronouncing words; vertigo and impaired ability to walk; weakness or paralysis on both sides

BRAIN CHANGES

SENSES, MOTION, EMOTIONS, & MEMORY

The senses become less acute with age through changes in the sense organs themselves as well as changes in the brain. Minimum levels of stimulation, called thresholds, are required before the brain perceives a sensation. With age, thresholds rise, requiring greater stimulation before sensations register. Sensations that occur below the threshold may seem not to occur at all. In addition, the aging brain suffers a decline in working memory, making it more prone to overstimulation and distraction. An overabundance of lights and sounds can overwhelm the elderly brain and impair cognition. That's why driving, especially in heavy traffic, becomes much more difficult with age.

Eyes and ears, the primary means of gathering information about the world, suffer the most dramatic ravages of age. As they lose their edge, they affect the brain's ability to perform at peak efficiency.

VISION

Many eye diseases are common among the elderly. These include glaucoma, cataracts, and macular degeneration. But ordinary wear and tear cause widespread damage to vision. Muscles that control the expansion and contraction of the pupils grow less efficient with age, and the pupils tend to be much more constricted, letting in less light. Furthermore, the lenses of mature eyeballs start to discolor and lose clarity. As they grow foggy, the lenses scatter light and create a noticeable glare at nighttime. In combination, the discoloration and tightening dramatically reduce the amount of light that hits the retinas of people older than 70. The eyes of the elderly also tend to be dry and at risk for infection because lachrymal glands that lubricate them lose their edge. As vision declines, nearly everyone older than age 55 needs corrective lenses at least part of the time.

Elderly lenses are prone to cataracts, which reduce clarity through the buildup of proteins that fold and harden. Some cataracts occur at birth, but more commonly they arise from aging, smoking, long exposure to sunlight, and diabetes

+ CATARACTS +

CATARACT SURGERY dates back at least 2,500 years old to India, where a cataract-clouded lens was pushed to the back of the eye. In ancient Rome, cataracts were broken up by needle, much like today's ultrasound procedure. Surgical removal dates to a 1748 operation in France. Today, a step not available then inserts a replacement plastic lens to sharpen vision.

BREAKTHROUGH

GRAEME CLARK grew up in Australia wanting to relieve the burden of his father's profound deafness, but he knew the condition was almost always impossible to cure.

Conventional hearing aids amplify sounds so their vibrations register on the hairs in the cochlea. However, the profoundly deaf have missing or damaged hairs, and thus little or no mechanism for converting vibrations into electrical signals the brain can interpret.

As a research professor at the University of Melbourne in 1967, Clark began to imagine a way around the problem: Why not send stimuli via electrodes

CROSS REFERENCE: See "Sights & Sounds," PAGE 106

A South Korean man reenacts an ancient exam for status as a venerated civil servant.

mellitus. The fogging of the lenses deprives the brain of sharp images through loss of acuity.

When vision becomes impaired over short distances, acceleration of mental decline may result. A 2005 study by the University of Texas Medical Branch in Galveston revealed that of 2,000 elderly Mexican Americans who were tested, those with poor vision at short distances tended to exhibit a steeper decline in mental functioning over a seven-year period. Why that's so isn't clear, but logic suggests the lack of clear vision for reading and performing eye-hand coordination would limit the ability to do brain-strengthening exercises. Lack of visual stimulation also might depress the workings of neural circuitry.

HEARING

Ears also suffer abuses through age. Few people with healthy hearing suffer much damage through childhood and adulthood. However, by age 60 or so, auditory deterioration almost always becomes noticeable. The spiral organ (of Corti) loses its hair receptors faster than they can be replaced. Over the course of a lifetime, loud noises, disease, and drugs destroy some of the 40,000 hair cells of the spiral organ. Hair cells usually get replaced, but at such a slow rate that they don't provide a return of total hearing function. Doctors estimate that if humans

BREAKTHROUGH

directly into the auditory nerve? "Probably as many as 99 percent of scientists around the world initially said that it was not feasible," Clark said.

Clark's work paid off in 1978, though, with the first surgical implantation of a cochlear device, sometimes called a "bionic ear." The U.S. Food and Drug Administration approved implants for adults in 1985 and for children in 1990; today, the fastest-growing demographic for recipients is children under age five.

Each device has two parts. An external system collects and transmits sounds to an internal system, which translates them into electronic pulses. Those pulses directly stimulate the auditory nerve, which transfers them to the brain's auditory complex for decoding as sounds. Cochlear implants and similar devices thus provide access to ranges of sounds that help patients recognize speech and other auditory stimuli.

could live to be 140, they would lose all hearing receptors. The ability to hear high-pitched sounds is the first auditory function to disappear. Once considered a disease of old age, the loss of hearing in the high registers now is appearing in younger and younger patients, thanks to the world becoming a noisier place.

As the inner ear also helps the body maintain balance, deterioration of the structures of the inner ear contributes to instability. That's one reason the elderly are more prone to falling.

Elderly volunteers who submit to tests of vision and hearing tend not to perform as well as younger subjects. In matters of visuospatial skills, an elderly brain exhibits some decrease in depth perception, spatial skills, and the rapid identification of complex geometric shapes. The impairment is mild enough that it may be caused by a decline in vision or even a lack of interest in the testing process.

Weakened eyesight may help cause recognition and recall problems as the brain ages.

Some skills of speech perception deteriorate with age through changes in the brain. The biggest change is the common loss of the aging brain's ability to hear high-frequency sounds, such as a faint, high-pitched voice. Another change relates to the slowing of the brain's ability to process information, which affects memory. If an elderly person has a memory or learning impairment stemming from auditory decline, he or she would learn more efficiently by reading information rather than having it imparted through lecture or discussion. The extra effort to recognize spoken words interferes with the encoding of memory, whereas visual stimulation, which follows a different series of brain circuits, may be unhampered.

Aging muscles lose mass, but Isaac and Isabel Ortiz still enjoy dancing in Sacramento.

SMELL & TASTE

Taste and smell are closely related, with much of the appreciation of food attributed to its aroma. Both decline with age. The number of taste buds begins to fall off in middle age, and the remaining buds lose some of their mass. Saliva production also decreases, which can make swallowing difficult and affect the taste of food. The sense of smell declines most dramatically after age 70, possibly because of the loss of nerve endings in the nose. According to a 1997 study by Duke University psychologist

+ TASTE CHANGES +

TASTE BUDS begin to decrease in number and in size in middle age—in women, at around 40 to 50, in men, about a decade later. The sensation of taste may not decline after the loss, but if it does, the change usually occurs after 60. First to go are the sweet and salty tastes. Studies suggest the aging process probably has little impact on why taste declines with age. Smoking, disease, and environmental exposures are more likely culprits.

CROSS REFERENCE: See "Smell & Taste," PAGE 116

Susan Schiffman, more than three-quarters of people at least 80 years old reported major difficulty in perceiving and identifying odors. Elderly people on medication typically require two to fifteen times as much of an odor or taste as their younger counterparts before they detect its presence.

Loss of the pleasurable experience of food can lead to alterations in diet, sometimes with significant results. The affected elderly may be at risk of vitamin deficiencies, malnutrition, accidental poisonings, and lowered immunity to diseases. Failure to detect appropriate levels of sugar or salt in food could have disastrous consequences for diabetics and those suffering from high blood pressure.

Why the nerves responsible for taste and smell decline with old age isn't clear. Some studies suggest that beyond the loss of taste buds, normal aging has little impact on smell and taste. Changes could be the result of disease, smoking, and environmental stimuli.

TOUCH

Similarly, it's not clear whether changes in touch occur through simple aging or as a result of disorders common to the elderly. Decreased blood flow, for example, may negatively affect the sensitivity of touch receptors, as may dietary deficiencies such as lack of enough vitamin B_1. One change that does

A 95-YEAR-OLD TRACK STAR

Leland McPhie, throwing a javelin, has won hundreds of medals since returning to track.

LELAND MCPHIE remembers observing a track competition in 1994 and an official inviting him to compete in the 50-meter dash. McPhie ran—and won. Not bad for a man who was 80 and taking part in the San Diego Senior Olympics. The race rekindled McPhie's competitive spirit, dormant for 60 years.

McPhie, born in Salt Lake City, Utah, in 1914, taught himself to pole vault and competed in high school from 1929 to 1933. Using a bamboo pole, he set a record in 1935 of 12 feet, 10 inches, at San Bernardino Valley College. Competing for San Diego State University in 1937, he led the team to a conference title. Since returning to track McPhie has won more than 200 gold medals. At 95 he threw a 35-pound mass nearly 22 feet at the U.S. Masters Indoor Track & Field Championship in March 2009. Asked by a San Diego newspaper when he might stop, McPhie answered, "Not until I have to."

He gave up competitive sprints after having open-heart surgery but still competes in six events, crediting his longevity to good diet and mental exercise—and getting out of bed every morning.

appear to be directly related to the aging process is a slight increase among some elderly people in their ability to detect fine-touch sensations. This may result from the thinning of the skin with age.

As touch thresholds rise, many elderly people experience a drop in temperature sensitivity. They may have trouble telling the difference between warm and hot, and between cool and cold. This puts them at increased risk of hyperthermia and hypothermia, serious medical conditions in which body temperatures are too high or too low, as well as localized frostbite and burns.

MOTION

How the brain and body move also changes with age. Voluntary motion requires the conscious action of nerves on skeletal muscles. Aging increases the amount of connective tissue in skeletal muscles but decreases the number of muscle fibers. Muscle proteins begin to degrade more rapidly than they can be replaced by age 30, and by age 75 muscles have lost half their mass, causing a decrease in strength and body weight. Aging muscles also suffer a decrease in the number of junctions with nerve cells, called motor units. With fewer neurons to release acetylcholine, which makes muscles contract, elderly muscles are prone to weakness. Doctors previously believed that the decline of motor units occurred as a natural part of aging. Now many believe the loss occurs as a result of inactivity. Strength training has been shown to improve muscle mass, leading to more endurance and a lowered likelihood of falls that can cause devastating injuries to brittle bones. A 1994 study at Tufts University found that increasing muscle mass by 10 percent through exercise led to a more than 100 percent improvement in strength. Exercises that emphasize balance, such as tai chi, can not only increase strength but also decrease debilitating falls, the likelihood of which rises with age-related changes in the neural pathways connecting the ear, brain stem, and cerebral cortex.

FAST FACT Dementia patients often are more confused at sundown. The reasons aren't clear.

Among all the muscle groups, smooth muscle tissues, such as the gastrointestinal tract, remain remarkably trouble free from the effects of aging. When problems do arise, it's often in response to an external irritant.

Aging also affects involuntary motions. Nerve impulses don't travel as efficiently along the conduction fibers located at the back of the spinal cord. When an elderly person starts to stumble, the brain must recalculate the body's position in space and issue a host of orders to skeletal muscles to readjust to avoid a fall. Slowed response times mean the calculations may take too long to prevent a tumble.

An elderly man looks at old photos. Depression and loneliness are common among the elderly.

CROSS REFERENCE: See "Brain in Action," PAGE 136

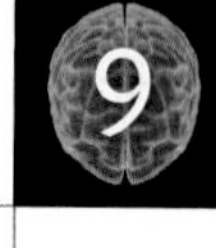

Taste and smell decline with age, yet food and drink can remain a pleasurable part of life.

STATES OF MIND

Changes in sleep patterns are part and parcel of growing old. As the brain ages, the amount of sleep it demands each night tends to remain relatively stable. What changes is the amount of time spent in the deepest stages of sleep. The duration of deep sleep slips from 20 to 25 percent of the sleep total for 30-year-olds to only 5 to 10 percent for most 70-year-olds—and sometimes, none at all. Furthermore, the typical 20-year-old takes only 8 minutes to fall asleep, while the typical octogenarian requires 18 minutes.

Sleep disturbances are common among the elderly. They include extreme difficulty in falling asleep or staying asleep, breathing problems, leg movements that cause the sleeper to wake up, and dysfunctional patterns of sleeping that include awaking at odd hours. Elderly sleepers suffer two ways: They typically react more strongly to external stimuli, which causes them to wake up. But even in sensory deprivation chambers, the elderly are prone to waking up in the middle of the night without having any lights or sound to blame for the interruption.

Alcohol and tobacco contribute to sleep problems. Alcohol affects the elderly brain more strongly because of lowered metabolic rates. A nightcap can bring on feelings of relaxation and ease the pathway to sleep, but too much alcohol robs the sleeper of the benefits of deep sleep by reducing or eliminating REM and its beneficial dreams. After the initial, fitful stages of sleep, alcohol continues

STAYING SHARP

IF YOU FEEL GOOD about growing older, chances are you'll have that feeling a long time. Researchers who examined personality traits and longevity in 1998 discovered that those who view aging as a positive experience outlive their sourpuss peers by an average of seven and a half years. That makes attitude a more effective life-extending trait than, say, a regular exercise regimen, which adds about three years.

The research began in 1975 with 660 people age 50 and older in Oxford, Ohio. They were asked detailed questions about their attitudes toward aging, including whether they agreed with statements such as "I am as happy now as I was when I was younger." Twenty-three years later, the study checked to see which subjects had died, and when. The results surprised Yale University researcher Becca Levy, one of the principal investigators. Even after controlling for sex, race, loneliness, self-reported health, and income level, the subject's view on aging remained a powerful predictor of longevity. Further studies reveal that the will to live, optimism, stress, and control over one's circumstances all correlate to longer life.

But that doesn't necessarily mean positive attitudes directly delay death. It's possible that some other variable, such as genetics, produces both good feelings and longer life. So, what's a person with a bad attitude to do? Martin E. P. Seligman, author of several books on optimism, said it may be possible to reorient people to adopt a more optimistic view of life, which is healthy if it's grounded in reality. At the University of Pennsylvania, Seligman trains a group of freshmen to boost their optimism as a way to reduce their stress. He said they suffer fewer illnesses throughout college than their peers.

Dementia patient Herb Winokur snoozes at breakfast with his caregiver and grandson. His daughter and son-in-law provide care in their home.

to interfere by causing frequent awakenings. Tobacco use makes it harder to both fall asleep and wake up. Among women, smoking has been linked to drowsiness during the day; among men, it's likely to cause unpleasant dreams.

To improve the sleep experience, the elderly should follow regular patterns of going to bed and getting up. Sleep should never be forced, as concentration leads to arousal. Instead, mental and physical exercises should be avoided before bedtime, and alcohol, tobacco, and caffeine intake minimized. If sleep still won't come, it's a good idea to get out of bed and read until feeling sleepy. Slavishly staying in bed and staring at the ceiling during periods of unwanted wakefulness won't do much good.

FAST FACT Tai chi has been shown to reduce tension and depression among all age groups.

Melatonin, a hormone that occurs naturally in the brain, is an effective sleep inducer in small amounts. The elderly brain is particularly sensitive to it because the pineal gland, which produces melatonin, makes less of it as it ages. A two-milligram, controlled-release melatonin capsule is likely to increase the length and quality of sleep without the nasty side effects possible with use of hypnotic prescription medication.

Sleep disturbances can have profound consequences. Lack of proper sleep has been associated with declining memory, weakened concentration, and impaired ability to function during the day. Sleepy

CROSS REFERENCE: See "Brain at Rest," PAGE 186

seniors are more likely to have an accident, fall down, or suffer from chronic fatigue.

Once awake, the healthy elderly brain has great difficulty concentrating completely—entering the hyperattentive zone required to, say, play complex video games or drive a race car. Studies at the Neuroscience Imaging Centre of the University of California at San Francisco found that older people take longer to complete simple tasks because they struggle to filter out irrelevant information.

AGING & EMOTION

Aging brains suffer a decrease in the production of neurotransmitters and the number of their receptor sites. Lowered levels of some neurotransmitters may cause the remaining ones to have an undue influence on mood. This makes older brains more susceptible to depression and anxiety—so much so that neurologist Richard Restak is convinced that most cases of depression occurring late in life stem from biochemical imbalances instead of psychological fears about mortality, decreased mobility, and impaired health. Between 15 percent and 25 percent of people more than 65 years old are seriously depressed. Nearly one-fifth of all suicides occur in this age group, with the rate for white males more than twice that of American adolescents. These alarmingly high rates put a premium on the early detection and treatment of depression among the elderly. Unfortunately, a correct diagnosis often is delayed because signs of depression are mixed and masked by physical ailments and injuries.

+ ON THE ROAD +

SOME FOLKS DRIVE just fine in their senior years. Most, however, experience a decline in their automotive abilities, especially after age 70. Hearing and vision tend to grow duller, narrowing the ability to gather information about traffic; and the executive function of the prefrontal lobes loses a bit of its processing speed and working memory as neurons shrink with age. That makes it harder to take in and process stimuli from the road and longer to make decisions such as whether to hit the brake or the accelerator.

One group of neurotransmitter receptors that thins with age is particular to dopamine. Synaptic docking sites for this transmitter decrease in density in the frontal lobes. Having fewer dopamine receptors eventually causes frontal lobe dysfunctions, including a decline in working memory. As science expands knowledge of neurotransmitters' roles in maintaining homeostasis in the brain, researchers have begun creating drugs that replace depleted chemicals to reverse some of aging's effects. Behavioral and chemical treatments are available for many of the disturbances to which the elderly brain is susceptible, including anxiety, insomnia, depression, alcoholism, and sexual dysfunctions. The elderly also typically suffer loss of appetite and a decreased enjoyment of normal activities during bouts of depression.

In small amounts, alcohol has been shown to have beneficial effects on the health of the elderly. A few glasses of beer or wine each week may increase blood flow to the brain and provide intellectual benefits. Heavy use of alcohol does just the opposite, affecting a host of cognitive abilities. Alcohol abuse affects up to 10 percent of elderly people who live at home and up to 40 percent of those in nursing homes. It takes an especially heavy toll on the elderly because metabolism decreases with age—the beer or two that causes mild impairment of a middle-aged adult is likely to have a more powerful effect on someone much older, particularly when the alcohol interacts with prescription medication. Alcohol, alone or with drugs, can disrupt judgment, memory, and thinking, and in the worst cases mimic some of the symptoms of Alzheimer's disease. Among both sexes, the loneliness of old age can contribute to a drinking problem.

To stave off loneliness, the elderly who maintain friendships and make new ones have a distinct advantage. Many studies have shown that morale increases when an older person spends time with friends, even more so than with family. Restak suspects this is because time with friends provides a break from the routines of daily life, as well as an opportunity to confide worries and fears that may not get communicated to family. This is perhaps one of the reasons that women, raised in a culture that encourages them to be more at ease with self-disclosure, tend to outlive men.

LEARNING & DECISION-MAKING

For centuries, observers have known that people process information more slowly as they age. But slowness doesn't mean failure. The aging brain, if not subject to disease or disorder, maintains its ability to process information. The elderly themselves usually are aware of the situation. They may say, "I have the answer on the tip of my tongue," taking an extra moment or two to recall it. As a trade-off for their decline in speed, the elderly tend to consider problems more carefully and then offer a well-reasoned response. Their brain changes to compensate for decreases in reaction time with more considered application of the information they recall.

"The experience of the older person enables him or her to put information into a context or a specialized-knowledge-based framework," Restak said. Elderly workers who are mentally sharp have the capacity to lose some of their functional cognitive capacity and still retain skill and intelligence levels on a par with their average but younger counterparts. Such individual cases argue against mandatory retirement ages; people age at different rates and maintain—or lose—mental skills at different rates, too.

+ MUSIC THERAPY +

WHEN WORDS ALONE cannot reach dementia patients, sometimes music therapy can. Familiar tunes, especially from the patient's teenage and young adult years, can bypass damaged language retrieval centers and make emotional connections. One *New York Times* writer told in 2009 of a World War II veteran with Alzheimer's disease who couldn't say a single word but danced with his wife for a half hour to Frank Sinatra songs from the 1940s. Appreciation for music fights dementia's encroachments because it's stored and processed more widely than plain speech.

MEMORY

One area of mental function that appears to be concretely affected by age is memory. Stereotypes of absent-minded elderly people are common in everyday life and in the mass media. But much of the common view of memory impairment among the aged may be exaggerated. Difficulty recalling facts and autobiographical data may be clouded by anxiety, depression, and other negative states, which are more common among the elderly than young people, and all of them interfere with memory.

Nevertheless, when controlling for variables that could skew memory experiments, it becomes clear that aging weakens memory. Normal degradation of memories occurs both for events of the recent past and those that may have occurred decades earlier.

Furthermore, the elderly lose a degree of working memory, the "mental desktop" that allows them to hold and manipulate information for a few seconds.

PET scans of elderly brains reveal how they differ in memory tasks from younger brains. Both young and old brains activate the hippocampal regions in both hemispheres to begin the process of encoding memories. However, elderly brains showed lower levels of activation than youthful brains during experiments aimed at testing their recognition of a set of 32 faces they had been asked to memorize. Not surprisingly, the elderly test subjects had lower accuracy levels in the recognition test. Researchers believe the lessened activation in the hippocampus points to greater difficulty in encoding memories.

The elderly brain may attempt to force the recall of uncertain information by calling on the frontal lobes to assist in memory, but PET scans reveal the older brain has more trouble activating these lobes, particularly in the left hemisphere. Problems with memory need not be cause for concern until they become serious enough to interfere with one's life. Then they may be ripe for discussion with a doctor.

ALZHEIMER'S DISEASE

Alzheimer's disease is a condition that slowly attacks the entire brain and its ability to function. Impairment of memory is almost always the first symptom of Alzheimer's disease. It is the most common form of dementia, afflicting most dementia patients. It targets the portions of the brain that are crucial to remembering, thinking, and

Keeping her brain sharp, 69-year-old Miyoko Izumikawa learns Internet stock trading during a free computer school for the aged in Tokyo.

FAST FACT Alzheimer's disease accounts for 55 percent to 60 percent of dementias. Genes on chromosomes 21, 19, 14, and 1 have been linked concretely with an increased risk of being diagnosed with disease.

reasoning. As these regions deteriorate, the patient eventually loses a lifetime of skills and becomes more and more reliant on the help of others. In the final stages, the disease turns the cognitive regions of the brain into a wasteland, spiraling the patient toward a vegetative state and eventual death.

How and why Alzheimer's attacks the brain remain something of a mystery. New treatments are under way to slow its advance, and lifestyle regimens can decrease the risk of getting the disease and delay its onset. However, more than a century after its discovery and description by German psychiatrist Alois Alzheimer, the disease that bears his name remains one of the scariest diagnoses a patient and family can receive.

STAGES & SYMPTOMS

Stage one of Alzheimer's disease may be hard to diagnose because of the slow advance of symptoms, which may look at first like an extreme form of the ordinary mental slowness common among the elderly. In this stage, Alzheimer's patients forget things that happen during the day that they normally would remember—such as what they had for lunch and who joined them. In the second stage, Alzheimer's patients struggle to come up with words. They also suffer a decline in their ability to perform abstract thinking and acts of imagination, and begin to suffer emotional downturns that decrease their interest in people and events. The third and final stage results in such total disruption of intellectual abilities that memories disappear, leaving the personality destroyed and the patient unable to recognize loved ones.

The brain of Alzheimer's patients shows telltale signs. Although it is normal for the brain to shrink somewhat as it ages, Alzheimer's patients lose about 10 percent to 15 percent of their mass. They also lose many more neurons than would be accounted for in normal aging, including up to 25 percent in certain cortical regions. Alzheimer's patients also show vastly more destruction of neurons in the hippocampus, crucial to memory formation, than would be expected through normal aging.

A striking marker for Alzheimer's disease is its accumulation of plaques and tangles. Dense bundles of fibers formed by a molecule called tau inside the neurons alter them from orderly patterns of neural connections to chaotic bird's nests of twists and turns. Outside the afflicted cells lie fatty globs of plaque. The accumulation of plaques and tangles makes neurons shrink and disappear. When the loss reaches a critical point, cognition suffers. Scientists used to think that Alzheimer's disease caused the creation of the plaques and tangles. Now they believe the buildup of debris leads to the symptoms

BREAKTHROUGH

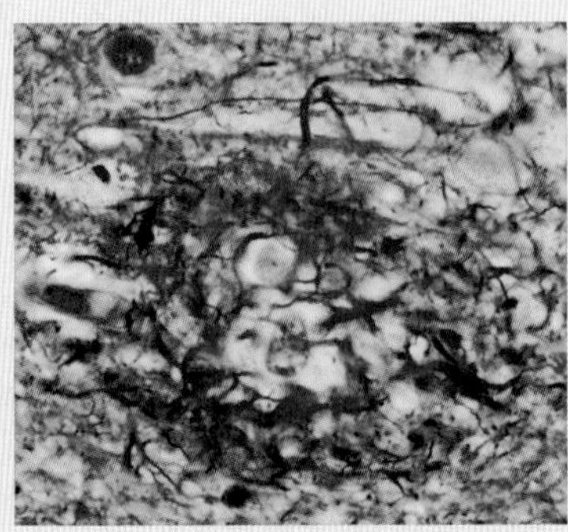

"SHE SITS on the bed with a helpless expression. What is your name? *Auguste.* Last name? *Auguste.* What is your husband's name? *Auguste, I think.* Your husband? *Ah, my husband.* She looks as if she didn't understand the question."

So opens the first entry in the long-lost file of a woman identified only as Auguste D, rediscovered in a German archive after 90 years. The interview occurred November 26, 1901. The physician who took notes was neuropsychiatrist Alois Alzheimer. Alzheimer described Auguste's symptoms as progressive cognitive impairment, hallucinations, delusions, and

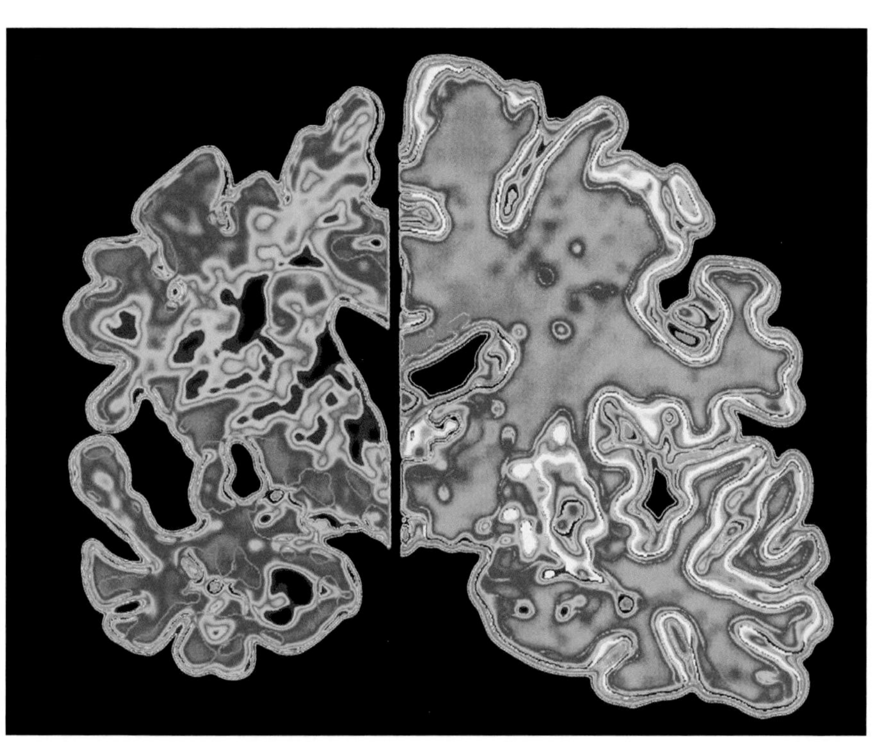

Vertical slices of a normal brain, right, and an Alzheimer's brain show shrinkage in the latter.

of the disease. A key chemical building block was pinpointed in 1984 with the discovery of beta-amyloid protein in the plaque. Beta-amyloid builds up gradually in every brain but accelerates in Alzheimer's patients. Research has focused on ways to slow or reverse the buildup, as well as to target defective genes that encode for excessive beta-amyloid production.

PREVENTION STRATEGIES

One thread of research suggests that anti-inflammatory drugs, such as ibuprofen, may lessen the risk of developing Alzheimer's because the presence of sticky plaques pushes glial cells into an inflammatory response that worsens the plaques' impact. Another thread focuses on creating a vaccine that would trick the body's immune system into targeting beta-amyloid as a foreign substance that must be destroyed.

Although there is no cure, the Food and Drug Administration has approved five drugs to treat the symptoms of Alzheimer's by improving cognitive functions. Four—Reminyl, Exelon, Aricept, and Cognex—work by delaying the destruction of acetylcholine, which is significantly lowered in the brain of Alzheimer's patients. The fifth, Namenda, offers protection to brain cells from an overabundance of the neurotransmitter glutamate, necessary for normal brain function but toxic in high amounts.

Trials of a drug in Great Britain showed promise in 2008. Patients with mild and moderate Alzheimer's who were given a daily capsule of Rember had the progress of their disease slowed by as much as 81 percent. During the 19 months of the clinical trial, patients showed no significant mental decline. The pill prevents the formation of new tangles in neurons and loosens those that exist. It may be available by prescription in a few years.

BREAKTHROUGH

psychosocial incompetence. He kept up with her case as her mental abilities further declined and she died in 1906. That year, Alzheimer described what he found in Auguste's brain during autopsy, including neurofibrillary tangles, cortical shrinkage, and plaque buildup (shown left) between neurons. It was the first description of a form of dementia that became known as Alzheimer's disease.

Alzheimer's description of the physical causes of Auguste's disease altered the way science viewed the aging brain. In place of the belief that senility was a normal part of growing older, Alzheimer linked a degradation of the mind to a physical disease. That set off a race to refine the disease's description and seek ways to slow its progress—or even find a cure. A century ago, relatively few people lived into their 70s and 80s. Today's extended life spans have helped to make Alzheimer's a bigger problem.

LIVING LONGER

OLDER BRAINS, BETTER BRAINS

Increasing life expectancy is graying much of the world, including France, site of this residential facility for the elderly.

CROSS REFERENCE: See "Maturity," PAGE 92

FAILURE OF ANY of the body's vital organ leads inevitably to death, but the legal definition of death itself rests on the irreversible cessation of brain functions. When the heart or lungs give out, death occurs because the brain becomes deprived of life-giving oxygen.

The brain, therefore, occupies a special position among the organs of the body: Life continues as long as the brain still works. Furthermore, the condition of the brain is the single most important factor in determining how long a person will live. When it comes to longevity, neuroscientist Richard Restak says, "the brain is not just one organ among many but *the pivotal determiner of how long we will live.*"

FAST FACT: Brain and nervous system disorders are the leading cause of disability in the U.S.

LIFE SPANS

How long can human life be extended, assuming a healthy brain continues to function? Nobody can be sure. The oldest authenticated record for a human life span is 122 years, achieved by a Frenchwoman named Jeanne Calment who died in 1997. That's well over twice the average life expectancy of children born in the late 19th century. For thousands of years, until the early 1900s, people could expect to live, on average, only into their 40s. Better prenatal care, access to medical treatments such as antibiotics, improved diet, and lifestyle changes including exercise, weight control, and abstainance from tobacco have steadily increased the average life span. A baby born in America in 2006 can be expected, on average, to live 77.7 years, a record high, according to the National Center for Health Statistics.

SCHOOLS OF THOUGHT

Theories of aging fall into two main categories. First are the so-called stochastic theories, which perceive aging as the result of random events that cannot be predicted precisely. This lack of perfect prediction explains why on average it's a good idea to give up tobacco because smoking increases the risk of cancer and other life-threatening illnesses, but practically everyone knows someone who smoked well into old age. When applied to the cells of the brain and body, stochastic principles suggest that the accumulation of damage over time from normal interaction with the environment causes cells to become less efficient and eventually to die. The second group of theories hold that aging occurs in response to the expression of a series of aging-related genes. The breakdown of cells over a lifetime becomes a kind of programmed senescence built in to the very fabric of life.

Both ideas appear to be at least partially true. No matter how high the average life expectancy rises, some people will fall short

NUNS OF MANKATO

Study of nuns' brains (kept as tissue samples) may lead to ways to stave off dementia.

SISTER MARY LOUISE PIHALY, 93, has trouble remembering the names of some of her fellow nuns at the School Sisters of Notre Dame in Mankato, Minnesota. But she stays active, reads, and stays positive. An ongoing study of the nuns, started in 1986, suggests that an upbeat temperament is key not only to living longer but also to staving off dementia. Neurology professor David Snowden began studying the nuns and others in their order around the country after discovering that many live well into their 90s and beyond. He realized the advantages of studying a group with so much in common: The nuns kept detailed records, and their shared diet, environment, unmarried status, health care, and other conditions eliminated many variables. After Snowden's retirement, others continued the research, including Kevin Lim and Karen Santa Cruz of the University of Minnesota. About 50 nuns remain in the study.

Research on 670 brains has found that folic acid may play a role in preventing Alzheimer's. And the complexity of the nuns' writings six decades ago appears to correlate with avoidance of Alzheimer's. The study also has linked mental stimulation with longevity and clarity of mind. Santa Cruz noted that 10 to 15 brains in the so-called Nun Study collection show signs of diseases, even though those nuns didn't exhibit symptoms while alive. The University of Minnesota has agreed to digitally scan tissue samples from each brain and make them available to researchers online.

+ WORLD'S OLDEST +

JEANNE CALMENT is recognized by the *Guinness Book of World Records* for the longest authenticated life. She died in Arles, France, on August 5, 1997, at age 122. She remained tart-tongued and funny, even as her memory and hearing began to fail. According to the World Health Organization, life expectancy is longest for women in Japan, where a girl born today will likely reach age 86. For men, the tiny European country Monaco offers the greatest life expectancy, 81. American life expectancy is 81 years for women and 76 for men. The oldest American whose age could be verified was Sarah Knauss of Pennsylvania, who died in 1999 at 119 years, 97 days.

because of illness, accident, and other unpredictable events. And no matter how healthy a person is, sooner or later he or she will hit what appears to be an upper limit for the life of the human species.

Evidence for the biological limit for each species comes from a series of studies conducted in the 1960s by microbiologist Leonard Hayflick. He spread a layer of embryonic human cells onto a growth culture in a sterile dish. When the cells divided enough to cover the culture, Hayflick removed a few cells and deposited them on a new culture in another dish. He repeated this process to see how many times the cells would divide. Division slowed after 35 iterations. By the 50th division, the skin cells had completely halted their divisions. Hayflick found that each species has its own maximum number of cell divisions. For mice, it's 15 or so; for the long-lived Galápagos turtle, it's 110 or more. This Hayflick limit suggests the existence of a genetic clock that chimes midnight after a set number of cell divisions.

The number of divisions isn't constant for all tissues. Bone marrow and skin cells divide continually over a lifetime, and taste buds are constantly being replaced. On the other hand, nearly all of the neurons in the brain do not divide at all—the cells that you have as a child are the same ones you have in old age. The Hayflick limit, therefore, may strongly influence the longevity of other vital organs but have little direct impact on the brain itself.

However, all cells are subject to chemical reactions with a class of molecules they encounter every day called free radicals. These molecules have an unstable electric charge because they contain at least one unpaired electron. Free radicals seek to achieve a neutral electrical charge by stealing electrons from nearby molecules. In the body, free radicals commonly take electrons from the walls of neurons and other cells, and more critically from cellular DNA and from the energy-producing cell bodies known as mitochondria. When a free radical steals an electron, it alters the electrical charge of the donor, which in turn seeks to rectify its own imbalance. That sets off a chain reaction that can cause extensive damage over the life of a cell.

FAST FACT The Centers for Disease Control reports that owning a pet helps blood pressure and cholesterol and tryglyceride levels.

In the 1950s, biogerentologist Denham Harman of the University of Nebraska Medical School proposed that aging occurs as a natural consequence of free radical damage. "Measures that slow initiation of damaging free-radical reactions by mitochondria may increase both the maximum and functional life spans," Harman said. According to this theory, antioxidants such as vitamins C and E, which soak up free radicals, as well as the body's own ability to repair damaged DNA may slow the aging process.

AGING & THE BRAIN

The aging process affects the brain in a variety of ways, from the decline of memory, to slower processing of thoughts, to the falloff in the brain's processing of high-frequency sounds. Along with aging, diseases often take their toll on cognition.

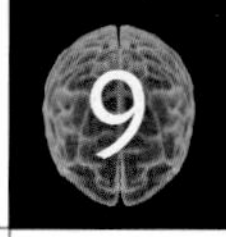

A third factor that affects cognitive ability is its continued use. Lack of mental fitness precipitates a decline in mental function just as lack of physical exercise makes muscles atrophy.

To keep the brain going stronger longer requires a regimen of exercise. All body organs except the brain decline with the wear and tear of age. The brain, on the other hand, *improves* with use. Some scientists believe that mental stimuli improve neurons' ability to repair their damaged DNA, while others point to the strengthening of neuronal connections as circuits enjoy repeated stimulation. Still others favor the explanation that the brain's plasticity creates new circuits that, with repeated use, contribute to the survival of their constituent neurons.

Whatever the explanation, neurons stay healthy when they are challenged with new stimuli in old age, whereas those given too much rest degenerate and contribute to unhealthy aging.

FORMING NEW NEURONS

In the brain of a baby developing in the uterus, unspecialized cells in the brain grow into hundreds of particular kinds of neurons, each appropriate to its location and function. The undifferentiated, embryonic forms are called stem cells, and they are found in several locations throughout the body of an adult. Given the right stimulation, these cells can grow into virtually anything.

Until 1998, scientists believed the adult brain contained no stem cells and that no neurons could

Maintaining strong friendships and social connections is one simple way to keep the brain going strong.

form after early childhood. However, the discovery that year of stem cells in the hippocampus, and the subsequent location of stem cells in the olfactory bulb and caudate nucleus, touched off a frenzy of speculation about ways they might be induced to create replacement neurons. Neurons damaged by disease or injury could have their functions taken over by new, healthy neurons grown from stem cells and transplanted into the brain or replaced by new neurons generated within the brain itself. Ideally, medical research could lead to ways for the brain to heal itself.

Thousands of new cells grow in the hippocampus every day. Many die soon afterward. Recent studies with rodents have shown that the more the test subject learned, the more new neurons survived in the hippocampus. Antidepressant therapy also seems to encourage the growth and survival of new neurons. Likewise, exercise in combination with the creation of the mood-elevating chemical beta-endorphin by the hypothalamus and pituitary gland has been shown to increase the creation and survival rate of new hippocampal neurons in mice. Other factors act to inhibit neurogenesis. Among the most notable is stress, which already has been associated with the higher risk of a variety of diseases. As science learns more about factors influencing the formation of new neurons, it hopes to gain a measure of control.

FAST FACT Formation of brain cells aiding in song memory was found in the 1980s in adult songbirds.

One initial challenge is to get stem cells to grow into particular kinds of neurons. "You can't just put stem cells somewhere in the brain and expect them to take over the function of missing or damaged cells," said neurodegenerative disease researcher Fred Gage of the Salk Institute in La Jolla, California. "You have to figure out the requirements for turning these cells into neurons."

Gage's research has focused on finding the right chemical stimuli to turn a laboratory dish of stem cells into mature, electrically active nerve cells. A breakthrough occurred in 2007 when teams of researchers in Wisconsin and Japan took ordinary human skin cells and transformed them into stem cells by inserting genes that reprogrammed them.

Gage envisions taking skin cells from Alzheimer's patients and turning them into neurons so they can be carefully watched for signs of developing the disease. Gage's approach is slightly different in that instead of inserting genetic material, he aims to turn on inactive genes in live skin cells to create the transformation.

REPAIRING THE DAMAGE

Another thread of research aims not to implant neurons grown from stem cells, but rather to activate precursor cells already in the brain to migrate to damaged areas, change into the necessary replacement neurons and heal the injury. This method, called self-repair, avoids the difficulties of controlled

BREAKTHROUGH

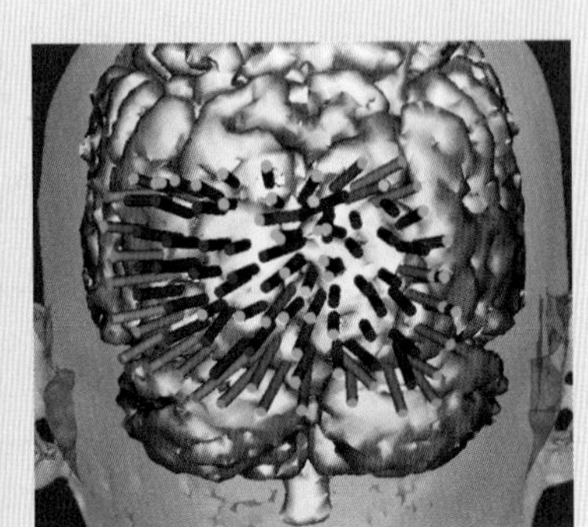

MICROSOFT co-founder Paul Allen put up $100 million in 2003 to completely map the brain of a mouse. Three years later, the Allen Brain Atlas went online with a 3-D database showing where each of more than 21,000 genes is expressed in the brain of the common *Mus domesticus.*

The research team at Seattle's Allen Institute for Brain Science captured 85 million images of stained mouse samples. The researchers chose the mouse because more than 90 percent of its brain has a direct counterpart to the genetic library of the human brain. It's also easy to obtain lots of mice from the

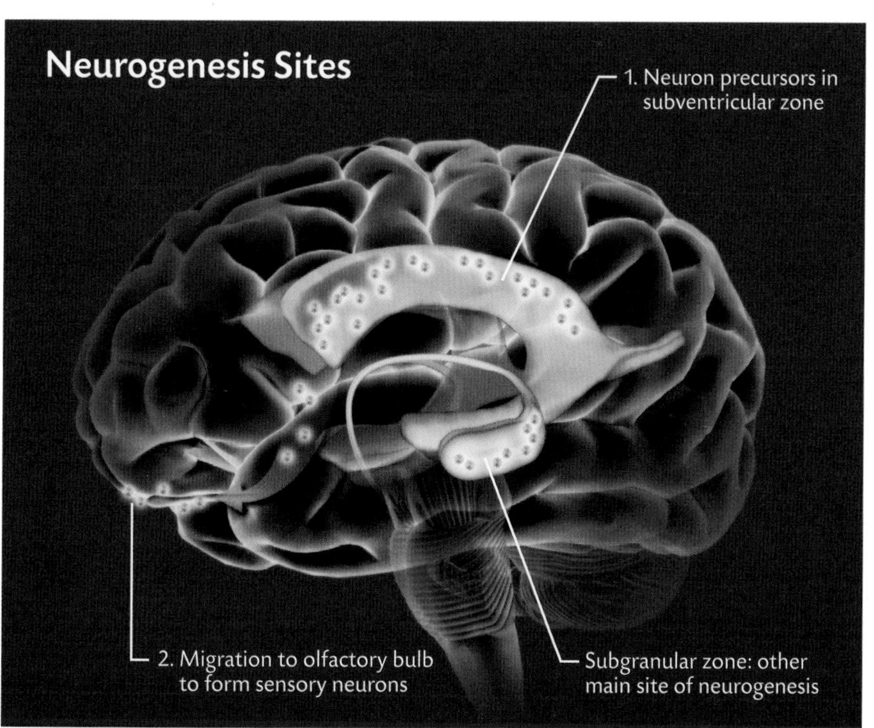

Neurogenesis sites *(brown and bright blue)* send cell precursors to the olfactory bulb *(purple)*.

growth in the lab. However, to get stem cells to migrate and change as desired, researchers need to learn much more about the chemical instructions that control neuron development.

Some success already has been obtained with mice. Jeffrey Macklis of Children's Hospital in Boston has induced the birth of neurons in mature mice brains. Precursor cells in the brains' fluid-filled ventricles moved toward areas of cortical damage and developed into mature neurons that acted just like those already there. The new neurons sent out axons and dendrites to connect and function like the cells they replaced. If such self-repair could be induced in humans, degenerative diseases that destroy or damage neurons could be treated by replacing the defective regions with healthy substitutes. The treatment probably would be restricted in scope because of limits on how many neurons could be replaced, as well as the complexity of the disorder. Parkinson's disease, for example, results from deficiencies in dopamine-producing neurons in a compact region of the substantia nigra. That would make it a more likely candidate for neurogenesis therapy than an illness that affects widespread circuits containing different kinds of neurons. Among the latter illnesses is Alzheimer's, whose widespread effects don't provide as clear a target for self-repair.

Still, the prospects of finding new ways to target formerly untreatable diseases thrills neuroscientists around the world. "Stem cell research and regenerative medicine are in an extremely exciting phase right now. We are gaining knowledge very fast and many companies are being formed and are starting clinical trials in different areas," said Sweden's Jonas Frisén.

BREAKTHROUGH

same gene pool, which eases the process of mapmaking.

To begin making an atlas of the expression and interaction of the human brain's genes, the team mapped the folds of the cerebral cortex as it developed from infant to mature adult. Researchers hope to find the early signatures of brain disorders ranging from William's syndrome (marked by developmental delays and overly friendly personalities) to autism and schizophrenia. Many diseases have their roots in mutations and the expression of specific genes too soon or too late for optimal brain development. Comparing genetic expression in healthy and anomalous brains is likely to suggest new ways to treat or prevent diseases. With the completion of human atlases, researchers around the world will be working from the same maps when refining what science knows about the human brain.

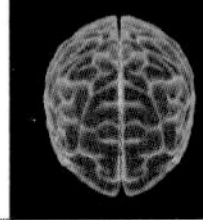

EPILOGUE

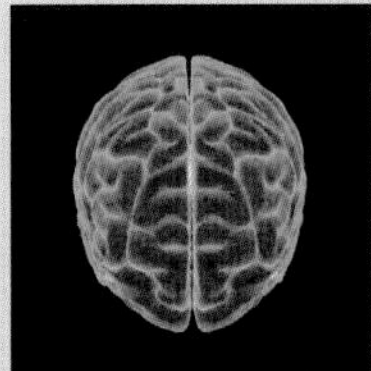

FUTURE OF THE BRAIN

THE BRAIN, that most complicated object, is not ready to reveal all of its secrets. Yet the application of the energy of the human mind, turned inward toward the brain itself, has begun to peel away layers of mystery. With greater understanding come new possibilities in creating artificial intelligence and enhancing human cognition through biological and mechanical means, including nootropic drugs, brain implants, and virtual reality.

Joey Chaos, a humanoid robot developed by Hanson Robotics, gets an adjustment to the wires in its head.

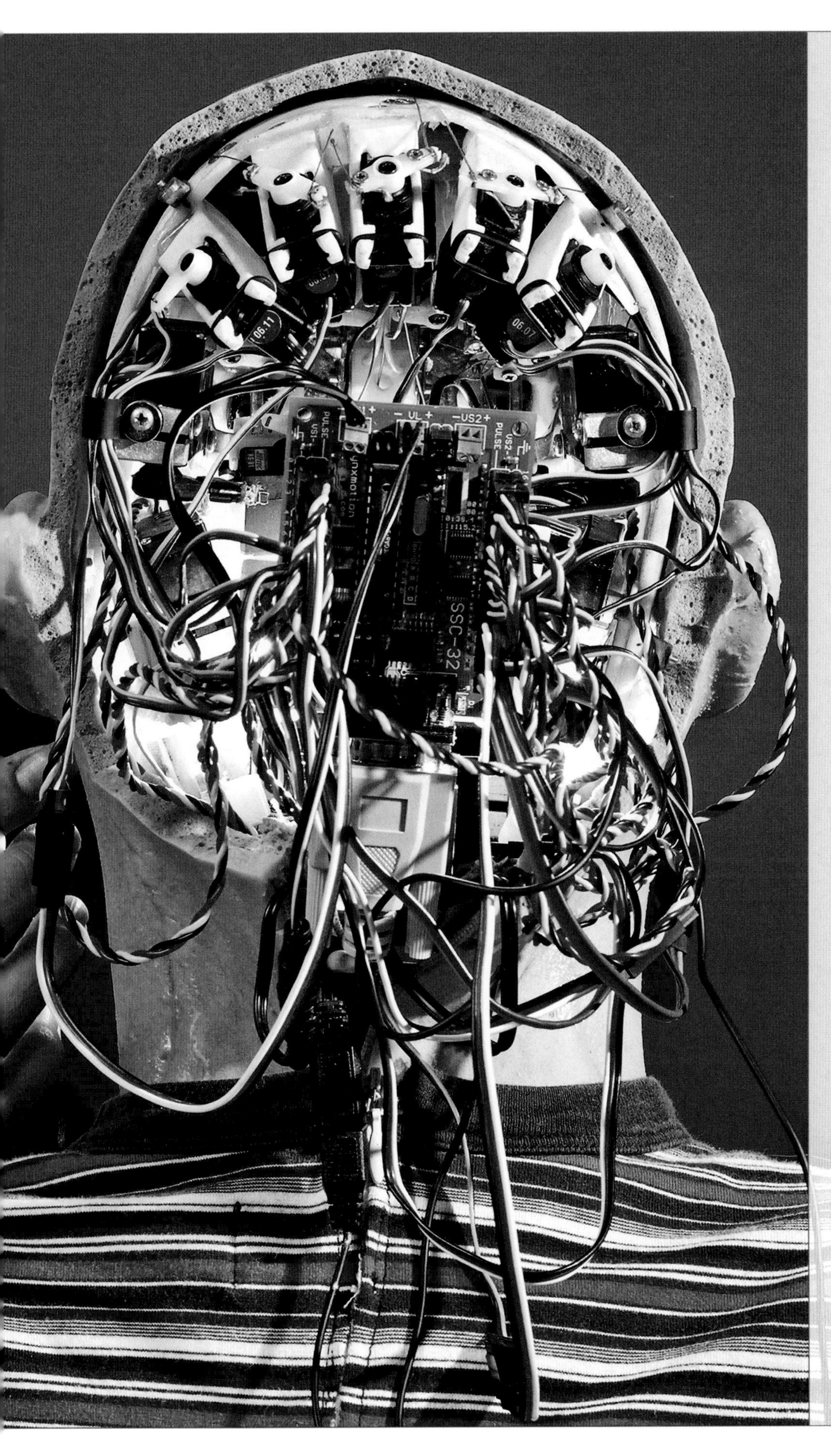

NEW INTELLIGENCE

THE PAST FEW years have seen startling advances in treatment of paralysis and Alzheimer's, memory modification, mind-machine interfaces, and prosthetic replacements for damaged eyes and ears. As the pace of discovery quickens, changes of the coming decades can only be guessed at.

DIFFERING VIEWS

Re-creating the brain's intelligence with computers is a controversial area as visions of the future of research on artificial intelligence tend to separate into two tracks. One path leads to a sort of heaven on Earth, the other to a purgatory of mankind's own making. The former comes from a group of scientists who foresee a utopian future when machines become self-aware and use the breadth, depth, and speed of their intelligence to attend to the Earth's many unresolved problems. The most prominent among these seers is Raymond Kurzweil, a pioneering researcher in artificial intelligence. In his 2005 book *The Singularity Is Near: When Humans Transcend Biology,* Kurzweil predicted that intelligent machines will reach a pivotal moment in their development in the year 2045, not only exceeding human intelligence but becoming capable of redrawing the plans for intelligent machines to make them exponentially more powerful. His use of the term "singularity" harks back to the publication of a paper in 1993 by science fiction writer and computer scientist Vernor Vinge, who likened the importance of the ever accelerating pace of technological change beginning in the late 20th century to the dawn of human life on Earth.

> **+ COMPUTER BIRTH +**
>
> AN IOWA STATE COLLEGE physics professor wrestling with how to build the first electronic digital computer drove his car to clear his head until he reached a bar 139 miles from campus. After getting a drink, John Atanasoff had his vision: The machine would run on base 2, use condensers, and compute by direct action. Atanasoff and graduate student Clifford Berry built a prototype in 1939 and working model in 1942. It weighed several hundred pounds.

The results of the singularity of artificial intelligence would be unpredictable, Kurzweil said, but he holds to the idea that intelligent machines will serve humanity rather than challenge it. Hybrids of biological and silicon-based intelligence will become possible, he said, and one day the contents of a human brain will be transferable into a metallic brain, like a CD-ROM uploading software into a computer. Such transfers would open the door to the possibility of functional immortality: Any time the brain (whether human or silicon) began to wear out in one form, it could be saved to another.

Some in the computer industry and at futuristic think tanks are not quite ready to plunk down the initial payment on an immortality machine. Foremost is Bill Gates, cofounder of Microsoft Corporation, who says nobody alive today will live to see the day when computers become sophisticated enough to pass for human. Also urging caution is Kevin Kelly, an editor at *Wired* magazine, who said, "People who predict a very utopian future always predict that it is going to happen before they die." Still, Kelly believes that someday—he's not sure when—the Earth's many computers may exhibit signs of intelligence developing through their myriad interconnections. It's the same idea that drives the plot of the *Terminator* movies: A military project becomes so sophisticated through its network of computers that it becomes self-aware and goes to war on the human race. That is one of the darker visions of the second track of humanity's continued exploration of the mind—a future in which science does more harm than good.

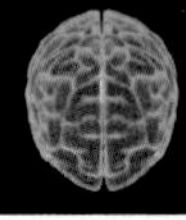

THE HISTORY OF AI

The idea of artificial intelligence dates at least to the early 19th century and included both demonic and benign forms in its inception. Mary Shelley wrote of the creation of artificial life in her 1818 novel *Frankenstein; or, The Modern Prometheus.* Transfixed by the processes that separated death from life, brilliant chemistry student Victor Frankenstein created a living man from inanimate matter, but then fled in horror from his own creation. His creature sought to be loved and accepted; it turned violent only when faced with continual rejection by the one who had made it. Three years after Shelley's book appeared in print, British inventor, mathematician, and mechanical engineer Charles Babbage created a design for the first Difference Engine, the dinosaur ancestor of the modern computer, which relied on moving parts instead of electronic circuitry.

Computers have gotten much more sophisticated since Babbage's day. Gordon Moore, a co-founder of the Intel computer chip company, famously described in 1965 how the speed of computer processing would repeatedly double itself every year or two. "Moore's Law" has come true; the world has grown accustomed to seeing sophisticated computerized devices arrive every year to supplant those that came before.

FAST FACT AI researcher Hugo de Garis names robotics as this century's key political issue.

A watershed occurred in 1997, when the IBM chess-specific computer Deep Blue, a high-performance machine that could

A microchip in a human brain interacts with surrounding neurons in an artist's vision of a functioning implant.

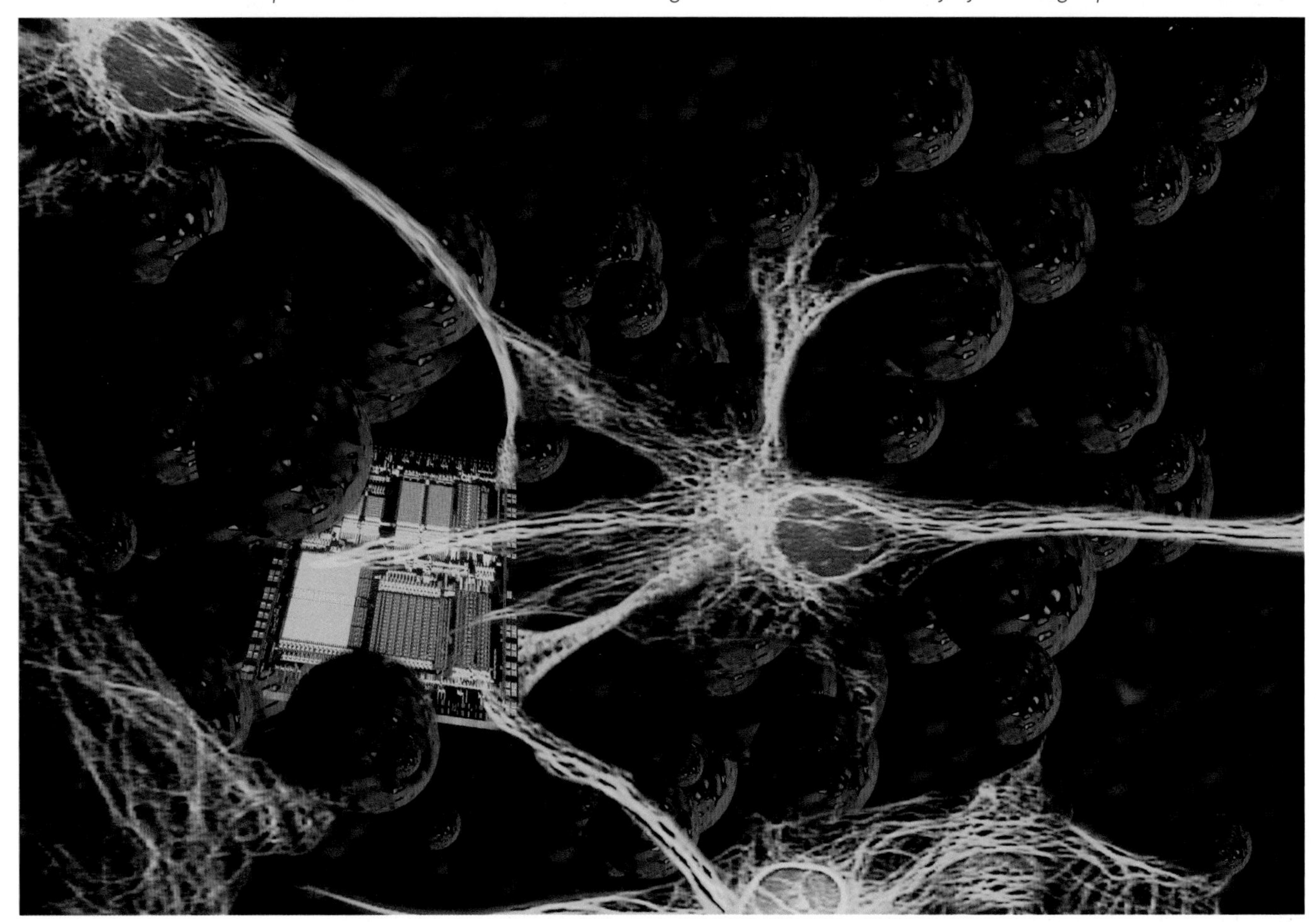

Two actors, one human and one a Mitsubishi Heavy Industries robot, perform at Japan's Osaka University.

explore up to 200 million positions in a second, defeated world chess champion Garry Kasparov. IBM said Kasparov could calculate only three positions per second. Fortunately, Deep Blue felt no satisfaction or joy—or at least, expressed none—at the defeat of the world's greatest human opponent.

COMPUTER-HUMAN HYBRIDS?

But could it? Today, agencies ranging from NASA to Google are looking at new applications for artificial intelligence. Creations capable of walking, talking, and interacting with people are making artificial intelligence one of the hottest ideas in the high-tech industry.

Human-machine hybrids serve as an intermediate step toward the dawn of some singularity. And they're already under way. Subretinal implants are being developed at universities and private corporations to restore some measure of sight to the blind. Myoelectrical prostheses, which pick up on tiny electrical signals in the skin of amputees' residual limbs, are making artificial arms and hands move with amazing precision. And cochlear implants have spread throughout the world in the past few years to improve the hearing of patients who not long ago would have had little or no hope of sound recognition.

COMPUTER-HUMAN INTERACTION

Computer programs are not only helping individual patients, they also are taking over greater and greater portions of human interactions. Robots are under development to track facial signs of expression that indicate state of mind and change their responses accordingly. Recently, a professor of mechanical engineering at Vanderbilt University created a robot that plays with autistic children and decodes their emotional states from data including heart rate, galvanic skin response, and temperature. Researcher Nilanjan Sarkar, who hopes to improve human-robot interactions, aims to help autistic kids learn social skills.

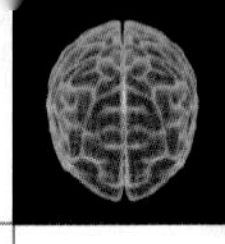

How far into the future will it be before computer holograms or sophisticated automatons take on the role of interacting with people in a store, government office, or business? Or a day when intelligent machines capable of gathering data from a medical patient, independent of the oversight of a doctor, spit out a diagnosis, prescription—and a bill?

Maybe that day is not far off. Rudimentary robot doctors have been at work for several years in a handful of hospitals, including Johns Hopkins Medical Center in Baltimore and the University of California at Davis Medical Center in Sacramento. Fleets of so-called rounding robots, produced by InTouch Health of California, visit patients in hospital rooms. Doctors working remotely use joysticks to send the 200-pound, 5-foot-tall robots to check up on their rounds. The robots have flat video screens that project the image of the inquiring doctor, and video cameras and microphones in place of eyes and ears. These robots don't replace human physicians but make it easy for doctors to gather information from far away. Half of the patients responding to a survey at Johns Hopkins said they would prefer a visit from a telerounding robot operated by their own physician to a visit from a real doctor whom they did not know well.

Robots that collect information at the command of humans are a far cry from artificially intelligent machines that could one day replace doctors themselves. Machines may evolve to take on tasks ranging from medicine to mechanical repair, but it isn't clear that they will supplant the creative power of the human mind.

So far, computers have excelled in doing simple tasks over and over, with great speed and singleness of purpose. The human brain typically lags in such abilities because it grows distracted and tired. The brain also fails to encode and recall information with the totality and clarity of machines. However, its higher functions were not designed to work by continually applying themselves to one task at a time. Nor does the brain work well when it retains or forgets everything it encounters.

The human brain excels in its power to adapt. It thinks, compares, learns, synthesizes, and makes new connections and new discoveries in ways that computers are only beginning to approach. Of course, that may be true only up to a point. The 2004 movie *I, Robot* addressed the emerging differences upon the approach of a singularly moment when the main character, a police officer who hates intelligent machines, confronts an exceptional robot. The officer asks with a sneer whether a robot could compose a beautiful symphony or paint a masterpiece. The robot replies, "Can you?"

FAST FACT Playwright Karel Capek (1890-1938) coined the word *robot* from *robota,* the Czech term for laborer. His play *R.U.R. (Rossum's Universal Robots)* had its debut in Prague in 1921.

+ THREE LAWS +

IN THE 20TH century, writer Isaac Asimov envisioned androids being controlled by "three laws of robotics": Robots (1) cannot injure a human or through inaction allow a human to come to harm; (2) must obey humans unless doing so conflicts with the first law; and (3) must protect themselves unless doing so conflicts with the first or second law. While many find these "laws" intriguing, they already are moot—robots have put some humans out of work, which is a form of harm.

SOCIAL SHIFTS CHANGING COMMUNICATION, CHANGING THE BRAIN

NEW TECHNOLOGY doesn't always lead to a better future, especially for the body and the brain, both of which extensive use of cell phones, the Internet, virtual reality, and an explosion of high-tech devices are changing. A 2009 British study published in *Biologist,* the journal of the Institute of Biology, linked a number of significant health risks to the decline in face-to-face social interactions. These include high blood pressure, stroke, narrowing of the arteries, a decline in immune system potency, and dementia. "One of the most pronounced changes in the daily habits of British citizens is a reduction in the number of minutes per day that they interact with another human being," said Aric Sigman, the author of the study.

Among his findings: fewer instances of people touching or hugging have led to a decline in levels of the neuropeptide oxytocin in the brain. Oxytocin has been linked to maintenance of a healthy heart, and its lack may correlate to higher incidence of cardiovascular disease. Mental functions also suffer when social experiences decline. Regular social interaction decreases the risk of dementia and increases cognitive functioning, the study found. "It is clear that this is a growing public health issue for all industrialised countries," Sigman said.

Communication through technology instead of face to face has become the most significant factor supporting society's growing physical estrangement, Sigman said. Both at home and elsewhere, people have become socially disengaged because of their earphones, mobile phones, and computers.

+ WINTER WONDER +

A VIRTUAL REALITY GAME called SnowWorld, designed for young burn patients, distracts them from their pain. Patients don a high-tech helmet that cuts them off from fear-inducing sights and sounds in the hospital. In their virtual world, they career along canyons of ice and throw snowballs at woolly mammoths, snowmen, and penguins. The mother of a six-year-old burn victim described the game as a lifesaver. "It doesn't totally take the pain away, but we know pain has a lot to do with your mind. And when your mind is centered on something else, it's as good as anesthesia," she said. SnowWorld was designed by Hunter Hoffman of the University of Washington. He chose virtual reality because young children cannot take many adult drugs, and a "cool" world of snow and ice offsets the sensations that burn victims suffer as their wounds receive treatment.

A GAMER'S MIND

The effect of video games on the developing mind of young children is uncertain. Positive effects may include the reinforcement of skills such as mastery of math, language, eye-hand coordination, and cooperation among multiple users. Harmful effects may include repetitive motion injuries, eyestrain, and social isolation. It's possible that playing computer games rewires users' brain to see the world in new kinds of ways. Gamers react to multiple variables and overlapping events, instead of working to solve problems linearly, one at a time. Development of integrated responses to complicated stimuli may cause video game players to have a more holistic view of causes and effects. Who knows? A holistic approach nurtured in the mind of video gamers may predispose them to solve the kinds of complex problems that dominate medicine, economics, and politics.

FAST FACT Atari's Battlezone (1980) is widely considered the first virtual reality game.

MARKETING & PROMOTION

Decreases in personal intereractions have yielded tactics for savvy marketers to take advantage of when promoting their products. With increased demands on our attention, it is even more important for advertisements to make a strong impact and stick in the brain. One

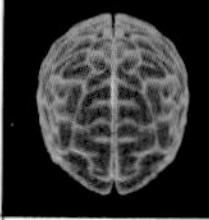

of the strongest tools, the earworm, harks back to the days of jingles. Researchers know certain kinds of songs get stuck in the head of the listener. Professor James Kellaris of the University of Cincinnati College of Business Administration describes the effect of these upbeat melodies with repetitive lyrics as a "cognitive itch" or "earworm." As many as 99 percent of the human race is susceptible to getting an earworm, he said. Kellaris's research intrigues advertisers and pop musicians looking for jingles and melodies that listeners will have a hard time forgetting, making them susceptible to marketing messages.

Marketing researchers are also turning to neuromarketing and using brain scans to dissect the subconscious feelings, desires, and motivations of people exposed to mass media advertising campaigns. A three-year study reported by Martin Lindstrom in his book *Buy.ology* revealed that purchasing decisions are much more emotional than rational. As the scans pinpoint positive reactions, advertisers tailor their pitches to take advantage of biochemical reactions in the brain to sell their products. The next product you buy may be consciously packaged to appeal to your brain at a subconscious level.

Image manipulation is another field where publicists and consultants are finding better ways to communicate their messages. In an experiment, photographs of political candidates were digitally morphed to incorporate some features from the faces of test subjects who were asked their reactions to the candidates. The photos, which included 60 percent of the politicians' faces and 40 percent of the

During the 2008 vice presidential debates, analysts monitored the candidates' every gesture and expression to see how it registered with the audience.

Residents of Lebanon talk on mobile phones. Much of new technology aims to enhance the power of communication.

respondents' faces, received favorable reviews. When combined with detailed demographic databases that can personalize appeals for money and votes, morphed images are predicted to have an increased role in political advertising.

GAMING & THE BRAIN

Virtual reality will continue to expand its role in education and entertainment. The University of Minnesota has embarked on a five-year study, financed by the National Institutes of Health, to examine how the brain learns by playing video games. The research team, led by psychology professor Paul Schrater, has found that games that require identifying and shooting targets improve specific neural skills such as shooting precision and peripheral vision. Video games generally improve the speed of processing stimuli from the senses and then responding. Fast-paced action games improve visual skills, including keeping track of multiple, independently moving objects, movement outside the focus of vision, and resolution of objects amid crowded scenes. Schrater warns that improvements don't happen overnight. "People that gain these skills play a minimum of five hours [a day] for a year or more," he said. "These skills are coming at costs in time."

It's not just game addicts who are learning through computer interfaces. In December 2008, the first revision of the U.S. Army field manual for training and full spectrum operations since the attacks of September 11, 2001, describes video gaming as a key way to replicate "an actual operational environment." Using computer technology to simulate a battlefield, soldiers can learn military skills without actually having to enter combat.

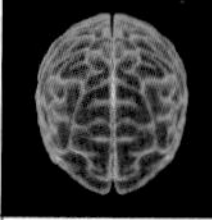

Potential recruits also get exposure to combat skills through free, downloadable video games such as America's Army, released in 2002. A third of cadets at West Point are believed to have played the game before entering the academy.

Department of Defense computer programmers also have developed software called DARWARS Ambush, a multiplayer game that shows the dangers of improvised explosive devices, ambushes, and other concerns related to the occupation of hostile countries such as Iraq and Afghanistan. Upgrades allow trainers to make the missions progressively more difficult and to give personalized feedback to players. Nonviolent components teach skills such as language and how to respond to culturally specific situations. Col. Casey Wardynski, who helped create the America's Army game, says it is getting results beyond game-specific abilities. A 2007 Rochester University study found a general improvement in cognitive skills among players who used the game a few hours each day for a month. With concrete results apparent, the Army has decided to institutionalize virtual reality gaming as part of its training.

+ VIRTUAL DOCTOR +

BLUECROSS BLUESHIELD of Hawaii began offering virtual hospital visits to members in 2009. Patients enrolled in the Online Healthcare Marketplace can log on at any time, type medical requests, and see a menu of specialists who can make a virtual visit. Physicians can communicate by phone, email, or videoconference. The system allows responding doctors instant access to their patients' medical records through a secure Web site.

FAST FACT

In 2009, the U.S. Army began operating islands in Second Life, a virtual platform with millions of users. Through their avatars, visitors to the Army's islands can learn about life as a soldier in a relaxed online environment.

SOCIAL NETWORKING

Virtual reality has the potential to reconnect people who have been cut off from social lives. While researchers have known for some time that social networking Web sites such as Facebook have attracted senior citizens with their ease of making friends and carrying on conversations, they now are investigating whether these virtual friendships provide the same sustaining benefits as flesh-and-blood relationships.

"One of the greatest challenges or losses that we face as older adults, frankly, is not about our health, but it's actually about our social network deteriorating on us, because our friends get sick, our spouse passes away, friends pass away, or we move," Joseph F. Coughlin, director of the AgeLab at the Massachusetts Institute of Technology, told the *New York Times*. New technology will help seniors stay connected because "the new reality is, increasingly, a virtual reality," he said. "It provides a way to make new connections, new friends and new senses of purpose." Internet networks create feelings of empowerment because elderly users can take control of their social lives without having to ask someone to drive them or help them. In addition, the tasks of posting to Web sites and communicating with friends online challenges minds that otherwise might not receive such stimulation.

VIRTUAL REALITY

The elderly and disabled could benefit from the development of brain-computer interfaces that give them more physical freedom. Technology under development in Europe allows people to turn on lights, change TV channels, and open electrically controlled doors just by thinking about the actions. Futuristic "smart homes" could be wired to take advantage of the technology. So far, users must wear EEG caps to monitor their brain waves and identify the distinctive patterns associated with thoughts about performing tasks. Mind-controlled electric wheelchairs, with users training in virtual reality before taking the real thing for a spin, are a possible offshoot of such technology.

BETTER BRAINS

IMPROVING THE MIND THROUGH SCIENCE

ONE OUTGROWTH of the new neuroscience is greater understanding of how cognitive functions can be enhanced. It is no longer a fantasy to think of drugs that could improve memory, intelligence, or some other mental ability. Such drugs are called nootropics, a word coined by psychologist Corneliu E. Giurgea in 1964 by combining the Greek words *noos* ("mind") and *tropein* ("to turn or bend"). They change the availability of neurochemicals in the brain, increase the amount of oxygen available to neurons, and stimulate the spread of neural connections.

The Pentagon gave IBM nearly five million dollars in 2008 to try to mimic a cat's brain.

Four mental capacities in particular are targeted by pharmaceutical companies that are researching nootropics: concentration, focus, memory, and mental endurance. Improvements in focus and concentration would keep the mind on task, minimizing distractions and maximizing the benefits of attention. Memory-improving drugs would enhance encoding of memories and help strengthen their links to previous information, effectively expanding the brain's storehouse of knowledge.

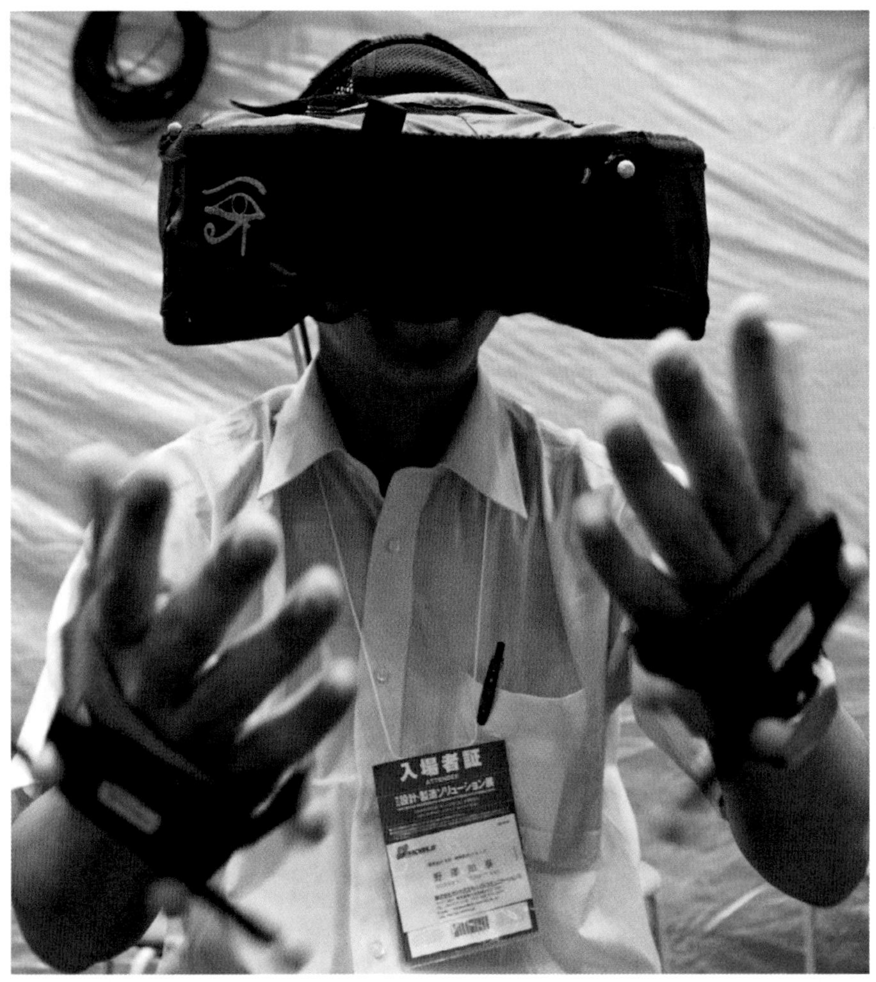

A three-dimensional virtual reality system gets a tryout at an expo in Tokyo in 2008.

RISKS OF IMPROVEMENT

While drugs have been available for years to improve these brain functions, they have side effects that may compromise their use. For example, stimulants such as amphetamines improve cognitive functions but require larger and larger doses to attain the same effects over time. New generations of stimulants that lack the worst side effects are likely to become popular in the next few decades. (Currently, modafinil—trade name Provigil—is allowing military personnel, long-distance drivers, and students to stay awake and alert for long periods of time.) Their use overrides the normal, diurnal cycle of sleeping and wakefulness, leading to unusually prolonged concentration. Such drugs convey a competitive edge, and isn't hard to imagine the use of attention enhancers expanding in the workplace, as employees who are angling for an advantage over their co-workers in a highly competitive economy turn to medications that permit them to do better work for longer periods. One possible consequence would be the further entrenchment of a workaholic culture in which people have less time, or less incentive, to relax, spend time with friends and family, and naturally recharge their reserves of energy.

ETHICAL ISSUES

As drugs continue to evolve, one can imagine a day when they will artificially enhance mental performance in ways similar to the impact of steroids and other drugs that boost energy and strength among athletes. Should students, for example, be allowed to take such drugs before sitting down to an entrance exam for graduate school? Neurologist Richard Restak says he favors maintaining a "level playing field" by strictly controlling the use of nootropics in competitive mental exercises, such as nationalized standardized tests. "Students are already required to show proof of identity prior to the exams, so why not also require proof that their performance on the exam isn't going to be influenced by cognition-enhancing drugs?" he asks. Certain drugs would be exempt, including prescription medications for students with attention deficit disorders, he says, but otherwise students could submit a saliva sample when they picked up their test packet.

In the next few decades, nootropics also are likely to expand their entry into the realm of mood control and memory. Drugs could eliminate unwanted feelings, such as grief, guilt, and shame. Or they could improve desired feelings, such as confidence. A nootropic pill may someday actually do what public radio host Garrison Keillor long has touted as one of the best attributes of his imaginary brand of biscuits: giving shy people "the strength to get up and do what needs to be done."

+ CUP OF JOE +

CUTTING-EDGE RESEARCH is developing new medicines every day, but something as old-fashioned as a cup of coffee could be the next big thing for Alzheimer's patients. In July 2009, the *Journal of Alzheimer's Disease* published findings from the Florida Alzheimer's Disease Research Center, which found that caffeine significantly decreased levels of the protein linked to Alzheimer's disease in the brains and blood of mice exhibiting symptoms.

READING BRAIN SCANS

Instead of artificially stimulating particular states of mind through nootropic drugs, brain-scanning techniques such as fMRI could someday be enlisted to detect desirable personality qualities for particular jobs. Findings of a high social intelligence would be useful for jobs requiring an extroverted personality, such as sales clerk. On the other end of the personality spectrum, scans could select the best personality types for combat roles in the armed forces.

Such scanning is already possible. Tests carried out by the National Institute of Mental Health examined neural circuit activation levels in Special Forces soldiers as they examined pictures of faces. The soldiers believed they were trying to tell the sex of the people in the pictures. In reality, their fMRI scans measured the activation levels in brain regions associated with strong negative emotions as they witnessed faces registering anger and fear. Compared with volunteers in a control group, the soldiers exhibited higher activation levels in the amygdala, a site linked to fear, as well as the inferior frontal cortices and anterior cingulate regions, both associated with emotional regulation. The pairing of emotion and control suggests the soldiers reacted strongly to the pictures but then mastered their emotions, as would be beneficial in the chaos and trauma of combat. The implications are significant for creating the most effective troops: Applicants for special training could be prescreened for the emotional aptitude required for the job.

SELECTING MEMORIES

Soldiers also would be good candidates for treatment by memory modification. The discovery that recalling memories alters them as they are returned to long-term storage holds out hope that psychologically damaging memories, such as those associated with post-traumatic stress disorder, could be given effective medical treatment. Scientists reported in the online edition of *Nature Neuroscience* that if painful memories could be altered when called up from storage, they might lose their emotional punch when recalled at a later date. That's a tall order, given that memories encoded with strong emotional content, such as those associated with death and loss, are among the most powerful. "Even the most effective treatments [for PTSD] only eliminate fearful responding, leaving the original fear memory intact, as [shown by] the high percentages of relapse after apparently successful treatment," the *Nature Neuroscience* article said. "Once emotional memory is established, it appears to last forever."

Beta blockers, widely used as anti-anxiety drugs because of their ability to interfere with the effects of adrenaline and thus lower blood pressure, have shown promise in eliminating strong emotional content from memory formation. In a 2002 pilot study at Harvard Medical School, a team led by Roger Pitman gave the beta blocker propranolol to trauma victims. Those victims exhibited fewer PTSD symptoms three months later than did a control group that received a placebo. Facts could be dredged up from memory as easily as before, but they did not reactivate the initial emotional response. Therapeutic applications are obvious: Victims of violence could shed themselves of the emotional baggage of their ordeals but, under the proper treatment, retain knowledge of essential details that would be useful for criminal proceedings, litigation, or military intelligence.

Daniel Sokol, a lecturer in medical ethics at St George's, University of London, noted that removing bad memories "will change our personal identity since who we are is linked to our memories. It may perhaps be beneficial in some cases, but before eradicating memories, we must reflect on the knock-on effects that this will have on individuals, society and our sense of humanity."

+ 3-D PATIENT +

SCIENTISTS AT THE University of Calgary have created a giant computer-animated human to help physicians visualize how the body metabolizes medicine. Inside the university's $1.5 million virtual reality room, visitors wearing special 3-D goggles can watch an aspirin as it is swallowed and absorbed by the bloodstream, goes to work, and is flushed out of the body via the kidneys. Biochemist Christoph Sensen and colleagues named their "patient" CAVEman, for Automated Virtual Environment. It models 3,000 body parts via high-definition holography. Sensen envisions a future in which doctors could wade into a computer projection of a patient's internal organs to analyze a 3-D rendering of a cancerous tumor. Or, he said, computer simulations could model the progress of Alzheimer's disease or diabetes in detail, helping scientists find cures.

BETTER RECALL

Work on memory enhancement also is proceeding at universities on both coasts. Research led by Theodore Berger at the University of Southern California and Samuel Deadwyler at Wake Forest University in North Carolina has focused on creating an artificial hippocampus in lab rats. The hippocampus, crucial to the formation of new memories, suffers damage in Alzheimer's patients. Electronically bypassing the hippocampus through the use of a prosthetic implant holds out hope of restoring the ability to make new memories in brain-damaged patients. If the project succeeds, it could lead to enhanced memory and greater understanding of the mechanics of higher cognitive functions.

In their rat experiments, Deadwyler and Berger temporarily stunned the rats' hippocampus with drugs. When the rats pressed a bar, it sent an electrical charge through a microchip that they believe sent out the same electrical signals that

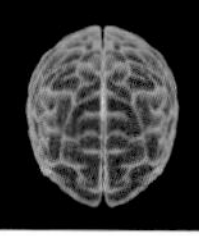

War veterans, such as injured Gunnery Sgt. Kenneth Sargent, might benefit from trauma research that potentially could include memory editing.

the stunned hippocampus would have. They hope to learn more about how memories are formed. Greater understanding of the means of encoding information could lead to downloading of information from machine to brain—the kind of memory upgrades suggested in the science fiction movies *Total Recall* and *The Matrix*. In the former, false memories implanted by a computer give one the illusion of having gone places and done things, creating virtual vacations or visions of other worlds. In the latter, enhanced physical skills are available as neural plug-ins containing the proper combinations of storage and recall.

Some researchers believe such knowledge must await not only deeper understanding of the brain but also the development of more sophisticated computers. Henry Markram, a neuroscientist at the Swiss Federal Institute of Technology in Lausanne, heads the Blue Brain Project, a supercomputer-based exploration of the brain's structure that began in 2005. The project aims to reverse-engineer the brain at the level of individual cells and molecules. It started with rat brains with the goal of moving on to the more cortically enhanced human variety. Teasing out the neural functions that arise from molecules and cells in a computer model most likely won't be possible until computers are available with a thousand times the current supercomputers' best processing power. Markram looks forward to the wonders that await: "I think there will be a conceptual breakthrough that will have significant implications for how we think of reality," he said.

BRAIN MEDICINE

ADVANCES IN TREATMENT

ADVANCEMENTS in treating mental disorders will emerge from the precision with which the brain will be scanned and understood. Imaging techniques will tease out moment-by-moment processes in the firing of neural circuitry. Chemical reactions in the synapses and the neurons will become more fully understood. Already, some imaging techniques are not only expanding the understanding of neuroscience but raising intriguing questions about the future.

MAGNETIC FUTURE

Magnets pose interesting possibilities. The brain creates its own tiny magnetic fields. A new, noninvasive process called transcranial magnetic stimulation, or TMS, subtly alters the brain's magnetic processes without resorting to drugs or surgery. Coils of magnets in the shape of a figure eight are placed on the scalp next to the region targeted for stimulation. As the magnets are turned on and off repeatedly, they can inhibit or excite nearby

A confocal light micrograph depicts the blood-brain barrier, which screens out many drugs that otherwise could treat tumors.

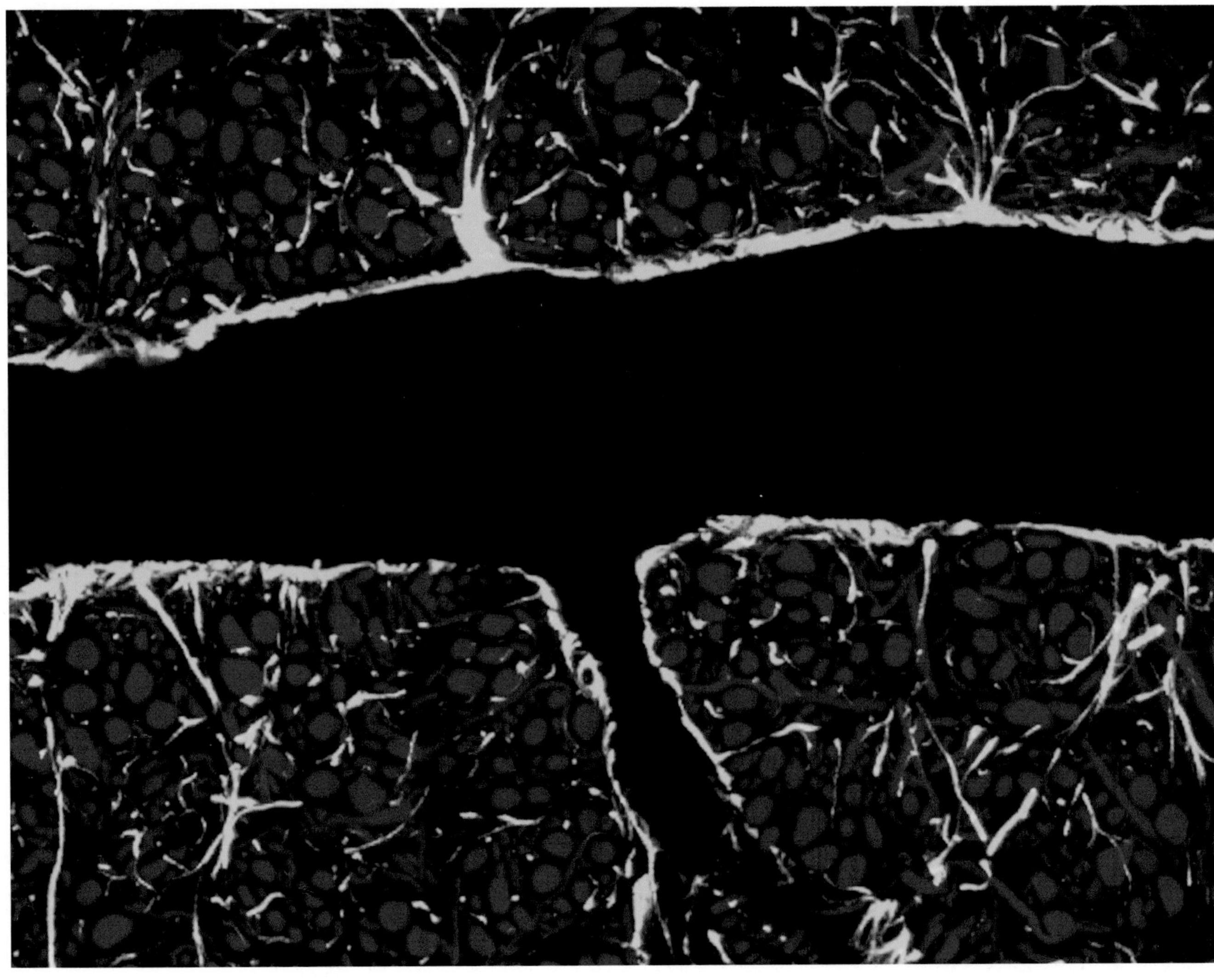

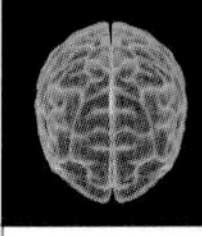

synaptic activity. When high-frequency bursts occur more than five times a second, they excite activity in the cerebral cortex and hold out promise for alleviating the effects of depression, Parkinson's disease, and other neurological disorders without the side effects of electric shocks or powerful drugs. Neurons in the left prefrontal cortex, where electrochemical activity tends to be dampened among the depressed, have sparked with greater fire in response to TMS.

Scientists at the National Institute of Mental Health also have found that magnetic treatments directed toward hyperactivity in the frontal lobes can lessen the grip of obsessive-compulsive disorder. At lower frequencies and slower pulses, TMS slows instead of hastens cortical activation. It may therefore work to ease the symptoms of illnesses such as the electric storms of epilepsy and the hallucinations of schizophrenia.

When applied to the prefrontal lobes, TMS has been shown to enhance the speed and agility of cognitive processing. The TMS bursts act like a localized jolt of caffeine, but nobody knows for sure how the magnets actually do their work. Some scientists suspect that high-frequency magnetic energy may boost the firing rate of brain cells, prompting them to more quickly and easily form the connections required for solving problems or learning new skills. Or it's possible that the magnets merely increase blood flow and thereby make more oxygen available to individual cells.

+ ROBORAT +

IN 2002, physiologist John K. Chapin unveiled "Roborat." The rat carried a radio antenna, a microprocessor, and a video camera. Radio signals stimulated areas of the rat's brain sensitive to left- and right-whisker touch, as well as pleasure circuits of the medial forebrain bundle. Controllers watching its video feed trained the rat to turn left or right by simulating whisker sensations and then rewarding the rat's pleasure circuits.

IMPLANTS

Mechanical implants in brains are likely to provide an expanding, alternative method of boosting brain power and treating symptoms of disease. Just like computers, microscopic implants are expected to get smaller and more powerful. Radio-powered microchips may be able to fire up long-dormant nerve fibers and restore movement to paralytics; silicon simulacra deposited next to damaged neurons in brains and spinal cords may be able to repair them or replace signals no longer produced in deadened regions.

Cochlear implants already help restore hearing lost through the disappearance of sensitive hair cells in the inner ear. Now, prosthetic devices are beginning to do similar work for vision. William Dobelle, who created an artificial vision system at his Dobelle Institute in New York City, implanted electrodes inside the brain's visual regions. The electrodes communicate with a tiny digital camera mounted on a pair of eyeglasses and a portable computer. Light patterns gathered electronically through the glasses get converted directly into stimuli that can be unscrambled at some level by the visual cortex. A patient of Dobelle's named Jens, blind as a result of two accidents, had enough visual sensation restored that he could move through rooms and navigate a car around trash bins in a parking lot at the institute's lab. Dobelle, who died in 2004, foresaw refinements that would allow blind drivers to venture into real traffic. He said he designed the system to improve patients' mobility but expected rapid advances to "provide the possibility that the patients will be able to scan the Internet and watch television."

The U.S. Army has given scientists at the University of California at Irvine a four-million-dollar

FAST FACT Probably fewer than 35 percent of people with Alzheimer's disease or other dementias have that specific diagnosis appear in their medical records, forestalling efforts for appropriate treatment.

grant to explore brain-machine connections that would establish "synthetic telepathy" and could have military or medical applications. According to the university, the interface would use a noninvasive imaging technology such as electroencephalography to translate thoughts into messages. For example, a soldier could think about a particular message. Computer technology would detect the brain waves involved in forming the thought and then decode them. The decoded thoughts then would be transmitted toward the intended receiver of the communication.

The system would take extensive training to be effective. At first, it would require communicators to focus on simple messages that the system would be taught to recognize—"Go," "Fire," "Help," and so forth. As it develops, the interface might be able to handle more complicated communications. It also could be adapted to help stroke and paralysis patients. When motor skills involved in speech have been compromised, a system that verbalizes thoughts would reestablish communication and reconnect patients with their world.

FAST FACT The ancient Greeks believed that lodestones, or magnetic rocks, had medicinal powers.

Simple communications sent by brain waves directly into computers have already begun. A University of Wisconsin biomedical engineering doctoral student named Adam Wilson took it as a personal challenge when he heard someone on the radio say how cool it would be if you could post messages to the Internet social networking site Twitter simply by thinking. In spring 2009, Wilson demonstrated he could do just that during an interview with CNN. Wearing a red cap fitted with electrodes, Wilson posted two messages, or "tweets," to Twitter through the power of thought. "Go Badgers," said the first, a reference to the Wisconsin mascot. The second said, "Spelling with my brain." Wilson accomplished the feat by concentrating on one letter at a time to "type" his message, then posted it to the Internet by focusing on the word "twit" on his computer screen. It took only a few days to write the software to connect the existing technology of reading brain waves to a Twitter interface. Wilson sent his first brain-created tweet to his adviser, assistant professor Justin Williams.

Posting messages to a social communication network through the power of concentration is more than a neat trick. It could be adapted to help paralytics, such as those with "locked-in syndrome," communicate. People with amyotrophic lateral sclerosis (ALS), a brain stem stroke, or a high spinal cord injury would benefit.

NEW CELLS

Like tiny mechanical implants, genetically engineered cells are treating a variety of brain disorders, and no doubt will do even more in the future. Designed to produce agents to speed or retard biochemical reactions in the brain, these cellular implants help the brain and body heal themselves. Encouraging results are emerging for treatment of cancers, eye diseases, spinal cord injuries, blood and muscle diseases, Parkinson's disease, and ALS, also known as Lou Gehrig's disease. The process involves taking cells from patients, changing a tiny bit of their genetic code, and then returning the cells to their host. Several treatments have advanced to human trials and produced encouraging results.

Alzheimer's disease patients are among those likely to benefit from cellular implants. Researchers at Harvard Medical School in 2007

+ VISION IMPLANTS +

RESEARCHERS AT THE University of Utah, led by ophthalmology professor Richard Normann, are refining and shrinking implants in the brain's visual cortex to restore vision of those with damage to their retinas or optic nerves. Their goal is to create visual networks sophisticated enough to read print. They believe human-engineered methods of treating profound blindness will move more quickly to clinical trials than biological approaches.

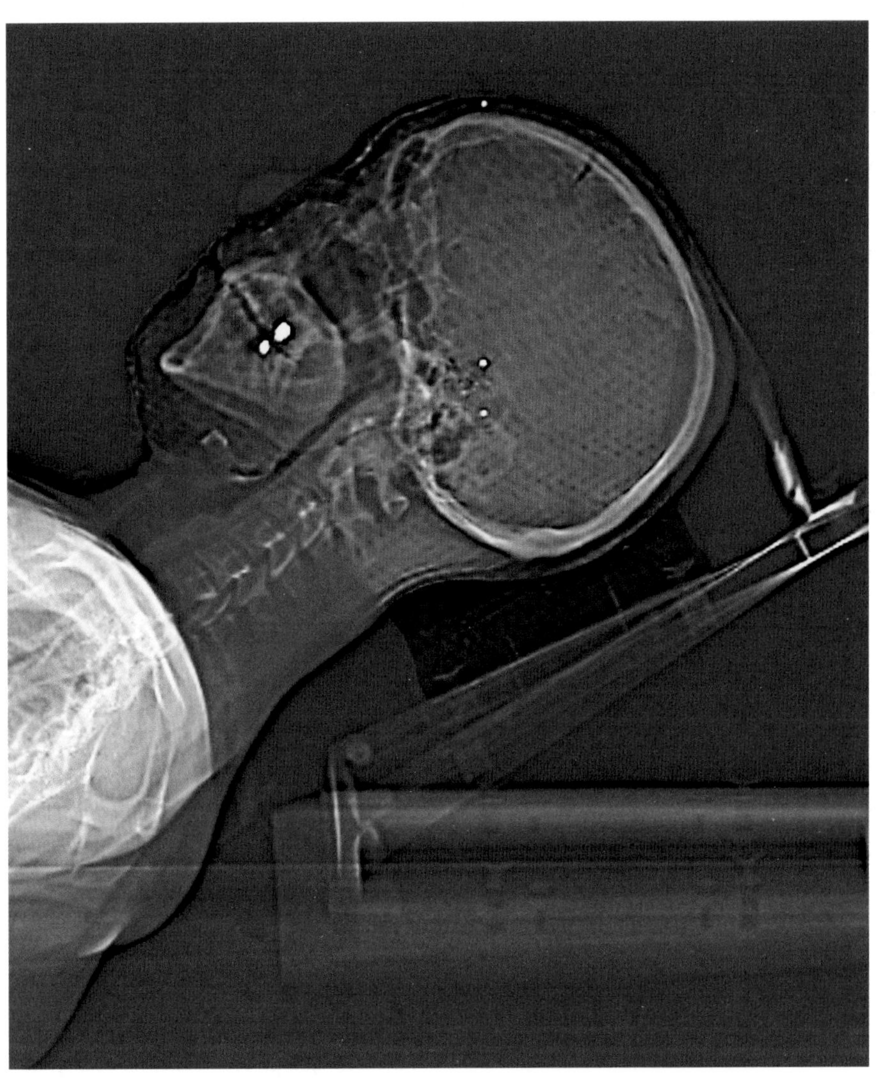

A colored CT scan reveals how radiotherapy is used to treat a brain cancer.

implanted a human gene that made mice rapidly develop the telltale neural plaques of Alzheimer's disease. When the mice received implants of cells genetically altered to produce an enzyme called neprilysin that targets the beta-amyloid compounds that form the plaques, the plaques dissolved and left the mice's brain clear. The implants caused no observable adverse side effects, the Harvard researchers said. Whether the technique might work in humans is unknown. The human brain is much larger than the mouse brain, and would require much higher concentrations of the enzyme-producing cells to cover the greater volume. One option for solving that problem is to place the re-engineered cells where they would circulate the amyloid-dissolving chemicals into the cerebrospinal fluid that bathes the brain or the blood that circulates to every cell.

BETTER DIAGNOSES

Other breakthroughs appear on the horizon for Alzheimer's. Soon, a person who wants to know the likelihood of getting the disease will take a simple blood test to search for markers, much as cholesterol tests today can indicate heightened risk of heart disease. Early identification of the likelihood of getting Alzheimer's would set up the patient to get drugs and therapies to slow or stop the mental decline associated with it. Vaccines to prevent the accumulation of amyloid plaque in the brain are possible, too. An experimental vaccine called ACC-001 spurs the body to create antibodies to attack beta-amyloid. If the body treats harmful plaques as invasive agents, it could target them for removal by activating its own immunity system. Other vaccines aim to directly remove the plaques from the brain.

The Blanchette Rockefeller Neurosciences Institute announced in May 2009 that it hoped to have a commercial version of an early Alzheimer's detection test available in 12 to 18 months. A home test might be available in three years. The test requires a collection of skin cells. The cells are grown in a glass dish and combined with an enzyme that signals the presence of Alzheimer's. The test proved to be 98 percent accurate upon autopsy, which is the only surefire way to diagnose the disease.

NEXT STEPS

A JUMP INTO THE UNKNOWN

RAYMOND Kurzweil, the futurist who predicts a singularity of human-machine interaction by the year 2045, expects tiny mechanical devices and software programs to greatly improve the function of the human brain. He takes as his starting point the world created in the trilogy of *Matrix* movies, which envision sentient computer programs, virtual reality indistinguishable from the real world, and the ability to download directly to the brain such skills as martial arts and the ability to fly a helicopter.

Kurzweil expects all of those developments to occur in the next three to four decades. He believes the foundation is being laid now, as computing power continues its explosive growth. Within two decades, he says, scientists will be able to reverse-engineer the brain and duplicate its functions in silicon, and computers will be as intelligent as people and virtually indistinguishable from their biological creators. And the advances will continue. "By 2050, $1,000 of computing will equal one billion human brains," he said.

DIRECT INPUT

He expects computers to have the capability of performing the tricks seen in *The Matrix*—perhaps even better than onscreen. In the movies, the hero has a thick cable inserted into his brain stem to download information. More likely, Kurzweil said, data could be downloaded by wireless connections or through the work of nanobots, microscopic machines that could enter the bloodstream and interface with the brain from the inside.

+ TURING TEST +

BRITISH MATHEMATICIAN and computer scientist Alan Turing proposed a futuristic "imitation game" in 1950 as a way to determine when machines have gained intelligence. One person puts questions to a machine and a second person, both hidden. They answer via text. After five minutes, the interrogator must declare which respondent is the human. Machines will win the Turing Test when the questioner cannot tell.

The microtechnology isn't ready yet, but researchers are on the verge of practical development. Already, the U.S. Department of Defense is at work on "Smart Dust," a cluster of robots one millimeter in diameter that can be dropped from a plane to perform reconnaissance on the ground. The aim is to create devices that gather information, communicate with each other, and share their intelligence with their controller at a remote site. Similar devices for deployment inside the bloodstream also are under development. If such nanobots could be deployed inside the brain and scan and communicate with an external computer, they could provide precisely detailed brain scans and interact with the brain at the cellular level.

Kurzweil foresees not only communication between implanted nanobots and external machines but also between nanobots and the brain itself. Neuron transistors now have the simple ability to detect the firing of a neuron or to cause it to fire. More sophisticated versions may take up positions inside the brain and create a form of virtual reality every bit as "real" as the outside world by blocking stimulations from the senses and replacing them with other signals. When such virtual reality is hooked to a computer network, it could be shared. It may not be a fantasy for groups of people to feel as if they were actually living inside computer games. Kurzweil sees no reason why the nanobots couldn't enhance emotional experiences or alter perception of one's own body.

TALKING TO THE BRAIN

Finally, Kurzweil envisions the ability to transfer knowledge from machines to the brain, and vice versa. Hybrid minds of neuronal and electronic components may expand the number of connections in the brain beyond the limits of biology. Under this scenario, a person who wants to learn to speak French would merely access the lexicon and rules of grammar as stored in a data bank. Or he or she could download the knowledge and motor skills to fly the attack helicopter that saves the hero in the first *Matrix* movie.

PREDICTIONS

Kurzweil's scenarios peer a few decades into the future. What lies beyond that? Evolutionary biologist Richard Dawkins says it's the question he is asked most frequently, and his short answer is: Nobody can make a clear prediction about the future of the body and brain. Genetic modifications, biomechanical implants, human impact on the planet, and unforeseen variables in human evolution make the prediction just too murky to qualify as science.

Undaunted, MSNBC science editor Alan Boyle asked scientists to project possibilities based on observable trends. Biological changes evolve in our body in response to the continued spread of toxins in the biosphere. Poisons found in pesticides and industrial solvents have hastened biochemical reactions in humans, bringing on higher incidences of breast cancer and lower sperm counts, as well as earlier entry into puberty. Might humans evolve in response to the spread of toxins that mimic hormones and neurochemicals?

FAST FACT American Film Institute rates *2001: A Space Odyssey's* computer, Hal, a top movie villain.

In an artist's rendering of some future operation, mechanical driller nanobots clear an artery of a blood clot.

Another possibility is the benevolent enhancement of the brain and body. "If you look at the superheroes of the '30s and the '40s, just about all of the technologies they had exist today," said Joel Garreau, author of *Radical Evolution*. If enhancements become common, Garreau foresees the possibility of humanity splitting into three groups: those who can afford enhancements and choose to use them; those who can afford them but eschew their use for moral reasons; and those who cannot afford access to enhancement. Such a scenario could create a world of haves and have-nots, as well as all of the political, ethical, and religious turmoil that would imply. Possible improvements might include treatments that could keep workers in their prime well past the age of 100—at a cost of opening up jobs for younger replacement workers.

A POSTBIOLOGICAL ERA

Others believe that humans will evolve in tandem with robots. If robots become extremely sophisticated, the question may arise of who might assimilate whom. Artificially intelligent robots would most likely have the wherewithal to reproduce and improve themselves. That might lead to the evolution of two separate species, one human and the other mechanical. Or robots may interact with humans in ways that improve human life.

Futurist Hans Moravec chooses the word *postbiological* to describe the key evolutionary changes of the coming century, when machines will evolve into beings as complex as people, and then move further into some transcendent state that science can only guess at. Humans should be proud when such machines refer to themselves as our descendants, he said.

"Unleashed from the plodding pace of biological evolution, the children of our minds will be free

Evolution's path from past to present leads to an uncertain future, but one likely to include sophisticated, mechanical children of the human mind.

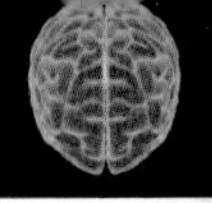

to grow to confront immense and fundamental challenges in the larger universe," Moravec wrote in the 1988 book *Mind Children: The Future of Robot and Human Intelligence.* "We humans will benefit for a time from their labors, but . . . like natural children, they will seek their own fortunes, while we, their aged parents, silently fade away."

FAST FACT University of the South Pacific algorithms let robotic cars "talk" to avoid collisions.

Creating artificially intelligent offspring that may someday supplant the human race doesn't bother Moravec; to choose not to develop such mind children would be the same as humanity turning its back on its potential, and "by my standards we would be tragic failures," he said. Given proper control, humanity can oversee the bounds of robotic so their fundamental motivation is to serve. "Like dutiful children caring for aging parents, these machines could provide a long, luxurious retirement for biological humanity. Few people will object to being offered steadily greater wealth for steadily less labor," Moravec said.

THE VIEW FROM HERE

Science has the power to improve the brain and body. The goal is to help people live lives that are healthy, meaningful, and long. The shape of technology that will help get humanity closer to that goal, and its impact on the shape of life itself, aren't clear. But neuroscience stands prepared to explore and eventually understand vast, uncharted regions. Over the next several decades the brain will go through amazing modifications, with technology driving each discovery. Humanity's challenge will be to harness the power of change while respecting the unique power of each mind. And while it has inspired both lofty visions and nightmares of the future, science ultimately will take the human brain places we cannot yet imagine.

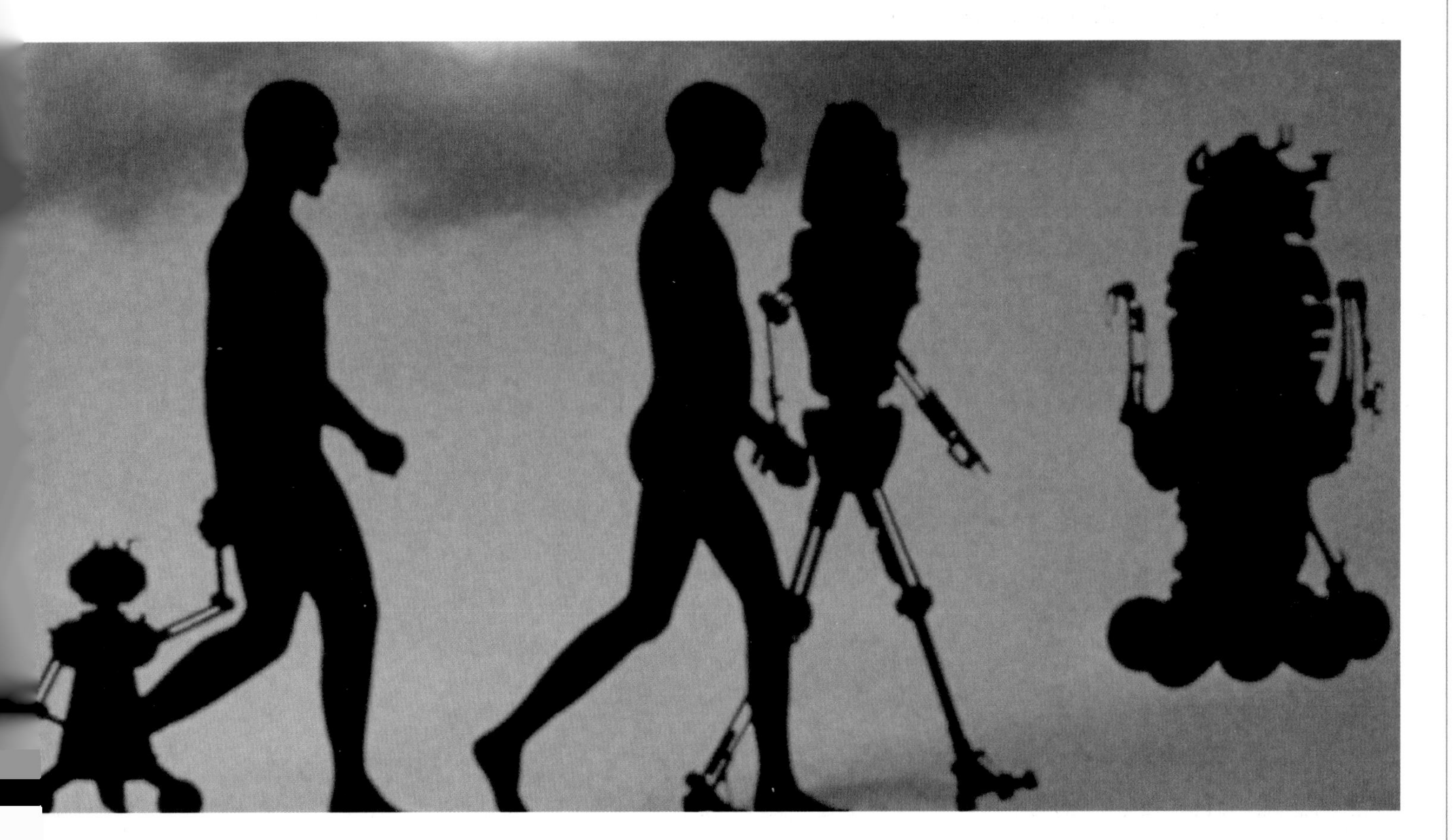

GLOSSARY

ABSENCE SEIZURES. Mild epileptic seizures in which consciousness is lost and facial muscles twitch briefly. Most occur in young children and disappear by age ten. Formerly known as petit mal seizures. (Chapter Two)

ACETYLCHOLINE. Neurotransmitter that causes muscles to contract. (Chapter Five)

ADRENAL GLANDS. Produce hormone and regulate metabolism and blood flow and volume. Play an active role in the fight or flight response. (Chapter Three)

ALIEN HAND. A rare neurological condition in which one hand acts independently of an individual's conscious control. (Chapter Six)

ALPHA WAVES. Brain waves common at times of relaxed alertness. Important state for learning and using new information. (Chapter Six)

ALZHEIMER'S DISEASE. The most common cause of dementia, primarily affecting memory, thinking, and reasoning. Nearly all brain functions are affected. (Chapter Nine)

AMUSIA. Neurodegenerative condition affecting the ability to understand or express music. Often occurs with aphasia. (Chapter Four)

AMYGDALA. An almond-shaped section of the forebrain. This component of the limbic system plays a central role in response to fear and terror. (Chapter One)

ANTERIOR CINGULATE CORTEX. Region of the brain associated with error recognition and inhibition. (Chapter Seven)

ANTERIOR TEMPORAL CORTEX. This region of the temporal cortex stores facial memory and plays a key role in facial recognition and identification. (Chapter Eight)

ANTEROGRADE AMNESIA. Loss of the ability to create memories of events and experiences following a trauma that causes amnesia. (Chapter Eight)

APHASIA. Neurodegenerative condition affecting the ability to understand or express language, generally as a result of stroke or similar brain trauma. (Chapter Four)

APRAXIA. Disorder of the nervous system in which an individual is unable to preform learned coordinated muscle movements although both muscles and senses work properly. (Chapter Four)

ARACHNOID. The weblike middle layer of the three membranes surrounding the brain and spinal cord. (Chapter One)

ARCUATE FASCICULUS. A bundle of nerve fibers connecting Broca's and Wernicke's areas. (Chapter Eight)

ASCENDING RETICULAR FORMATION. Part of a branch of interconnected nuclei located in the brain stem. Responsible for the waking state. (Chapter Six)

ASSOCIATIVE LEARNING. A process in which learning occurs by associating an action with its consequence. (Chapter Eight)

ATHEROSCLEROSIS. A blood vessel disease characterized by hardening of artery walls due to the buildup of fatty deposits. (Chapter Nine)

ATHETOSIS. A movement disorder, linked to an overactive basal ganglia, characterized by slow continuous movement. (Chapter Five)

ATTENTION DEFICIT HYPERACTIVITY DISORDER (ADHD). A common disorder in which individuals have difficulty concentrating and multitasking and are easily distracted. Affects 4 to 5 percent of children and may continue into adolescence and adulthood. (Chapter Three)

ATTENTIONAL BLINK. The brain's inability to detect a new target object flashed milliseconds after the first. Generally observed during testing involving rapid presentation of visual stimuli. (Chapter Six)

AUTISM. One of a group of developmental disorders characterized by significant communication and social interaction impairments and unusual interests and behaviors. (Chapter Seven)

AUTONOMIC NERVOUS SYSTEM. Involuntary nervous system. Consists of visceral motor fibers that activate the heart, digestive tract, and glands. (Chapter Two)

AXON. The electrically sensitive fiber of a neuron responsible for the transmission of information away from the nerve cell. (Chapter One)

AXON TERMINALS. The knoblike endings of the terminal branches of an axon. (Chapter One)

BASAL GANGLIA. A group of nuclei consisting of the caudate nucleus, putamen, and globus pallidus. Believed to play a role in movement regulation and coordination. (Chapter Five)

BEHAVIORAL INHIBITION. A group of responses that develop in some children faced with new people or situations, who may express fear or shyness, and may cry or seek comfort in a familiar person. (Chapter Three)

BEHAVIORISM. The science of predicting and controlling behavior through use of stimuli and conditioned responses. (Chapter Three)

BETA-AMYLOID. A protein that forms the characteristic plaques found in the brains of Alzheimer's patients. (Chapter Nine)

BETA BLOCKERS. A class of drugs that lower blood pressure by blocking the effects of epinephrine (adrenaline). (Chapter Five)

BETA WAVES. Brain waves associated with periods of active mental states, such as problem solving and critical thinking. (Chapter Six)

BIPOLAR DISORDER. A neurological condition characterized by extreme mood swings from mania to depression. (Chapter Seven)

BLOOD-BRAIN BARRIER. Membranes in the brain's blood vessels with limited permeability that inhibit the transfer of many substances from the blood into brain tissues. (Chapter One)

BRAIN DEATH. The brain's lack of electrochemical activity and loss of function. A brain-dead individual cannot recover. (Chapter Six)

BROCA'S AREA. Region in the left frontal cortex of the brain responsible for motor movements in the production of speech. (Chapter Eight)

BRODMANN AREAS. Fifty-two sites located and mapped on the cerebral cortex by neuroscientist Korbinian Brodmann in 1906. Damage to a particular area manifests itself in a distinct, predictable way. (Chapter Two)

CATARACTS. A clouding of the eyes' natural lens due to protein buildup. (Chapter Nine)

CENTRAL FISSURE. Groove separating the frontal lobe from the parital lobe. (Chapter One)

CENTRAL NERVOUS SYSTEM. The brain and spinal cord. This control hub integrates incoming sensory information and issues motor responses. (Chapter Two)

CEREBELLUM. Region of the brain that is most responsible for producing smooth, coordinated muscle movement. (Chapter Five)

CEREBRAL CORTEX. The outermost layer of the brain, responsible for creativity, planning, language, and perception. (Chapter Two)

CEREBROSPINAL FLUID. Plasmalike fluid that cushions and protects the brain and spinal cord and provides nourishment to brain tissues. (Chapter One)

CERVICAL NERVES. Eight pairs of spinal nerves that issue from the first seven vertebrae and supply movement and feeling to the arms, neck, and upper chest. (Chapter Five)

CHEMORECEPTORS. Neural receptors that respond to the presence of chemicals. (Chapter Two)

CHOLINESTERASE INHIBITOR. A common type of anti-dementia drug that slows the breakdown of the neurotransmitter acetylcholine. (Chapter Nine)

CHOREA. A movement disorder characterized by involuntary, irregular, jerking motions of the limbs and trunk. (Chapter Five)

CHRONIC TRAUMATIC ENCEPHALOPATHY (CTE). A degenerative brain disease found in individuals with a history of frequent concussions. Characterized by depression, memory loss, aggression, confusion, and the early onset of dementia. (Chapter Five)

CHUNKING. The technique of dividing a large amount of information into smaller groups to facilitate memorization. (Chapter Eight)

CIRCADIAN RHYTHM. Any pattern an organism's body follows in an approximately 24-hour period, such as the sleep-wake cycle in humans. (Chapter Six)

CLASSICAL CONDITIONING. A process of behavioral training in which a previously neutral stimulus evokes a particular response through repeated pairing with a stimulus that naturally evokes it. (Chapter Eight)

COCCYGEAL NERVE. Single pair of nerve cells that issue from the coccyx, or tailbone. Supplies feeling to the skin between the coccyx and the anus. (Chapter Five)

COCHLEAR HAIR CELLS. Hearing receptors located in the inner ear that allow the processing of sound. They bend in response to vibrations entering the ear and transmit signals to the brain stem via the audatory nerve. (Chapter Four)

COGNITIVE RESERVE THEORY. Suggests that building and strengthening neural connections through mental stimulation may offset dementia symptoms. (Chapter Nine)

COGNITIVE UNCONSCIOUSNESS. Takes in most of the information that allows individuals to consciously act within and know their environment. (Chapter Five)

COMA. A deep level of unconsciousness in which an individual cannot be awakened and does not respond to stimuli. (Chapter Six)

COMPLEX REGIONAL PAIN SYNDROME. A chronic condition causing swelling and pain in a limb and differences in skin coloration and temperature. Believed to be caused by a dysfunction of the central or peripheral nervous system. (Chapter Six)

CONCUSSION. A high-velocity impact injury to the brain that may interfere with movement, balance, speech, and memory and can have short- and long-term effects. (Chapter Five)

CONES. Photoreceptive cells on the retina that provide color vision. They respond to the wavelengths of red, blue, and green. (Chapter Four)

CONFABULATION. A disorder in which an individual unintentionally fabricates occurrences to fill gaps in his or her memories, believing them to be accurate. (Chapter Eight)

CORPUS CALLOSUM. A thick band on nerve fibers that connects and allows for communication between the left and right hemispheres of the brain. (Chapter Six)

CORTICAL DEMENTIA. Dementia induced by damage to the cerebral cortex. Results in impaired social and behavioral skills, thinking, memory, and language. (Chapter Nine)

CORTICOTROPIN RELEASING HORMONE. Released by the hypothalamus, increasing anxiety, putting the body on alert, and causing adrenaline and the stress hormone cortisol to be released. (Chapter Two)

CORTISOL. A hormone released by the adrenal cortex in periods of extended stress. Has anti-inflammatory properties. (Chapter Six)

CRANIAL NERVE. Carries information to and from the brain. Except for the vagus nerve, all terminate in the head and neck. (Chapter Two)

CRANIUM. The fused bones that encase and protect the brain and the organs for equilibrium and hearing; skull. (Chapter One)

CREUTZFELDT-JAKOB DISEASE. A rare and fatal degenerative brain disorder characterized by rapid progressive dementia. (Chapter Nine)

DELTA WAVES. Brain waves that occur most often during sleep or in periods of unconsciousness. Common in newborns. (Chapter Six)

DEMENTIA. Term for a group of symptoms caused by disease, infection, or trauma resulting in a loss of mental functions, interfering with normal daily life. (Chapter Nine)

DENDRITES. Branching fibers of axons that act as receptors of information. These receive messages from other neurons and deliver them to the main body of the nerve cell. (Chapter One)

DIABETES INSIPIDUS. A condition caused by a lack of sufficient amounts of antidiuretic hormone (ADH) in the body. Patients experience extreme thirst and frequent urination. (Chapter Two)

DIABETES MELLITUS. A condition caused by a lack of insulin, resulting in heavy blood sugar loss through urination. (Chapter Two)

DISSOCIATIVE FUGUE. A psychiatric disorder, often induced by stress, that is characterized by amnesia of self or personality. (Chapter Eight)

DOMAINS. Specific cortical areas that have been linked to specific motor and sensory functions by PET scans and fMRIs. (Chapter Two)

DURA MATER. The tough outermost layer of the three membranes surrounding the brain and spinal cord, located directly beneath the cranium. (Chapter One)

DYSCALCULIA. A learning disability that is characterized by severe difficulty in understanding math. (Chapter Eight)

DYSGRAPHIA. A learning disability affecting an individual's ability to write. This may affect both fine motor hand control and idea processing. (Chapter Eight)

DYSLEXIA. A learning disability of neurological origin that impairs the ability to process language, leading to difficulties in spelling, reading, and writing. (Chapter Eight)

DYSTHYMIA. A chronic depression less severe than major depression, characterized by occurrence of depressive symptoms nearly daily for over two years. (Chapter Seven)

ECTODERM. Outermost layer of cells of a developing embryo that becomes the skin and nervous system. (Chapter Three)

ELABORATION. A memory technique in which new information is associated with previously learned material, aiding in long-term storage. (Chapter Eight)

EMBOLIC STROKE. Stroke occurring when fatty deposits, detached from an artery wall in another region of the body, lodge in an artery of the brain, cutting off blood supply. (Chapter Nine)

EMOTION. A spontaneously occurring mental state characterized by strong feeling and often accompanied by physiological and behavioral changes. (Chapter Seven)

EMOTIONAL CONTAGION. The process of emotion transfer from one individual to another by the unconscious mimicking of the emotional behavior of others in the vicinity. (Chapter Seven)

ENDODERM. Innermost layer of cells of a developing embryo that becomes the digestive tract. (Chapter Three)

ENDORPHINS. Proteins that reduce the perception of pain. (Chapter One)

ENGRAM. A term for the physical trace that memory formation may leave on participating neurons. (Chapter Eight)

ENKEPHALINS. Natural pain suppressants that inhibit the discharge of pain-exciting neurotransmitters. (Chapter Two)

ENVIRONMENTAL DEPENDENCY SYNDROME (EDS). A neurological condition in which an individual feels compelled to mimic others' actions or to use tools within their environment. (Chapter Six)

EPINEPHRINE. Also called adrenaline. Primary hormone produced by the adrenal medulla. This works with norepinephrine to put body systems on alert. (Chapter Three)

EPISODIC MEMORY. A type of declarative memory that consists of stored autobiographical remembrances of personal experiences. (Chapter Eight)

EPITHELIAL CELLS. Cells that form the tastebuds. Consist of supporting, basal, and taste receptor cells. (Chapter Four)

ESTROGEN. Female sex hormone produced in the adrenal glands and ovaries. These feminizing hormones are found in both sexes, but in a much higher quantity in women. (Chapter Seven)

EXPLICIT MEMORY. Memories that are consciously recalled. (Chapter Eight)

EXTEROCEPTORS. Sensory neurons that respond to external stimuli and are the basis for the five senses. (Chapter Four)

FETAL ALCOHOL SYNDROME. A range of developmental disorders and birth defects caused by excessive drinking during pregnancy. Children may be born with abnormal features and malformed organs and are at risk of mental retardation. (Chapter Three)

FIGHT OR FLIGHT RESPONSE. Triggered by the sympathetic branch of the automatic nervous system, the brain's response to flee from or defend itself against a potential or perceived threat. (Chapter Two)

FISSURES. The deepest inward folds or grooves of the brain. (Chapter One)

FOVEA. Center of the retina. This region contains the highest concentration of photoreceptors and is the area of highest visual acuity. (Chapter Four)

FREE RADICALS. Molecules in the body with an unstable electric field. They take an electron from a neighboring molecule, making it unstable and creating a chain reaction that may result in cell damage. (Chapter Nine)

GAMMA WAVES. Brain waves that occur continuously in all states except for nondreaming sleep. Believed to promote various brain functions, especially memory. (Chapter Six)

G-FACTOR. Short for "general intelligence factor." A psychological measure of cognitive mental abilities. (Chapter Two)

GLAUCOMA. An eye condition that develops as fluid pressure builds inside the eye. This may damage the optic nerve, leading to vision impairment or blindness. (Chapter Nine)

GLOMERULI. Cells located in the olfactory bulb that respond to particular odors. (Chapter Four)

GLUTAMATE. Excitatory neurotransmitter prevalent in the central nervous system. (Chapter Three)

GONADOTROPHIN-RELEASING HORMONES. A hormone produced in the hypothalamus that induces the production of luteinizing and follicle-stimulating hormones in the pituitary gland. (Chapter Seven)

GYRI (SING., gyrus). The outward, elevated folds of the cerebral cortex. (Chapter One)

HAPTIC TOUCH. The physical manipulation of objects in order to determine tactile qualities such as texture, hardness, and shape. (Chapter Four)

HAYFLICK LIMIT. The term given to the discovery that cell division occurs a finite amount, the number of which may determine an organism's life span. (Chapter Nine)

HEMIBALLISMUS. Uncontrolled movement of the limbs as a result of damage to the basal ganglia. (Chapter Five)

HEMISPHERECTOMY. The surgical removal of one of the hemispheres of the brain. (Chapter Three)

HEMORRHAGIC STROKE. Stroke that occurs when a brain artery ruptures, generally as a result of high blood pressure. (Chapter Nine)

HIPPOCAMPUS. Region of the brain aiding in converting new information to long-term memory. (Chapter Six)

HOMEOSTASIS. A state of equilibrium referring to the body's ability to remain internally stable while external environments vary. (Chapter Two)

HUNTINGTON'S DISEASE. A hereditary condition causing degeneration of the neurons in the basal ganglia and cerebral cortex. It is ultimately fatal. (Chapter Five)

HYPERALGESIA. An increased sensitivity to pain. (Chapter Two)

HYPERTHYMESTIC SYNDROME. A condition in which an individual has a superior autobiographical memory. (Chapter Eight)

HYPOCRETIN. A neurotransmitter that promotes wakefulness. (Chapter Six)

HYPOTHALAMUS. Brain region located directly above the brain stem. The center of emotional response. Regulates body temperature, hunger, thirst, and sleeping. (Chapter One)

IMPLICIT MEMORY. Memory recalled unconsciously during physical activity. (Chapter Eight)

INSULA. Region of the cerebral cortex responsible for the recognition and perception of disgust in others. (Chapter Seven)

INTERNEURONS. Neurons confined to the brain and spinal cord that integrate information between motor and sensory neurons. (Chapter One)

INTEROCEPTORS. Sensory neurons that register internal stimuli, such as chemical changes. (Chapter Four)

INTRALAMINAR NUCLEI. Located in the thalamus, nuclei are responsible for creating the brain's gamma waves. (Chapter Six)

JACKSONIAN EPILEPSY. A type of epilepsy characterized by predicable fits confined to certain parts of the body. (Chapter Five)

JET LAG. A circadian rhythm disruption caused by rapid long-distance flights. Irritability, fatigue, digestive problems result. (Chapter Six)

KETAMINE. A drug that blocks NMDA receptors in the central nervous system. It is used as an anesthetic and in some cases to treat chronic severe pain. (Chapter Six)

KORSAKOFF'S PSYCHOSIS. A form of amnesia, often caused by severe alcoholism, in which an individual is unable to form or store new memories and is much given to confabulation. (Chapter Eight)

LATERAL GENICULATE. Located in the thalamus, this serves as a visual relay center for signals in transit to the visual cortex in the occipital lobes. (Chapter Four)

LATERAL PONTINE TEGMENTUM. A region of the pons responsible for inducing REM sleep. (Chapter Six)

LEPTIN. A hormone produced by fat cells that helps to regulate metabolism and food consumption. (Chapter Five)

LEVATOR ANGULI ORIS. Facial muscle that raises the angle of the mouth. (Chapter Seven)

LEVATOR LABII SUPERIORIS. Facial muscle in the upper lip and cheek that raises the upper lip when one is smiling. (Chapter Seven)

LEWY BODY DEMENTIA. The second most common form of degenerative dementia; characterized by abnormal structures in certain areas of the brain. (Chapter Nine)

LIPOFUSCIN. A naturally occurring fatty brown pigment that accumulates in cells as the body ages. (Chapter Nine)

LOCKED-IN SYNDROME. A neurological condition resulting in loss of voluntary muscle movement in all regions of the body except the eyes, though cognitive awareness and reasoning remain normal. (Chapter Five)

LONGITUDINAL SULCUS. A deep groove that separates the cerebral hemispheres. (Chapter One)

LONG-TERM POTENTIATION (LTP). The strengthening of neural connections, considered to be the cellular basis of memory. (Chapter Seven)

LUMBAR NERVES. Five pairs of spinal nerves that supply the lower back, fronts of the legs, and feet. (Chapter Five)

MACULAR DEGENERATION. A chronic eye disease characterized by the degeneration of the eye tissue responsible for clear center vision. (Chapter Nine)

MAJOR DEPRESSION. The most severe form, characterized by profound interference with normal daily activities. May occur as a single episode after significant trauma, or repeatedly through life. (Chapter Seven)

MECHANORECEPTORS. These receptors create nerve impulses when their shape is deformed by a mechanical force such as pressure or touch. (Chapter Two)

MEDIAL TEMPORAL LOBE. Region of the brain that includes the hippocampus and amygdala. This area is crucial to the formation, storage, and organization of memory. (Chapter Eight)

MEDULLA OBLONGATA. Part of the brain stem, it connects the spinal cord to higher brain centers. Controls heartbeat and respiration. (Chapter One)

MELATONIN. A hormone produced in the pineal gland that helps regulate the sleep-wake cycle. (Chapter Seven)

MESODERM. The middle layer of cells in a developing embryo. This forms the muscles, skeleton, heart, and genitalia. (Chapter Three)

MIRROR NEURONS. Neurons that fire during a familiar action and when thinking of or observing others performing it. (Chapter Five)

MITRAL CELLS. Neurons in the olfactory bulb that refine and amplify signals from glomeruli and relay the information to the olfactory tract. (Chapter Four)

MONOAMINES. A class of neurotransmitters that includes dopamine, serotonin, and adrenaline. (Chapter Six)

MONOAMIDE OXIDASE INHIBITORS (MAOIS). Antidepressant drugs that boost mood by preventing monoamide oxidase from metabolizing serotonin, norepinephrine, and dopamine. (Chapter Seven)

MOTION BLINDNESS. Loss of the ability to detect changes in movement. Motion appears as a series of differing still images. (Chapter Five)

MOTOR HOMUNCULUS. A diagram linking body parts to the corresponding region of the motor cortex, with sizes of body parts shown in proportion to the number of neural connections. (Chapter Five)

MOTOR NEURONS. Neurons carrying impulses away from the central nervous system to activate muscles and glands. (Chapter One)

MYELIN SHEATH. An insulating and protective wrapping of fatty tissue that surrounds axons and increases the speed of the transmission of nerve impulses. (Chapter One)

NARCOLEPSY. The inability to regulate sleep-wake cycles. Linked to the absence of or lowered amounts of hypocretin in the brain. (Chapter Six)

NEGATIVITY BIAS. A psychological phenomenon by which individuals react more strongly to unpleasant situations and stimuli than to their positive counterparts. (Chapter Seven)

NEURAL DARWINISM. A term coined by Nobel laureate Gerald Edelman, this describes the process in which neurons that receive constant simulation grow and those that do not atrophy. (Chapter Three)

NEURAL GROOVE. In a developing embryo, during the second stage of brain and spinal cord development, this occurs as the neural plate begins to fold inward. (Chapter Three)

NEURAL PLATE. Formed during the third week of embryonic development as the ectoderm thickens. The first stage of brain and spinal cord formation. (Chapter Three)

NEURAL TUBE. Formed as the neural groove fuses together. Occurs by day 22 in embryonic development. (Chapter Three)

NEUROBLASTS. Primitive nerve cells. (Chapter Three)

NEUROGLIA. Glial cells; these brain cells insulate, guide, and protect neurons. (Chapter One)

NEUROLEPTICS. Antipsychotics, a class of drugs generally used to treat schizophrenia and other psychotic disorders. (Chapter Three)

NEURON. Nerve cell; a central nervous system cell that generates and transmits information from nerve impulses. (Chapter One)

NEURONAL MIGRATION. Process where new neurons created in the prefrontal cortex relocate to other parts of the brain and assume new tasks. This occurs during the final months of fetal development. (Chapter Three)

NOCICEPTOR. Pain receptor that responds to potentially harmful stimuli. (Chapter One)

NODES OF RANVIER. Regular gaps in the myelin sheath occurring along the coated axons. (Chapter One)

NONASSOCIATIVE LEARNING. Learning that occurs through repeated exposure to a stimulus without the result of either positive or negative consequences. (Chapter Eight)

NOREPINEPHRINE. A neurotransmitter and adrenal medullary hormone, this works with epinephrine to activate the sympathetic branch of the autonomic nervous system. Also involved in arousal and the regulation of sleep and mood. (Chapter Three)

NUCLEUS ACCUMBENS. Region of the brain associated with feelings of pleasure and reward. (Chapter Six)

OBSESSIVE-COMPULSIVE DISORDER (OCD). An anxiety disorder characterized by intrusive, unwanted thoughts (obsessions) and/or strong urges to perform countering actions (compulsions). (Chapter Five)

OLFACTORY EPITHELIUM. A fluid-coated patch of sensory cells that is able to detect odor. It is located on the roof of the nasal cavity. (Chapter Four)

OPERANT CONDITIONING. A process of behavioral training in which a voluntary action is reinforced through reward or diminished through punishment. (Chapter Eight)

ORBICULARIS OCULI. Facial muscles responsible for crinkling the eyes when an individual smiles. (Chapter Seven)

OTOLITHS. Crystals of calcium carbonate located in the otolithic membrane in the ear. Playing a vital role in balance, these structures both detect gravity and aid in the awareness of the head's spatial orientation. (Chapter Four)

OXYTOCIN. A hormone produced in the pituitary gland that is released during pregnancy and intercourse, promoting trust and pair bonding. (Chapter Seven)

PAPEZ CIRCUIT. A system of interconnected brain regions, including hippocampus, hypothalamus, and cingulate gyrus, that participates in short-term memory formation and emotional processing. (Chapter Eight)

PARALLEL PROCESSING. The transmission of information through the body whereby one neuron excites multiple others and several paths are utilized at once. (Chapter Two)

PARASYMPATHETIC BRANCH. This branch of the autonomic nervous system is responsible for relaxing the body. It lowers heart rate and blood pressure and reduces breathing rate. (Chapter Two)

PATELLAR TENDON. Connects the kneecap to the shinbone and aids in leg extension. Site of the knee jerk reflex test. (Chapter Five)

PERCEPTION. The interpretation of the meaning of a stimulus. (Chapter Four)

PERIPHERAL NERVOUS SYSTEM. The nerves that branch out of the brain and spinal cord. (Chapter Two)

PERSISTENT VEGETATIVE STATE. A condition in which the brain maintains functions necessary to keep the body alive, but not cognitive function. (Chapter Six)

PHENYLETHYLAMINE (PEA). A neurotransmitter found in small amounts in the brain and in foods such as chocolate. This releases dopamine into the limbic system, creating feelings of pleasure. (Chapter Seven)

PHEROMONES. Chemicals produced by insects and animals that transmit messages to or affect the behavior of other individuals. (Chapter Four)

PHOBIA. An unreasonable fear that may cause avoidance and panic. (Chapter Seven)

PHONEME. The smallest sound element in a language that can be altered to change the meaning of a word. These have no meaning on their own. (Chapter Three)

PHONEME CONTRACTION. The loss of the ability to hear and differentiate all language sounds. Occurs between the age of six months and one year. (Chapter Three)

PHOTORECEPTORS. Neurological receptors that react to light energy. (Chapter Two)

PHRENOLOGY. Pseudoscience popular in the 19th century believing that personal characteristics and mental abilities can be derived from knots and knobs of the skull. (Chapter Two)

PIA MATER. The innermost cerebral membrane, this thin layer of connective tissue clings to every dip and curve of the cerebral cortex. (Chapter One)

PLANUM TEMPORALE. Region of the brain associated with speech and sign language comprehension. Larger in the left hemisphere in two-thirds of the population. (Chapter Three)

PLASTICITY. The brain's ability to reshape neural interactions. (Chapter One)

PONS. Part of the brain stem that serves as a bridge between the medulla and midbrain and aids the medulla in respiratory regulation. (Chapter One)

PRECENTRAL GYRUS. Area of the cerebral cortex containing the primary motor complex and thus responsible for body movement. Located on the frontal lobe of each hemisphere. (Chapter One)

PREFRONTAL CORTEX. Brain region located in the anterior frontal lobe. Responsible for reasoning, planning, judgment, empathy, abstract ideas, and conscience. (Chapter One)

PRIMARY DEMENTIA. A dementia, such as Alzheimer's, that does not occur as a result of another disease. (Chapter Nine)

PROGRESSIVE DEMENTIA. Dementia that becomes worse through time. (Chapter Nine)

PROPRIOCEPTORS. Sensory neurons that are responsible for the sense of self, and the awareness of body position and movement. (Chapter Four)

PROSODY. Additional meaning imparted to language through the rhythm and intonation of speech. (Chapter Seven)

PROSOPAGNOSIA. A condition also known as face blindness in which an individual is unable to recognize a person by his or her facial features or to differentiate between faces. (Chapter Eight)

PRUNING. A natural process of the brain where weak neural connections die off. Occurs on a large scale during fetal development and during teenage years, and on a lesser scale through adulthood. (Chapter Three)

PSUEDOBULBAR PALSY. A condition in which an individual is unable to voluntarily control the muscles of his or her face. (Chapter Seven)

RAPHE NUCLEI. A cluster of neurons in the pons, medulla, and midbrain primarily responsible for production of serotonin. (Chapter Nine)

RAPID EYE MOVEMENT SLEEP (REM). The fifth stage of sleep, characterized by high levels of activity in the cerebral cortex. Stage in which dreaming occurs. (Chapter Six)

RECALL. Memory process that involves retrieving previously stored information. (Chapter Eight)

RECOGNITION. Retrieval process of memory that consists of the identification of learned items. (Chapter Eight)

REFLEX. An automatic and uncontrolled reaction to stimuli. (Chapter Two)

RESPONSIBLE FOR BODY MOVEMENT. Located on the frontal lobe of each hemisphere.

RETINA. The sensory membrane that lines the interior of the eye. Composed of several layers, including one of rods and cones, it receives the image formed by the lens, converts it to signals, and transmits it to the optic nerve. (Chapter Four)

REVERSIBLE DEMENTIA. Dementia brought on by an unrelated, treatable condition. (Chapter Nine)

RISORIUS MUSCLES. Facial muscles responsible for the lateral movement of the corners of the mouth. (Chapter Seven)

RODS. Photoreceptors that register dim light. Located on the retina, these allow vision at low-light levels. (Chapter Four)

ROOTING. A reflex that causes an infant to turn toward facial stimulation. Believed to facilitate feeding. (Chapter Four)

SACRAL NERVES. Five pairs of spinal nerves that issue from the sacrum below the lower back. These nerves supply the backs of the legs and sexual organs. (Chapter Five)

SCHIZOPHRENIA. A chronic neurological disease of distorted thoughts and perceptions. This affects both men and women and usually surfaces during adolescence or young adulthood. Symptoms include delusions and hallucinations. (Chapter Three)

SEASONAL AFFECTIVE DISORDER (SAD). A type of depression that occurs most often with the onset of winterr. (Chapter Seven)

SECONDARY DEMENTIA. Dementia resulting from injury or another disease. (Chapter Nine)

SELECTIVE SEROTONIN REUPTAKE INHIBITORS (SSRIS). Drugs that inhibit serotonin reabsorption. Used to treat depression, anxiety disorders, obsessive-compulsive disorder, and eating disorders. (Chapter Five)

SEMANTIC MEMORY. Stored knowledge of general facts and data. (Chapter Eight)

SENSATION. The brain's registration and awareness of a stimulus. (Chapter Four)

SENSORY NEURONS. Neurons that send impulses from the skin and parts of the body to the central nervous system. (Chapter One)

SEPARATION ANXIETY. The distress found in some young children at the departure of a parent or the introduction to a stranger.

SEPTUM. Region of the brain associated with orgasm. (Chapter Seven)

SERIAL PROCESSING. Transmission of information along a direct chain of neurons; a single neuron is excited at a time. (Chapter Two)

SEROTONIN. Inhibitory neurotransmitter that plays a role in sleep, mood regulation, memory and learning. (Chapter Six)

SLEEP APNEA. A sleep disorder in which an individual frequently stops breathing for short periods of time. (Chapter Six)

SLEEP REGULATORY SUBSTANCES. Proteins that accumulate in cerebrospinal fluid during wakefulness. Induce sleep upon reaching threshold levels. (Chapter Six)

SOMATIC NERVOUS SYSTEM. Voluntary nervous system. Sends signals from the central nervous system to the skeletal muscles. Usually under conscious control. (Chapter Two)

SPINAL NERVES. The nerves branching out from and relaying information to and from the spinal cord. (Chapter Two)

STEM CELL. An unspecialized cell with the ability to grow and develop into other types of cells and tissues. (Chapter Nine)

STIMULUS. A change in the environment that evokes a reaction from the brain. (Chapter Four)

SULCI (SING., sulcus). Inward folds of the cerebral cortex, more shallow than fissures. (Chapter One)

SUBCORTICAL DEMENTIA. A dementia affecting the lower regions of the brain, resulting in changes in movement and emotion as well as memory problems. (Chapter Nine)

SUBSTANTIA NIGRA. A cluster of cells at the base of the midbrain responsible for the production of dopamine. (Chapter Nine)

SUPERIOR COLLICULUS. Located in the midbrain, this region adjusts the head and eyes to achieve maximal visual input. (Chapter Four)

SUPERIOR TEMPORAL SULCUS. Brain region containing the neural networks responsible for motion detection/analysis. (Chapter Five)

SYLVIAN FISSURE. Groove separating the parietal lobe from the temporal lobe. (Chapter One)

SYMPATHETIC BRANCH. A branch of the autonomic nervous system that puts the body on alert and supplies it with energy in response to fear or excitement. (Chapter Two)

SYNAPSE. Tiny gap between the axon terminals of two neurons through which communication occurs. (Chapter One)

SYNESTHESIA. A condition in which the stimulation of one sense is simultaneously percieved by another sense or senses. (Chapter Four)

TASTANTS. Chemicals that stimulate the sensory cells in taste buds. (Chapter Four)

TAU. A protein that aids in the formation of microtubules, which help transport nutrients within neurons. Becomes irregular in Alzheimer's disease. (Chapter Nine)

TESTOSTERONE. Male sex hormone produced primarily in the testes. This masculinizing hormone is found in both sexes, but in much higher quantity in men. (Chapter Seven)

THERMORECEPTORS. Neuroreceptors that register changes in temperature. (Chapter Two)

THETA WAVES. Experienced between waking and sleeping, during prayer, daydreaming, creativity and intuition. Thought to promote learning, memory. (Chapter Six)

THORACIC NERVES. Spinal nerves that issue from the 12 vertebrae of the upper back. They serve the trunk and abdomen. (Chapter Five)

THROMBOTIC STROKE. Occurs when an artery supplying blood to the brain gets clogged by fatty deposits on its inner walls, cutting off blood supply. (Chapter Nine)

TINNITUS. Hearing disorder characterized by chronic ringing or clicking in the ears in an otherwise silent environment. (Chapter Four)

TISSUE PLASMINOGEN ACTIVATOR (TPA). A clot-destroying drug that, when administered soon after a stroke occurs, improves chances of full recovery or minimizing damage. (Chapter Nine)

TONIC-CLONIC SEIZURES. The most severe epileptic seizures, often causing loss of bowel and bladder control, tongue biting, and strong convulsions. Formerly known as grand mal seizures. (Chapter Two)

TOURETTE'S SYNDROME. Neurological disorder characterized by repetitive and involuntary motor tics and vocalizations. (Chapter Five)

TRANSECT. To completely sever. (Chapter One)

TRANSIENT ISCHEMIC ATTACK (TIA). A stroke occurring when an artery serving the brain becomes temporarily blocked. Often precedes more severe, acute stroke. (Chapter Nine)

TRIGEMINAL CRANIAL NERVE. Connects the face to the brain stem. Responsible for facial sensation and motor control. (Chapter Five)

TRIUNE BRAIN. A 1967 theory on the evolution of the brain suggested by Paul MacLean in which there are three separate areas of the brain representing evolutionary development. (Chapter Three)

TROPHIC FACTORS. Proteins that promote the survival, function, and growth of neurons, and that are responsible for the correct wiring of neurons during brain development. (Chapter Three)

TRYPTOPHAN. An essential amino acid that aids in the body's production of serotonin and vitamin B_3. Prevalent in turkey and dairy products. (Chapter Six)

TYMPANIC MEMBRANE. The eardrum, which conducts sound wave vibrations from the ear canal to the bones of the middle ear. (Chapter Four)

UMAMI. The fifth flavor, linked to foods containing glutamate and aspartate. (Chapter Four)

VAGUS NERVE. One of the primary communications pathways between the brain and the body's major organs. (Chapter Seven)

VENTRAL PALLIDUM. One of the brain's main reward circuits; associated with attachment and stress relief. (Chapter Seven)

VESTIBULAR ORGAN. Nonauditory portion of the inner ear. Responsible for the body's detection of the spatial orientation and movement of the head and for maintaining balance and posture. (Chapter Four)

VESTIBULAR REFLEX. Automatic adjustment to the muscle tone in the body and neck to maintain the posture of the head. (Chapter Five)

VESTIBULOOCULAR REFLEX. Causes automatic adjustments to the eye muscles to maintain a stable gaze regardless of head movement. (Chapter Five)

VISUOSPATIAL MEMORY. A type of declarative memory, allowing the remembrance of the location of objects in space. (Chapter Eight)

VOMERONASAL ORGAN (VNO). A small olfactory organ found in most vertebrates that detects pheromones and sends corresponding signals to the brain. (Chapter Seven)

WERNICKE'S AREA. Brain area located in the posterior region of the temporal lobe; responsible for ability to understand and produce intelligible speech. (Chapter Eight)

WILLIAM'S SYNDROME. A rare genetic condition characterized by mental retardation or learning difficulties, an overly friendly personality, and a distinctive facial appearance. (Chapter Nine)

WORKING MEMORY. Where information is stored temporarily in the brain. Also called short-term memory. (Chapter Two)

ZYGOMATICUS MUSCLES. Facial muscles involved in smiling. These lift the corners of the mouth. (Chapter Seven)

ZYGOTE. A fertilized egg. (Chapter Three)

INDEX

Boldface indicates illustrations.

B

C

D

E

F

G

H

P

T

FURTHER READING

BOOKS

Aamodt, Sandra, and Sam Wang. ***Welcome to Your Brain: Why You Lose Your Car Keys but Never Forget How to Drive and Other Puzzles of Everyday Life.*** Bloomsbury, 2008.

Bauby, Jean-Dominique. ***The Diving Bell and the Butterfly***, trans. Jeremy Leggatt. Alfred A. Knopf, 1997.

Buettner, Dan. ***The Blue Zones: Lessons for Living Longer From the People Who've Lived the Longest.*** National Geographic Society, 2008.

Damasio, Antonio. ***The Feeling of What Happens: Body and Emotion in the Making of Consciousness.*** Harcourt, 1999).

—. ***Looking for Spinoza: Joy, Sorrow, and the Feeling Brain.*** Harcourt, 2003.

Della Sala, Sergio. ***Tall Tales About the Mind & Brain: Separating Fact From Fiction.*** Oxford University Press, 2007.

Dement, William C., and Christopher Vaughan. ***The Promise of Sleep: A Pioneer in Sleep Medicine Explores the Vital Connection Between Health, Happiness, and a Good Night's Sleep.*** Delacorte Press, 1999.

Edelman, Gerald M. ***Second Nature: Brain Science and Human Knowledge.*** Yale University Press, 2006.

—. ***Wider Than the Sky: The Phenomenal Gift of Consciousness.*** Yale University Press, 2004.

Finger, Stanley. ***Minds Behind the Brain: A History of the Pioneers and Their Discoveries.*** Oxford University Press, 2000.

—. ***Origins of Neuroscience: A History of Explorations Into Brain Function.*** Oxford University Press, 1994.

Gaddes, William H., and Dorothy Edgell. ***Learning Disabilities and Brain Function: A Neuropsychological Approach.*** Springer-Verlag, 1994.

Goldberg, Elkhonon. ***The Executive Brain: Frontal Lobes and the Civilized Mind.*** Oxford University Press, 2001.

Jamison, Kay Redfield. ***Touched With Fire: Manic-Depressive Illness and the Artistic Temperament.*** Free Press, 1993.

—. ***An Unquiet Mind: A Memoir of Moods and Madness.*** Alfred A. Knopf, 1995.

Kurzweil, Ray. ***The Singularity Is Near: When Humans Transcend Biology.*** Viking, 2005.

Marieb, Elaine N., and Katja Hoehn. ***Human Anatomy & Physiology***, 7th ed. Benjamin Cummings, 2007.

Ratey, John J. ***A User's Guide to the Brain: Perception, Attention, and the Four Theaters of the Brain.*** Pantheon Books, 2001.

Restak, Richard. ***The Brain.*** Bantam Books, 1984.

—. ***The Mind.*** Bantam Books, 1988.

—. ***Mozart's Brain and the Fighter Pilot: Unleashing Your Brain's Potential.*** Harmony Books, 2001.

—. ***Mysteries of the Mind.*** National Geographic Society, 2000.

—. ***The Naked Brain: How the Emerging Neurosociety Is Changing How We Live, Work, and Love.*** Harmony Books, 2006.

—. ***The New Brain: How the Modern Age Is Rewiring Your Mind.*** Rodale, 2003.

—. ***Older & Wiser: How to Maintain Peak Mental Ability for as Long as You Live.*** Thorndike Press, 1998.

—. ***Poe's Heart and the Mountain Climber: Exploring the Effect of Anxiety on Our Brains and Our Culture.*** Harmony Books, 2004.

—. ***Receptors.*** Bantam Books, 1994.

—. ***The Secret Life of the Brain.*** Joseph Henry Press, 2001.

Sacks, Oliver. ***The Man Who Mistook His Wife for a Hat: And Other Clinical Tales.*** Summit Books, 1985.

—. ***Migraine.*** Vintage Books, 1999.

Schwartz, Jeffrey M., and Sharon Begley. ***The Mind and the Brain: Neuroplasticity and the Power of Mental Force.*** Regan Books/HarperCollins, 2002.

Taylor, Jill Bolte. ***My Stroke of Insight: A Brain Scientist's Personal Journey.*** Viking, 2006.

Thayer, Robert E. ***Calm Energy: How People Regulate Mood With Food and Exercise.*** Oxford University Press, 2001.

Trimble, Michael R. ***The Soul in the Brain: The Cerebral Basis of Language, Art, and Belief.*** Johns Hopkins University Press, 2007.

van Campen, Cretien. ***The Hidden Sense: Synesthesia in Art and Science.*** MIT Press, 2008.

Zimmer, Carl. ***Soul Made Flesh: The Discovery of the Brain, and How It Changed the World.*** Free Press, 2004.

GAMES

Allen, Robert, Philip Carter, and Ken Russell. ***The Mensa Mind Games Pack.*** Carlton Books, 2005.

Brain Age: Train Your Brain in Minutes a Day! (Dr Kawashima's Brain Training: How Old Is Your Brain?). Nintendo DS, 2006.

Crossword Dice. Levenger Company.

Goldberg, Elkhonon. ***Brain Games: Lower Your Brain Age in Minutes a Day.*** Publications International, 2008.

Visual Brainstorms. ThinkFun.

WEBSITES

The Aging Brain

Alzheimer's Association, ***http://www.alz.org***
A leading voluntary health organization specializing in Alzheimer's disease care, support, and research

Alzheimer's Disease Education and Referral Center, ***http://www.nia.nih.gov/alzheimers***
Branch of the National Institute on Aging that focuses on Alzheimer's research and education

Elder Wisdom Circle, ***http://www.elderwisdomcircle.org/***
A nonprofit organization employing more than 600 Elders who provide free and confidential advice to people of all ages

Brain Health

National Institute of Mental Health, ***http://www.nimh.nih.gov/index.shtml***
U.S. government institute devoted to the understanding and treatment of mental illnesses

National Institute of Neurological Disorders and Stroke, ***http://www.ninds.nih.gov/***
U.S. government organization that conducts and supports research on the more than 600 disorders that can affect the brain and nervous system

Brain Research

Society for Neuroscience, ***http://www.sfn.org/***
Nonprofit organization of scientists and physicians who study the brain. Website features neuroscience publications created for the general public

The Whole Brain Atlas, ***http://www.med.harvard.edu/AANLIB/home.html***
Online resource for images of the brain and the central nervous system

The Future

Computer History Museum, ***http://www.computerhistory.org/***
Museum that seeks to preserve the history of computing, including the inventors, machines, software, business and competitive environments, and social implications

KurzweilAI.net, ***http://www.kurzweilai.net/***
Site focused on the exponential growth of biological and mechanical intelligence and their potential merger

General Health & Wellness

Centers for Disease Control and Prevention, ***http://www.cdc.gov/***
Branch of the U.S. Department of Health and Human Services that distributes reliable, credible information on public health and safety

Medline Plus, ***http://medlineplus.gov/***
Information compiled from the U.S. National Library of Medicine, the National Institutes of Health, and other government agencies and health-related organizations. Contains extensive information about drugs, an illustrated medical encyclopedia, interactive patient tutorials, and the latest health news

WebMD, ***http://www.webmd.com/***
Website providing medical information and tools for managing a healthy lifestyle

PHOTO CREDITS

Cover: Don Farrall/Getty Images.
Half Title Page: (LE), UpperCut Images/Getty Images; (CTR), Sebastian Kaulitzki/Alamy; (RT), Hank Morgan – Rainbow/Science Faction. Title Page: MedicalRF.com. Foreword: Sebastian Kaulitzki/Shutterstock.

Chapter One: 0-1, UHB Trust/Getty Images; 1, page icon, MedicalRF.com; 2, 3D4Medical.com/Getty Images; 4, Dr. Jeremy Burgess/Photo Researchers, Inc.; 5, Sheila Terry/Photo Researchers, Inc.; 6, Bettmann/CORBIS; 7, Sandra Baker/Alamy; 8, (c) The British Library Board. All Rights Reserved. 781.b.3, frontispiece; 9, SPL/Photo Researchers, Inc.; 10, Illustration © 2009 Scientific Publishing Ltd., Elk Grove Village, IL, USA; 11, Portrait of Santiago Ramon y Cajal (1852-1934) 1906 (oil on canvas), Sorolla y Bastida, Joaquin (1863-1923)/ Diputacion General de Aragon, Zaragoza, Spain/Index/The Bridgeman Art Library; 13 (UP), Michael Delannoy/Visuals Unlimited; 13 (CTR), Sebastian Kaulitzki/Alamy; 13 (LO), Phototake, Inc./Alamy; 14, MedicalRF.com/CORBIS; 15, SPL/Photo Researchers, Inc.; 16, Dr. Torsten Wittmann/Photo Researchers, Inc.; 18, Matthias Kulka/CORBIS; 19, Visuals Unlimited/CORBIS; 20, Illustration © 2009 Scientific Publishing Ltd., Elk Grove Village, IL, USA; 21, Digital Vision/Getty Images; 22, BSIP/Photo Researchers, Inc.; 23, Library of Congress, #3b49887; 24, Edward Kinsman/Photo Researchers, Inc.; 25, SSPL/The Image Works; 26, DAJ/Getty Images; 27, SSPL/The Image Works; 28, ISM/Phototake; 29, AP Photo/Niklas Larsson; 30, MedicalRF.com/Getty Images; 31, Hank Morgan - Rainbow/Science Faction.

Chapter Two: 32-33, J. Meric/Getty Images; 34, Barry Blackman/Getty Images; 35, Illustration © 2009 Scientific Publishing Ltd., Elk Grove Village, IL, USA; 36, Science Museum/Science & Society Picture Library; 38, Beverly Joubert/NationalGeographicStock.com; 42, Illustration © 2009 Scientific Publishing Ltd., Elk Grove Village, IL, USA; 43, Taylor S. Kennedy/NationalGeographicStock.com; 44, Maximilian Weinzierl/Alamy; 45, Mark Thiessen, NGS; 46, James Balog/Getty Images; 47, Robert Harding Picture Library Ltd./Alamy; 48, Scott Rick, Assistant Professor of Marketing, University of Michigan; 49, Eric Gevaert/Shutterstock; 50 (UP), John Burcham/NationalGeographicStock.com; 50 (LO), Bettmann/CORBIS; 52, Dimas Ardian/Getty Images; 53, Don W. Fawcett/Photo Researchers, Inc.; 54, 3D4Medical.com/Getty Images; 56, Gary Carlson/Photo Researchers, Inc.; 57, Lazar/Shutterstock; 58, Chris McGrath/Getty Images; 59, Al Lamme/Phototake; 60, AP Photo/Martin Meissner; 61, SIM Laboratories/Photo Researchers, Inc.; 62, Cameramannz/Shutterstock; 63, Bochkarev Photography/Shutterstock.

Chapter Three: 64-65, Atlaspix/Alamy; 66, Mansell/Time & Life Pictures/Getty Images; 67, Dr. Keith Wheeler/Photo Researchers, Inc.; 68, Science Source/Photo Researchers, Inc.; 70, Travelib Prime/Alamy; 71, Stella/fStop/Alamy; 72, Dorling Kindersley/Getty Images; 74, SIU/Peter Arnold, Inc.; 75, Dr. G. Moscoso/Photo Researchers, Inc.; 76, Peter Cade/Getty Images; 77, Lee O'Dell/Shutterstock; 78, Sarah Leen/NGS Image Collection; 79, Guigoz/Steiner/Photo Researchers, Inc.; 80, Magdalena Szachowska/Shutterstock; 81, Archives of the History of American Psychology - The University of Akron; 82, Tom Grill/CORBIS; 83, Medical Body Scans/Photo Researchers, Inc.; 84, Patrick Sheandell O'Carroll/Getty Images; 85, BSIP/Photo Researchers, Inc.; 86, AP Photo/Rogelio V. Solis; 88, Richard Seagraves/Getty Images; 89, Martin Lladó/Gaia Moments/Alamy; 91, China Photos/Getty Images; 92, James Cavallini /Photo Researchers, Inc.; 94, W. Eugene Smith/Time & Life Pictures/Getty Images; 96, Jose Luis Pelaez, Inc./Blend Images/CORBIS.

Chapter Four: 98-99, Susana Vera/Reuters; 100, Richard Nowitz/NationalGeographicStock.com; 101, NASA/JPL; 102, C. Wells Photographic/Shutterstock; 103, UpperCut Images/Getty Images; 104, Sarah Leen/NGS Image Collection; 105, Aleksandar Kamasi/Shutterstock; 106, Illustration © 2009 Scientific Publishing Ltd., Elk Grove Village, IL, USA; 107, Mary Evans Picture Library/Alamy; 108, Cordelia Molloy/Photo Researchers, Inc.; 109, Omikron/Photo Researchers, Inc.; 110, Wellcome Department of Cognitive Neurology/Photo Researchers, Inc.; 111, Art Montes De Oca/Getty Images; 112, Library of Congress, Prints and Photographs Division, #3a02026; 113, Illustration © 2009 Scientific Publishing Ltd., Elk Grove Village, IL, USA; 115, Maxell Corporation of America/Manhattan Marketing Ensemble; 116, Junker/Shutterstock; 117, John M. Daugherty/Photo Researchers, Inc.; 118, Eric Tadsen/iStockphoto.com; 119, Omikron/Photo Researchers, Inc.; 121, BSIP/Photo Researchers, Inc.; 122, LWA-Dann Tardif/CORBIS; 123, Photo Researchers, Inc.; 124, Alfred Wekelo/Shutterstock; 125, Andrew Newberg, M.D., University of Pennsylvania; 126, Carlyn Iverson/Photo Researchers, Inc.; 127, Leïla Bellon/Photo Researchers, Inc.; 128, Jiang Dao Hua/Shutterstock; 129, Fiona McDougall, OneWorld Photo; 130, Library of Congress, Prints and Photographs Division, #3b26065; 131, Bob Krist/CORBIS.

Chapter Five: 134-135, Gustoimages/Photo Researchers, Inc.; 136, ISM/Phototake, Inc.; 137, Sidney Moulds/Photo Researchers, Inc.; 138, AM-Studio/Shutterstock; 139, David Grossman/Photo Researchers, Inc.; 140, Alex Wong/Getty Images; 141, Paul Spinelli/Getty Images; 142, Sally and Richard Greenhill/Alamy; 143, The Print Collector/Alamy; 144, Mary Evans Picture Library/Alamy; 145, Peter Gardiner/SPL/Photo Researchers, Inc.; 146, David Madison/Getty Images; 147, Zephyr/Photo Researchers, Inc.; 148, iofoto/Shutterstock; 149, Carolyn M. Carpenter/Shutterstock; 152, John W. Karapelou, CMI/Phototake; 154, Karina Maybely Orellana Rojas/Shutterstock; 155, Thomas Deerinck/Getty Images; 156, Chris Cole/Getty Images; 157, CORBIS; 158, Chip Somodevilla/Getty Images; 159, Rick Gomez/CORBIS; 160, AP Photo/Michael Dwyer; 161, Mary Evans Picture

Library/Alamy; 162, AP/Wide World Photos; 163, WDCN/Univ. College London/Photo Researchers, Inc.; 164, Pulse Picture Library/CMP Images/Phototake; 165, ISM/Phototake; 166, Dr. Torsten Wittmann/Photo Researchers, Inc.; 167, Glenn Asakawa/University of Colorado.

Chapter Six: 168-169, Alfred Pasieka/Photo Researchers, Inc.; 170, Tino Soriano/NationalGeographicStock.com; 171, Victorian Traditions/Shutterstock; 172, RTimages/Shutterstock; 173, Joel Sartore/NationalGeographicStock.com; 174, SVLumagraphica/Shutterstock; 175, Jodi Cobb/NGS Image Collection; 178, Bildarchiv Preussischer Kulturbesitz/Art Resource, NY; 179, Justin Guariglia/NationalGeographicStock.com; 180, Rick Wood/Journal Sentinel, Inc.; 181, AP Photo/File, Andre Rau/Elle; 182, Kalina Christoff, UBC Cognitive Neuroscience of Thought Laboratory; 183, AP Photo/Sigmund Freud Museum; 184, El Greco/Shutterstock; 185, Justin Guariglia/CORBIS; 186, Paul Chesley/NationalGeographicStock.com; 187, British Museum/Art Resource, NY; 188 (UP), Hank Morgan/Photo Researchers, Inc.; 188 (CTR), Hank Morgan/Photo Researchers, Inc.; 188 (LO), Hank Morgan/Photo Researchers, Inc.; 189, Davis Meltzer/NGS Image Collection; 190-191, Visuals Unlimited/CORBIS; 191, Bruce Benedict/Transtock/CORBIS; 192, Louie Psihoyos/Science Faction/CORBIS; 193, Davis Meltzer/NGS Image Collection; 194, MediVisuals/Photo Researchers, Inc.; 195, Louie Psihoyos/Science Faction; 196, Banque d'Images, ADAGP/Art Resource, NY; 197, Philippe Garo/Photo Researchers, Inc.; 198, Kairos, Latin Stock/Photo Researchers, Inc.; 199, AP Photo/Keystone/Walter Bieri; 200 & 201 (LO), Rosenbloom MJ and Pfefferbaum A (2008): Magnetic resonance imaging of the living brain: Evidence for brain degeneration among alcoholics and recovery with abstinence. Edited by E. V. Sullivan and A. Noronha in *Alcohol Research & Health* 31 (4):363-376; 201 (UP), Visuals Unlimited/CORBIS; 202, Reprinted by permission from Macmillan Publishers Ltd: New Insights Into the Genetics of Addiction, Ming D. Li, Margit Burmeister, Nature Reviews Genetics 10, 225-231 (31 March 2009) doi:10.1038/nrg2536 Progress; 203, Alex Saberi/Shutterstock.

Chapter Seven: 204-205, Hill Street Studios/Blend Images/CORBIS; 206, Gary Hershorn/Reuters/CORBIS; 207, The Image Works; 208, David H. Seymour/Shutterstock; 209, Ariel Skelley/Getty Images; 210, Tek Image/Photo Researchers, Inc.; 211, James Cavallini/Photo Researchers, Inc.; 212, AP Photo/Mark Wilson; 214, WDCN/Univ. College London/Photo Researchers, Inc.; 215, Uli Wiesmeierr/CORBIS; 216, Anson0618/Shutterstock; 217, Geoff Oliver Bugbee; 219, Nic Bothma/epa/CORBIS; 220 (LE), BSIP/Photo Researchers, Inc.; 220 (CTR), BSIP/Photo Researchers, Inc.; 220 (RT), BSIP/Photo Researchers, Inc.; 221, Tom McHugh/Photo Researchers, Inc.; 222 (UP), Dennis Kunkel Microscopy, Inc./Visuals Unlimited; 222 (CTR), M I (Spike) Walker/Alamy; 222 (LO), Sidney Moulds/Photo Researchers, Inc.; 223, LookatSciences/Phototake, Inc.; 224, Jose Luis Pelaez/Getty Images; 225, Alfred Pasieka/Photo Researchers, Inc.; 227, David Patrick Valera/Getty Images; 228, Chris Johns, NGS; 229, Christopher Furlong/Getty Images; 230, Wellcome Department of Cognitive Neurology/Photo Researchers, Inc.; 231, Vincent van Gogh/Bridgeman Art Library/Getty Images; 232 (UP), Cyberonics, Inc. via Getty Images; 232 (LO), Andy Crump, TDR, World Health Organization/Photo Researchers, Inc.

Chapter Eight: 234-235, Arlene Jean Gee/Shutterstock; 236, Carol Donner/Phototake; 237, Joel Sartore/NationalGeographicStock.com; 238, Mario Savoia/Shutterstock; 239, North Wind Picture Archives/Alamy; 241, © 2009 Suzanne Corkin, reprinted with permission of The Wylie Agency LLC; 242, AP Photo/Walla Walla Union-Bulletin, Jeff Horner; 243, Hypnotic/Universal Pictures/The Kobal Collection/Endrenyi, Egon; 244, AP Photo/Princeton University; 246, Illustration © 2009 Scientific Publishing Ltd., Elk Grove Village, IL, USA; 247, Heidi and Hans-Jurgen Koch/Minden Pictures/NationalGeographicStock.com; 248, Fred R. Conrad/The New York Times; 248-249, Albert Tousson/Phototake; 250, Jaimie Duplass/Shutterstock; 251, AP Photo/Jeff Chiu; 252, Rosalind Rosenberg; 253, Bliznetsov/Shutterstock; 254, NASA; 255, AP Photo/Jodi Hilton, POOL; 256, AP Photo/Al Grillo; 257, AP Photo/John Froschauer; 258, AP/Wide World Photos; 259, ISM/Phototake; 260, Bettmann/CORBIS; 261, Rick Friedman/CORBIS; 262, Allen R. Braun, M.D., NIDCD/NIH ; 263, Will & Deni McIntyre/Photo Researchers, Inc.; 264, Rue des Archives/The Granger Collection, New York; 265, Kevin Winter/Getty Images for AFI.

Chapter Nine: 266-267, Paul A. Souders/CORBIS; 268, Hinata Haga/HAGA/The Image Works; 269, Journal-Courier/Clayton Salter/The Image Works; 270, Getty Images for Nintendo DS; 271, Scott Camazine/Photo Researchers, Inc.; 272, Dennis Hallinan/Alamy; 273, Zephyr/Photo Researchers, Inc.; 274, Jeff Zelevansky/Reuters/CORBIS; 275, Kevin J. Miyazaki/Redux Pictures; 277, AP Photo/Jonathan Fredin; 278 (UP), Cordelia Molloy/Photo Researchers, Inc.; 278 (LO), AJPhoto/Photo Researchers, Inc.; 279, Jung Yeon-Je/AFP/Getty Images; 280, Joe Jaszewski/Getty Images; 281, Christine Stone; 282, Karen Kasmauski/Science Faction/CORBIS; 283, Tim Graham/Getty Images; 284, Ed Kashi/CORBIS; 285, David McLain/Aurora/Getty Images; 286, Silvano Audisio/Shutterstock; 287, Yoshikazu Tsuno/AFP/Getty Images; 288, Dr. Cecil H. Fox/Photo Researchers, Inc.; 289, Alfred Pasieka/Photo Researchers, Inc.; 290, Patrick Landmann/Getty Images; 291, Karen Kasmauski/CORBIS; 292, Chris Rank/CORBIS; 293, VEER Barry Rosenthal/Getty Images; 294, MIT AI Lab/Surg. Planning Lab, BWH/Photo Researchers, Inc.; 295, Henning Dalhoff/Bonnier Publications/Photo Researchers, Inc.

Epilogue: 296-297, Volker Steger/Photo Researchers, Inc.; 299, Mike Agliolo/Photo Researchers, Inc.; 300, Yoshikazu Tsuno/AFP/Getty Images; 303, AP Photo/Paul Sakuma; 304, Joseph Barrak/AFP/Getty Images; 306, Kim Kyung-Hoon/Reuters/CORBIS; 309, Charles Ommanney/Getty Images; 310, C.J. Guerin, PhD, MRC Toxicology Unit/Photo Researchers, Inc.; 313, Zephyr/Photo Researchers, Inc.; 315, Erik Viktor/Photo Researchers, Inc.; 316-317, Francesco Reginato/Getty Images.

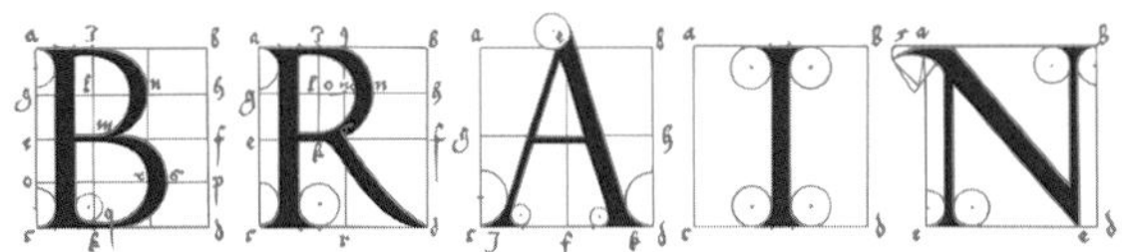

THE COMPLETE MIND

Michael S. Sweeney
Foreword by Richard Restak, M.D.

Published by the National Geographic Society
John M. Fahey, Jr., *President and Chief Executive Officer*
Gilbert M. Grosvenor, *Chairman of the Board*
Tim T. Kelly, *President, Global Media Group*
John Q. Griffin, *Executive Vice President; President, Publishing*
Nina D. Hoffman, *Executive Vice President; President, Book Publishing Group*

Prepared by the Book Division
Barbara Brownell Grogan, *Vice President and Editor in Chief*
Marianne R. Koszorus, *Director of Design*
Susan Tyler Hitchcock, *Senior Editor*
Carl Mehler, *Director of Maps*
R. Gary Colbert, *Production Director*
Jennifer A. Thornton, *Managing Editor*
Meredith C. Wilcox, *Administrative Director, Illustrations*

Staff for This Book
Amy Briggs, *Editor*
Susan Blair, *Illustrations Editor*
Kristin Sladen, *Illustrations Editor*
Melissa Farris, *Art Director*
Cameron Zotter, *Designer*
Al Morrow, *Design Assistant*
Judith Klein, *Copy Editor*
Caitlin Mertzluff, *Researcher*
R. Gary Colbert, *Production Project Manager*

Manufacturing and Quality Management
Christopher A. Liedel, *Chief Financial Officer*
Phillip L. Schlosser, *Vice President*
Chris Brown, *Technical Director*
Nicole Elliott, *Manager*
Rachel Faulise, *Manager*

The National Geographic Society is one of the world's largest nonprofit scientific and educational organizations. Founded in 1888 to "increase and diffuse geographic knowledge," the Society works to inspire people to care about the planet. It reaches more than 325 million people worldwide each month through its official journal, *National Geographic,* and other magazines; National Geographic Channel; television documentaries; music; radio; films; books; DVDs; maps; exhibitions; school publishing programs; interactive media; and merchandise. National Geographic has funded more than 9,000 scientific research, conservation and exploration projects and supports an education program combating geographic illiteracy. For more information, visit nationalgeographic.com.

For more information, please call 1-800-NGS LINE (647-5463) or write to the following address:

National Geographic Society
1145 17th Street N.W.
Washington, D.C. 20036-4688 U.S.A.

Visit us online at www.nationalgeographic.com

For information about special discounts for bulk purchases, please contact National Geographic Books Special Sales: ngspecsales@ngs.org

For rights or permissions inquiries, please contact National Geographic Books Subsidiary Rights: ngbookrights@ngs.org

ISBN: 978-1-4262-0547-7
ISBN: 978-1-4262-0548-4 (deluxe)

Printed in the U.S.A.

09/RRDW/1